CELLULAR ENDOCRINOLOGY IN HEALTH AND DISEASE

CELLULAR ENDOCRINOLOGY IN HEALTH AND DISEASE

Edited by

ALFREDO ULLOA-AGUIRRE

*Universidad Nacional Autónoma de México and Instituto Nacional de Salud Pública,
México D.F., México*

P. MICHAEL CONN

*Departments of Internal Medicine and Cell Biology-Biochemistry, Texas Tech University Health Sciences Center,
Lubbock, TX, USA*

ELSEVIER

AMSTERDAM • BOSTON • HEIDELBERG • LONDON
NEW YORK • OXFORD • PARIS • SAN DIEGO
SAN FRANCISCO • SINGAPORE • SYDNEY • TOKYO

Academic Press is an imprint of Elsevier

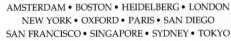

Academic Press is an imprint of Elsevier
32 Jamestown Road, London NW1 7BY, UK
225 Wyman Street, Waltham, MA 02451, USA
525 B Street, Suite 1800, San Diego, CA 92101-4495, USA

British Library Cataloguing-in-Publication Data
A catalogue record for this book is available from the British Library.

Library of Congress Cataloging-in-Publication Data
A catalog record for this book is available from the Library of Congress.

ISBN : 978-0-12-408134-5

For information on all Academic Press publications
visit our website at elsevierdirect.com

Typeset by MPS Limited, Chennai, India
www.adi-mps.com

Working together
to grow libraries in
developing countries

www.elsevier.com • www.bookaid.org

Contents

12. Bone as an Endocrine Organ

GERARD KARSENTY

13. Regulation of Steroidogenesis

ANDREW A. BREMER, WALTER L. MILLER

14. Adipose Tissue as an Endocrine Organ

NICOLAS MUSI, RODOLFO GUARDADO-MENDOZA

15. Insulin-Secreting Cell Lines: Potential for Research and Diabetes Therapy

SHANTA J. PERSAUD, ASTRID C. HAUGE-EVANS,
PETER M. JONES

16. Architecture and Morphology of Human Pancreatic Islets

ALVIN C. POWERS

17. Computational Models to Decipher Cell-Signaling Pathways

ANNE POUPON, ERIC REITER

List of Contributors

Ana Aranda Instituto de Investigaciones Biomédicas "Alberto Sols," Madrid, Spain, Consejo Superior de Investigaciones Científicas and Universidad Autónoma de Madrid, Madrid, Spain

Richard Bertram Florida State University, Tallahassee, FL, USA

Andrew A. Bremer Vanderbilt University, Nashville, TN, USA

Maria Luisa Brandi University of Florence, Medical School, Florence, Italy

Sally A. Camper University of Michigan Medical School, Ann Arbor, MI, USA

Nancy Carrasco Yale University School of Medicine, New Haven, CT, USA

Luisella Cianferotti University of Florence, Medical School, Florence, Italy

P. Michael Conn Departments of Internal Medicine and Cell Biology-Biochemistry, Texas Tech University Health Services Center, Lubbock, TX, USA

Constanza Contreras Jurado Instituto de Investigaciones Biomédicas "Alberto Sols," Madrid, Spain, Consejo Superior de Investigaciones Científicas and Universidad Autónoma de Madrid, Madrid, Spain

Lique M. Coolen University of Mississippi Medical Center, Jackson, MS, USA

Pascale Crépieux CNRS, Nouzilly, France, François Rabelais University, Tours, France

Francisco Dominguez INCLIVA, Instituto Universitario IVI (IUIVI), Valencia University, Valencia, Spain

Shereen Ezzat University of Toronto, Princess Margaret Hospital, and the Ontario Cancer Institute, Toronto, ON, Canada

Laurine Gagniac CNRS, Nouzilly, France, François Rabelais University, Tours, France

Nathalie Gallay CNRS, Nouzilly, France, François Rabelais University, Tours, France

Peter D. Gluckman University of Auckland, Auckland, New Zealand

Karen Gomez-Hernandez University of Toronto, Princess Margaret Hospital, and the Ontario Cancer Institute, Toronto, ON, Canada

Arturo E. Gonzalez-Iglesias Florida State University, Tallahassee, FL, USA

Robert L. Goodman West Virginia University, Morgantown, WV, USA

Rodolfo Guardado-Mendoza University of Guanajuato, León, Mexico

Florian Guillou CNRS, Nouzilly, France, François Rabelais University, Tours, France

Mark A. Hanson University of Southampton, Southampton, UK

Astrid C. Hauge-Evans King's College London, London, UK

Tomohiro Ishii Department of Pediatrics, School of Medicine, Keio University, Tokyo, Japan

Peter M. Jones King's College London, London, UK

Gerard Karsenty Columbia University Medical Center, New York, NY, USA

Michael N. Lehman University of Mississippi Medical Center, Jackson, MS, USA

Felicia M. Low University of Auckland, Auckland, New Zealand

Olaia Martínez-Iglesias Instituto de Investigaciones Biomédicas "Alberto Sols," Madrid, Spain, Consejo Superior de Investigaciones Científicas and Universidad Autónoma de Madrid, Madrid, Spain

Jan M. McAllister Pennsylvania State University College of Medicine, Hershey, PA, USA

Walter L. Miller University of California, San Francisco, CA, USA

Bhavi Modi Virginia Commonwealth University, Richmond, VA, USA

Nicolas Musi Barshop Institute for Longevity and Aging Studies and Geriatric Research, San Antonio, TX, USA

Juan Pablo Nicola Yale University School of Medicine, New Haven, CT, USA

Aurea Orozco Instituto de Neurobiologia, Universidad Nacional Autónoma de México (UNAM), Quéretaro, México

María Inés Pérez Millán University of Michigan Medical School, Ann Arbor, MI, USA

Shanta J. Persaud King's College London, London, UK

Anne Poupon INRA, Nouzilly, France, CNRS, Nouzilly, France, François Rabelais University, Tours, France

Alvin C. Powers Vanderbilt University Medical Center, Nashville, TN, USA

Eric Reiter INRA, Nouzilly, France, CNRS, Nouzilly, France, François Rabelais University, Tours, France

Ludivina Robles-Osorio Facultad de Medicina, Universidad Autónoma de Quéretaro, Quéretaro, México

Lidia Ruiz-Llorente Instituto de Investigaciones Biomédicas "Alberto Sols," Madrid, Spain, Consejo Superior de Investigaciones Científicas and Universidad Autónoma de Madrid, Madrid, Spain

Carlos Simon INCLIVA, Instituto Universitario IVI (IUIVI), Valencia University, Valencia, Spain

Juan Carlos Solís-S Facultad de Medicina, Universidad Autónoma de Quéretaro, Quéretaro, México

Jerome F. Strauss III Virginia Commonwealth University, Richmond, VA, USA

Toru Tateno University of Toronto, Princess Margaret Hospital, and the Ontario Cancer Institute, Toronto, ON, Canada

Manuel Tena-Sempere University of Córdoba, Córdoba, Spain, Instituto de Salud Carlos III, Córdoba, Spain, Instituto Maimónides de Investigaciones Biomédicas (IMIBIC)/Hospital Universitario Reina Sofia, Córdoba, Spain

Judith L. Turgeon University of California, Davis, USA

Alfredo Ulloa-Aguirre Universidad Nacional Autónoma de México and Instituto Nacional de Salud Pública, Mexico D.F., Mexico

Carlos Valverde-R Instituto de Neurobiologia, Universidad Nacional Autónoma de México (UNAM), Quéretaro, México

Michael D. Walker Weizmann Institute of Science, Rehovot, Israel

Dennis W. Waring University of California, Davis, USA

Preface

The last two decades have witnessed tremendous advances in endocrinology — specifically in our knowledge of how endocrine cells govern their own function, as well as the functions of other cells. The intricate mechanisms that control the biosynthesis and secretion of hormones by endocrine cells, as well as the cellular responses to stimuli from a large variety of hormones and chemical signals, are becoming better understood — thanks to improvements in the techniques used to explore cell and molecular biology; these include microscopy, recombinant DNA technology, cell micromanipulation, and all "OMICS" fields. We are now able to identify many of the genetic, biochemical and structural responses that are regulated by environmental cues and endogenous stimuli. We are only starting to understand the intracellular and intercellular networks that maintain homeostasis of the whole organism, but it is clear that that regulatory networks have their genesis in the response of the individual cell.

This book is our attempt to provide an understanding of how endocrine glands function by integrating information resulting in biological effects on both local and systemic levels. The book explores and dissects the function of a number of cell systems, including those whose function as part of the endocrine was not obvious until recently, among these, the bone and the adipose tissue. To this end, the editors selected authors based on their research contributions and their ability to express their thoughts clearly.

The editors want to express appreciation to the authors for providing contributions in a timely fashion and to the staff at Elsevier for helpful input.

Alfredo Ulloa-Aguirre
P. Michael Conn

Thyroid Hormone Receptors and their Role in Cell Proliferation and Cancer

Olaia Martínez-Iglesias, Lidia Ruiz-Llorente, Constanza Contreras Jurado and Ana Aranda

Instituto de Investigaciones Biomédicas "Alberto Sols," Madrid, Spain, Consejo Superior de Investigaciones Científicas and Universidad Autónoma de Madrid, Madrid, Spain

THYROID HORMONE ACTION

The important physiological actions of the thyroid hormones (THs) are mediated by binding to the nuclear thyroid hormone receptors (TRs). The thyroid gland produces predominantly thyroxine (T4), but triiodothyronine (T3) is the most active TH, since it has a higher affinity by the receptors.[1] THs are released by the thyroid gland to the bloodstream and they enter the cells through the adenosine triphosphate (ATP)-dependent monocarboxylate transporters MCT8 and MCT10 and the organic anion transporter proteins (OATPs).[2] The amount of T3 available for binding to the nuclear receptors is regulated by cell-specific expression of selenoenzymes deiodinases (DIOs). DIO1 and DIO2 catalyze the conversion of T4 to T3 in target tissues, increasing intracellular levels of the active hormone, while DIO3 causes hormone inactivation since it converts T4 and T3 by inner ring deiodination to the inactive metabolites reverse T3 (rT3) and T2, respectively.

TRs belong to the superfamily of nuclear receptors and act as ligand-dependent transcription factors.[3] Several TR protein isoforms are generated by promoter use or alternative splicing of the primary transcripts of the *TRα* and *TRβ* genes. The TRα1, TRβ1 and TRβ2 are the main hormone-binding isoforms and their relative levels of expression vary among cell types and at different developmental stages, suggesting that they could have organ-specific functions. In the case of TRβ, TRβ1 is more widely expressed, while the expression of TRβ2 is restricted to the anterior pituitary, and some neural cells.[4,5] Studies with genetically modified mice have shown that TRα and TRβ can substitute for each other to mediate some actions of the thyroid hormones but they can also mediate isoform-specific functions.[6]

As shown in Figure 1.1, TRs are composed of several functional domains. The *N*-terminal

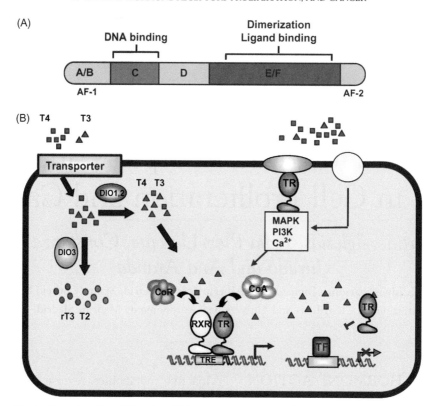

FIGURE 1.1 **Mechanism of action of the thyroid hormone receptors.** (A) Schematic representation of a thyroid hormone receptor, showing the different functional domains. (B) Thyroxine (T4) and triiodothyronine (T3) enter the cell through transporter proteins such as MCT8 and 10 or OATPs. Inside the cells, deiodinases (DIO1,2) convert T4, to the more active form T3. DIO3 produces rT3 and T2 from T4 and T3, respectively. T3 binds to nuclear thyroid hormone receptors (TRs) that regulate transcription by binding, generally as heterodimers with the retinoid X receptor (RXR), to positive or negative thyroid hormone response elements (TREs) located in regulatory regions of target genes. Activity is regulated by an exchange of corepressor (CoR) and coactivator (CoA) complexes. TRs can also regulate the activity of genes that do not contain a TRE through "cross-talk" with other transcription factors (TF) that stimulate target gene expression. Binding of T3 to a subpopulation of receptors located outside the nuclei can also cause rapid "non-genomic" effects through interaction with adaptor proteins, leading to stimulation of signaling pathways. T4 can also bind to putative membrane receptors such as integrin αVβ3 inducing mitogen activated protein kinase (MAPK) activity.

region (A/B) contains a constitutive ligand-independent transcriptional activation domain, the autonomous activation function 1 (AF-1). This region is followed by the DNA-binding domain (DBD), or region C. The DBD is the most conserved region among the nuclear receptors and is composed of two zinc fingers. In each zinc finger, four invariable cysteines coordinate tetrahedrically with one zinc ion. Amino acids required for discrimination of the

thyroid hormone response element (TRE) are present at the base of the first finger in a region termed the "P box," and other residues of the second zinc finger that form the so called "D box" are involved in dimerization. Through the DBD the receptors interact with the major groove of DNA. A hinge domain, or D region, connects the DBD with the E region or ligand-binding domain (LBD), also responsible for dimerization. This hinge domain contains residues essential for interaction with corepressors. Crystallographic analysis has shown that the LBDs are formed by 12 α-helices, and the C-terminal helix (H12) encompasses the ligand-dependent transcriptional activation function, or AF-2.

TRs regulate gene transcription by binding, preferentially as heterodimers with retinoid X receptors (RXRs), to short DNA binding motifs, called thyroid hormone response elements or TREs, which are located in regulatory regions of target genes.[7] TREs are composed of two copies of the AGG/TTCA motif. They can be configured as palindromes (Pal), inverted palindromes (IPs), or direct repeats spaced preferably by four non-conserved nucleotides (DR4). Although TRs can bind to their response elements as monomers or homodimers, heterodimerization with RXR strongly increases the affinity for DNA and transcriptional activity.

Transcriptional regulation by these receptors is mediated by the recruitment of coactivators and corepressors.[3,8,9] In the absence of ligand, TRs can act as constitutive repressors when bound to TREs, due to their association with corepressors such as NCoR (nuclear receptor corepressor) or SMRT (silencing mediator of retinoic and thyroid receptor). NCoR and SMRT belong to multicomponent repressor complexes that contain histone deacetylases (HDACs) and cause chromatin compaction and consequently transcriptional inhibition.[10] NCoR and SMRT are related both structurally and functionally. They contain three autonomous repressor domains (RD) and a receptor interacting domain (the CoRNR motif) located toward the carboxyl terminus. Transcriptional repression by the corepressor-bound receptors appears to be mediated by the recruitment of HDACs to the target gene. HDAC1 or 2 (class I deacetylases) are recruited to the first RD of the corepressors via the adaptor mSin3 protein, and the RD3 has been demonstrated to repress transcription by directly interacting with class II deacetylases (HDACs 4, 5 and 7).

In addition, a repressor complex containing the corepressors, HDAC3 and transducin beta-like proteins (TBL1 or TBL1R) appears to be required for repression by TR. Although a receptor CoR box, located within the hinge region, is essential for interaction of receptors with the corepressors, the CoRNR motif does not interact directly with residues in this region, but docks to a hydrophobic groove in the surface of the LBD at H3 and 4.

Hormone binding induces a conformational change in the receptor that allows the release of corepressors and allows the recruitment in a sequential manner of coactivator complexes. The stronger change observed in the receptors upon ligand binding is the position of H12.[11] This helix projects away from the body of the LBD in the absence of ligand. However, upon hormone binding H12 moves in a "mousetrap" model being tightly packed against H3 or 4 and making direct contacts with the ligand. This change generates a hydrophobic cleft responsible for interaction with coactivators.[12] A glutamic acid residue in H12 and a lysine residue in H3, which are conserved throughout the superfamily of nuclear receptors, interact directly with the coactivator and form a charge clamp that stabilizes binding. Consequently, mutation of these residues abolishes coactivator binding and causes the loss of thyroid hormone-dependent transcriptional activation.[13] Since the coactivator binding surface overlaps with that involved in corepressors interaction, coactivator and corepressor binding is mutually exclusive. Some coactivators belong to ATP-dependent chromatin-remodeling complexes, others are part of complexes that induce post-translational modifications of histones, such as acetylation or arginine methylation, and others interact with the basic transcriptional machinery causing the recruitment of RNA polymerase II to the target promoter. Binding of the coactivators causes chromatin decompaction and transcriptional activation.

In addition to causing ligand-dependent transcriptional activation, TRs can also repress gene transcription in a hormone-dependent manner. In some cases, this repression is associated with binding to negative TREs (nTREs). Although the properties of nTREs are not yet well known, these elements are often located very close to the transcriptional start site,[3] and corepressors and deacetylase activity appear to be involved in hormone-dependent negative regulation.[14] TRs can also regulate the expression of genes that do not contain a TRE by positive or negative interference with the activity of other transcription factors or signaling pathways, a mechanism referred to as transcriptional crosstalk.[3] Thus, we have shown that TRs can antagonize AP-1,[15,16] cyclic AMP (cAMP) response element-binding protein (CREB),[17,18] or NF-kB-mediated transcription.[19,20] In this case, the receptors do not bind directly to the DNA recognition elements for these transcription factors in the target gene, but can be tethered to these binding motifs via protein-to-protein interactions. This type of transcriptional crosstalk between transcription factors and nuclear receptors has been shown to be critical for regulation of many cellular functions, including anti-inflammatory and anti-proliferative actions of nuclear receptor ligands.[21–23] Finally, thyroid hormones can elicit rapid non-genomic effects initiated at the cell membrane that can lead to stimulation of kinase pathways. These actions could be mediated by a fraction of membrane-associated nuclear receptors, or by occupancy of putative membrane receptors, such as integrin $\alpha V \beta 3$, which would bind T4 preferentially.[9] Figure 1.1B illustrates the main aspects of thyroid hormone actions on cells.

TRS AND CANCER

The first evidence linking TRs with cancer was the finding that TRα is the cellular counterpart of the v-erbA oncogene of the avian erythroblastosis virus (AEV), a retrovirus that causes erythroleukemia and sarcoma in chickens. v-ErbA acts as a constitutive dominant-negative of TRs since it contains mutations that abolish ligand binding, recruitment of coactivators and hormone-dependent transcriptional stimulation, while maintaining the ability to bind corepressors.[24] There is also evidence that reduced TR expression and/or alterations in TR genes are common events in human cancer.[25] In particular, decreased TR levels as well as somatic mutations in TR genes are frequently present in breast cancers and aberrant TRs have been found in more than 70% of human hepatocarcinomas. Most of these mutants have been shown to act as dominant-negative inhibitors of TR activity,[26] suggesting that the native receptors could act as tumor suppressors and that loss of expression and/or function of this receptor could result in a selective advantage for cell transformation and tumor development. In agreement with this idea it has been shown that TRs could function as tumor suppressors in a mouse model of metastatic follicular thyroid carcinoma.[27]

INHIBITION OF TUMOR CELL PROLIFERATION BY THE THYROID HORMONE RECEPTORS

T3 blocks proliferation of N2a neuroblastoma cells which express TRβ1 (N2a-β cells). Our results have shown that T3 coordinately regulates the expression of several genes that play a key role in cell cycle control. Thus, the hormone induces a rapid down-regulation of the c-*myc* gene, a decrease of *cyclin* D1 transcription, and an induction of the cell cycle inhibitors p27Kip1 and p21Cip.[28–30] Furthermore, gene expression analysis indicates a decreased expression of other cyclins (F, T1, D1 and B2), cyclin-dependent kinase 4

(CDK4) and other cell cycle components such as Wee1 and Cdc20 after T3 treatment.[31]

The c-*myc* oncogene plays an important role in cell cycle progression and different signals that arrest cell growth suppress expression of c-Myc. Transcription of the c-*myc* gene is controlled by several promoters, and a block in transcriptional elongation appears to be essential in the regulation of c-*myc* gene expression. Sequences known to function as a polymerase II pausing region are located immediately downstream of the P2 promoter, and a binding site for the transcriptional repressor CTCF maps precisely within this region of polymerase II pausing and release. Interestingly, we have demonstrated the existence of a nTRE in this region. This element binds TR-RXR heterodimers and is adjacent to the TCTF binding site. Furthermore, a c-*myc* promoter fragment containing binding sites for both transcription factors confers repression by T3 when located downstream of an heterologous promoter, indicating that the receptor in cooperation with CTCF causes premature termination of transcription, decreasing c-*myc* mRNA levels.[32]

One of the molecular events required for cell cycle progression is the inactivation by hyperphosphorylation of retinoblastoma protein family. Accordingly, we found that T3-mediated growth arrest of neuroblastoma cells is associated with hypophosphorylation of the retinoblastoma proteins pRb and p103. This modification is catalyzed by cyclin-dependent kinases (CDKs), whose activity is regulated by different mechanisms including their association with cyclins and with cyclin kinase inhibitors (CKIs). As indicated above, p27Kip1 and p21Cip levels increase upon incubation of N2a-β cells with T3. The strong and sustained increase of p27Kip1 by the hormone is secondary both to augmented levels of p27Kip1 mRNA and to a longer half-life of the CKI, indicating that transcriptional and post-transcriptional mechanisms are involved in T3-induced CKI induction. The increased levels of

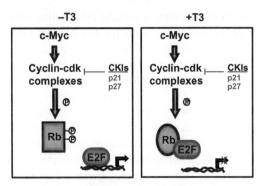

FIGURE 1.2 **Inhibition of proliferation of neuroblastoma N2a-β cells by T3.** In the absence of T3, retinoblastoma proteins (Rb) are phosphorylated by cyclin–cdk complexes and E2F transcription factors are free to bind to genes important for progression through the cell cycle. T3 inhibits expression of the c-*myc* protooncogen and reduces expression of genes encoding several cyclins, among them the *cyclin* D1 gene, while reducing expression of cyclin-dependent kinases such as CDK4. In addition, the hormone increases the levels of the cyclin kinase inhibitors p21 and p27. These changes lead to a reduced activity of cyclin–cdk complexes and to hypophosphorylation of Rb. Under these conditions E2F factors remain bound to Rb and cell cycle progression is blocked.

p27Kip1 lead to a significant increase in the amount of CKI bound to CDK2 and to a marked inhibition of the kinase activity of the cyclin/CDK2 complexes. As a consequence of these changes, retinoblastoma proteins are hypophosphorylated in T3-treated N2a-β cells and progression through the restriction point in the cell cycle is blocked[30] (Figure 1.2).

THE THYROID HORMONE RECEPTOR ANTAGONIZES RAS-INDUCED PROLIFERATION

Ras oncoproteins are small guanosine triphosphate (GTP)-binding proteins that play a crucial role in normal and malignant cell proliferation. Oncogenic mutations in the *ras* gene result in a constitutively active protein that is present in at least 30% of human tumors and can efficiently transform most immortalized rodent cells.[33]

Ras activation induces, among others, the activation of the Ras/mitogen-activated protein kinase (MAPK) signaling pathway, which is a key mediator for mitogenic signaling and Ras-induced transformation.[34] In this pathway, activation of the MAPK extracellular signal-regulated kinase 1/2 (Erk1/2) leads to phosphorylation of transcription factors of the Ets family, or to activation of downstream kinases such as Rsk or Msk,[35,36] which then phosphorylate other transcription factors, among them b-Zip factors of the CREB/ATF family.

Cyclin D1 is one of the main targets for the proliferative, transforming and tumorigenic effects of the *ras* oncogene.[37] In N2a-β cells expression of oncogenic Ras increases Cyclin D1 levels and T3 reverses significantly this induction. In parallel, Ras increases proliferation of N2a-β cells and T3 inhibits this response. Furthermore, the inhibitory effect of T3 on proliferation is significantly reversed after overexpression of Cyclin D1, showing that the repression of Cyclin D1 expression by T3 plays an important role in the mechanism by which the hormone represses Ras-mediated proliferation.[28] In transient transfection experiments with reporter genes containing the *cyclin* D1 gene promoter we have observed that T3 blocks induction of *cyclin* D1 promoter activity by *ras* not only in neuroblastoma cells but also in human hepatocarcinoma cells, in murine fibroblasts, and in rat tumor pituitary cells, indicating that T3-dependent repression on Ras-mediated transcription is a rather general effect. The v-*src* oncogene also stimulates transcription of *cyclin* D1 in a Ras-dependent manner and T3 also antagonizes this response. T3 represses expression of the *cyclin* D1 gene in response to the *ras* oncogene through proximal promoter sequences that do not contain a TRE but contain a CRE (cyclic AMP response element). The CRE constitutively binds b-Zip factors such as CREB and ATF-2 and, accordingly, neither Ras nor the hormone alters the abundance of the factors that bind this motif. However, activation of these

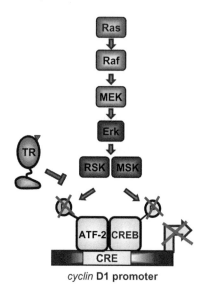

cyclin **D1 promoter**

FIGURE 1.3 **Model of the regulation of** *cyclin* **D1 transcription by** *ras* **and the thyroid hormone receptors.** Oncogenic Ras stimulates the Ras-MAPK pathway leading to activation Rsk or Msk kinases, which phosphorylate transcription factors such as CREB or ATF-2. These factors bind to a CRE motif located in the proximal promoter of the *cyclin* D1 gene, and their activation stimulates transcription. The liganded thyroid hormone receptor antagonizes activation of the MAPK pathway and interacts with the b-Zip factors blocking their activation by Ras and transcription of the *cyclin* D1 gene.

transcription factors by Ras is blocked by TR in a T3-dependent manner (Figure 1.3). In addition to antagonizing the MAPK pathway and the activation of downstream kinases such as Rsk2 or Msk, TRs can interact directly with ATF-2 and with CREB inhibiting its phosphorylation.[17]

THE THYROID HORMONE RECEPTORS ANTAGONIZE TRANSFORMATION AND TUMORIGENESIS BY ONCOGENIC RAS

Since TRs can inhibit Ras-dependent proliferation and *cyclin* D1 transcription, we also examined the possibility that they could repress Ras mediated cellular transformation and tumor growth. To prove this hypothesis, we analyzed formation of transformation foci in NIH-3T3 fibroblasts transfected with oncogenic Ras in the presence or absence of TRs. The results obtained showed that the transforming capacity of the *ras* oncogene is strongly decreased in TR-expressing fibroblasts, although TRβ1 appears to have a stronger anti-transforming activity than the α1 isoform. Not surprisingly, TRs were also able to antagonize fibroblast transformation by v-*src*. Furthermore, the inhibition of transformation by TRβ1 was lost after *cyclin* D1 over-expression, indicating that downregulation of this cyclin is also involved in the anti-transforming effects of the receptor. To analyze whether TRs could act as suppressors of tumor formation by the *ras* oncogene in mice, NIH-3T3 cells expressing in a stable manner oncogenic Ras alone or in combination with TRα1 or TRβ1 were injected into the flanks of immunodeficient nude mice. Whereas large tumors developed in mice injected with fibroblasts expressing Ras alone, tumor formation was blocked in mice injected with fibroblasts co-expressing the oncoprotein and TRβ1. Co-expression of oncogenic Ras with TRα1 abolished tumor formation but caused a strong delay in the appearance of tumors.[28] Furthermore, the tumors formed in the presence of TRα1 presented a more differentiated phenotype, as demonstrated by an increased presence of collagen and a more fusiform morphology of the cells. Therefore, TRs could play a relevant role as suppressors of *ras*-dependent tumors, and although both isoforms suppress tumor growth, TRβ1 appears to exert a stronger anti-tumorigenic effect *in vivo*.

FUNCTIONAL DOMAINS INVOLVED IN TR ANTAGONISM OF Ras RESPONSES

To analyze the mechanisms and receptor domains involved in the antagonism of Ras-induced transcription, we examined the effect of various TRβ1 mutants on *cyclin* D1 promoter activity in HepG2 and NIH-3T3 fibroblasts[13] (Figure 1.4A). Mutants E457Q in H12 or K288I in H3 still presented a significant activity to antagonize the activation of the *cyclin* D1 promoter by oncogenic Ras and to reduce Cyclin D1 protein levels. This indicates that residues in H3 and H12 that are essential for coactivators recruitment and ligand-dependent transactivation on a TRE are not required for inhibition of Ras responses by the receptor. TRβ1 mutants in the hinge domain were used to analyze the role of corepressors in this antagonism. A receptor containing the triple mutation AHT-GGA in the CoR box that abolishes interaction with corepressors[38] and ligand-independent repression on a TRE, did not block *cyclin* D1 promoter stimulation by Ras, indicating that corepressors could play a role in the transcriptional

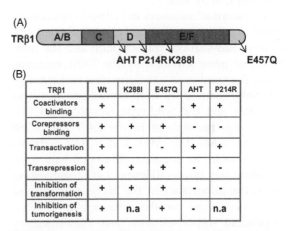

FIGURE 1.4 **Thyroid hormone receptor domains involved in blocking the responses to the *ras* oncogene.** (A) Scheme of TRβ1 showing the position of the mutations used. (B) Effect of the different receptor mutants on T3-dependent in vitro recruitment of coactivators and corepressors in "pull-down" assays, transactivation of a reporter plasmid containing a TRE in transient transfection assays, transrepression of the response of the *cyclin* D1 promoter to oncogenic Ras, inhibition of formation of transformation foci in NIH-3T3 fibroblasts transfected with the oncogene and inhibition of tumor growth in nude mice injected with Ras-transformed fibroblasts. n.a, not analyzed.

antagonism of TRβ1. In addition, mutation P214R in a residue preceding H1, homologous to that present in a v-erbA transformation deficient mutant (td359),[39] also blocks interaction with corepressors and is unable to antagonize Ras-mediated transcription in a T3-dependent manner (Figure 1.4). Both corepressor mutants were, however, capable of inducing T3-dependent transcription, demonstrating that Ras antagonism is mechanistically different from classical TR actions on transcriptional stimulation. The hinge mutants that were unable to mediate transrepression of the *cyclin* D1 promoter also lost the ability of antagonizing ATF-2 activation by the oncogene, reinforcing the notion that corepressors play a role in the antagonistic effects of TRβ1 on Ras-dependent transcription. This is supported by the finding that overexpression of NCoR or SMRT further represses T3-dependent *cyclin* D1 transcription by Ras.

Transformation assays in NIH-3T3 fibroblasts showed a strict parallelism between the antagonism of Ras-induced transcription and the inhibition of transformation by the TRβ1 mutants. Thus, the number of transformation foci was significantly reduced by T3 in fibroblasts coexpressing Ras and the wild-type receptor and this action was maintained in the E457 and K288 mutants, while the AHT and P214R mutants were unable to mediate T3-dependent antitransforming effects (Figure 1.4B), suggesting again the importance of the hinge domain and the corepressors in the antagonism of Ras-dependent responses. This was confirmed by the finding that the inhibition of formation of transformation foci by T3 was more marked after co-transfection of TRβ1 with SMRT or NCoR. The relevant role of endogenous corepressors on the antitransforming effects of TRβ1 was demonstrated with siRNA experiments that revealed that NCoR downregulation counteracted to a significant extent the inhibitory effect of T3 on cellular transformation, whereas SMRT depletion had a weaker effect. Furthermore,

NCoR knockdown increased foci formation by Ras in the absence of the transfected receptor, indicating that this corepressor can act as an endogenous inhibitor of fibroblast transformation.[13] By contrast, the oncogenic TR mutants such as v-erbA or PV that act as dominant-negatives blocking the antiproliferative effects of the wild-type receptors do not suppress Ras function although they bind corepressors constitutively, indicating that ligand binding is also required for this receptor action. This is further suggested by the finding that another dominant-negative receptor, a mutant TRα in three conserved lysine residues in the carboxy-terminal extension of the DBD that, unexpectedly, displays a strongly reduced ligand binding affinity, is unable to antagonize transformation by the *ras* oncogene in a T3-dependent manner.[40]

To analyze *in vivo* the role of corepressors and coactivators on suppression of tumor formation by TRβ1, nude mice were injected with NIH-3T3 fibroblasts expressing the oncoprotein alone or in combination with wild-type TRβ1 or the AHT and E457Q mutants. Tumor formation was blocked in mice injected with cells coexpressing Ras and either the native receptor or the E457 mutant, while the AHT mutant did not inhibit tumor development. These results suggest that corepressors (but not classical coactivators) are essential not only for the antitransforming effects of TRβ1, but also for inhibiting tumor formation in *vivo* (Figure 1.4B).

TRβ1 SUPPRESSES TUMOR INVASION AND METASTASIS

To analyze the role of TRs in tumor progression and metastatic growth, we re-expressed TRβ1 in hepatocarcinoma SK-hep1 (SK) cells and breast cancer MDA-MB-468 (MDA) cells, which have lost receptor expression, generating the SK-TRβ and MDA-TRβ cell lines. The competence of cancer cells to proliferate in the absence of a solid substrate is required for the acquisition of an invasive and metastatic phenotype and we observed that TRβ1 inhibited colony formation by these cells in soft agar and blocked cell growth in suspension under rocking conditions. This could decrease survival of TRβ1-expressing cancer cells in the bloodstream. In addition, TRβ1 expression strongly reduced invasion in matrigel assays and this could inhibit cell intravasation and extravasation of the tumor cells. Proliferation, survival and cell invasion also depends on the response of the tumor cell to autocrine and paracrine growth factors. We found that epidermal growth factor (EGF) or insulin-like growth factor-1 (IGF-1) induced proliferation of parental hepatocarcinoma and breast cancer cells and that this response was lost in cells expressing TRβ1. The receptor blocks the mitogenic action of these growth factors by reducing expression of IGFIR, EGFR or ErbB3 receptors and by suppressing activation of ERK and PI3K signaling pathways that are critical for cell proliferation and invasion.[41] In accordance with our observations that TRs can inhibit Ras-mediated responses downstream of ERK, TRβ1 expression blocked the activation of transcription factors such as ELK1 or ATF-2 by EGF and IGF-1 in hepatocarcinoma and breast cancer cells. Furthermore, TRβ1 inhibited TGFβ-dependent proliferation. Many actions of this transforming growth factor are mediated by SMAD activation, but TGFβ can also increase MAPK and PI3K activity, and TRβ1 blocked stimulation of these pathways by TGFβ. Therefore, TR antagonizes stimulation of signaling pathways by growth and transforming growth factors that play a key role in invasion and tumor progression.

Genes that are relevant for metastatic progression have been identified. Among them, high levels of expression in primary tumors of the prostaglandin-synthesizing enzyme cyclooxygenase 2, the transcriptional inhibitor ID1, the chemokine receptors CXCR4, CCR6 and CCR1, the protooncogen c-Met, or some metalloproteases is associated with high risk of metastasis and poor prognosis in patients. Strikingly, we found that TRβ1 coordinately downregulated the expression of these prometastatic genes, suggesting a common molecular mechanism for this receptor action. The finding that inhibitors of MAPK or PI3K inhibit their expression in parental cells suggests that antagonism of these pathways by TRβ1 could participate in their transcriptional repression,[41] although the gene elements and molecular mechanism of repression by the receptor remain to be determined.

When parental, as well as TRβ-expressing MDA and SK cells, were inoculated into nude mice, it was observed that TRβ1 reduced tumor proliferation (Figure 1.5) retarding tumor growth, and enhanced expression of epithelial markers while reducing expression of mesenchymal markers. In addition, TRβ1 increased the necrotic area of the tumors and significantly reduced angiogenesis. These changes are compatible with reduced invasion and tumor growth. Indeed, whereas tumors formed by the parental hepatocarcinoma and breast cancer cells are highly infiltrative, TRβ1-expressing cells caused the appearance of tumors to be more compact and surrounded by a pseudocapsule of collagen and inflammatory cells (Figure 1.6). Most of these tumors do not infiltrate the adjacent tissues and they do not originate distant nodular metastasis. To analyze the effect of the receptor in formation of experimental metastasis, parental and TRβ1-expressing cells were injected into the tail vein of nude mice. Examination of the lungs at necropsy showed that TRβ1 had a potent inhibitory effect on metastasis formation (Figure 1.7). The incidence of metastasis, their size, the number of metastases per lung, and the area of the lungs affected by metastatic growth were markedly reduced in mice injected with TRβ1-expressing cells. Extravasation was also strongly decreased in TRβ1-expressing cells, indicating that the receptor has anti-metastatic

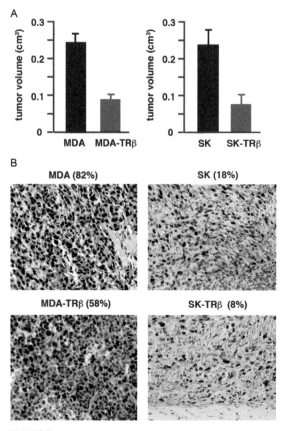

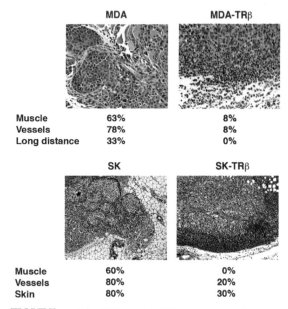

FIGURE 1.6 TRβ1 inhibits tumor invasion. Hematoxylin and eosin staining of the tumors formed by parental MDA-MB-468 breast cancer cells and SK-hep1 hepatocarcinoma cells shows that most tumors infiltrate surrounding tissues. In contrast, tumors formed by TRβ1-expressing cells are more compact and less infiltrative. The percentage of tumors invading the indicated tissues and forming spontaneous long distant metastasis is indicated.

FIGURE 1.5 TRβ1 inhibits tumor growth. MDA-MB-468 breast cancer cells and SK-hep1 hepatocarcinoma cells were stably transfected with an empty vector or with TRβ1 and inoculated into the fat mammary pad or the flank of nude mice, respectively. (A) Mean tumor volume 25 days after inoculation. (B) The tumors were excised and tumor cell proliferation was assessed by immunohistochemistry with the proliferation marker Ki67.

HYPOTHYROIDISM RETARDS TUMOR GROWTH BUT INCREASES METASTASIS DEVELOPMENT

activity by blocking not only the ability of cancer cells to proliferate and colonize the lung parenchyma but also by reducing cancer cell extravasation.[41] The inhibitory effect of TRβ1 on tumor invasiveness and formation of metastasis observed in the mice is compatible with the reduced invasion and the repression of expression of pro-metastatic genes observed in the cultured cells.

In contrast with the well-accepted role of TRs as tumor suppressors, no clear association between thyroidal status and cancer has been found in humans. Thus, hypothyroidism is more prevalent in hepatocarcinoma patients and might be a possible risk factor for liver cancer,[42] but hypothyroid patients have been reported to present both a higher and a reduced incidence of breast carcinomas.[43,44] These confounding effects could be secondary to the important metabolic changes associated with hypothyroidism rather than to direct

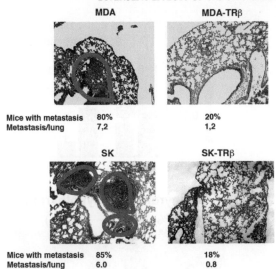

MDA   MDA-TRβ

| Mice with metastasis | 80% | 20% |
| Metastasis/lung | 7,2 | 1,2 |

SK   SK-TRβ

| Mice with metastasis | 85% | 18% |
| Metastasis/lung | 6.0 | 0.8 |

FIGURE 1.7   **TRβ1 inhibits formation of experimental metastasis.** Parental and TRβ1-expressing cells were injected into the tail vein of nude mice and formation of lung metastasis was analyzed 30 days later. The figure illustrates representative hematoxylin and eosin staining of lungs showing the appearance of large nodular metastasis (delineated by a continuous line) in mice injected with parental MDA-MB-468 or SK-hep1 cells, which are absent in the TRβ1-expressing cells. The percentage of injected animals bearing metastatic lesions in the lungs, as well as the mean number of metastasis/lung is indicated.

binding of the thyroid hormones to TRs in the tumor cells, as they often present inactivating mutations. Therefore, we compared tumor growth, invasion, and formation of metastasis in control and hypothyroid nude mice injected with both parental and TRβ1-expressing SK and MDA cells. We found that tumor growth was retarded in hypothyroid mice inoculated with both parental and TRβ1-expressing cells (Figure 1.8). The reduced tumor volume in hypothyroid hosts correlated with a lower proliferation index, reduction of Cyclin E expression and increased necrosis of the tumors. In addition, tumors developed in hypothyroid mice had a strong reduction of epithelial markers and a more mesenchymal phenotype, which could facilitate invasion and dissemination of the tumor cells to distant organs.

Indeed, we found that hypothyroidism increased the number of invasion fronts of the tumors, the infiltration to neighboring tissues such as muscle, blood and lymph vessels or skin and the formation of distant metastasis in lung, liver or bone (Table 1.1). Furthermore, formation of lung metastasis after inoculation of the cancer cells into the tail vein of the hypothyroid nude mice was also markedly enhanced with regard to the metastatic growth observed in normal hosts, again both in parental and TRβ1-expressing cells (Figure 1.9). Therefore, hypothyroidism appears to favor a permissive tissue microenvironment for cancer metastasis. The increased malignancy of tumors formed by the parental hepatocarcinoma and breast cancer cells that do not express TRs in hypothyroid mice, suggest that changes in the stromal cells, most likely secondary to the important metabolic changes associated with hypothyroidism, rather than a direct effect of the hormone on the cancer cells, could be responsible for the increased tumor aggressiveness. In summary, normal thyroid hormone levels appear to favor growth of primary xenografts, but they also block tumor cell dissemination and metastasis formation.[45] These divergent effects could help to explain the contradictory reports on the influence of hypothyroidism in human tumors. Furthermore, because our results show that similar effects are observed independently of the presence or absence of TR in the cancer cells, it would be expected that thyroidal status could impact tumor progression, even in tumors in which TRs are deleted or mutated, a common event in human cancer.

DIVERGENT EFFECTS OF TRs ON NORMAL AND TRANSFORMED CELL PROLIFERATION

The actions of THs are highly pleiotropic, and the various TR isoforms might have

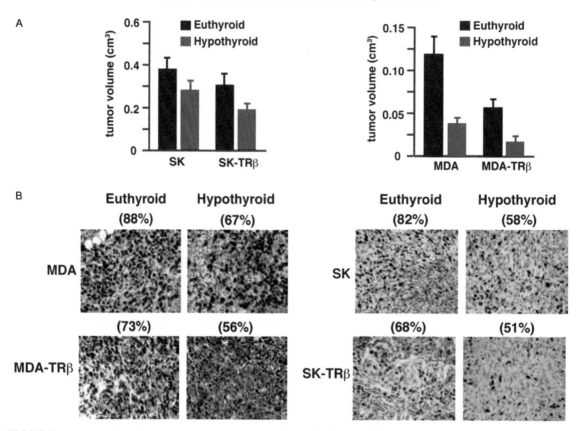

FIGURE 1.8 **Hypothyroidism reduces tumor growth.** MDA-MB-468 and SK-hep1 cells were stably transfected with an empty vector or with TRβ1 and inoculated into euthyroid and hypothyroid nude mice. (A) Mean tumor volume 7 weeks after inoculation. (B) The tumors were excised and tumor cell proliferation was assessed by immunohistochemistry with the proliferation marker Ki67.

opposing effects on cell proliferation depending on the cell type, the cellular context or the transformation status (see ref 5 for a recent review). For instance, in agreement with our results in hepatocarcinoma cells showing that expression of TRβ1 retards tumor growth and inhibits metastasis formation in nude mice, it has been found that THs induce a rapid regression of carcinogen-induced hepatic nodules in rodents, reducing the incidence of hepatocarcinoma and lung metastasis in a TRβ-dependent manner.[46−48] Paradoxically, TH administration can cause liver hyperplasia in animals.[49] Although the adult liver has a

low replicative activity, hepatocytes can proliferate in response to partial surgical hepatectomy. It has been shown that hypothyroidism delays liver regeneration after partial hepatectomy,[50] and we have demonstrated that TRβ promotes hepatocyte proliferation in response to hepatectomy, since knockout (KO) mice lacking TRβ or TRα1 and TRβ show a significant retardation in the restoration of liver mass. In the absence of TRs there is a retarded initiation of proliferation accompanied by an important but transient apoptotic response that does not occur in normal mice. These changes are linked to increased nitrosative

TABLE 1.1 Hypothyroidism Enhances Tumor Invasion.

Cells Inoculated	Host Mice	Muscle	Vessels	Bone	Skin	Lung	Liver
MDA	Euthyroid	40%	58%	n.d	n.d	n.d	n.d
	Hypothyroid	60%	82%	30%	n.d	n.d	n.d
MDA-TRβ	Euthyroid	30%	30%	n.d	n.d	n.d	n.d
	Hypothyroid	50%	50%	n.d	n.d	n.d	n.d
SK	Euthyroid	60%	80%	n.d	80%	0%	0%
	Hypothyroid	100%	100%	n.d	100%	30%	30%
SK-TRβ	Euthyroid	0%	20%	n.d	20%	n.d	n.d
	Hypothyroid	0%	50%	n.d	40%	n.d	n.d

Abbreviation: n.d, not detected.
Parental and TRβ1-expressing MDA-MB-468 and SK-hep1 hepatocarcinoma cells were inoculated into euthyroid or hypothyroid nude mice. Tumors were excised 8 weeks later and the percentage of tumors invading the indicated tissues for each condition was scored

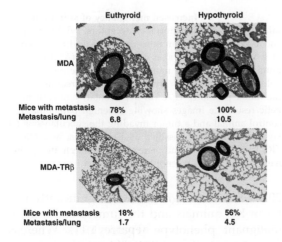

FIGURE 1.9 **Hypothyroidism enhances formation of metastasis.** Parental and TRβ1-expressing MDA-MB-468 cells were injected into the tail vein of nude mice and formation of lung metastasis was analyzed 30 days later. The figure illustrates representative hematoxylin and eosin staining of lungs in the different groups. The percentage of injected animals bearing metastatic lesions in the lungs, as well as the mean number of metastasis/lung is indicated.

stress, resulting from a drop in the levels of asymmetric dimethylarginine (ADMA), a potent physiological inhibitor of nitric oxide synthase (NOS) activity.[51] Therefore, TRs

appear to have opposite effects on normal hepatocytes with low proliferative activity and on tumor hepatic cells that proliferate rapidly, although the mechanism responsible for these differences is still unknown.

The skin is also a target for TR-mediated cell proliferation. Thyroid dysfunction is associated with skin pathologies in patients,[52,53] and topical application of TH stimulates epidermal proliferation and dermal thickening and accelerates wound healing in rodents.[54,55] Using genetically modified mice we have observed that the effects of TH on skin proliferation are mediated through interactions with both TRα1 and TRβ.[56,57] We found reduced keratinocyte proliferation and decreased hyperplasia in response to topical application of 12-O-tetradecanolyphorbol-13-acetate (TPA) or retinoids in the epidermis of mice lacking TRs and also in hypothyroid mice (Figure 1.10). Reduced proliferation in TR KO mice correlates with increased expression of cyclin-dependent kinase inhibitors (CKIs) in the interfollicular epidermis, with strongly reduced Cyclin D1 expression in the keratinocytes of the basal layer and with increased production of pro-inflammatory cytokines,

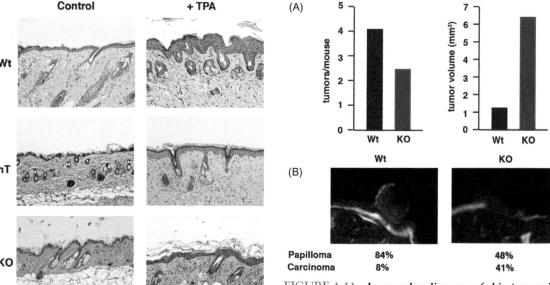

FIGURE 1.10 **The liganded thyroid hormone receptor is required for normal keratinocyte proliferation.** Control mice (Wt), mice made hypothyroid (hT) by treatment with anti-thyroidal drugs and genetically modified mice lacking the thyroid hormone binding isoforms TRα1 and TRβ (knockout, KO), were treated topically for 4 days with the tumor promoter TPA or with the solvent (control). Hematoxylin and eosin staining shows the reduced skin hyperplasia found in KO and hypothyroid animals.

FIGURE 1.11 **Increased malignancy of skin tumors in mice lacking thyroid hormone receptors.** Wild-type mice and TRα1/TRβ knockout mice were subjected to a protocol of two-stage chemical skin carcinogenesis. (A) The mean number of tumors per mouse as well as the mean tumor volume was measured after 30 weeks of treatment with the tumor promoter TPA. (B) Representative nuclear magnetic resonance images showing a benign papilloma in a control animal and a tumor invading the subcutaneous fat layer, histologically classified as a carcinoma, in a TRα1/TRβ knockout mice. The percentage of benign and malignant tumors for each genotype is indicated.

which is associated with enhanced phosphorylation of p65/NF-κB and STAT3 transcription factors.[56,57]

In a protocol of two-stage chemical skin carcinogenesis benign papillomas and malignant tumors can be induced on the backs of mice after exposure to the initiating carcinogen 7,12-dimethylbenzaanthracene (DMBA), and subsequent chronic treatment with TPA. The finding that TR KO mice show a strongly reduced epidermal proliferation in response to the tumor promoter suggests that they may be less sensitive to skin carcinogenesis. Indeed, we have observed that TRα1/TRβ develop fewer tumors than normal mice when subjected to the carcinogenesis protocol. However, after

20 weeks, tumor growth was significantly faster in KO animals and the tumors had a more malignant phenotype (Figure 1.11). Whereas most tumors found in wild-type mice were typical well-differentiated papillomas, in KO mice almost half of the tumors were classified as *in situ* carcinoma or squamous cell carcinomas (SCCs). Therefore, TR deficiency seems to inhibit benign tumor formation at early stages of skin carcinogenesis and increases malignization at later stages, indicating again that these receptors can have divergent effects on cell proliferation and malignant transformation. Supporting the notion that TRβ could act as a suppressor of tumor progression, we also found that TRβ expression can be detected in

normal skin of wild-type mice, but receptor expression was strongly reduced in the papillomas and was totally lost in SCCs.[41]

Acknowledgements

The laboratory of Ana Aranda is supported by Grants BFU2011-28058 from Ministerio de Economía y Competitividad, RD12/0036/0030 from the Fondo de Investigaciones Sanitarias, and S2011/BMD-2328 (TIRONET) from the Comunidad de Madrid.

References

1. Flamant F, Baxter JD, Forrest D, Refetoff S, Samuels H, Scanlan TS, et al. International union of pharmacology. Lix. The pharmacology and classification of the nuclear receptor superfamily: Thyroid hormone receptors. *Pharmacol Rev* 2006;**58**:705–11.
2. Visser WE, Friesema EC, Visser TJ. Minireview: thyroid hormone transporters: the knowns and the unknowns. *Mol Endocrinol* 2011;**25**:1–14.
3. Aranda A, Pascual A. Nuclear hormone receptors and gene expression. *Physiol Rev* 2001;**81**:1269–304.
4. Lazar MA. Nuclear receptor corepressors. *Nucl Recept Signal* 2003;**1**:e001.
5. Pascual A, Aranda A. Thyroid hormone receptors, cell growth and differentiation. *Biochim Biophys Acta* 2013;**1830**:3908–16.
6. Flamant F, Gauthier K. Thyroid hormone receptors: the challenge of elucidating isotype-specific functions and cell-specific response. *Biochim Biophys Acta* 2013;**1830**:3900–7.
7. Yen PM, Ando S, Feng X, Liu Y, Maruvada P, Xia X. Thyroid hormone action at the cellular, genomic and target gene levels. *Mol Cell Endocrinol* 2006;**246**:121–7.
8. Brent GA. Mechanisms of thyroid hormone action. *J Clin Invest* 2012;**122**:3035–43.
9. Cheng SY, Leonard JL, Davis PJ. Molecular aspects of thyroid hormone actions. *Endocr Rev* 2010;**31**:139–70.
10. Privalsky ML. The role of corepressors in transcriptional regulation by nuclear hormone receptors. *Annu Rev Physiol* 2004;**66**:315–60.
11. Wagner RL, Apriletti JW, McGrath ME, West BL, Baxter JD, Fletterick RJ. A structural role for hormone in the thyroid hormone receptor. *Nature* 1995;**378**:690–7.
12. Feng W, Ribeiro RC, Wagner RL, Nguyen H, Apriletti JW, Fletterick RJ, et al. Hormone-dependent coactivator binding to a hydrophobic cleft on nuclear receptors. *Science* 1998;**280**:1747–9.
13. Garcia-Silva S, Martinez-Iglesias O, Ruiz-Llorente L, Aranda A. Thyroid hormone receptor beta1 domains responsible for the antagonism with the ras oncogene: role of corepressors. *Oncogene* 2011;**30**:854–64.
14. Wang D, Xia X, Liu Y, Oetting A, Walker RL, Zhu Y, et al. Negative regulation of tsh{alpha} target gene by thyroid hormone involves histone acetylation and corepressor complex dissociation. *Mol Endocrinol* 2009;**23**(5):600–9.
15. Perez P, Palomino T, Schonthal A, Aranda A. Determination of the promoter elements that mediate repression of c-fos gene transcription by thyroid hormone and retinoic acid receptors. *Biochem Biophys Res Commun* 1994;**205**:135–40.
16. Perez P, Schonthal A, Aranda A. Repression of c-fos gene expression by thyroid hormone and retinoic acid receptors. *J Biol Chem* 1993;**268**:23538–43.
17. Mendez-Pertuz M, Sanchez-Pacheco A, Aranda A. The thyroid hormone receptor antagonizes creb-mediated transcription. *Embo J* 2003;**22**:3102–12.
18. Sanchez-Pacheco A, Palomino T, Aranda A. Negative regulation of expression of the pituitary-specific transcription factor ghf-1/pit-1 by thyroid hormones through interference with promoter enhancer elements. *Mol Cell Biol* 1995;**15**:6322–30.
19. Chiloeches A, Sanchez-Pacheco A, Gil-Araujo B, Aranda A, Lasa M. Thyroid hormone-mediated activation of the erk/dual specificity phosphatase 1 pathway augments the apoptosis of gh4c1 cells by downregulating nuclear factor-kappab activity. *Mol Endocrinol* 2008;**22**:2466–80.
20. Lasa M, Gil-Araujo B, Palafox M, Aranda A. Thyroid hormone antagonizes tumor necrosis factor-alpha signaling in pituitary cells through the induction of dual specificity phosphatase 1. *Mol Endocrinol* 2010;**24**:412–22.
21. De Bosscher K, Vanden Berghe W, Haegeman G. The interplay between the glucocorticoid receptor and nuclear factor-kappab or activator protein-1: molecular mechanisms for gene repression. *Endocr Rev* 2003;**24**:488–522.
22. Rosenfeld MG, Lunyak VV, Glass CK. Sensors and signals: a coactivator/corepressor/epigenetic code for integrating signal-dependent programs of transcriptional response. *Genes Dev* 2006;**20**:1405–28.
23. Pascual G, Glass CK. Nuclear receptors versus inflammation: mechanisms of transrepression. *Trends Endocrinol Metab* 2006;**17**:321–7.
24. Wolffe AP, Collingwood TN, Li Q, Yee J, Urnov F, Shi YB. Thyroid hormone receptor, v-erba, and chromatin. *Vitam Horm* 2000;**58**:449–92.
25. Aranda A, Martinez-Iglesias O, Ruiz-Llorente L, Garcia-Carpizo V, Zambrano A. Thyroid receptor: roles in cancer. *Trends Endocrinol Metab* 2009;**20**:318–24.

26. Chan IH, Privalsky ML. Thyroid hormone receptors mutated in liver cancer function as distorted anti-morphs. *Oncogene* 2006;**25**:3576−88.

27. Zhu XG, Zhao L, Willingham MC, Cheng SY. Thyroid hormone receptors are tumor suppressors in a mouse model of metastatic follicular thyroid carcinoma. *Oncogene* 2010;**29**:1909−19.

28. Garcia-Silva S, Aranda A. The thyroid hormone receptor is a suppressor of ras-mediated transcription, proliferation, and transformation. *Mol Cell Biol* 2004;**24**:7514−23.

29. Garcia-Silva S, Perez-Juste G, Aranda A. Cell cycle control by the thyroid hormone in neuroblastoma cells. *Toxicology* 2002;**181−182**:179−82.

30. Perez-Juste G, Aranda A. The cyclin-dependent kinase inhibitor p27(kip1) is involved in thyroid hormone-mediated neuronal differentiation. *J Biol Chem* 1999;**274**:5026−31.

31. Bedo G, Pascual A, Aranda A. Early thyroid hormone-induced gene expression changes in n2a-beta neuroblastoma cells. *J Mol Neurosci* 2011;**45**:76−86.

32. Perez-Juste G, Garcia-Silva S, Aranda A. An element in the region responsible for premature termination of transcription mediates repression of c-myc gene expression by thyroid hormone in neuroblastoma cells. *J Biol Chem* 2000;**275**:1307−14.

33. Bos JL. Ras oncogenes in human cancer: a review. *Cancer Res* 1989;**49**:4682−9.

34. Calvo F, Agudo-Ibanez L, Crespo P. The ras-erk pathway: understanding site-specific signaling provides hope of new anti-tumor therapies. *Bioessays* 2010;**32**:412−21.

35. Anjum R, Blenis J. The rsk family of kinases: emerging roles in cellular signalling. *Nat Rev Mol Cell Biol* 2008;**9**:747−58.

36. Arthur JS. Msk activation and physiological roles. *Front Biosci* 2008;**13**:5866−79.

37. Gille H, Downward J. Multiple ras effector pathways contribute to g(1) cell cycle progression. *J Biol Chem* 1999;**274**:22033−40.

38. Horlein AJ, Naar AM, Heinzel T, Torchia J, Gloss B, Kurokawa R, et al. Ligand-independent repression by the thyroid hormone receptor mediated by a nuclear receptor co-repressor. *Nature* 1995;**377**:397−404.

39. Damm K, Beug H, Graf T, Vennstrom B. A single point mutation in erba restores the erythroid transforming potential of a mutant avian erythroblastosis virus (aev) defective in both erba and erbb oncogenes. *EMBO J* 1987;**6**:375−82.

40. Sanchez-Pacheco A, Martinez-Iglesias O, Mendez-Pertuz M, Aranda A. Residues k128, 132, and 134 in the thyroid hormone receptor-alpha are essential for receptor acetylation and activity. *Endocrinology* 2009;**150**:5143−52.

41. Martinez-Iglesias O, Garcia-Silva S, Tenbaum SP, Regadera J, Larcher F, Paramio JM, et al. Thyroid hormone receptor beta1 acts as a potent suppressor of tumor invasiveness and metastasis. *Cancer Res* 2009;**69**:501−9.

42. Reddy A, Dash C, Leerapun A, Mettler TA, Stadheim LM, Lazaridis KN, et al. Hypothyroidism: a possible risk factor for liver cancer in patients with no known underlying cause of liver disease. *Clin Gastroenterol Hepatol* 2007;**5**:118−23.

43. Beatson G. On the treatment of inoperable cases of carcinoma of the mamma: suggestions for a new method of treatment, with illustrative cases. *Lancet* 1896;**2**:104−7.

44. Cristofanilli M, Yamamura Y, Kau SW, Bevers T, Strom S, Patangan M, et al. Thyroid hormone and breast carcinoma. Primary hypothyroidism is associated with a reduced incidence of primary breast carcinoma. *Cancer* 2005;**103**:1122−8.

45. Martinez-Iglesias O, Garcia-Silva S, Regadera J, Aranda A. Hypothyroidism enhances tumor invasiveness and metastasis development. *PLoS One* 2009;**4**: e6428.

46. Ledda-Columbano GM, Perra A, Loi R, Shinozuka H, Columbano A. Cell proliferation induced by triiodothyronine in rat liver is associated with nodule regression and reduction of hepatocellular carcinomas. *Cancer Res* 2000;**60**:603−9.

47. Perra A, Kowalik MA, Pibiri M, Ledda-Columbano GM, Columbano A. Thyroid hormone receptor ligands induce regression of rat preneoplastic liver lesions causing their reversion to a differentiated phenotype. *Hepatology* 2009;**49**:1287−96.

48. Perra A, Simbula G, Simbula M, Pibiri M, Kowalik MA, Sulas P, et al. Thyroid hormone (t3) and trbeta agonist gc-1 inhibit/reverse nonalcoholic fatty liver in rats. *FASEB J* 2008;**22**:2981−9.

49. Francavilla A, Carr BI, Azzarone A, Polimeno L, Wang Z, Van Thiel DH, et al. Hepatocyte proliferation and gene expression induced by triiodothyronine in vivo and in vitro. *Hepatology* 1994;**20**:1237−41.

50. Malik R, Mellor N, Selden C, Hodgson H. Triiodothyronine enhances the regenerative capacity of the liver following partial hepatectomy. *Hepatology* 2003;**37**:79−86.

51. Lopez-Fontal R, Zeini M, Traves PG, Gomez-Ferreria M, Aranda A, Saez GT, et al. Mice lacking thyroid hormone receptor beta show enhanced apoptosis and delayed liver commitment for proliferation after partial hepatectomy. *PLoS One* 2010;**5**:e8710.

52. Paus R. Exploring the "thyroid-skin connection": concepts, questions, and clinical relevance. *J Invest Dermatol* 2010;**130**:7−10.

53. Slominski A, Wortsman J. Neuroendocrinology of the skin. *Endocr Rev* 2000;**21**:457–87.
54. Safer JD, Crawford TM, Holick MF. Topical thyroid hormone accelerates wound healing in mice. *Endocrinology* 2005;**146**:4425–30.
55. Safer JD, Fraser LM, Ray S, Holick MF. Topical triiodo-thyronine stimulates epidermal proliferation, dermal thickening, and hair growth in mice and rats. *Thyroid* 2001;**11**:717–24.
56. Contreras-Jurado C, Garcia-Serrano L, Gomez-Ferreria M, Costa C, Paramio JM, Aranda A. The thyroid hormone receptors as modulators of skin proliferation and inflammation. *J Biol Chem* 2011;**286**:24079–88.
57. Garcia-Serrano L, Gomez-Ferreria MA, Contreras-Jurado C, Segrelles C, Paramio JM, Aranda A. The thyroid hormone receptors modulate the skin response to retinoids. *PLoS One* 2011;**6**:e23825.

The Molecular Cell Biology of Anterior Pituitary Cells

Arturo E. Gonzalez-Iglesias and Richard Bertram

Florida State University, Tallahassee, FL, USA

INTRODUCTION

"Do not stop to question whether these ideas are new or old, but ask, more properly, whether they harmonize with nature." **Marcello Malpighi**

The dawn of neuroendocrinology is coincident with the beginning of the tale of a remarkable gland: the pituitary gland. Around AD 170 Galen postulated that the "pituita" (from the Greek *ptuo*, "to spit," and the Latin *pituita*, "mucus") secreted waste products (phlegm, one of the four humors of the body) from the brain into the nasal cavities. Nineteen centuries later, we have found not only that Galen's concept was far from the truth but also that a fascinating – yet still primitive – understanding of the gland has emerged.

The pituitary is one of the two elements that make up the hypothalamo–hypophysial unit, the joint anatomical structure comprising hormone-producing neurons and cells by which the brain regulates the vital functions of the body. Indeed, a key regulator of body homeostasis during development, stress, and other physiological processes, the pituitary gland acts as a double interpreter, mediating the talk between the brain and the peripheral organs, and integrating their respective cues as well as those of its own (local autocrine and paracrine factors). Being functionally and anatomically connected to the hypothalamus by the median eminence (ME) via the infundibular stalk,[1] the pituitary has two embryologically and functionally distinct divisions: the neurohypophysis (or neural lobe) and the adenohypophysis (anterior pituitary and intermediate lobes). The remarkable molecular and cellular aspects of the biology of the cells that constitute the anterior pituitary is the focus of this chapter.

ANTERIOR PITUITARY: ORGANIZATION, CELL TYPES, HORMONES AND FUNCTIONS

The anterior pituitary is an endocrine gland responsible for secreting hormones that regulate a wide range of functions. These hormones

Cellular Endocrinology in Health and Disease.
DOI: http://dx.doi.org/10.1016/B978-0-12-408134-5.00002-0

are synthesized and released by distinct groups of polygonal endocrine cells that are organized as interlacing cords and lined up on an anastomizing web of capillary vessels (the secondary capillary plexus) derived from the hypophysial artery. The cytoplasm of these cells contains granules of stored hormone that are released by exocytosis. The endothelial cells of the capillaries are fenestrated to facilitate the exchange of molecules between the endocrine cells and the blood, which not only bring in the hypothalamic and peripheral factors (through the long portal vessels and hypophysial arteries, respectively) that regulate the activity of the gland but also carry the released pituitary hormones away into the general circulation.[1] In addition, an extensive web of interconnected folliculostellate (FS) cells surround the endocrine cells. These cells regulate both the interaction of neighboring endocrine cells and the exchange of molecules between them and the capillaries. FS cells represent about 5−10% of the anterior pituitary cell population.

The traditional view of the pituitary holds that there are five endocrine cell types that are responsible for synthesizing six anterior pituitary hormones (Table 2.1). For each cell type, several immortalized cell lines have been developed, characterized and used extensively.[2]

Somatotrophs, which synthesize and release growth hormone (GH), are the major endocrine cell type in the anterior pituitary and constitute 40−50% of its cell population. They are localized predominantly to the lateral portions of the anterior lobe. Somatotroph function is primarily regulated by hypothalamic factors: growth hormone-releasing hormone (GHRH) produced by neurons in the arcuate nucleus is stimulatory, whereas somatostatin (STT) produced by neurons in the periventricular nucleus is inhibitory. STT suppresses both basal and GHRH-induced GH release, having no effect on GH synthesis. GH production and secretion also receives inhibitory feedback from the major target of GH, insulin-like

growth factor-I from the liver. Somatotrophs express receptors for many other regulators of GH synthesis and release, including ghrelin, pituitary adenylate cyclase-activating peptide, thyroid hormone, glucocorticoids, insulin and endothelins. GH is secreted from the somatotrophs at a pulse frequency of about 1−2 h with a half-life that ranges between 6−20 min, and the pattern exhibits gender differences. In the case of males, the pulses are much larger early at night, whereas in females the pattern is more irregular and the pulses tend to be more uniform throughout the day. The pattern of GH release appears to be driven by the rate at which GHRH is released from the arcuate nucleus neurons.

GH is also called somatotropin (*soma*, "body") because of its profound and widespread anabolic effects throughout the body. In its absence, growth is stunted. Although virtually every tissue responds to some degree, skeletal muscle cells, liver, and chondrocytes (cartilage cells) are particularly sensitive to GH levels. Though the metabolic effects are direct actions of GH, it is now apparent that most, if not all, of the anabolic effects of GH are mediated by the production of a family of peptide hormone intermediaries known as insulin-like growth factors (IGFs) which are secreted by the liver, cartilage, muscle, and other tissues where they can act locally in a paracrine or autocrine fashion. GH, acting through the IGFs, stimulates protein synthesis, cell growth and a positive nitrogen balance, leading to an increase in lean body mass and a decrease in body fat. Many hours must elapse after administration of GH before its anabolic, growth-promoting effects become evident.

Thyrotrophs comprise approximately 5% of the anterior endocrine cell population and are typically spread over the anteriomedial and lateral portions of the gland. Thyrotrophs synthesize and secrete thyroid-stimulating hormone (TSH), also known as thyrotropin, which is controlled by central and peripheral

TABLE 2.1 Hypothalamic Neurohormones and Anterior Pituitary Endocrine Cell Types and Hormones

Hypothalamic Releasing or Inhibiting Neurohormone	Target Cell in Anterior Pituitary Gland	Anterior Pituitary Hormone Release Stimulated or Inhibited	Target Anterior Pituitary Hormone
Growth hormone releasing hormone (GHRH)	Somatotroph	Growth hormone (GH)	Multiple somatic tissues
Somatostatin (STT)		GH	Multiple somatic tissues
Thyrotropin releasing hormone (TRH)	Thyrotroph	Thyroid stimulating hormone (TSH)	Stimulates synthesis and release of thyroid hormone by thyroid gland
Corticotropin releasing hormone (CRH)	Corticotroph	Adrenocorticotrophic hormone (ACTH)	Stimulates synthesis and release of steroids by adrenal cortex
Gonadotropin releasing hormone (GnRH)	Gonadotroph	Follicle stimulating hormone (FSH) and luteinizing hormone (LH)	Stimulates synthesis and release of gonadal steroids, gametogenesis, ovulation
Dopamine	Lactotroph	Prolactin	Milk synthesis and secretion by the mammary gland

regulators. The dominant stimulatory control of thyrotroph function and TSH secretion is exerted by the hypothalamic neurohormone thyrotropin-releasing hormone (TRH) released by neurons of the paraventricular nucleus. TSH is secreted in pulses lasting for 2–3 h with a nocturnal surge before sleep. Once the sleep phase begins, TSH release is curtailed. The half-life of the hormone in blood is approximately of 1 h.

TSH is a glycoprotein with a molecular weight of 28,000 and consists of a heterodimer of two subunits (α and β) that are tightly associated by noncovalent forces and encoded by separate genes. Although both subunits are required for receptor binding and hormone action, the β subunit confers biological specificity to the TSH molecule, as the α subunit is also a component of the anterior pituitary gonadotropin hormones luteinizing hormone (LH) and follicle-stimulating hormone (FSH). The essential actions of TSH are those exerted on the thyroid gland, where it promotes growth and differentiation of the gland and stimulates all steps in the secretion of the thyroid hormones

thyroxine (T4) and triiodothyronine (T3). These steps include glandular uptake of iodide, its organification, the completion of thyroid hormone synthesis, and the subsequent release of thyroid gland products. T3 and T4 act at the hypothalamic and pituitary levels to block the secretion of TSH via feedback inhibition. Fasting decreases thyrotroph responsiveness to TRH, while exposure to cold increases it.

Corticotrophs constitute about 15% of the adenohypophysial endocrine cell population and are scattered throughout the anterior lobe in adult animals, but are primarily found in the anteromedial part of the gland. They synthesize proopiomelanocortin (POMC) and release its proteolytic derivatives, adrenocorticotropin hormone (ACTH), α-melanocyte-stimulating hormone (α-MSH), lipotropic hormone (LPH) and endorphins. The main releasing factors for are the hypothalamic corticotropin-releasing hormone (CRH) and arginine-vasopressin (AVP), which acts in synergy with CRH to potentiate hormone release. Glucocorticoids, secreted by the adrenal cortex, are the major physiological inhibitor. ACTH levels in plasma exhibit a circadian rhythm, with a peak in the early morning followed by a gradual decline during the night, and is controlled by clock neurons in the suprachiasmatic nucleus that synapse with CRH-producing neurons in the hypothalamic paraventricular nucleus. ACTH also exhibits faster, ultradian and hourly rhythms due to feedback between corticotrophs and adrenal cells.[3]

As its name implies, one of the major actions of ACTH is the promotion of growth of adrenal cortex cells. ACTH effects on these cells are also necessary for both basal and stress-induced secretion of glucocorticoids and aldosterone. Its half-life of 10 min allows for rapid adjustments of circulating levels of glucocorticoids. ACTH, CRH, and glucocorticoids are the main effector molecules of the hypothalamo–pituitary–adrenal (HPA) axis, the system that activates to help the organism adapt and cope with different stress cues (hypoglycemia, anesthesia, surgery, trauma, hemorrhage, infection, pyrogens) and psychiatric disorders like anxiety or depression.

Gonadotrophs constitute about 10–15% of the anterior pituitary endocrine cells and are localized throughout the pars distalis and most of the pars tuberalis of the anterior lobe. They form intimate contacts with lactotroph cells, with which they have extensive cell-to-cell (paracrine) interactions. Gonadotrophs synthesize and release two hormones essential to the growth and function of the gonads in both genders (hence their common designation as gonadotropins): luteinizing hormone (LH) and follicle-stimulating hormone (FSH). Synthesis and release of both hormones are stimulated by gonadotropin-releasing hormone (GnRH), which is secreted in a pulsatile manner by neurons that are dispersed within the mediobasal hypothalamus and preoptic areas. In addition, other hypothalamic factors such as gonadotropin-inhibiting hormone, vasopressin, substance P, as well as feedback of gonadal factors (estrogens, progestogens, androgens and inhibin) contribute to the regulation of gonadotropins. The release of LH and FSH occurs in phase with pulses of GnRH, with intervals that range from 30 to 60 min depending on the species and, for females, the stage of the ovarian cycle. During the ovarian cycle, the levels of LH and FSH in plasma correlate strongly with the gonadotropin content of the pituitary, and both are highest just before ovulation.

LH, with a molecular weight of 28,000, and FSH, with a molecular weight of 33,000, are glycoproteins with similar structures. Each is composed of the common pituitary hormone α subunit, also found in TSH, and a unique β subunit that confers them hormonal specificity. The carbohydrate moiety of the latter subunit also enhances their half-life (1 h for LH and about 3 h for FSH) and is critical for receptor binding and biological responses, and its modification allows considerable variation of the bioactivity of secreted LH and FSH molecules in different physiological conditions. LH stimulates the interstitial cell lines of male and female gonads (Leydig and thecal cells) mainly to secrete androgens (particularly testosterone), whereas FSH stimulates testicular Sertoli and ovarian granulosa cells to secrete estrogens (particularly estradiol) and a variety of protein products essential to spermatogenesis and oogenesis, respectively. During the initiation of the reproductive cycle of females, FSH acts on primary follicles to stimulate growth of the granulosa cells. Once the follicular phase is advanced and draws near to ovulation, LH also acts in female granulosa cells to promote progesterone production.

Lactotrophs make up about 15–25% of the adenohypophysial endocrine cell population and are a particularly non-homogeneous group of endocrine cells scattered throughout the anterior pituitary. A significant number are also found in the posterior medial portion of the gland. They synthesize and release prolactin (PRL), a 198 amino-acid protein that owes its name to its role in milk production during lactation in mammals. PRL has structural similarity and a comparable half-life (20 min) to GH. Like somatotrophs, lactotrophs inherently have high secretory activity due to their spontaneous electrical activity, though the former cells are not as active as lactotrophs in terms of their secretory activity. Thus, unlike all other anterior pituitary hormones, the physiological control of PRL is predominantly via tonic hypothalamic inhibition mediated by dopaminergic neurons located primarily in the arcuate nucleus. Therefore, a drop in the levels of dopamine (DA) in portal plasma often translates into an increase of the PRL-releasing activity of these cells. Hormone release by lactotrophs can also be stimulated by an extensive range of factors such as TRH, oxytocin, vasoactive intestinal peptide (VIP), angiotensin II, endothelin-1, serotonin and estrogens, but none of them have been established as a physiologically relevant PRL-releasing hormone. Like other trophic hormones, PRL secretion rises at night, and release occurs in episodic pulses.

In addition to its roles in the growth of alveolar breast cells and milk production, prolactin inhibits GnRH-induced release of gonadotropins in the anterior pituitary as well as the actions of gonadotropins on the gonads. This may be the mechanism by which it prevents ovulation in lactating women and normal sperm production in males. Indeed, PRL facilitates the release of DA from the ME and thus acts in a negative feedback loop to inhibit its own secretion. Many other roles have been described for prolactin, including control of sexual behavior, induction of maternal behavior in pregnancy, and support of the corpora lutea during pregnancy and pseudopregnancy (in rodents). Similarly to ACTH, PRL release is stimulated by different stressor cues, including insulin-induced hypoglycemia, infection, surgery, anesthesia, and fear. Exercise and stimulation of the nipples both stimulate prolactin

secretion, an effect that is thought to be mediated by oxytocin.

DISEASES OF THE ANTERIOR PITUITARY

Diseases of the pituitary can be divided into two major categories of pituitary disturbances, namely, hyperactivity (termed hyperpituitarism) and hypoactivity (hypopituitarism).[4] We focus on those that arise from disorders of the anterior pituitary. Because the function of the anterior pituitary is based on the specialized action (and interaction) of its different endocrine cells, the type of clinical response to each pituitary condition varies with the type of cell that is affected.

Hypopituitarism

With an annual estimated incidence (number of new cases per population in a given time period) of 4.2 per 100,000 and prevalence (proportion of the total number of cases to the total population) of 45.5 per 100,000, hypopituitarism might be caused by either an inability of the gland itself to produce hormones or an insufficient supply of hypothalamic-releasing hormones. It is causally associated with pituitary tumors (61%), non-pituitary lesions (9%), and non-cancerous causes (30%), including perinatal insults, genetic causes, trauma and idiopathic cases. Often, mutations in genes encoding single hormones (or the receptor for their cell-specific hypothalamic releasing factor) result in single pituitary hormone deficiency. Unless successfully treated, hypopituitarism is chronic and lifelong, and in cases of shortage of ACTH or TSH it can cause life-threatening events and lead to increased mortality. If there is decreased secretion of most pituitary hormones, the term panhypopituitarism is used.

ACTH DEFICIENCY This causes adrenal atrophy and ACTH-receptor downregulation, low blood pressure, low blood sugar level, fatigue, weight loss, and low tolerance for stress.

TSH DEFICIENCY Features of TSH deficit include underactive thyroid, fatigue, cold intolerance, weight gain, constipation, hair loss, dry skin and cognitive slowing.

LH, FSH DEFICIENCIES In premenopausal women, absent or infrequent menstrual cycles, infertility, vaginal dryness, dyspareunia (painful sexual intercourse), loss of libido, and loss of some female characteristics occur. In men, these deficiencies are associated with impotence, shriveling of testes, decreased sperm production, infertility, loss of libido, and loss of some male characteristics. In childhood, deficiencies result in delayed or missing onset of puberty.

GH DEFICIENCY The hallmarks of insufficient GH release are decreased muscle mass and strength, visceral obesity, fatigue, decreased quality of life, and impairment of attention and memory; there is stunted growth and dwarfism in children.

PRL DEFICIENCY Inability to produce breast milk after childbirth occurs in some women. However, elevated prolactin concentrations sometimes occur in hypopituitarism because of disruption of inhibitory signals by the hypothalamus, causing lactation, tenderness of the breast, and suppression of gonadotropins, leading to symptoms of hypogonadism.

Hyperpituitarism

Hyperpituitarism is the primary hypersecretion of pituitary hormones. It typically results from a pituitary tumor or adenoma, which represent from 10–25% of all intracranial neoplasms. Most pituitary adenomas are benign, functional, and secrete a hormone that produces clear symptoms characteristic of their condition. The four most common types of adenoma-related hyperpituitarism are prolactinoma, corticotropinoma (Cushing's disease), somatotropinoma (gigantism), and null cell adenomas that do not secrete hormones. Since the enlargement of the anterior pituitary gland can damage the optic nerves and compress the hypothalamus, some of its common symptoms are headache, loss of side or peripheral vision, and hyposecretion of neighboring anterior pituitary hormones. Clinically active pituitary adenomas affect approximately one in 1000 of the general population but are rare in children.

PROLACTINOMA This is an adenoma of lactotroph cells and the most common type of pituitary tumor (30% of pituitary adenomas). In women, high blood levels of prolactin often cause changes in menstruation (periods may become irregular or disappear) and those who are not pregnant or nursing may begin producing breast milk (galactorrhea). Because of the hyperprolactinemia-induced inhibition of the gonadotropin axis, women with hyperprolactinemia also exhibit symptoms of gonadotropin deficit, such as loss of libido, dyspareunia, vaginal dryness and hypogonadism. In males, the most common symptom of prolactinoma is impotence, loss of libido, hypogonadism, oligospermia, and diminished ejaculate volume.

CORTICOTROPINOMA Also known as Cushing's disease, this is a tumor of corticotroph cells that accounts for 20% of pituitary adenomas. The symptoms reflect the presence of excess cortisol or ACTH and include weight gain, high blood pressure, poor short-term memory, irritability, extra fat around the neck, a round and ruddy face, fatigue, and poor concentration. Women also exhibit menstrual irregularity and hirsutism (abnormal hair growth).

SOMATOTROPINOMA This is a tumor of somatotroph cells that represents 15% of pituitary adenomas. The resulting hypersecretion of GH in adults causes acromegaly (an overgrowth of the terminal parts of the skeleton such as the nose, mandible, hands and feet). When it presents in children and adolescents it causes gigantism; disruption of sexual maturation is common, either because of hormone hypersecretion or because of manifestations caused by compression of hypothalamic connections by the adenoma.

NULL CELL ADENOMAS These are dysfunctional tumors that account for 25% of pituitary adenomas. Because they do not secrete hormones, symptoms are restricted to headache, increased intracranial pressure and visual field defects.

GONADOTROPHIC ADENOMAS These are tumors of gonadotroph cells that account for 10% of pituitary adenomas. Although functional, they are usually clinically silent.

THYROTROPHIC ADENOMAS These are rare (less than 1%) but have the particular characteristic of being plurihormonal, since they produce the common glycoprotein α-subunit, prolactin, and the specific β-subunits of LH and FSH.

PHENOTYPIC PROFILE OF ANTERIOR PITUITARY CELL TYPES: DIFFERENT, YET SOMEWHAT SIMILAR

The anterior pituitary endocrine cell types are classically defined on the basis of the expression of a specific hormone and its corresponding mRNA, and this is determined by various lineage-specific transcription factors.[5] Although two-thirds of the population of each anterior pituitary cell type exclusively express their respective hormone at both mRNA and protein levels, the remainder coexpress the mRNAs of two to four different hormones.[6] For example, a GH cell can also express the mRNAs of PRL, TSH-β, LH-β and POMC. In addition, a significant fraction of the non-hormone producing cells contain multiple hormone mRNAs that fail to be translated into the mature protein. Thus, at the mRNA level, the various anterior pituitary hormones are shared by a fraction of cells that express multilineage phenotypes. These multiple mRNA-containing (or polyhormonal) cells are thought to be progenitor cells or "reserve" cells that upon appropriate signals will terminally differentiate depending on developmental and/or physiological needs. The origin and biological relevance of these cells are currently a matter of intense investigation and debate.

The current paradigm for hypothalamic control of anterior pituitary hormone secretion holds that each hypothalamic-releasing hormone modulates the release of a single pituitary hormone by acting on a single cell type. However, when attempts were made to characterize cell phenotypes on the basis of whether their intracellular calcium concentration ($[Ca^{2+}]_i$) responded to a specific hypothalamic releasing hormone, many cells were found to respond to two or more of these agents. Furthermore, hormone release assays in single living cells revealed that multi-responsiveness also exists with respect to hormone release. Such "paradoxical" hormone release by a noncorresponding hypothalamic releasing hormone had, in fact, been observed in a number of *in vitro* and *in vivo* studies using normal and pathological cells. Indeed, a growing body of evidence suggests that 30–40% of normal pituitary cells remain in a multipotential state, able to respond to as many as four different hypothalamic secretagogues.[7] In some pituitary adenomas, these multi-responsive cells can constitute up to 80% of the endocrine cells. Moreover, much like the polyhormonal cells described above, multi-responsive cells are not restricted to any particular cell type. Thus, different cell types can also share receptors to allegedly cell-specific hypothalamic neurohormones. In fact, the mRNA of the receptor for the classic hypothalamic GH inhibitory hormone, somatostatin, is found among all five major cell types of the anterior pituitary. The type 1 CRH receptor (CRH-R1) is not exclusively expressed by corticotrophs; subsets of lactotrophs, gonadotrophs and thyrotrophs also express it in different degrees. It has also been shown that about 35% of somatotrophs bind biotinylated GnRH, the stimulating hormone for gonadotrophs. It has been hypothesized that estrogens and possibly activin stimulate the expression of GnRH receptors in pre-existing GH cells at midcycle, rendering them capable of fully responding as a gonadotroph to help support the GnRH-mediated proestrus surge of LH and FSH release. Additionally, retention of GHRH receptors would allow supporting the GH needs of the reproductive system at this time.[8]

Whether these multi-responsive cells derive from an independent multipotential subset or from pre-existing differentiated cells (termed *transdifferentiation*), they provide a cellular basis for anterior pituitary plasticity. Two observations are consistent with this. The first one is that multi-responsive cells are maximally abundant at puberty, a time signed by huge endocrine and physiological transformation. Secondly, multifunctional cells seem to be more abundant in females than in males, suggesting not only that the female pituitary may be more plastic than the male pituitary, but also that the hormonal changes associated with the female reproductive cycle may promote transdifferentiation.

Unstimulated anterior pituitary cells can also be classified by their biophysical and biochemical properties, more specifically, on their patterns of electrical activity and intracellular

Ca^{2+} dynamics.[9,10] A general feature of cultured anterior pituitary cells is that their membrane potential (V_m) oscillates between potentials of -60 to -50 mV. When V_m reaches the threshold level, pituitary cells fire action potentials (APs), a feature called spontaneous electrical activity and is observed in 15–80% of the cells, depending on the cell type and of cultural and/or recording conditions. The firing of APs causes transients of intracellular Ca^{2+} concentration ($[Ca^{2+}]_i$) that in turn reflect the pattern of electrical activity. Two patterns of electrical activity are typically observed in anterior pituitary cells. The first, termed axonal-type AP spiking, is typically found in gonadotrophs,[11] thyrotrophs,[12] and largely in corticotrophs,[13] and is characterized by sharp single APs that are short in duration (less than 100 ms), with spike frequencies of about 0.7 Hz and amplitudes of more than 60 mV. Axonal-type AP spiking is associated with small-amplitude $[Ca^{2+}]_i$ transients that range from 20 nM to 70 nM and low basal hormonal release in these cells. The second pattern, termed pseudo-plateau bursting and characteristic of cultured somatotrophs[14] and lactotrophs,[15] exhibits broader V_m oscillations in the form of a depolarizing plateau with superimposed bursts of small amplitude APs that usually do not reach 0 mV. Such bursts have a longer duration (several seconds) than gonadotroph APs, so that the burst frequency is significantly lower (about 0.3 Hz), and result in an oscillatory increase in $[Ca^{2+}]_i$ of high amplitude that ranges from 0.3 to 1.2 μM and is sufficient to maintain high and steady hormonal (GH and prolactin) release. Although the typical patterns of spontaneous electrical activity are useful for the identification of the different cell populations of the anterior pituitary, they should not be used as the sole criteria, since the electrical activity of these cells often experience spontaneous reversible transitions between the two modes.

HETEROGENEITY WITHIN ANTERIOR PITUITARY CELL TYPES: SIMILAR, YET SOMEWHAT DIFFERENT

Although different cell types have similarities, it is also apparent that similar cells (that is, cells within a cell type population) may present striking differences. It has been long recognized that each of the different types of anterior pituitary cells are functionally heterogeneous, which is thought to be related to structural heterogeneity. One of the best known examples is found among gonadotrophs,[16,17] which are heterogeneous in size (cell area ranging from $30-170 \, \mu m^2$), morphology (large rounded, small oval and angular stellate cells), physical density (allowing their separation using sedimentation techniques at unit gravity), ultrastructural characteristics (type I gonadotrophs characterized by dilated rough ER (RER) cisternae and secretory granules of 200 nm and $300-700$ nm in diameter, type II gonadotrophs by flattened RER cisternae and $200-250$ nm secretory granules, type III gonadotrophs by a stellate shape and secretory granules of $220-250$ nm), hormone content (small gonadotrophs seem to store only one of the gonadotropins, whereas most of the larger cells either store both LH and FSH, or FSH alone) and responsiveness (variable capacity to bind and respond to GnRH). These gonadotroph subpopulations may account for the differential regulation of LH and FSH secretion *in vivo*.

The other anterior pituitary cell types exhibit heterogeneity of these cellular features as well. In the porcine pituitary, somatotrophs comprise two morphologically distinct subpopulations of low- (LD) and high-density (HD) cells, separable by Percoll gradient, that respond differently to hypothalamic regulators. In LD somatotrophs, somatostatin inhibits GHRH-induced GH secretion, whereas somatostatin alone stimulates GH release from HD somatotrophs.

Functional heterogeneity of corticotrophs displays striking sexual dimorphism. In males, the corticotrophs are of the orthodox phenotype, that is, monohormonal (storing only ACTH) and monoreceptorial (responding only to CRH). Their female counterparts are made of about equal parts of orthodox and multifunctional cells. Sexual dimorphism and functional heterogeneity are even more striking in thyrotrophs, which are mostly polyhormonal in both genders, but only female thyrotrophs co-store GH and/or ACTH in addition to prolactin and/or LH. Among lactotrophs, three morphological types in rodents have been defined by electron microscopy. Type I "classical" lactotrophs contain large irregular-shaped electron-dense secretory granules (diameter $300-700$ nm), type II cells contain numerous medium sized spherical electron-dense granules (diameter $100-250$ nm) and type III cells contain small (<100 nm) spherical granules. Though thought to represent different stages of cell maturity, these morphologically-defined lactotroph subpopulations have functional differences, as these subtypes are differentially sensitive to hypothalamic and local regulators of prolactin secretion.[18]

Several substates of a particular pattern of electrical activity can be found within a given anterior pituitary cell type. In somatotrophs, the burst period of somatotrophs ranges from 2 to 10 s, with longer active phase duration associated with the slower bursting, which results in increased Ca^{2+} influx through voltage-gated Ca^{2+} channels (VGCC) and higher amplitude Ca^{2+} oscillations. An explanation of the heterogeneity of the active phase duration has been recently provided.[14,19]

Lactotrophs also display considerable variability in their spontaneous and receptor-controlled patterns of electrical activity and Ca^{2+} dynamics. The authors have recently addressed the heterogeneity in the $[Ca^{2+}]_i$ responses of lactotrophs to TRH.[20] Responses were evaluated in the absence of extracellular Ca^{2+} to prevent Ca^{2+} influx during the agonist challenge and remove one potential source of heterogeneity. Figure 2.1A shows thirteen Ca^{2+} traces from individual lactotrophs responding to the same TRH application, exhibiting considerable variability. In contrast, Figure 2.1B−E shows traces from four different lactotroph cells subjected to two consecutive TRH applications. In each cell, the response to the second TRH pulse was very similar to that of the first pulse. Thus, during the time course of the observations, heterogeneity in

the Ca^{2+} response to TRH could be observed among lactotroph cells, with uniformity of response within single cells to multiple TRH applications. Of the many potential variables that could result in the observed heterogeneity of the Ca^{2+} responses to TRH, the authors have found that variability in the rate of Ca^{2+} extrusion through the plasma membrane (k_{PMCA}) largely accounts for it. Since TRH may modulate the activity of the plasma membrane Ca^{2+}-adenosine triphosphatase (ATPase) pump, it might also contribute to the observed k_{PMCA}

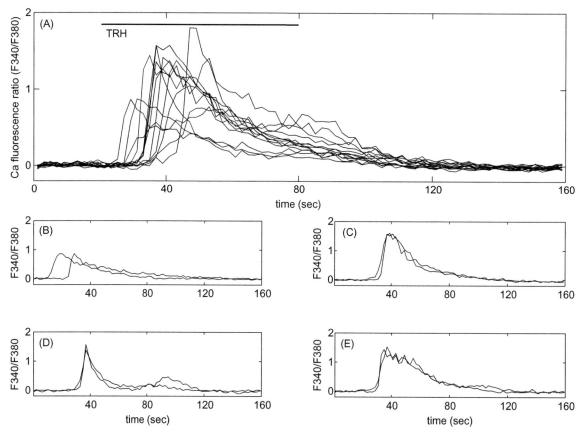

FIGURE 2.1   **Heterogeneity of the Ca^{2+} response to thyrotropin-releasing hormone (TRH) in pituitary lactotrophs.** In all panels, the x-axis shows time in seconds, and the y-axis shows the Ca^{2+} fluorescence ratio (F_{340}/F_{380}). (A) Thirteen individual Ca^{2+} traces from the same experiment. Extensive variation exists in the peak, area, latency of response and decay rate. (B−E) Examples of single cell Ca^{2+} traces from the same experiment showing very similar responses to two successive challenges of TRH, 30 min apart. Original figure published in ref[20].

heterogeneity. We also found that variability in the $[Ca^{2+}]$ in the endoplasmic reticulum ($[Ca^{2+}]_{ER}$) is likely a key element of heterogeneity. In addition, our results did not show a positive correlation between the peak and decay rate of the intracellular Ca^{2+} transients, suggesting that variations in the G_q/IP_3 signaling pathway are not the main source of heterogeneity.

PATTERNS OF ANTERIOR PITUITARY CELL SIGNALING AND SECRETION

Molecular Mechanisms Underlying Excitability and Basal Secretion

In excitable anterior pituitary cells, the resting membrane potential is controlled by classical inward rectifier K^+ (K_{ir}), ether-a-go-go-related gene (ERG) and TWIK-related (TREK-1) K^+-conducting channels. Because the V_m of anterior pituitary cells ranges between -50 to $-60\,mV$, other depolarizing conductances must contribute to maintaining the resting potential. Recent findings indicate that tetrodotoxin (TTX)-insensitive Na^+-conducting channels, along with low-voltage-activated transient (T)-type voltage-gated Ca^{2+} channels, not only participate in the control of the resting potential but also act as pacemaking currents underlying the spontaneous activity frequently observed in isolated cells.[10] In physiological (*in vivo*) conditions, TTX-sensitive voltage-gated Na^+ channels may play a role in the production of action potentials in hyperpolarized cells. However, high voltage-activated, dihydropyridine-sensitive long lasting (L)-type Ca^{2+} channels account for the bulk of the conductance during spike depolarization as removal of extracellular Ca^{2+} and addition of VGCC blockers abolish electrical activity in the majority of the endocrine pituitary cells without affecting their resting membrane potential. Finally, the concerted activity of delayed-rectifying K^+ channels with both BK (large conductance) and SK (small conductance) types of Ca^{2+}-activated K^+ channels repolarize the V_m during the downstroke of an AP. In bursting cells, BK channels are also functionally important.

The distinctive patterns of electrical activity and Ca^{2+} dynamics exhibited by different types of pituitary cells are thought to arise from differences in the expression levels of ion channels such as those involved in bursting and repolarization. For example, BK channels are expressed at higher levels in somatotrophs and lactotrophs (which exhibit pseudo-plateau bursting activity) than in gonadotrophs (which display continuous, axonal-type AP spiking).[21] This correlation between BK channel expression levels and pattern of electrical activity suggests that larger BK conductance favors the generation of bursting activity and global Ca^{2+} signals. Indeed, blockade of BK channels in pituitary somatotrophs can switch the activity pattern of these cells from bursting to spiking, greatly reducing the amplitude of $[Ca^{2+}]_i$ oscillations.[22] In other words, it is possible to convert the firing phenotype of somatotrophs to that of gonadotrophs by reducing their BK conductance. The reciprocal conversion is also true: addition of an artificial BK conductance to cultured gonadotrophs through the dynamic clamp technique changes their activity pattern from spiking to bursting.[19] Mathematical modeling predicted that the activation time constant of the BK conductance is important. BK activation must be fast to promote bursting; if too slow, then the BK current does not promote bursting and instead takes on its traditional inhibitory role on electrical activity.[19] Faster BK activation kinetics limit spike amplitude, preventing full activation of the repolarizing delayed rectifier K^+ current, so membrane potential weakly oscillates around a depolarized level[14,22,23] before falling back toward rest, which results in bursting, increased Ca^{2+} influx and global Ca^{2+} signals.

Unlike classical neuronal synapses, neuroendocrine and endocrine cells require sustained depolarization to trigger the exocytotic pathway. In anterior pituitary cells, VGCC are open only briefly during the short time of a single AP, and the elevated $[Ca^{2+}]_i$ is localized to the nanodomains that form at the inner mouths of open channels. With longer durations and smaller burst amplitudes, VGCC stay open much longer and significant Ca^{2+} influx ensues, resulting in the summation of individual $[Ca^{2+}]_i$ nanodomains that generate a global Ca^{2+} signal. Therefore, differences in the ability of spontaneous firing patterns to generate global $[Ca^{2+}]_i$ signals will determine their basal secretory activity For example, the basal release of GH and prolactin is much higher than that of LH, which is consistent with the pseudo-plateau bursting activity of somatotrophs and lactotrophs that result in high-amplitude (global) $[Ca^{2+}]_i$ signals and the short-duration APs of gonadotrophs that evoke low-amplitude (local) $[Ca^{2+}]_i$ signals.

Receptor-Modulated Hormone Release and Voltage-Gated Ca^{2+} Influx through cAMP-Dependent Pathways

A number of hypothalamic neurohormones and peripheral hormones control AP-driven Ca^{2+} dynamics and Ca^{2+}-dependent hormone release in anterior pituitary cells through modulation of cAMP-dependent pathways. These bind and activate a G protein-coupled receptor (GPCR) to modulate the basal activity of adenylyl cyclase (AC, the enzyme responsible for synthesizing cAMP) and, in turn, Ca^{2+} entry. Those that promote AP-driven Ca^{2+} dynamics and Ca^{2+} dependent secretion bind to GPCRs coupled to the stimulatory class of G proteins (G_s) to increase AC activity through the activation of the α subunit of the G_s protein, whereas those that reduce Ca^{2+} signaling and secretion bind to GPCRs coupled to the inhibitory class of G proteins $(G_{i/o})$ to decrease AC activity.

The stimulatory G_s-signaling pathway is triggered by CRH receptors in corticotrophs, GHRH receptors in somatotrophs and VIP/pituitary adenylate cyclase-activating polypeptide (PACAP) receptors in lactotrophs, somatotrophs and FS cells. Activation of the G_s-operated GPCRs causes plasma membrane depolarization, increased electrical activity and Ca^{2+} entry. The type of $[Ca^{2+}]_I$ response is a plateau elevation of $[Ca^{2+}]_I$ or an increase in the frequency and/or amplitude of $[Ca^{2+}]_I$ transients as elevated cAMP levels promote, directly or indirectly, electrical activity and voltage-gated Ca^{2+} influx (VGCI). The direct pathway consists of the activation of a background Na^+ conductance that results upon binding of cAMP to hyperpolarization-activated and cyclic nucleotide-regulated (HCN) channels. HCN channels are expressed in somatotrophs,[24] lactotrophs,[25] gonadotrophs and thyrotrophs,[12] and probably corticotrophs.[26] Because they are nonselective cation channels, they are likely to play a role in the initiation of the pacemaker depolarization. The indirect pathway consists of the activation of cAMP-dependent kinases (PKAs) that induce phosphorylation-mediated modulation of the function of several plasma membrane ion channels. In corticotrophs, CRH-induced PKA activation inhibits Kir channels to promote slow depolarization and enhanced excitability. In somatotrophs, GHRH-stimulated PKA-mediated phosphorylation results in the opening of not only T-type and L-type Ca^{2+} channels, but also TTX-insensitive voltage-gated Na^+ channels, leading to the upstroke of a voltage spike.

The inhibitory $G_{i/o}$-signaling pathway is triggered in lactotrophs by dopamine D2 and endothelin-1 ETA receptors, and in both lactotrophs and somatotrophs by somatostatin sst1, sst2 and sst5 receptors. Other receptors linked to this pathway and expressed by anterior pituitary cells include those for adenosine, γ-aminobutyric acid (GABA), serotonin, melatonin and neuropeptide Y. Activation of this pathway opposes the actions mediated by the G_s-

signaling pathway resulting in membrane hyperpolarization, silencing of electrical activity and inhibition of Ca^{2+} entry and Ca^{2+}-dependent hormone secretion. $G_{i/o}$-mediated inhibitory actions, which can be irreversibly blocked by application of pertussis toxin, comprise two major signaling branches. The first one stems from the $G_{i/o}$ α-subunit-mediated inhibition of AC activity, downregulating all cAMP-stimulated effects on electrical activity, Ca^{2+} entry and secretion. The second is due to the activation of $G_{i/o}$ βγ dimers that activate Kir3 (also known as G protein-gated inwardly rectifying K^+ channels, GIRK) and inhibit L-type Ca^{2+} channels in a cAMP/PKA-independent fashion, leading to hyperpolarization and cessation of AP firing. Interestingly, three of the four mammalian Kir3 channels are specifically induced by estradiol in lactotrophs in proestrus,[27] underlying dopamine effects that are only observed in this stage of the cycle: strong Kir conductance and robust hyperpolarization, the latter playing a critical role in the prolactin secretory rebound that follows dopamine withdrawal.[28] A novel mechanism has been recently described for endothelin ETA and dopamine D2 receptors in lactotrophs. In addition to the $G_{i/o}$ class of G proteins, these receptors couple to G_z proteins, a subfamily of $G_{i/o}$ proteins that are insensitive to pertussis toxin. Activation of the α subunit and βγ dimers of G_z by these receptors potently block VGCI and prolactin release by inhibiting AC activity and the exocytotic machinery responsible for secretion.[29,30]

Receptor-Modulated Hormone Release through Ca^{2+}-Mobilizing Pathways

All anterior pituitary cell types express Ca^{2+}-mobilizing GPCRs. Examples in gonadotrophs include the receptors for GnRH, endothelins, PACAP and substance P; in lactotrophs such receptors include those for acetylcholine, angiotensin II, TRH, oxytocin, ATP, endothelin, serotonin, galanin and substance P; somatotrophs express ghrelin and endothelin receptors; in corticotrophs, AVP and norepinephrine receptors; and in thyrotrophs, TRH (the main thyrotroph secretagogues) and endothelin elicit Ca^{2+} mobilization. When activated, these Ca^{2+}-mobilizing GPCRs couple to the G_q protein class of heterotrimeric G proteins, eliciting the dissociation of the $α_q$ subunit that triggers phospholipase C-mediated phosphoinositide hydrolysis resulting in the formation of inositol 1,4,5 triphosphate (IP_3) and diacylglycerol (DAG) (Figure 2.2). IP_3 binds to IP_3 receptors expressed in the membrane of the endoplasmic reticulum (ER), the primary storehouse of Ca^{2+} in most cells, causing a mobilization of Ca^{2+} out of this compartment and leading to a large and fast increase of $[Ca^{2+}]_I$.[31]

Two different Ca^{2+} signaling patterns can be observed after the initial $[Ca^{2+}]_I$ spike. The first one, termed "biphasic," is a non-oscillatory pattern found in lactotrophs, somatotrophs and thyrotrophs where the transient $[Ca^{2+}]_I$ spike is followed by a slow decline to a plateau level that is above basal. However, some cells exhibit "monophasic" responses where only the spike or the plateau is observed. A key condition of the biphasic Ca^{2+} response is that the IP_3Rs are opened continuously throughout the time of agonist application. The microdomain of Ca^{2+} that forms near the mouth of the IP_3 receptor during the Ca^{2+} spike is high enough to trigger exocytosis. Furthermore, the transient $[Ca^{2+}]_I$ surge activates the small-conductance Ca^{2+}-activated K^+ channels (SK) and hyperpolarizes the plasma membrane, terminating any previous electrical activity.[10,32] As the Ca^{2+} is removed from the cytosol, by means of plasma membrane ATPase pumps and Na^+/Ca^{2+} exchangers as well as sarco-endoplasmic reticulum Ca^{2+} ATPase (SERCA) pumps, the SK channels close and the membrane depolarizes. Depletion of the ER Ca^{2+} store also provides a signal for the activation of transient receptor potential canonical (TRPC) channels and other store-operated Ca^{2+} channels

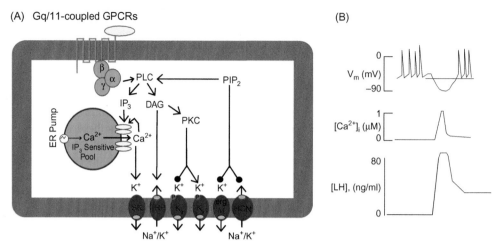

FIGURE 2.2   (A) Mechanism of receptor-induced electrical activity and Ca^{2+} signaling by activated G_q- coupled, Ca^{2+}-mobilizing receptors. (B) Schematic representation of a biphasic response typically observed in gonadotrophs upon GnRH-induced stimulation of electrical activity (top), cytosolic Ca^{2+} (middle) and hormone secretion (bottom). The generation of a biphasic response reflects the tonic activation of IP_3 receptors (see text for details).

(SOCs) important for depolarization. Diacylglycerol may directly activate TRP channels or may inhibit Kir currents through protein kinase C (PKC)-dependent phosphorylation (Figure 2.2). In addition to these events, downregulation of an M/ERG channel via PLC-mediated PiP_2 depletion leads to a sustained depolarization phase that activates voltage-gated Ca^{2+} channels, further depolarizing the cell and initiating single spiking or bursting (Figure 2.2). The resulting Ca^{2+} entry refills the ER Ca^{2+} store, enabling continued Ca^{2+} signaling.

In contrast to the biphasic Ca^{2+} response described above, GnRH-stimulated gonadotrophs and norepinephrine-stimulated corticotrophs engage in an oscillatory Ca^{2+} response after the initial transient of $[Ca^{2+}]_I$. Interestingly, oscillations in IP_3 are not required to generate oscillatory Ca^{2+} release in gonadotrophs. The IP_3 receptor itself is the source of the oscillation, since it is activated by cytosolic Ca^{2+} on a fast time scale and inhibited on a longer time scale. The delayed inhibition of the channel produces oscillations.[33] This is a true ER-mediated oscillation, as it can be produced even when the membrane potential is clamped at a voltage sufficiently high to allow Ca^{2+} influx that refills the ER between each bout of Ca^{2+} release.[34] In the unclamped cell, a key feature of the oscillatory response is the antiphasic pattern of electrical activity and Ca^{2+} release due to the SK channel-mediated inhibitory action of each Ca^{2+} pulse on the plasma membrane (see ref[10]). Once $[Ca^{2+}]_I$ returns to a low level following the Ca^{2+} pulse, firing resumes. The electrical activity and secretion are out of phase; the former serves to refill the ER Ca^{2+} store that periodically releases Ca^{2+} and evokes secretion during sustained stimulation.

PARACRINE (LOCAL) REGULATION AND INTERCELLULAR SIGNALING WITHIN THE ANTERIOR PITUITARY

More than 100 bioactive substances are expressed by cells of the anterior pituitary and can act within the gland through specific receptors, allowing these messengers to exert a local regulatory function during specific physiological states. Depending on the cellular location of both the signaling molecule and its respective receptor, such messengers may act either on the same cell (autocrine control), an adjacent cell (juxtacrine control), a neighboring cell (paracrine control) or even within the same cell where the signaling molecule is produced without being secreted (intracrine control). Such interactions are highly context-dependent and constitute the cellular basis of locally controlled anterior pituitary plasticity. In general, they either promote or inhibit hormone release and cell proliferation, and are switched on/off when hormonal outputs need to be adapted to changing demands of the organism, such as during mating, pregnancy, lactation, stress, inflammation, immune responses, starvation and circadian rhythms. We will briefly highlight physiologically relevant paracrine and autocrine interactions that have been substantiated with reasonable evidence. Comprehensive reviews on this topic have been recently published.[17,35]

Well-established autocrine mechanisms in lactotrophs include the stimulatory loops of VIP, galanin and TGF-α as well as the inhibitory loops of TGF-β1 and endothelins. In gonadotrophs, the autocrine loop made up by activin B (stimulatory) and follistatin (inhibitory), along with that of inhibin (inhibitory), constitute one of the fundamental mechanisms for selective regulation of FSH expression and secretion, as changes in GnRH pulse frequency cause changes of expression of these three modulators that in turn change the FSH:LH ratio.[36] Nitric oxide (NO) is also an important autocrine/paracrine modulator of gonadotrophs as both these cells and FS cells express nitric oxide synthase (NOS). NO may play both inhibitory and stimulatory roles, and this might depend on the cellular source of NO, as NO from FS cells seems to inhibit GnRH-stimulated LH release whereas gonadotroph

NO stimulates basal LH and FSH secretion. PACAP also plays autocrine roles in gonadotrophs where it is specifically expressed in proestrus to stimulate LH release through interaction with the gonadotroph NO system and GnRH signaling pathways. The purine ATP has been shown to be costored with hormones in secretory granules and cosecreted in all endocrine cell types of the anterior pituitary,[37] but autocrine roles have only been consistently shown in gonadotrophs where it potently stimulates basal LH release. In corticotrophs, expression of CRH and AVP (which has also been detected in all endocrine cell types except somatotrophs) and their receptors underlies an autocrine mechanism by which CRH and AVP may contribute to the well-established potentiation of ACTH release by hypothalamic CRH and AVP. In somatotrophs, ghrelin has been shown to sensitize the cell to GHRH at early postnatal age (and possibly at puberty).[38] In thyrotrophs, stimulatory and inhibitory autocrine roles for leptin and neuromedin B, respectively, have been shown to be important in the control of TSH secretion during adaptation to nutritional status.

The majority of the paracrine interactions reported so far are those of gonadotrophs with different endocrine cell types of the anterior pituitary, particularly lactotrophs, somatotrophs and corticotrophs. Among these, one of the first described is the GnRH-stimulated prolactin release by lactotrophs,[39] an effect that is mediated by the release of a still uncharacterized molecule from immature postnatal gonadotrophs.[40] In immature cells, GnRH also elicits a biphasic hormone release response on somatotrophs that begins with inhibition of growth hormone secretion during GnRH application and is followed by a rapid rebound secretion of GH that slowly returns to basal levels. These interactions might be partially related to developmental roles that GnRH may have on these cell types. Some of the candidate paracrine factors from gonadotrophs thought to be involved in the development of lacto-trophs are the glycoprotein hormone α-sub-unit, the growth factor TGF-α and prolactin-releasing peptide (PrRP). Stimulatory paracrine interactions must be counterbalanced with inhibitory ones for the gland to meet and maintain homeostasis throughout different physiological stages; one good example of neg-ative interaction between gonadotrophs and lactotrophs is that mediated by calcitonin (and calcitonin-like peptides). Calcitonin immunore-activity is primarily located in gonadotrophs that are associated with *cup-shaped* lactotrophs (described later in this section). It inhibits basal and TRH-stimulated prolactin synthesis and release as well as lactotroph mitosis. This para-crine effect is likely mediated by the release of TGF-β1 by lactotrophs, which in turn inhibits lactotroph proliferation as well as prolactin expression and secretion.[41] Consistent with this role, estradiol, a well-known promoter of lactotroph secretory activity and proliferation, negatively regulates calcitonin expression in gonadotrophs. Positive and negative interac-tions have also been described between gonadotrophs and corticotrophs; calcitonin gene-related peptide accounts for the former whereas the natriuretic peptides (ANP and CNP) and adrenomedullin (a calcitonin pep-tide family member) account for the latter. Since stress responses are attenuated in preg-nancy and lactation, it is likely that the inhibi-tory tone of natriuretic peptides on corticotroph function is upregulated during these states of elevated estradiol levels in plasma as estrogens are known to upregulate natriuretic peptide expression in heart tissue.

Gonadotrophs can also be paracrine targets themselves. Although inhibition of ovulation during lactation is achieved primary by endorphin-mediated inhibition of hypotha-lamic GnRH neurons triggered by the suck-ling stimulus, local inhibitory cues from lactotrophs and corticotrophs to gonado-trophs at the anterior pituitary may contribute to the suckling-induced negative influence on ovulation. Increased release of β-endorphin and galanin by corticotrophs and lactotrophs, respectively, may contribute to inhibit preovulatory LH secretion. In pregnancy, estrogen-induced high galanin release by lac-totrophs underlies increased lactotroph activity and growth as well as decreased LH release. During stress and undernutrition, stress-induced activation of the hypothalamo−pituitary−adrenal (HPA) axis inhibits the hypothalamo−pituitary−gonadal (HPG) axis. At the pituitary level, this nega-tive influence of the HPA onto the HPG axis may be mediated by the negative paracrine signals of corticotroph β-endorphin on GnRH-induced gonadotroph LH release.

At the heart of the structure and function of the anterior pituitary gland is the key sup-portive and dynamic role of the FS cells. These cells are also excitable and are thought to coordinate activity of endocrine cells. They form two microanatomical structures that may have a large impact on pituitary cell physiology. Located in the center of the hor-monal cell cord, they are often arranged in clusters and form small follicles in rats, but are larger in humans and some other species. In the follicles, numerous microvilli protrude and some cilia are present. Follicle-forming FS cells are polarized. At the apical pole, border-ing the follicle, they form tight junctions among each other, although not always fully sealed, and, more laterally, junctions of the "zonula adherens" type (desmosomes). The basolateral side makes contact with the endo-crine cells and with other FS cells, and extends processes that end on the basal mem-brane surrounding the cell cords. The role of follicles remains unclear but the structures are thought to be involved in intercellular trans-port of metabolic products and ions. A second group of FS cells extends long processes between the hormonal cell types within each glandular cell cord. Although these processes form intercellular junctions among each other, mostly of the zonula adherens-type, they are also electrotonically coupled through gap junctions as shown by rapid propagation of Ca^{2+} currents over long distances in the gland.[42] On this basis, it is hypothesized that these cells coordinate the activity of endocrine cells. In support of this is the finding of a correlation between the number of gap junctions and reproductive maturation in the rat. Interestingly, estradiol seems to increase FS network connectivity, as a steep rise in gap junction number is observed at the end of pregnancy and during lactation. Moreover, in the estrous cycle, connectivity is highest during proestrus and estrus, and it has been recently suggested that this increased connectivity plays a significant role in the preovulatory LH surge.[43] Some FS cells make intimate foot processes with the basal membrane of the extra-vascular spaces at the periphery of the cell cords. In some species FS cells located in the periphery of the cell cords are juxtaposed in a way that they form sinusoid-like spaces. Intercellular lacunae are also often seen between endocrine cells. These lacunae, along with the sinusoid-like spaces surrounded by FS cells and perivascular spaces are thought to form a micro-channel system within the pituitary, through which hormones, local factors, nutrients, ions and waste products can circulate. The three-dimensional architecture of FS cells is under developmental control: in the infant rat, the FS follicles are elongated and participating FS cells have a columnar shape without cellular extensions and displaying very few junctions. At the onset of puberty, they separate into smaller follicular units and start making extensions and junctions, especially tight junctions.

It is thus reasonable to hypothesize that FS cells play key roles in anterior pituitary adaptation to varying physiological conditions, including immune, nutritional and other stresses in which they may operate as critical interfaces in homeostatic mechanisms. Prolonged pituitary activation during immune stress, pregnancy, lactation, starvation and other conditions, may lead to excess production of specific hormones that in turn may result in inhibition of essential physiological processes. In such events, it is likely that FS cells provide a mechanism to circumvent this issue through their capability to blunt many stimulated activities in the anterior pituitary. FS cells release NO that in turn may stimulate guanylyl cyclase activity and increase cyclic guanosine monophosphate (cGMP) in different endocrine cell types to inhibit hormone release (Figure 2.3). In addition, FS cells are permissive for the mitogenic effect of estradiol on lactotrophs by releasing fibroblast growth factor-2 (FGF-2), and mediate the stimulatory action of bacterial endotoxin lipopolysaccharide (LPS), tumor necrosis factor α (TNF-α), VIP, PACAP and interleukin-1 (IL-1) on ACTH secretion by releasing IL-6 which act in corticotrophs in a paracrine fashion (Figure 2.3). FS cells may possibly modulate responses to immune stress as they express receptors for epinephrine, acetylcholine, angiotensin II, calcitonin and ATP that are known to regulate immune cell functions. Indeed, FS cells also express glucocorticoid receptors and have been shown to mediate the glucocorticoid fast negative feedback effect on ACTH (and also prolactin and GH) secretion via ATP-binding cassette (ABC) transporter-mediated externalization of annexin 1, which in turn acts on specific binding sites on endocrine cells to inhibit hormone release[44] (Figure 2.3).

PLASTICITY OF ANTERIOR PITUITARY CELL NETWORK ORGANIZATION AND FUNCTION

The architecture of the anterior lobe is based on interlacing strands of mostly polygonal cells separated by a tremendously developed

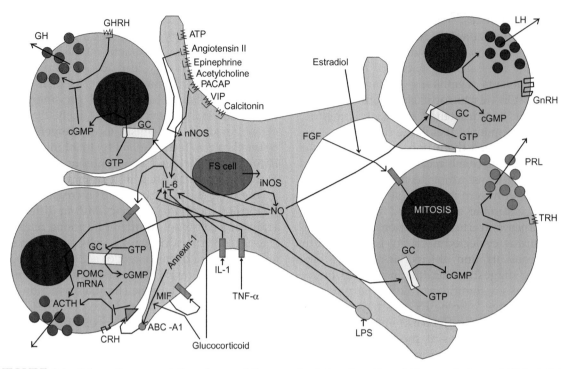

FIGURE 2.3   **Schematic representation of some of the paracrine interactions thought to occur between folliclostellate (FS) cells, lactotrophs, somatotrophs, gonadotrophs and corticotrophs.** →, stimulatory effect; |, inhibitory effect (see text for details).

array of sinusoidal capillaries. Early descriptions of the microanatomy of the adenohypophysis suggested that its various cell types are intermingled, forming cell networks (reviewed in refs[35,42]). Nakane later made the seminal observation that the different pituitary cell types are not distributed homogeneously over the various areas of the gland and within a specific cell cord.[45] In fact, a recent study using two-photon microscopy imaging of genetically engineered mouse somatotrophs expressing the enhanced green-fluorescent protein (EGFP) showed that GH-producing cells are interconnected via adherens junctions, seeming to form a "homotypic connected 3D cell continuum" that displays coordinated Ca^{2+} transients.[46] This cell system consists of numerous intercrossing strands of single GH cells, with larger clusters of GH cells

positioned at the intersection of the cords. Consistent with Nakane's observation, there are differences in the GH cell strand and cluster densities in different regions of the gland. Across the lifespan, the GH system architecture shows plasticity.[46] In prepubertal animals, the patterning of the GH cell system is similar between the lateral and the median zones of the gland. From puberty to adulthood there is a marked increase in the proportion of GH cells in clusters in the lateral portions of the gland than in the median zone surrounding the stalk. Interestingly, the volume-to-surface ratios of the GH cell system returns to prepubertal values in the lateral zones of aged mice, indicating that the plasticity of this cell system continues well into adulthood.[46] More recently, it has been shown that gonadotrophs exhibit significantly different distributions across physiological states,[47] so it is likely that other cell types share a similar cell system architecture and plasticity. In fact, Nakane had observed close associations between somatotrophs and corticotrophs and between gonadotrophs and lactotrophs. Some of the lactotrophs embraced the oval-shaped gonadotrophs with long cellular processes and Nakane named them "cup-shaped" PRL cells.[45] These cell—cell associations are thought to have functional consequences, not only in the cells involved in the association, but also in the pituitary gland as a whole. That the different cell types are both structurally and functionally interconnected is confirmed by the fact that genetic ablation of GH cells dramatically reduced the pituitary content of all hormones.[48]

Finally, the ample possibilities brought about by interconnectivity at the cellular network level may offer emergent properties that might be critical to anterior pituitary cells to adapt their physiology to suit the prevailing environmental conditions. As interconnected cell populations bestowed with the role of mounting large-scale responses to maintain homeostasis, experience-dependent plasticity may represent an inherent property of these cells. This concept has been recently illustrated in female mice that underwent one or more lactations to repeatedly and selectively stimulate activity of pituitary lactotrophs.[49] Throughout the lactotroph population, each lactation event induced alterations in functional connectivity through changes in structural connectivity mediated by differential homotypic lactotroph—lactotroph gap junctional contacts. Following weaning, the lactotroph population is able to maintain the pattern of functional connectivity for weeks to months, a hallmark of long-term experience-dependent plasticity that is reminiscent to that observed in neurons. Likewise, the lactotroph network retains functional connectivity through changes in the extent and strength of cell—cell communication, allowing repeated lactation demands to be met with evolved network dynamics and improved tissue output.[49] Thus, experience-dependent plasticity allows lactotrophs to adapt their hormone releasing activity upon repeated stimulation, a feature that is likely shared by other endocrine cells within the anterior pituitary and, possibly, throughout the body.

CONCLUDING REMARKS

Endocrine cells of the anterior pituitary have a remarkably varied signaling toolkit at their disposal which, in combination with GPCR expression profiles, and gap junctional electrical coupling, may allow incoming signals to be processed and propagated from cell to cell in a specific manner. Not only are these features subject to dynamic changes according to physiological needs, but also the morphological arrangement and functional connectivity of the different cell types of this gland can also go through profound remodeling. These fundamental features allow anterior pituitary cells to adapt their responses to environmental challenges and demands and thereby to meet their vital role in homeostatic control.

Acknowledgements

The authors wish to gratefully acknowledge Jose Arias-Cristancho for preparing Figures 2 and 3 of this chapter as well as the support provided by National Institute of Health grant DK-43200.

References

1. Gonzalez-Iglesias AE, Freeman ME. Brain control over pituitary gland hormones. In: Pfaff DW, editor. *Neuroscience in the 21st Century*. New York: Springer-Verlag; 2012.
2. Ooi GT, Tawadros N, Escalona RM. Pituitary cell lines and their endocrine applications. *Mol Cell Endocrinol* 2004;**228**:1—21.

3. Walker JJ, Terry JR, Tsaneva-Atanasova K, Armstrong SP, Mc Ardle CA, Lightman SL. Encoding and decoding mechanisms of pulsatile hormone secretion. *J Neuroendocrinol* 2010;**22**:1226–38.

4. Melmed S. Update in pituitary disease. *J Clin Endocrinol Metab* 2008;**93**:331–8.

5. Perez-Castro C, Renner U, Haedo MR, Stalla GK, Arzt E. Cellular and molecular specificity of pituitary gland physiology. *Physiol Rev* 2012;**92**:1–38.

6. Roudbaraki M, Lorsignol A, Langouche L, Callewaert G, Vankelecom H, Denef C. Target cells of gamma3-melanocyte-stimulating hormone detected through intracellular Ca^{2+} responses in inmature rat pituitary constitute a fraction of all main pituitary cell types, but mostly express multiple hormone phenotypes at the messenger ribonucleic acid level. Refractoriness to melanocortin-3 receptor blockade in the lacto-somatotroph lineage. *Endocrinology* 1999;**140**:4874–85.

7. Villalobos C, Nuñez L, Frawley S, García-Sancho J, Sánchez A. Multi-responsiveness of single anterior pituitary cells to hypothalamic-releasing hormones: a cellular basis for paradoxical secretion. *Proc Natl Acad Sci USA* 1997;**94**:14132–7.

8. Childs GV. Development of gonadotropes may involve cyclic transdifferentiation of growth hormone cells. *Arch Physiol Biochem* 2002;**110**:42–9.

9. Stojilkovic SS. Molecular mechanisms of pituitary endocrine cell calcium handling. *Cell Calcium* 2012;**51**:212–21.

10. Stojilkovic SS, Tabak J, Bertram R. Ion channels and signaling in the pituitary gland. *Endocr Rev* 2010;**31**:845–915.

11. van Goor F, Zivadinovic D, Martínez-Fuentes AJ, Stojilkovic SS. Dependence of pituitary hormone secretion on the pattern of spontaneous voltage-gated calcium influx. Cell type-specific action potential secretion coupling. *J Biol Chem* 2001;**276**:33840–6.

12. Kretschmannova K, Kucka M, Gonzalez-Iglesias AE, Stojilkovic SS. The expression and role of hyperpolarization-activated and cyclic nucleotide-gated channels in endocrine anterior pituitary cells. *Mol Endocrinol* 2012;**26**:153–64.

13. Liang Z, Chen L, McClafferty H, Lukowski R, MacGregor D, King JT, et al. Control of hypothalamic-pituitary-adrenal stress axis activity by the intermediate conductance calcium-activated potassium channel, SK4. *J Physiol* 2011;**589**:5965–86.

14. Tsaneva-Atanasova K, Sherman A, van Goor F, Stojilkovic SS. Mechanism of spontaneous and receptor-controlled electrical activity in pituitary somatotrophs: experiments and theory. *J Neurophysiol* 2007;**98**:131–44.

15. Gonzalez-Iglesias AE, Jiang Y, Tomic M, Kretschmannova K, Andric SA, Zemkova H, et al. Dependence of electrical activity and calcium influx-controlled prolactin release on adenylyl cyclase signaling pathway in pituitary lactotrophs. *Mol Endocrinol* 2006;**20**:2231–46.

16. Tougard C, Tixier-Vidal A. Lactotropes and gonadotropes. In: Knobil E, Neill JD, editors. *The physiology of reproduction*. New York: Raven Press Ltd; 1994. p. 1711–47.

17. Childs GV. Gonadotropes and lactotropes. In: Neill JD, editor. *Knobil and neill's physiology of reproduction*. 3rd ed. New York: Raven Press Ltd; 2006. . p. 1483–562..

18. Christian HC, Chapman LP, Morris JF. Thyrotrophin-releasing hormone, vasoactive intestinal peptide, prolactin-releasing peptide and dopamine regulation of prolactin secretion by different lactotroph morphological subtypes in the rat. *J Neuroendocrinol* 2007;**19**:605–13.

19. Tabak J, Tomaiuolo M, Gonzalez-Iglesias AE, Milescu LS, Bertram R. Fast-activating voltage- and calcium-dependent potassium (BK) conductance promotes bursting in pituitary cells: a dynamic clamp study. *J Neurosci* 2011;**31**:16855–63.

20. Tomaiuolo M, Bertram R, Gonzalez-Iglesias AE, Tabak J. Investigating heterogeneity of intracellular calcium dynamics in anterior pituitary lactotrophs usign a combined modelling/experimental approach. *J Neuroendocrinol* 2010;**22**:1279–89.

21. van Goor F, Zivadinovic D, Stojilkovic SS. Differential expression of ionic channels in rat anterior pituitary cells. *Mol Endocrinol* 2001;**15**:1222–36.

22. van Goor F, Li YX, Stojilkovic SS. Paradoxical role of large-conductance calcium-activated K+ (BK) channels in controlling action potential-driven Ca^{2+} entry in anterior pituitary cells. *J Neurosci* 2001;**21**:5902–15.

23. Tabak J, Toporikova N, Freeman ME, Bertram R. Low dose of dopamine may stimulate prolactin secretion by increasing fast potassium currents. *J Comput Neurosci* 2007;**22**:211–22.

24. Simasko SM, Sankaranarayanan S. Characterization of a hyperpolarization-activated cation current in rat pituitary cells. *Am J Physiol – Endocrinol Metab* 1997;**272**:E405–14.

25. Gonzalez-Iglesias AE, Kretschmannova K, Tomic M, Stojilkovic SS. ZD7288 inhibits exocytosis in an HCN-independent manner and downstream of voltage-gated calcium influx in pituitary lactotrophs. *Biochem Biophys Res Commun* 2006;**346**:845–50.

26. Tian L, Shipston MJ. Characterization of hyperpolarization-activated cation currents in mouse anterior pituitary, AtT20 D16:16 corticotropes. *Endocrinology* 2000;**141**:2930–7.

27. Christensen HR, Zeng Q, Murawsky MK, Gregerson KA. Estrogen regulation of the dopamine-activated GIRK channel in pituitary lactotrophs: implications for regulation of prolactin release during the estrous cycle. *Am J Physiol − Regul Integr Comp Physiol* 2011;**301**: R746−56.

28. Gregerson KA. Functional expression of the dopamine-activated K+ current in lactotrophs during the estrous cycle in female rats. *Endocrine* 2003;**20**:67−74.

29. Andric SA, Zivadinovic D, Gonzalez-Iglesias AE, Lachowicz A, Tomic M, Stojilkovic SS. Endothelin-induced, long lasting, and Ca^{2+} influx-independent blockade of intrinsic secretion in pituitary cells by Gz subunits. *J Biol Chem* 2005;**280**:26896−903.

30. Gonzalez-Iglesias AE, Murano T, Li S, Tomic M, Stojilkovic SS. Dopamine inhibits basal prolactin release in pituitary lactotrophs through pertussis toxin-sensitive and -insensitive signaling pathways. *Endocrinology* 2008;**149**:1470−9.

31. Berridge MJ. *Neuronal signalling. Cell signaling biology.* Portland Press Limited; 2012. p. 10.1−10.104

32. Li YX, Rinzel J, Keizer J, Stojilkovic SS. Calcium oscillations in pituitary gonadotrophs: comparison of experiment and theory. *Proc Natl Acad Sci USA* 1994;**91**:58−62.

33. Vergara LA, Stojilkovic SS, Rojas E. GnRH-induced cytosolic calcium oscillations in pituitary gonadotrophs: phase resetting by membrane depolarization. *Biophys J* 1995;**69**:1606−14.

34. Kukuljan M, Vergara L, Stojilkovic SS. Modulation of the kinetics of inositol 1,4,5-trisphosphate-induced $[Ca^{2+}]_i$ oscillations by calcium entry in pituitary gonadotrophs. *Biophys J* 1997;**72**(2 Pt 1):698−707.

35. Denef C. Paracrinicity: the story of 30 years of cellular pituitary crosstalk. *J Neuroendocrinol* 2008;**20**:1−70.

36. Kirk SE, Dalkin AC, Yasin M, Haisenleder DJ, Marshall JC. Gonadotropin-releasing hormone pulse frequency regulates expression of pituitary follistatin in folliculostellate cell-enriched primate pituitary cell cultures. *Endocrinology* 1994;**135**:876−80.

37. Stojilkovic SS, He ML, Koshimizu TA, Balik A, Zemkova H. Signaling by purinergic receptors and channels in the pituitary gland. *Mol Cell Endocrinol* 2010;**314**:184−91.

38. Szabo M, Cuttler L. Differential responsiveness of the somatotroph to growth hormone-releasing factor during early neonatal development in the rat. *Endocrinology* 1986;**118**:69−73.

39. Denef C, Andries M. Evidence for paracrine interaction between gonadotrophs and lactotrophs in pituitary cell aggregates. *Endocrinology* 1983;**112**:813−22.

40. Andries M, Denef C. Gonadotropin-releasing hormone influences the release of prolactin and growth hormone from intact rat pituitary *in vitro* during a limited period in neonatal life. *Peptides* 1995;**16**:527−32.

41. Sarkar DK, Kim KH, Minami S. Transforming growth factor-beta 1 messenger RNA and protein expression in the pituitary gland: its action on prolactin secretion and lactotropic growth. *Mol Endocrinol* 1992;**6**:1825−33.

42. Hodson DJ, Romano N, Schaeffer M, Fontanaud P, Lafont C, Fiordelisio T, et al. Coordination of calcium signals by pituitary endocrine cells in situ. *Cell Calcium* 2012;**51**:222−30.

43. Lyles D, Tien JH, McCobb DP, Zeeman ML. Pituitary network connectivity as a mechanism for the luteinizing hormone surge. *J Neuroendocrinol* 2010;**22**:1267−78.

44. Morris JF, Christian HC, Chapman LP, Epton MJ, Buckingham JC, Ozawa H, et al. Steroid effects on secretion from subsets of lactotrophs: role of Folliculo-Stellate cells and Annexin 1. *Arch Physiol Biochem* 2002;**110**:54−61.

45. Nakane PK. Classifications of anterior pituitary cell types with immuno-enzyme histochemistry. *J Histochem Cytochem* 1970;**18**:9−20.

46. Bonnefont X, Lacampagne A, Sanchez-Hormigo A, Fino E, Creff A, Mathieu MN, et al. Revealing the large-scale network organization of growth hormone-secreting cells. *Proc Natl Acad Sci USA* 2005;**102**:16880−5.

47. Alim Z, Hartshorn C, Mai O, Stitt I, Clay C, Tobet S, et al. Gonadotrope plasticity at cellular and population levels. *Endocrinology* 2012;**153**:4729−39.

48. Waite E, Lafont C, Carmignac D, Chauvet N, Coutry N, Christian H, et al. Different degrees of somatotroph ablation compromise pituitary growth hormone cell network structure and other pituitary endocrine cell types. *Endocrinology* 2010;**151**:234−43.

49. Hodson DJ, Schaeffer M, Romano N, Fontanaud P, Lafont C, Birkenstock J, et al. Existence of long-lasting experience-dependent plasticity in endocrine cell networks. *Nat Commun* 2012;**3**:605.

Sensing Calcium Levels: The Biology of the Parathyroid Cell

Luisella Cianferotti and Maria Luisa Brandi

University of Florence, Medical School, Florence, Italy

INTRODUCTION

Extracellular calcium concentration $[Ca^{2+}_o]$ in vertebrates is tightly regulated and has to be maintained within a narrow range. This guarantees the maintenance of multiple intracellular and extracellular metabolic processes in vertebrates, such as mineral and skeletal homeostasis, synaptic activity, muscle contraction, blood coagulation and others. Indeed, since the on-Earth environment is naturally poor in calcium and puts mammals at constant risk of calcium deficiency, systems that guarantee an adequate provision of calcium had to evolve and potentiate during adaptation to life on land, during the shift from an aquatic to a terrestrial environment. Indeed, the parathyroids present in tetrapods (i.e., animals possessing four extremities) are evolutionarily related to the gills in fish, which also express molecules capable of sensing small variations of $[Ca^{2+}_o]$.[1] Although fish do not need homeostatic systems for calcium regulation since they live in an environment that is naturally rich in this mineral (present in the concentration of 8 mM) and do not have distinct parathyroid glands, they are able to synthesize parathyroid hormone (PTH) transcribing two different genes.[2]

The parathyroids are small glands, numbering four or more, and originally took this name because of their close proximity to the thyroid, sharing the same location and vascular supply. The parathyroids originate from the endoderm of the third and the fourth pharyngeal pouches, with the contribution of neural crest cells in the pharyngeal arches, in a progression that includes the formation, migration and differentiation towards mature glands. Since their migration can sometimes arrest along their developmental route between the pharynx and the mediastinum, ectopic parathyroid gland may be found. The gene *Gcm2* (glial cell missing 2) is a key regulator of parathyroid development and it is expressed solely in this tissue.[3] If mutated, PTH-expressing cells fail to form in the parathyroid anlage. Other transcription factors, such as *Rae28*, *Hoxa3*, *Pax91* and *Pax9* are also important in the developing of functional

Cellular Endocrinology in Health and Disease.
DOI: http://dx.doi.org/10.1016/B978-0-12-408134-5.00003-2

parathyroids, since they induce and support *Gcm2* expression.

The parathyroid glands secrete parathyroid hormone (PTH). In response to hypocalcemia, the two major hormones that regulate calcium homeostasis, PTH, the biologically active form of vitamin D (1,25(OH)$_2$D or calcitriol) and calcitonin (CT) contribute to restore [Ca$^{2+}_o$] through coordinated actions on the kidney, bone and intestine, promoting calcium reabsorption, resorption and absorption, respectively. Besides the regulation of calcium metabolism, PTH, 1,25(OH)$_2$D together with circulating fibroblast growth factor 23 (FGF23) are also major determinants in phosphate homeostasis. This implies that feedback mechanisms exist for the coordinated actions of these molecules in the overall maintenance of mineral homeostasis. Indeed, the function of the parathyroid gland is regulated also by these other hormones through their action on specific receptors expressed on the parathyroid cell surface (Figure 3.1).

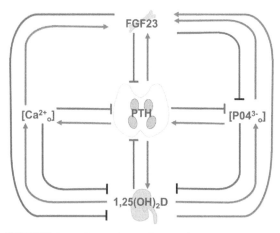

FIGURE 3.1   Systemic regulation of parathyroid function. PTH is directly regulated by mineral ions and main hormones involved in the regulation of calcium and phosphate homeostasis, i.e., FGF23 and 1,25(OH)$_2$D (in blue: inhibitory signals; in green: stimulatory signals; see text for further explanation).

In response to rapid decreases of [Ca$^{2+}_o$] (i.e., acute hypocalcemia), stored PTH is promptly released into the circulation, while in the setting of prolonged reductions of [Ca$^{2+}_o$] (i.e., chronic hypocalcemia) the parathyroids respond by increasing the synthesis of PTH. If the stimulus is long enough, the parathyroids increase also their size, so that hypertrophy and hyperplasia occur.

Many of the studies of parathyroid cell function have been limited by the impossibility of setting primary cultures of human parathyroid cells, since *in vitro* they rapidly lose key features which characterize them. In this view, epithelial *in vitro* models derived from the study of PTH expression and secretion could help in the study of the regulation of PTH expression and secretion and the regulation of the cell cycle of the parathyroid cell. The only clonal parathyroid epithelial cell line, named PT-r, expressing the parathyroid hormone-related peptide (PTHrP) and the calcium sensing receptor (CaSR) has recently been reported to express PTH gene.[4-6]

In this chapter, parathyroid cell biology and physiologic regulation will be described, with an additional brief overview of the principal mechanisms in human parathyroid pathophysiology, namely parathyroid hyperplasia. Mechanisms of parathyroid tumorigenesis responsible for states of primary hyperparathyroidism and of defective parathyroid function (i.e., hypoparathyroidism) will be not covered in this chapter.

BIOSYNTHESIS AND METABOLISM OF PTH

PTH is an 84-amino acid polypeptide (molecular weight (MW) about 9500 Da) stored and secreted by the parathyroid glands in response to relative decreases of [Ca$^{2+}_o$]. Circulating levels of PTH depend on the release of PTH from the pre-formed secretory granules and *de-novo* synthesis of PTH. While the first mechanism is mainly responsible for the rapid release of PTH in response to acute stimuli and occurs within minutes, the second system is activated thereafter, if the stimulus persists. Thus, the long-term replenishment of the secretory vesicles depends upon the transcription and availability of PTH mRNA. The biosynthesis of PTH involves several steps and is estimated to occur in less than one hour.

The human PTH gene is localized on chromosome 11 (11p15). While the rat PTH gene has only one functional TATA transcription starting site, the human and bovine PTH genes have two functional homologous TATA boxes each, which direct the transcription of two different PTH transcripts both in physiology (normal parathyroids) and in pathology (parathyroid hyperplasia or adenoma). PTH gene is composed by three exons separated by two introns. The coding region encompass the 5′ and 3′ untranslated regions, the pre(signal) peptide and the PTH itself, with exon 1 coding for the 5′ untranslated region, exon 2 for the signal peptide, and exon 3 for PTH itself and the 3′ untranslated region. PTH mRNAs belonging to different species conserve homology both in the translated and untranslated regions and intron/exon sequences.[7] In addition, there is substantial homology among mammalian PTH genes. Indeed, human and bovine PTH genes share 85% homology, while human and rat sequences share 75% homology. Since PTH-related peptide (PTHrP), which shares homology with PTH, is located on a similar region on chromosome 12, it is likely that that PTH and PTHrP derive from a common precursor by chromosomal duplication.

The human PTH promoter region contains a cyclic adenosine monophosphate (cAMP)-responsive element, an Sp1 element, and a nuclear factor-Y (NF-Y) binding site, necessary for Sp1 and NF-Y assembly, which strongly drives gene transcription. It also harbors specific DNA sequences, which mediate the negative regulation of PTH transcription by certain hormones (see below).

PTH gene is initially transcribed as a primary translational product, a 115-amino acid-long preproPTH, which is then cleaved to the 90-amino acid proPTH.[8] The pre-peptide of 25 amino acids appears to harbor properties of a signal peptide, with hydrophobic core, necessary for the transport within the rough endoplasmic reticulum and greatly conserved among the different species, bounded by charged amino acids at the *N-* and *C*-terminal ends. The cleavage of the pre-peptide by microsomial enzymes produces the 90-amino acid-long proPTH. Afterwards, the pre-peptide is quickly degraded and quite impossible to detect in intact cells. ProPTH is eventually converted in mature PTH within minutes by a subsequent proteolitic cleavage, which takes place in the Golgi apparatus. The 84-amino acid long PTH is the major stored and secreted and form of PTH. PTH 1–84 is the biologically active hormone, which is first packaged within secretory granules and then secreted upon several stimuli (see below). The *N*-terminus of PTH contains the biologically active 34-amino acid-long region involved in the binding to the specific receptor (PTH1R or type 1 PTH/PTHrP receptor), in common with the PTHrP, and signal transduction. PTH is greatly conserved among different species, while the *C*-terminal region is important in the secretory process.

Upon proper stimuli (i.e., hypocalcemia), PTH is released from the dense secretory granules and released into the circulation. Circulating PTH has a brief half-life (5 min) since it is quickly cleared by the liver and the kidney. The synthesis of mature PTH with the parathyroid cell depends directly on the amount of preproPTH generated, since the conversion from preproPTH to proPTH and eventually PTH is spontaneous. Thus, the limiting step under regulation of systemic hormones and mineral ions is the amount of preproPTH mRNA transcribed.

Intracellular degradation of mature PTH usually occurs so that PTH C-terminal and N-terminal fragments are also present, like PTH, within the parathyroid cell and can be released into the circulation. In respect to this, serum PTH is immunoheterogeneous.[9] The intracellular degradation of PTH may take place in two different ways. First, the secretory granules can fuse to lysosomes or to vesicles containing cysteine proteases. Serum PTH fragments (mainly C-terminal fragments), which lack the N-terminal portion, can also derive from peripheral catabolism taking place mainly in the liver, and they are eventually cleared by the kidney. It was initially believed that the C-terminal fragments were biologically inactive. Indeed, it has been recently demonstrated that these fragments are able to exert opposite effects relative to 1−84 or 1−34 PTH, likely binding to alternative receptors (supposedly C-PTH receptor) not yet cloned. In eucalcemic conditions, C-fragments constitute 20% of total circulating PTH molecules, whereas in response to hypocalcemia or hypercalcemia they can decrease or rise, respectively.

REGULATION OF PTH SECRETION AND TRANSCRIPTION

Both the acute and chronic PTH secretion are strictly regulated primarily by $[Ca^{2+}{}_o]$, but also by other mineral ions (phosphate, magnesium) and systemic hormones. The key limiting intracellular steps, which are under the control of systemic hormones and mineral ions, are depicted in Figure 3.2. Besides the intact biologically active form, the intracellular production of N-terminal and C-terminal fragments is also tightly regulated.

Moreover, calcium can also directly regulate the amount of C-terminal fragments generated by the liver. All in all, serum PTH concentration is the result of multiple hints in order to maintain $[Ca^{2+}{}_o]$ within the target range.

Calcium-Sensing Properties of the Parathyroids: The CaSR

The major regulatory signal for the rapid regulation of PTH secretion is $[Ca^{2+}{}_o]$.[10] This process is mainly mediated by the extracellular CaSR, which is mainly expressed on the membrane of cells belonging to organs classically involved in maintaining mineral homeostasis, such as the parathyroid chief cells, thyroidal C-cells, apical membrane of cells of the renal distal tubule, intestine and bone cells. Many insights in the physiology of the CaSR have derived from the study of diseases caused by mutations of the CaSR gene. Inactivating mutations of CaSR causes familial hypercalciuric hypercalcemia (FHH1, OMIM 14598), while activating mutations lead to autosomal dominant hypocalcemia (ADH, OMIM 601298).[11] The CaSR belongs to family C of the G protein-coupled receptors (GPCRs), which is likely present in the form of a homodimer on the cell surface. Disulfide bonds and non-covalent interactions mediate the formation of dimers. This process likely occurs in the rough endoplasmic reticulum, before the expression on the plasma membrane. Each monomer comprises a long (600 amino acid) N-terminal extracellular tail, 7 transmembrane domains, and a short C-terminal intracellular tail harboring several phosphorylation sites. Although a solved crystal structure is not available for CaSR,[12] it is conceivable that the receptor binds to Ca^{2+} in an extracellular tridimensional configuration, the so-called "Venus flytrap" (i.e., Venus flytrap domain, VFTD), which harbors Ca^{2+}-binding sites and folds autonomously.[13,14] The cysteine residues contained in the VFTD are essential for forming the disulfide bonds across the dimer interface. Moreover, this domain is critical in transducing CaSR signaling, since it translates the closing of the VFTD and twisting dimer interface into activation of the heptahelical domain upon Ca^{2+} binding.[15] Mutations in

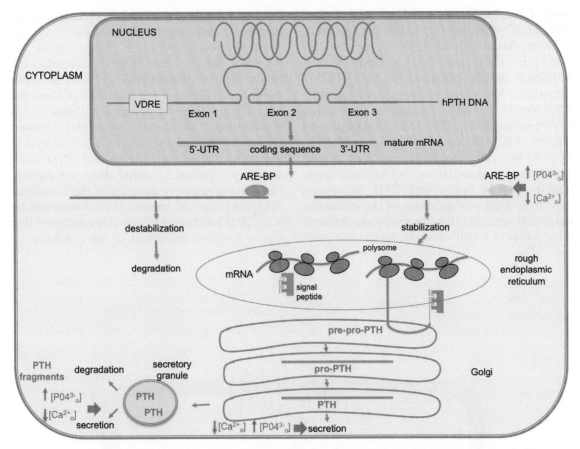

FIGURE 3.2 **Transcription, maturation, and secretion of PTH in the parathyroid chief cell and regulation of the limiting intracellullar steps by mineral ions.** The promoter region of *PTH* gene harbors a response element for 1,25(OH)$_2$D/VDR (VDRE), involved in the transrepression of PTH transcription. The stability of PTH mRNA is controlled by cytosolic proteins, i.e., AU-rich element-binding proteins (ARE-BP), which, upon binding to AREs, stabilize or destablized PTH mRNA. Stabilizing ARE-BPs are induced by hypocalcemia and hyperphosphatemia. Nascent preproPTH is transferred by a signal peptide to the Golgi apparatus, where final maturation to proPTH and eventually PTH occurs. Mature 1−84 PTH peptide is then stored in secretory granules, secreted upon proper stimuli or eventually degraded into fragments, which can be also released in the circulation, contributing to PTH immunoreactivity.

the cysteine-rich domain causes FHH1 and abrogate CaSR-mediated signaling. Indeed, after Ca^{2+} binding, the CaSR changes its conformation in the extracellular domain, which is then transmitted to the 7-pass transmembrane domain to favor interactions with G-protein subunits. Other divalent cations and other compounds are able to bind CaSR and, therefore, can modulate its activity.

Indeed, agents such as L-amino acids with aromatic side chains determine conformational changes in the receptor that render CaSR more sensitive to Ca^{2+}, and for this reason they are referred to as "calcimimetics" (see below). The class-C GPCRs transduce the signal through the activation of heterotrimeric G proteins (G$_{q/11}$, G$_{i/o}$ and G$_{12/13}$). The isoforms of G protein associated to the receptor make CaSR-mediated signaling upon ligand binding cell-type specific. In the parathyroids, CaSR-mediated activation of $G_{q/11}$ by divalent ions (i.e., Ca^{2+} and Mg^{2+}) leads to the activation of phospholipase C beta (PLCβ), consequent hydrolysis of phosphatidylinositol-4,5-biphosphate and formation of inositol 1,4,5-trisphosphate (IP3) and diacylglycerol (Figure 3.3). IP3 mobilizes calcium $[Ca^{2+}_i]$ from its intracellular stores. High $[Ca^{2+}_i]$ inhibits cAMP accumulation, which ultimately inhibits vesicle fusion and PTH exocytosis, maybe through modifications of the cytoskeleton. CaSR can also directly modulate intracellular levels of cAMP through interaction with $G_{i/o}$, which inhibits adenylate cyclase activity, thereby reducing intracellular cAMP. Moreover, CaSR activates phospholipase A2 (PLA2), which, in turn, increase the production of arachidonic acid and activates phosphatidylinositol 4-kinase that restores the intracellular levels of phosphatidylinositol-4,5-biphosphate. All these intracellular changes elicited by the binding of Ca^{2+} to CaSR lead to the quick release of pre-formed PTH from secretory granules into the circulation. Parathyroid-selective ablation of the α subunit of G_q abolishes the control of PTH secretion by $[Ca^{2+}_o]$. It has been recently demonstrated that loss of function mutations of the α subunit of

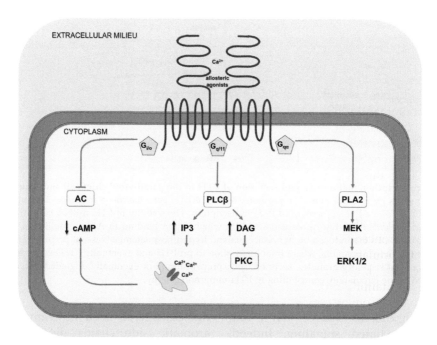

FIGURE 3.3 Schematic representation of calcium sensing receptor (CaSR)-mediated signaling in the parathyroid cell. Upon Ca^{2+} binding to the CaSR, the receptor interacts with heterotrimeric G proteins. The activation of $G_{q/11}$ results in the activation of phospholipase C beta (PLCβ), which, in turn, increase inositol trisphosphate (IP3) and dyacilglycerol (DAG) levels, thus increasing the mobilization of calcium from the intracellular store and activating protein kinase C (PKC). CaSR may also interact with $G_{i/o}$ and inhibit adenylate cyclase (AC) activity, thus decreasing directly cyclic adenosine monophosphate (cAMP) levels. CaSR signaling may also activate phospholipase A2 (PLA2), which activates the mitogen-activated protein kinase (MAPK) pathway, with increases in MEK, ERK and arachidonic acid formation, thus regulating the cell cycle.

G_{11}, referred to as *GNA11*, are responsible for of FHH type 2, while $G\alpha_{11}$ mutants with gain of function cause ADH type 2.[16] The intracellular tail of the CaSR harbors residues that can be phosphorylated or dephosphorylated. Phosphorylation of Thr-888 operated by protein kinase C (PKC) inhibits the mobilization of $[Ca^{2+}_i]$ induced by changes in $[Ca^{2+}_o]$. The activation of PKC by phorbol esters indirectly elicits PTH secretion by inducing phosphorylation of the CaSR. The same effect is obtained by means of mutations of Thr-888, which lead to substitutions of negatively charged amino acids mimicking a permanent phosphorylation at this site. Conversely, mutations of Thr-888, which abolish the phosphorylation site, cause a constitutive activation of the receptor with a chronic suppression of PTH secretion. Indeed, the CaSR appear to exist in two different states, depending on the phosphorylation of Thr-888. In the presence of low–normal $[Ca^{2+}_o]$ the receptor is mainly present in the phosphorylated state, while when $[Ca^{2+}_o]$ is high, it is predominantly dephosphorylated.

The CaSR is sensible to small changes of $[Ca^{2+}_o]$. When activated by small increases of serum calcium concentration, (above 1 mM, which represents the normal serum calcium concentration), it inhibits PTH secretion and calcium reabsorption in the renal distal tubule in order to decrease $[Ca^{2+}_o]$. An increase in serum calcium concentration has the additional effect of stimulating the secretion of CT, which lowers $[Ca^{2+}_o]$, even if CT plays a marginal role in maintaining mineral homeostasis in adult humans. Conversely, a decrease of serum calcium concentration will trigger PTH secretion and favor renal calcium reabsorption and calcium resorption from bone. PTH, in turn, will enhance renal 1α-hydroxylase activity and $1,25(OH)_2D$ formation in the kidney, thus favoring intestinal calcium absorption. All in all, the CaSR indirectly and directly regulates the production of the main hormones that control mineral fluxes between outer and inner cellular compartments (Figure 3.1). The synthesis of $1,25(OH)_2D$ is also directly regulated by calcium and phosphate, i.e., dampened down by hypercalcemia/hyperphosphatemia or elicited by hypocalcemia/hypophosphatemia.

When parathyroid cells are exposed to decreases of $[Ca^{2+}_o]$, PTH begins to increase within 1 min, reach a peak within 5–10 min and then slowly decreases more of 50% relative to the maximum level after one hour despite the persistence of hypocalcemia, while rapid correction of hypocalcemia causes PTH to decrease within minutes thereafter. The relationship between $[Ca^{2+}_o]$ and PTH is described by a steep inverse sigmoidal curve, according to which circulating PTH responds with large changes to small variation of $[Ca^{2+}_o]$. This ensures the maintenance of $[Ca^{2+}_o]$ within a narrow range.[10] This curve is characterized by the "set-point," namely the $[Ca^{2+}_o]$ at which PTH secretion is half-maximal. The $[Ca^{2+}_o]$ at set-point roughly corresponds to the $[Ca^{2+}_o]$ that is maintained *in vivo*. PTH then exerts its actions primarily on bone, promoting bone resorption, and on the kidney, promoting phosphaturia in the proximal tubule, calcium reabsorption in the distal tubule and enhancing the formation of biologically active vitamin D in the proximal tubule, as previously described. While the first two actions are rapid (occurring within minutes), the last one requires some hours. In view of this, the CaSR-mediated control of PTH secretion is primarily important in defending towards hypocalcemia. Conversely, direct and indirect actions of $[Ca^{2+}_o]$ in the kidney play a key role in protecting from hypercalcemia. $[Ca^{2+}_o]$ directly promotes calcium excretion acting on the CaSR in the cortical thick ascending limb (CTAL). In some species, a increase in $[Ca^{2+}_o]$ also mediates a CT-mediated inhibition of bone resorption, thus contributing to the rapid response against hypercalcemia. Indeed, mice devoid of PTH are still able to respond to increases in

$[Ca^{2+}_o]$ by means of increased calcium excretion, underlining the importance of these homeostatic responses. Thus, the "hormone" Ca^{2+} directly activates a short feedback loop directly controlling its own levels.

In parathyroid gland adenomas or carcinomas, abnormal parathyroid cells are characterized by decreased sensitivity to $[Ca^{2+}]_o$, resulting in an uncontrolled continuous release of PTH, with a subsequent increase in serum PTH levels and hypercalcemia.

Since the CaSR is a main regulator of parathyroid function, it has been considered also a major target for novel compounds potentially able to regulate mineral and skeletal metabolism through the modulation of CaSR-mediated signaling. In view of this, allosteric modulators of the CaSR have been developed in order to modulate PTH secretion.[17] Several compounds can modulate the three-dimensional structure of the CaSR, making it more or less sensitive to changes in $[Ca^{2+}_o]$. Activators of the CaSR (calcimimetics) lead to an inhibition of PTH secretion and stimulation of renal calcium excretion. They have been approved for the treatment of secondary hyperparathyroidism (SHPT) and as a medical therapy for primary hyperparathyroidism with severe hypercalcemia, where surgery is contraindicated. Moreover, they are currently being employed in other forms of hyperclacemia and some forms of hypophosphatemia. Conversely, inhibitors of CaSR-mediated signaling (calcilytics) can induce a transient increase in PTH secretion, which can mimic the anabolic effects on bone of PTH-intermittent therapy. Nonetheless, so far calcilytics have failed as anabolic therapy for osteoporosis, while they are currently being tested in conditions of hypocalcemia and hypercalciuria.[18] In animal models of parathyroid hyperplasia secondary to CKD, the administration of calcimimetics has been found to partially restore the expression both of CaSR and VDR on parathyroid cell surface.[19]

Phospate/FGF23 Regulation of the Parathyroids

Among PTH actions in the kidney, the direct inhibition of phosphate reabsorption and the $1,25(OH)_2D$-mediated enhancement of phosphate intestinal absorption and bone resorption contribute to the maintenance of phosphate homeostasis. Inorganic phosphate $(H_2PO_4^-, HPO_4^{2-})$, in turn, which is maintained within a narrow range in the circulation, is implicated in multiple cellular metabolic signaling processes and structures. Among the different functions, phosphate is also recognized as a major regulator of PTH secretion and synthesis and parathyroid cell division. Moreover, it has been demonstrated both *in vitro* and *in vivo* that phosphate actions on the parathyroids are independent of modifications in $[Ca^{2+}_o]$ and $1,25(OH)_2D.$[20] Nonetheless, the precise mechanisms by which phosphate is sensed by the parathyroids and the specific phosphate sensors have not been still identified.[21] Besides PTH, which has been demonstrated to serve also as a phosphaturic factor, the major regulator of serum phosphate levels is serum FGF23, a bone-derived phosphaturic hormone, which acts on a specific co-receptor complex, the FGF receptor 1 (FGF1)/Klotho. This receptor complex is mainly expressed in the renal distal tubule, but also in the parathyroids and other tissues, and is stimulated by phosphate and $1,25(OH)_2D$ increase. Although a specific structure for phosphate sensing, the *pho regulon*, has been identified in yeast and bacteria, a specific mammalian phosphate sensing system has not yet been characterized.[22] Nonetheless, it has been shown that phosphate-induced ERK-mediated apoptosis of hypertrophic chondrocytes could be prevented by means of inhibition of ERK phosphorylation. This, together with the crosstalk between different endocrine organs (mainly kidney—parathyroids—bone) in response to changes in phosphate, point towards the presence of a phosphate sensing mechanism also in mammalian tissues.

In chronic kidney disease (CKD), as renal function progressively decreases, PTH levels gradually increase and SHPT occurs. In this setting, phosphate retention, decreased renal production of 1,25(OH)$_2$D, hypocalcemia and their pathophysiological interplay are the major inducers of SHPT.[23] Phosphate is believed to play a key role in the pathogenesis of SHPT, both indirectly, through modifications of [Ca^{2+}$_o$], and directly. It was traditionally believed that hyperphosphatemia could affect PTH secretion through the establishment of a transient hypocalcemia that would have triggered hyperparathyroidism. Indeed, an increase in serum phosphate levels, as an effect of the decreased renal function, is directly "sensed" by the parathyroids and has the primary effect of increasing directly the synthesis of PTH and enhancing PTH mRNA stability.[24]In vitro studies have demonstrated that increasing phosphate concentrations have the ability to stimulate PTH secretion independently of changes in calcium concentration when administered parathyroid tissue slices or to whole rat parathyroid glands in culture, but not on isolated cells. The parathyroid response to changes in serum phosphate concentration occurs at different levels: secretion, gene expression and cellular proliferation. The mechanisms by which phosphate controls PTH secretion might involve the modulation of phospholipase A2 signaling, through arachidonic acid mobilization.

Dietary-induced hypophosphatemia has been shown to markedly decrease PTH gene expression to the same extent as diet-induced hypocalcemia induces increase in PTH mRNA levels. The mechanism of calcium and phosphate-induced modification on PTH mRNA levels are post-transcriptional. Indeed, calcium and phosphate induce changes in parathyroid cytosolic proteins, which are responsible for stabilizing mRNA through binding to instability regions

(cis elements) contained in the PTH mRNA 3′ untranslated region (UTR).[25] These proteins, such as AU-rich binding factor (AUF1) are induced both in conditions of hyperphosphatemia and hypocalcemia, and their expression is directly correlated to PTH mRNA levels in vivo since they stabilize the PTH transcript. Thus, phosphate directly regulates the stability of PTH mRNA.

The major hormone implicated in phosphate homeostasis is FGF23. Klotho has been demonstrated in the parathyroids by means of immunoblotting, reverse transcription polymerase chain reaction (RT-PCR), and immunoistochemistry, while it is not present in the surrounding tissues, such as the thyroid. The co-expression of FGFR1 makes the parathyroid a target organ for FGF23.[26] The first cues that FGF23 plays a role in the control of PTH secretion come from the characterization of transgenic mice overexpressing FGF23 or humans with hypophosphatemic rickets or tumor induced osteomalacia, diseases characterized by high levels of FGF23. In these models, in contrast to the expected hyperparathyroidism in the face of hypocalcemia and low levels of 1,25(OH)$_2$D, PTH secretion was blunted. The intraperitoneal administration in rats of a full-length FGF23 harboring mutations making it resistant to proteolytic cleavage, led to the development of hypophosphatemia and reduction in 1,25(OH)$_2$D levels, as expected, but also to a inhibition of serum PTH and PTH mRNA within the parathyroids. Since an increased phosphorylation of ERK1/2 was demonstrated in the parathyroids, a pathway activated by FGF23-activated MAPK-signaling, the PTH reduction was interpreted as a direct effect of FGF23 on the parathyroid cells. Blocking the MAPK pathway by means of an ERK1/2 phosphorylation inhibitor, prevented the FGF23-mediated reduction in PTH levels in vivo and in vitro in rat parathyroid tissue slices. Thus, FGF23 exerts an inhibitory effect on PTH synthesis and secretion both in vivo and *in vitro* through activation of MAPK signaling. Incubating primary cultures of bovine parathyroid cells with FGF23 further confirmed the inhibition of PTH expression in terms of mRNA. Since increased levels of 1α-hydroxylase mRNA could be demonstrated, it was speculated that the intracellular formation of $1,25(OH)_2D$ could mediate the inhibitory effect of FGf23 on PTH synthesis. Since in transgenic mice expressing human PTH gene in the parathyroids the intraperitoneal administration was still able to blunt PTH expression, it was inferred that FGF23 effects on the parathyroids are similar in men and rodents.

Vitamin D Regulation of the Parathyroids

The biologically active form of vitamin D, calcitriol or $1,25(OH)_2D$, is a potent inhibitor of PTH transcription.[27] The vitamin D receptor (VDR), a ligand-dependent transcription factor, is expressed in the parathyroids as abundantly as in the duodenum, the classic target site for calcitriol. Upon ligand binding, VDR heterodimerizes with the retinoid X receptor (RXR), translocates into the nucleus, binds to specific vitamin D response elements (VDREs) and activates target genes transcription. The inhibitory effect of $1,25(OH)_2D$ on PTH secretion was initially demonstrated *in vitro* in primary cultures of bovine parathyroid cells, where the administration of calcitriol markedly reduced PTH mRNA levels and, consequently, PTH secretion. This effect was confirmed *in vivo* in rats, where the administration of non-hypercalcemic doses of calcitriol reduced PTH mRNA levels in the parathyroids. Thereafter, the presence of a specific VDRE was demonstrated in the promoter region of *PTH* gene.[28] The sequence of this element and likely the recruitment of nuclear factors differ from that of other known upregulatory VDRE present in other genes. Indeed, the binding of 1,25 $(OH)_2D/VDR$ complex to this response element mediates a strong transcriptional repression of *PTH* gene transcription. Calcitriol may also interact with another promoter element, the E-box (CANNTG)-like motif, which represents another nVDRE first identified in the human 1α-hydroxylase promoter. In the absence of $1,25(OH)_2D$, a VDR interacting repressor (VDRI) binds the nVDRE and transactivates transcription through interaction with the histone acetylase (HAT) coactivator p300/CBP. When $1,25(OH)_2D$ binds to the VDR, the VDRI transactivation function is inhibited by protein–protein interaction, which induces dissociation from HAT and interaction with histone deacetylase (HDAC) mediating ligand-induced transrepression.[29]

Calcitriol may also enhance the expression of its own receptor – the VDR – mRNA in the parathyroids. This leads to a stronger and amplified inhibitory effect of $1,25(OH)_2D$ on PTH transcription under physiological conditions. This does not occur in the presence of hypocalcemia, where, despite the high levels, $1,25(OH)_2D$ does not increase VDR mRNA. This unexpected effect might be mediated by calreticulin, a calcium-binding protein contained in the endoplasmic reticulum, which also regulates gene transcription through the interaction of hormone response elements. Indeed, calreticulin, which is increased in the nuclear fraction in the setting of hypocalcemia, associates with the VDR/RXR through a protein–protein interaction, thus blocking the inhibitory effect of $1,25(OH)_2D$ on PTH transcription. This can explain why hypocalcemia is characterized by increased PTH mRNA levels despite high levels of $1,25(OH)_2D$ and why many patients with CKD and SHPT are resistant to treatment with $1,25(OH)_2D$ or its analogs.

To assess whether the $1,25(OH)_2D/VDR$ system is important *in vivo* in the control of PTH secretion directly or indirectly, through modification of mineral homeostasis, experiments were carried out in mice devoid of VDR, which display hypocalcemia, hypophosphatemia, secondary hyperparathyroidism and rickets. The correction of mineral metabolism in these mice by means of a diet high in calcium, phosphate and lactose, prevented the development of SHPT and parathyroid hyperplasia, leading to the concept that the $1,25(OH)_2D/VDR$ system is not essential (or is redundant) for the control of PTH secretion *in vivo*, nor for the control of parathyroid cell proliferation.

Calcitriol may also indirectly inhibit PTH synthesis by enhancing CaSR-mediated control of PTH secretion. Indeed, while the transcription of CaSR is not influenced directly by $[Ca^{2+}_o]$, it is regulated by $1,25(OH)_2D$ in the parathyroids as well as in other tissues (kidney, thyroidal C cells). VDRE have been identified in the two promoter regions of *CaSR* gene, yielding alternative transcripts. Upon ligand binding, the VDR upregulates CaSR mRNA expression, thus rendering the parathyroid cell more sensitive to changes in $[Ca^{2+}_o]$, which then evokes the events above described. Conversely, CaSR-mediated signaling upregulates the VDR mRNA in the parathyroids.

Despite the strong inhibitory effects of $1,25(OH)_2D$ on PTH synthesis, its use in SHPT has been limited because of the hypercalcemic effect. Several analogs, synthesized by modifications of the side chain of $1,25(OH)_2D$, such as oxacalcitriol, calcipotriol, and paracalcitol have been employed. These analogs keep the ability to evoke the same nuclear responses of $1,25(OH)_2D$, but they are administered in non-hypercalcemic doses. However, although they display *in vitro* a similar effect in inhibiting PTH gene expression and PTH secretion; this ability is blunted *in vivo*. Indeed, $1,25(OH)_2D$ remains the most effective in decreasing PTH levels *in vivo*, even if it is used at doses which do not cause hypercalcemia. Thus, the absolute and relative advantages of administering calcitriol analogs

instead of $1,25(OH)_2D$ are still debated and remain to be demonstrated.[30]

MAIN MECHANISMS OF PARATHYROID GLAND HYPERPLASIA IN SHPT

Parathyroid cells are terminally differentiated cells, which rarely undergo mitosis under physiologic conditions. This is one reason for the difficulty in maintaining parathyroid cells in culture.

Stimuli that are able to elicit PTH secretion are often able to stimulate growth of quiescent parathyroid cells and produce change in parathyroid anatomy (cellular hypertrophy and hyperplasia) if persisting constantly over time, since they trigger proteins (cyclins and cyclin dependent kinases) regulating the entry into the cell cycle. For instance, in chronic renal failure, hyperphosphatemia due to phosphate retention, low $[Ca^{2+}_o]$, low levels of $1,25(OH)_2D$, intestinal calcium malabsorption, which contribute to the genesis of secondary hyperparathyroidism, lead in the long term to the development of enlarged parathyroids, namely diffuse parathyroid hyperplasia, due to a multiclonal expansion of the parathyroid cells.[31] In addition to these systemic factors, local molecular alterations likely contribute to the genesis/worsening of this endocrine abnormality. Thus, reduction in the expression of the VDR, CaSR and FGFR1/Klotho on parathyroid cell surface, especially in the advanced stages of parathyroid hyperplasia, can render the parathyroids resistant to the modulation by $[Ca^{2+}_o]$, $1,25(OH)_2D$, and FGF23.[32,33] Since both $[Ca^{2+}_o]$ and calcitriol induce the transcription of their own receptors, the reduced expression of CaSR and VDR can be the result of the low $[Ca^{2+}_o]$ and calcitriol levels themselves. Altered modulation of PTH mRNA transcripts could also play a role in determining an abnormal control of parathyroid growth. In the establishment of hyperplasia during SHPT, at the beginning the parathyroids grow diffusely and polyclonally. Short-standing parathyroid hyperplasia can still be reverted since the opposite stimuli (i.e., increases in calcium, decreases in phosphate, or increases in calcitriol administration) may limit parathyroid cell growth. In advanced stages of parathyroid hyperplasia, some clusters of cells start proliferating at higher rate and form nodules, giving rise to nodular hyperplasia. These nodules are mainly monoclonal.[34]

In the nephrectomized, uremic, hyperphosphatemic rat model, the rapid development of parathyroid hyperplasia is associated to a robust expression of the cell-cycle inducer transforming growth factor alpha (TGFα) and its receptor, epidermal growth factor receptor (EGFR) and decreased expression of the cyclin kinase inhibitors p21 and p27. Conversely, phosphate restriction reduces the expression of TGFα and induces in the parathyroids the expression of the tumor suppressor gene p21, thus preventing parathyroid cell replication.[35] In humans, an enhanced expression of EGFR mRNA in CKD-associated parathyroid hyperplasia has been demonstrated. The administration of vitamin D metabolites increases the expression of p21 and decrease TGFα, thus preventing parathyroid cell proliferation and reverting SHPT.

In addition to the control of PTH secretion, $[Ca^{2+}{}_o]$ regulates parathyroid cell growth, differentiation and apoptosis through CaSR. Indeed, the activation of the CaSR in the parathyroids activates multiple signaling pathways and key proteins devoted to the control of the cell cycle, such as JNKs, MAPKs and ERKs. These effects seem to be mediated, at least in part, by the binding of the compartmentalized CaSR to filamin A, an intracellular scaffolding protein.[36] The importance of the CaSR signaling alterations in the establishment of parathyroid hyperplasia has been indirectly further demonstrated by the fact that in a rodent model of renal failure the administration of cinacalcet has been proven to prevent the development of parathyroid hyperplasia as well as to revert it once.[37]

CONCLUSIONS

Besides the CaSR-mediated signaling, which remains the major regulator of PTH secretion, other systemic and local actors play a role in regulating PTH gene expression, synthesis and secretion, both in the short and in the long term. Indeed, mineral metabolism is strongly influenced by the concerted modifications in PTH, 1,25(OH)$_2$D and FGF23, which contribute to regulate calcium and phosphate serum concentrations. These ions *per se* can influence parathyroid function in physiology and become mechanisms to sustain modifications in the anatomy of the parathyroids. The possibility of modulating CaSR-mediated signaling through allosteric modulators of CaSR, used alone or in combination with calcitriol or its analogs, represent a significant advance in the management of SHPT. Since the parathyroids represent a major target for phosphate and phosphate-regulating hormone FGF23, a future challenge is represented by the identification of the phosphate sensing mechanism in the parathyroid. This discovery could yield potential great opportunities in developing modulators of phosphate signaling in a similar way to what has happened with the CaSR. In view of this, the availability of a model of parathyroid cell culture would be useful to confirm previous results, to examine new mechanisms in the control of PTH transcription and secretion, and to test new compounds.

References

1. Okabe M, Graham A. The origin of the parathyroid gland. *Proc Natl Acad Sci USA* 2004;**101**:17716–9.

2. Gensure RC, Ponugoti B, Gunes Y, Papasani MR, Lanske B, Bastepe M, et al. Identification and characterization of two parathyroid hormone-like molecules in zebrafish. *Endocrinology* 2004;**145**:1634–9.

3. Ding C, Buckingham B, Levine MA. Familial isolated hypoparathyroidism caused by a mutation in the gene for the transcription factor GCMB. *J Clin Invest* 2001;**108**:1215–20.

4. Sakaguchi K, Santora A, Zimering M, Curcio F, Aurbach GD, Brandi ML. Functional epithelial cell line cloned from rat parathyroid glands. *Proc Natl Acad Sci USA* 1987;**84**:3269–73.

5. Ikeda K, Weir EC, Sakaguchi K, Burtis WJ, Zimering M, Mangin M, et al. Clonal rat parathyroid cell line expresses a parathyroid hormone-related peptide but not parathyroid hormone itself. *Biochem Biophys Res Commun* 1989;**162**:108–15.

6. Kawahara M, Iwasaki Y, Sakaguchi K, Taguchi T, Nishiyama M, Nigawara T, et al. Predominant role of 25OHD in the negative regulation of PTH expression: clinical relevance for hypovitaminosis D. *Life Sci* 2008;**82**:677–83.

7. Bell O, Silver J, Naveh-Many T. Parathyroid hormone, from gene to protein. In: Naveh-Many T, editor. *Molecular biology of the parathyroid.* Georgetown, TX: Landes Bioscience/Kluwer; 2005.

8. Habener JF, Amherdt M, Ravazzola M, Orci L. Parathyroid hormone biosynthesis. Correlation of conversion of biosynthetic precursors with intracellular protein migration as determined by electron microscope autoradiography. *J Cell Biol* 1979;**80**:715–31.

9. D'Amour P. Acute and chronic regulation of circulating PTH: significance in health and in disease. *Clin Biochem* 2012;**45**:964–9.

10. Brown EM. Role of the calcium-sensing receptor in extracellular calcium homeostasis. *Best Pract Res Clin Endocrinol Metab* 2013;**27**:333–43.

11. Hannan FM, Thakker RV. Calcium-sensing receptor (CaSR) mutations and disorders of calcium, electrolyte and water metabolism. *Best Pract Res Clin Endocrinol Metab* 2013;**27**:359–71.

12. Conigrave AD, Ward DT. Calcium-sensing receptor (CaSR): pharmacological properties and signaling pathways. *Best Pract Res Clin Endocrinol Metab* 2013;**27**:315–31.

13. Mun HC, Franks AH, Culverston EL, Krapcho K, Nemeth EF, Conigrave AD. The Venus Fly Trap domain of the extracellular Ca^{2+}-sensing receptor is required for L-amino acid sensing. *J Biol Chem* 2004;**279**:51739–44.

14. Kumar R, Thompson JR. The regulation of parathyroid hormone secretion and synthesis. *J Am Soc Nephrol* 2011;**22**:216–24.

15. Breitwieser GE. Minireview: the intimate link between calcium sensing receptor trafficking and signaling: implications for disorders of calcium homeostasis. *Mol Endocrinol* 2012;**26**:1482–95.

16. Nesbit MA, Hannan FM, Howles SA, Babinsky VN, Head RA, Cranston T, et al. Mutations affecting G-protein subunit α11 in hypercalcemia and hypocalcemia. *N Engl J Med* 2013;**368**:2476–86.

17. Cavanaugh A, Huang Y, Breitwieser GE. Behind the curtain: cellular mechanisms for allosteric modulation of calcium-sensing receptors. *Br J Pharmacol* 2012;**165**:1670–7.

18. Nemeth EF, Shoback D. Calcimimetic and calcilytic drugs for treating bone and mineral-related disorders. *Best Pract Res Clin Endocrinol Metab* 2013;**27**:373–84.

19. Mendoza FJ, Lopez I, Canalejo R, Almaden Y, Martin D, Aguilera-Tejero E, et al. Direct upregulation of parathyroid calcium-sensing receptor and vitamin D receptor by calcimimetics in uremic rats. *Am J Physiol Renal Physiol* 2009;**296**:F605–13.

20. Lopez-Hilker S, Dusso AS, Rapp NS, Martin KJ, Slatopolsky E. Phosphorus restriction reverses hyperparathyroidism in uremia independent of changes in calcium and calcitriol. *Am J Physiol* 1990;**259**:F432–7.

21. Bergwitz C, Jüppner H. Phosphate sensing. *Adv Chronic Kidney Dis* 2011;**18**:132–44.

22. Sabbagh Y. Phosphate as a sensor and signaling molecule. *Clin Nephrol* 2013;**79**:57–65.

23. Drüeke TB. Cell biology of parathyroid gland hyperplasia in chronic renal failure. *J Am Soc Nephrol* 2000;**11**:1141–52.

24. Silver J, Naveh-Many T. Phosphate and the parathyroid. *Kidney Int* 2009;**75**:898–905.

25. Kilav R, Bell O, Le SY, Silver J, Naveh-Many T. The parathyroid hormone mRNA 3′-untranslated region AU-rich element is an unstructured functional element. *J Biol Chem* 2004;**279**:2109–14.

26. Silver J, Naveh-Many T. FGF23 and the parathyroid. *Adv Exp Med Biol* 2012;**728**:92–9.

27. Silver J, Naveh-Many T. Vitamin D and the Parathyroids. In: Holick MF, editor. *Vitamin D: Physiology, molecular biology and clinical applications.* 2nd ed. New York: Humana Press; 2010. p. 235–54.

28. Demay MB, Kiernan MS, DeLuca HF, Kronenberg HM. Sequences in the human parathyroid hormone gene that bind the 1,25-dihydroxyvitamin D3 receptor and mediate transcriptional repression in response to 1,25-dihydroxyvitamin D3. *Proc Natl Acad Sci USA* 1992;**89**:8097–101.

29. Kim MS, Fujiki R, Murayama A, Kitagawa H, Yamaoka K, Yamamoto Y, et al. 1α,25(OH)2D3-induced transrepression by vitamin D receptor through E-box-type elements in the human parathyroid hormone gene promoter. *Mol Endocrinol* 2007;**21**:334–42.

30. Drüeke TB. Which vitamin D derivative to prescribe for renal patients. *Curr Opin Nephrol Hypertens* 2005;**14**:343–9.

31. Koizumi M, Komaba H, Fukagawa M. Parathyroid function in chronic kidney disease: role of FGF23-Klotho axis. *Contrib Nephrol* 2013;**180**:110–23 2012;**728**:92–9

32. Fukuda N, Tanaka H, Tominaga Y, Fukagawa M, Kurokawa K, Seino Y. Decreased 1,25-dihydroxyvitamin D3 receptor density is associated with a more severe form of parathyroid hyperplasia in chronic uremic patients. *J Clin Invest* 1993;**92**:1436–43.

33. Kifor O, Moore Jr FD, Wang P, Goldstein M, Vassilev P, Kifor I, et al. Reduced immunostaining for the extracellular Ca^{2+}-sensing receptor in primary and uremic secondary hyperparathyroidism. *J Clin Endocrinol Metab* 1996;**81**:1598–606.

34. Cunningham J, Locatelli F, Rodriguez M. Secondary hyperparathyroidism: pathogenesis, disease progression, and therapeutic options. *Clin J Am Soc Nephrol* 2011;**6**:913–21.

35. Dusso AS, Pavlopoulos T, Naumovich L, Lu Y, Finch J, Brown AJ, et al. p21(WAF1) and transforming growth factor-alpha mediate dietary phosphate regulation of parathyroid cell growth. *Kidney Int* 2001;**59**:855–65.

36. Hjälm G, MacLeod RJ, Kifor O, Chattopadhyay N, Brown EM. Filamin-A binds to the carboxyl-terminal tail of the calcium-sensing receptor, an interaction that participates in CaR-mediated activation of mitogen-activated protein kinase. *J Biol Chem* 2001;**276**:34880–7.

37. Miller G, Davis J, Shatzen E, Colloton M, Martin D, Henley CM. Cinacalcet HCl prevents development of parathyroid gland hyperplasia and reverses established parathyroid gland hyperplasia in a rodent model of CKD. *Nephrol Dial Transplant* 2012;**27**:2198–205.

The Biology of Pituitary Stem Cells

María Inés Pérez Millán and Sally A. Camper

University of Michigan Medical School, Ann Arbor, MI, USA

INTRODUCTION

The pituitary gland acts as the central endocrine regulator of growth, reproduction, metabolism and response to stress. In all vertebrates the adenohypophysis, or anterior lobe of the pituitary gland, contains specialized cell types that secrete different hormones into the hypophyseal portal system, and affects endocrine target organs. The cell types are known as lactotrophs, somatotrophs, thyrotrophs, corticotrophs, and gonadotrophs, and they secrete prolactin (PRL), growth hormone (GH), thyroid stimulating hormone (TSH), adrenocorticotropin (ACTH), and the gonadotropins, luteinizing (LH) and follicle stimulating (FSH) hormones, respectively. In rodents, each of these hormone-producing cell types can be detected at birth,[1] but the size of each population changes after birth in response to regulatory factors produced by the hypothalamus and feedback from target organs. There is some evidence that the pituitary can regenerate after tissue injury,[2] but more studies need to be done to define this process and the capabilities for each cell type. For example, mouse studies showed somatotroph recovery after genetic ablation, but lactotrophs did not regenerate significantly.[3,4] While the mechanisms of regeneration and cellular adaptation are not known, there are three proposed routes: proliferation of terminally differentiated cells; transdifferentiation of differentiated cells; and/or differentiation of progenitors/stem cells.

Stem cells were identified in adult organs with high regenerative capacity and/or turnover, including skin, liver, intestine and bone marrow.[5] In addition, stem cells were also found in organs considered to contain mostly post-mitotic, terminally differentiated cells, such as the brain[6] and the heart.[7] In all these organs, stem cells share three fundamental characteristics: capacity to proliferate and self-renew, differentiation potential, and ability to regenerate tissue after cell loss.

The pituitary gland is an organ with low cell turnover,[3] and while differentiated cells can re-enter the cell cycle, most hormone-producing cells are not dividing.[8] A great deal has been learned about anterior pituitary stem or progenitor cells in the last several years.[9] Many studies provide lines of evidence that support the presence of stem cells in the pituitary gland, including expression of markers or dye exclusion properties associated with stem cells in other tissues, and the ability of individual cells to differentiate *in vitro* into multiple cell types. In this review, we outline the varied approaches to identifying pituitary stem cells. To appreciate the current state of the art, we review the basic aspects of pituitary development and some of the foundation studies on pituitary progenitors. We also relate the knowledge of pituitary stem cells to the disease states of hypopituitarism and pituitary adenoma. We discuss areas of important, ongoing investigation and future challenges. These include defining the mechanisms necessary to preserve multi-potent cells in the niche, to induce differentiation, and to guide progenitors to adopt specific cell fates.

LOCATION OF PROLIFERATING AND DIFFERENTIATING CELLS DURING PITUITARY DEVELOPMENT

Fate mapping studies show that the pituitary gland derives from the anterior neural ridge, at the most anterior part of the developing embryo, and the hypothalamus develops from tissue immediately posterior it. As the head and oral cavity develop, the anterior neural ridge moves ventrally and posteriorly, maintaining contact with the tissue fated to become the hypothalamus. The oral ectoderm within the oral cavity invaginates to produce Rathke's pouch, and pinches off as cartilage shelves that form the palate move in. The adenohypophysis and pars intermedia (intermediate lobe) develop from Rathke's pouch, and the pars nervosa or posterior lobe and pituitary stalk develop from the overlying neural ectoderm. The neural ectoderm contains a signaling center that produces fibroblast growth factor (FGF) and other signaling molecules that stimulate proliferation of Rathke's pouch. The cells that are highly proliferative during development are located in

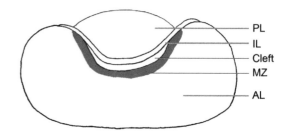

FIGURE 4.1 **Schematic representation of a coronal section of a mouse pituitary gland.** The rodent pituitary is composed of the anterior lobe (AL), intermediate lobe (IL), and posterior lobe (PL). The zone around the cleft is referred as the marginal zone (MZ, in red), and it is thought to contain the majority of the stem cells.

close proximity to the signaling center and line the cleft that is the remnant of the pouch. These cells are a densely packed epithelium, and as cells transition to differentiation, they delaminate from the epithelium and become the glandular parenchyma that will be the anterior lobe. In later stages of rodent gestation, early postnatal life, and in adults, a few scattered, proliferating cells are detectable in the parenchyma, and they are largely hormone negative. The zone of cells that line the cleft is referred to as the marginal zone, and it is thought to contain the majority of the stem cells (Figure 4.1). There are clearly progenitors in the parenchyma, and it is not clear whether the cells in each area have equivalent differentiation potential.

STRATEGIES FOR IDENTIFYING PITUITARY PROGENITORS

Chromophobes Can Differentiate into All Anterior Pituitary Hormone-producing Cells

In 1969, chromophobes, the cells that do not readily absorb histological stains, were proposed to be pituitary stem cells because of their behavior in transplant studies.[10] Chromophobes transplanted into the hypothalamus of hypophysectomized rats underwent proliferation and differentiation into mature basophils (thyro-trophs, gonadotrophs and corticotrophs) and acidophils (somatotrophs and lactotrophs). Further supporting evidence came from studies that induced chromophobes to differentiate into basophils and acidophils *in vitro*.[11,12]

Pituitary Progenitors Absorb a Fluorescent Dipeptide and Have Folliculostellate Cell Characteristics

Folliculostellate cells are agranular cells with star-like morphology, formed by long cytoplasmatic projections. They are located in the parenchymal tissue, usually around the marginal zone of the anterior pituitary gland. Folliculostellate cells are immunopositive for S100 and for glial fibrillary acidic protein (GFAP), and they constitute 5–10% of the pituitary cells in the adult gland. They are organized in a functional network with endo-crine cells, which they regulate in a paracrine manner by producing growth factors and cytokines. Their long cytoplasmic processes and gap junctions facilitate intercellular communication. They also act as scavenger cells with phagocytic activity.[13] Several studies suggested that folliculostellate cells could be a potential source of pituitary stem cells. Additional markers and functional tests are necessary to identify the subpopulation of folliculostellate cells that have stem cell or progenitor characteristics, such as the ability to form colonies *in vitro*.

Paul Thomas's group was the first to iden-tify an adult pituitary cell population able to form colonies *in vitro*.[14] He used fluorescence-activated cell sorting to select live, nucleated single cells that lacked propidium iodide, and plated the cells at low density. Most cells died, but colonies formed. Only 0.2% of anterior pituitary cells were able to form pituitary colony-forming cells. We recapitulated this in the lab using cholera toxin to select for cells

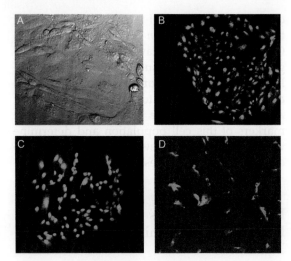

FIGURE 4.2 **Pituitary colony-forming cells in adult mice.** (A) Colony-forming pituitary cells in low-density culture colonies at day 4. These cells were positive for stem cells markers: SOX2 (B), SOX9 (C) and GFRa2 (D). Cell nuclei were counterstained with 4',6-diamidino-2-phenylindole (DAPI) in blue.

that cannot take up the toxin. These cells also formed colonies *in vitro* (Figure 4.2).

Pituitary colony-forming cells express S100 and GFAP, and they have the capacity of specific uptake of the fluorescent dipeptide β-Ala-Lys-Ne-AMCA(AMCA is (7-amino-4-methylcoumarin-3-acetic-N'-hydroxysuccinimide ester)).[15] Based on these results, the pituitary colony-forming cells were suggested to be a subpopulation of folliculostellate cells. Only AMCA-positive cells, which constitute 3.7% of the pituitary cells, were able to form pituitary colony-forming cells, but only 12.3% of them did, consistent with the apparent heterogeneity of the folliculostellate cell population.[14]

Angiotensin-converting enzyme is expressed in the marginal zone, which is proposed to comprise the niche and a source of precursor cells in the adult pituitary.[16] Cells sorted for angiotensin converting enzyme, but not the stem cell marker SCA1, enriched the AMCA-positive population of pituitary colony forming cells.[16] Moreover, 6 weeks after implantation of AMCA-positive, GH-negative cells, 3.3% of them differentiated *in vivo* and expressed GH.[17] These studies have confirmed the progenitor potential of a subpopulation of folliculostellate cells, based on their ability to form colonies *in vitro* and to differentiate *in vivo*. Additional studies are needed to define the markers of this cell population and determine whether it can self-renew and differentiate into other pituitary cell types.

Pituitary Progenitors in the Side Population That Exhibit Rapid Hoechst Dye Efflux

Using different approaches, Vankelecom and colleagues found adult pituitary cells with progenitor or stem cell characteristics.[12] One method is based on the concept that stem cells exclude harmful components, and thus exhibit rapid efflux of Hoechst dye, and the other method relies on clonal sphere formation. Cell sorting of bone marrow cells incubated with Hoescht 33342 reveals a side population of cells with rapid efflux that contains multipotential hematopoietic stem cells markers.[18] SCA1 is a phosphatidylinositol-anchored protein that is a member of the lymphocyte antigen 6 (Ly-6) family, and it is expressed in both fetal and adult hematopoietic stem cells. It has been used to enrich for stem cells in other tissues as well. The approach of Hoechst dye exclusion and fluorescence-activated cell sorting has been successful in many identifying stem cells in many tissues, including the pituitary gland.[12,19] In the pituitary, this side population is composed of cells expressing high and low levels of the stem cell marker SCA1, representing 60% and 40% of the population, respectively. Two pieces of evidence support the idea that pituitary progenitor cells are in the non-high SCA1 fraction. This latter group of cells also expresses transcription factors characteristic of Rathke's pouch progenitors,

including HESX1, PROP1, PAX6 and LHX4. More importantly, only non-high SCA1 cells can form spheres that can give rise to all endocrine cell types of the anterior pituitary.[20] This is in agreement with the demonstration that progenitors are confined to the angiotensin-converting enzyme positive, SCA1-negative fraction.[16]

The pituitary side population is composed of cells that express stem cells markers and can differentiate *in vitro*. Their capacity for self-renewal needs to be demonstrated. This field would also be advanced by identifying markers and cell sorting parameters that can purify the stem cells from the heterogenous side population.

Pituitary Progenitors in the SOX2-Positive Cell Population

A breakthrough in stem cell research came when Yamanaka and colleagues forced expression of a few transcription factors in differentiated cells and demonstrated that they were sufficient to induce pluripotency.[21] The expression of two of these transcription factors, SOX2 and SOX9, has been studied in the pituitary gland.

SOX2, a member of the SOXB1 subfamily of high-mobility group box transcription factors, is required for the maintenance of several stem cell populations in humans and rodents, including the central nervous system.[22] Robinson and colleagues showed that SOX2 is expressed in Rathke's pouch during development and in approximately 3% of adult pituitary cells, including many of the cells in the marginal zone and cells scattered in the parenchyma. SOX9 belongs to the SOXE family, and is a marker for stem cells in the pancreas, retina, and central nervous system.[23–26] In some organs, members of the SOXE family modulate the activity of SOXB1 family members by promoting differentiation along specific pathways.[27,28] During pituitary development in mice, SOX2-expressing cells are detectable at e12.5; they do not express SOX9, and they have a high proliferation rate. At e18.5 and in adult pituitaries the majority of SOX2-positive cells also express SOX9. These SOX2 +, SOX9 + cells can form pituispheres *in vitro* and differentiate into S100 expressing folliculostellate cells and all five types of hormone-producing cells of the adenohypophysis, suggesting that they are multipotent.[29] These cells have a low proliferation rate, suggesting that they may represent transient amplifying cells committed to a specific pituitary cell type. The small portion (1%) of SOX2-positive cells that are SOX9 negative, divide slowly, do not produce hormones, and are able to form secondary pituispheres *in vitro*. The Lovell Badge group recently demonstrated that SOX2 +, SOX9 + cells from embryos and adults give rise to pituitary endocrine cells, proving that they are progenitors. Furthermore, these adult stem cells can become mobilized and differentiate into the appropriate endocrine cell type in response to physiological demands. Estrogen administration in males and adrenal ablation causes SOX2 +, SOX9 + stem cells to proliferate and differentiate into somatotrophs and corticotrophs, respectively.[30] The challenge now is to investigate the mechanisms and stimuli that make these stem cells start differentiating and integrate to the right endocrine hormonal cell network. These findings are very promising and bring us closer to potential therapeutic strategies for pituitary diseases.

Pituitary Progenitors Expressing GFRa2 and PROP1: Rodent and Human

Alvarez and colleagues made an important contribution to the field by demonstrating expression of the stem-cell marker OCT4 in both human and rodent pituitaries, and examining expression of other key factors. While human pituitaries do not have a well-defined intermediate lobe, there are remnants of Rathke's pouch, and cells in these areas express OCT4. In addition, these cells express SOX2 and glial cell line derived neurotrophic factor receptor alpha 2 (GFRa2). GFRa2 is a glycosylphosphatidylinositol (GPI)-linked cell surface receptor that plays an important role in neuronal cell survival and differentiation. In the adult rodent pituitary gland, GFRa2 expression is detected in the marginal zone and in cells scattered throughout the adenohypophysis, representing 0.9% of the total cells. The GFRa2-positive cells are positive for the pituitary specific, homeodomain transcription factor PROP1. Mutations in PROP1 are the most common known cause of multiple pituitary hormone deficiency in humans, and the deficiency leads to progressive loss in hormone production.[30] GFRa2-positive cells are slowly proliferating and able to form spheres *in vitro*, can generate secondary pituispheres, and differentiate into the five hormone-producing pituitary lineages.[31]

Nestin—GFP Tagged Pituitary Progenitors

Marginal cells, that are lining the pituitary cleft, have been proposed to define a stem cell niche, as this epithelium is the remnant of the Rathke's pouch. Marginal cells are not granular; they have a poorly developed endoplasmic reticulum and an abundance of free ribosomes and polysomes (reviewed in ref[9]). The idea that marginal cells are stem cells came from the demonstration that nestin is expressed in cells lining the pituitary cleft adjacent to the marginal zone.[32] Using a genetic approach, an adult pituitary stem cell population was identified that expresses nestin and can generate all of the differentiated anterior pituitary cell types.[33] In this study, the nestin transgene expression marks a subset of cells in Rathke's pouch that do not express endogenous nestin, however.[34]

PITUITARY PROGENITORS AND DISEASE PATHOPHYSIOLOGY

Hypopituitarism: Failure to Produce and/or Retain Pituitary Progenitors

Pituitary hormone deficiency, or hypopituitarism, is the decreased (*hypo*) secretion of one or more of the hormones normally produced by the pituitary gland. The signs and symptoms of hypopituitarism vary, depending on which hormone deficiencies are present and on the underlying cause of the abnormality. Isolated growth hormone deficiency and multiple hormone deficiencies are prevalent, with an incidence of ~1 in 4000 live births. In many cases the deficiencies can be managed with injections of recombinant growth hormone, oral synthetic thyroid hormone, estrogen or testosterone therapy, and corticosteroid maintenance. It would be an incredible improvement in treatment if pituitary stem cell therapy could be developed that would be able to regenerate the pituitary endocrine lineages that are lacking in these patients. While this seems like a very distant possibility, an elegant study by Sasai and colleagues demonstrated that embryonic stem cells could be programmed to recapitulate Rathke's pouch formation and produce functional, differentiated corticotrophs.[35] Remarkably, transplantation of these induced corticotrophs into the kidney capsule of hypophysectomized mice was sufficient to rescue their stress response. The manipulations that guided this differentiation were developed from the knowledge that anterior pituitary development is stimulated by the neural ectoderm and regulated by WNT, Notch, BMP and FGF signaling. It would be especially exciting if protocols could be

developed that would reliably direct the development of other lineages, and that they could be adapted to induced pluripotent cells rather than embryonic stem cells.

Mutations in several transcription factors can cause hypopituitarism.[36] LHX4 and PROP1 expression have been associated with pituitary progenitor populations. They may be involved in regulating early pituitary progenitor proliferation and the transition to differentiation, respectively. LHX4 normally inhibits expression of the cell cycle inhibitor p21, but in *Lhx4* mutants p21 expression spreads throughout Rathke's pouch, cell proliferation is markedly reduced, and extensive apoptosis ensues very early in pituitary development.[37–39] Human patients with *PROP1* mutations have progressive hormone deficiencies that are usually first identified by growth insufficiency and reduced production of growth hormone, TSH, and gonadotropins. Pituitary hormone levels progressively decline, and eventually all anterior pituitary hormones may be lost, including ACTH.[30] It is tempting to speculate that the progression represents depletion of the stem cell pool, despite the fact that this evolution is not obviously mimicked in mice. The GH, TSH and PRL deficiencies are congenital and profound in *Prop1* mutant mice.[40] The organ size is normal in mutants at birth, but their pituitaries are obviously dysmorphic, apparently due to failure of the progenitor cells to delaminate from the marginal zone, mimicking a failed epithelial to mesenchymal transition.[41] The anterior pituitary proliferation rate is poor in late gestation *Prop1* mutants, and there are fewer proliferating cells located in the parenchyma. The pituitary gland normally grows markedly during the postnatal period, but in *Prop1* mutants there is very little additional growth, and apoptosis is evident in the marginal zone. This is consistent with a role for *Prop1* in regulating progenitor proliferation and transition to differentiation. Perhaps the species difference in disease manifestation is due to different developmental timing in PROP1 action OF progenitors. The mouse model provides an opportunity to identify *Prop1* target genes that affect progenitor proliferation, survival and delamination.

Stem Cells in Pituitary Adenomas

Pituitary adenomas are prevalent in humans. They represent up to 15% of all diagnosed intracranial neoplasms and are unexpectedly discovered in 10–20% of autopsies. Although they are benign tumors that very rarely metastasize, pituitary adenomas can cause compression symptoms, such as visual defects and hypopituitarism, and morbidity. Treatment can involve multiple invasive surgeries and radiation. There are some familial pituitary adenoma syndromes, but most tumors are sporadic and clonal. The common types have characteristics of differentiated cells such as hormone production and/or transcription factors expressed in committed cells. An exception to this is the craniopharyngioma, an aggressive childhood tumor that has progenitor characteristics.[42]

Markers of pituitary stem cells were studied in two different models of pituitary adenoma. First, SOX2-expressing cells were studied in a mouse model of adamantinomatous craniopharyngiomas (ACP) developed by Martinez-Barbera's group. Rare cells that overexpress β-catenin form clusters that have features of ACP, and express SOX2. SOX9 was found only in the cells surrounding these clusters.[42] The presence of SOX9-expressing cells around the cluster may have functional significance. In contrast, in human ACP they did not find SOX2-positive cells in the clusters. And, as in the mouse model SOX9 was expressed in cells surrounding the β-catenin clusters but was not seen within the cells clusters. Recently, the same group used transgenic mice to show that expression of β-catenin in SOX2+ cells

results in the formation of typical ACP pre-tumoral β-catenin-accumulating clusters and tumors. Interestingly, the tumor mass was not derived from the SOX2 population, suggesting a paracrine role of SOX2 in pituitary tumorigenesis, rather than being the cell of origin of the tumors.[43] The second model produces a different type of tumor. $Rb^{+/-}$ knockout mice with a nestin–GFP transgene develop hyperplastic pituitary nodules.[33] Haploinsufficiency for the tumor suppressor gene *Rb*, retinoblastoma, affects the intermediate lobe and causes melanocyte stimulating hormone (MSH) tumors in mice. None of the tumors cells were positive for GFP, however, GFP-positive cells that co-expressed SOX2 and LHX3 encapsulated the tumors. The relationship between the SOX2-positive cells and the initiation or promotion of growth is not clear. It is intriguing that both of these models have a core of tumor tissue surrounded by another abnormal layer expressing different markers. Perhaps there are paracrine interactions that influence growth of these adenomas. The different sites of SOX2 expression, inside the ACP-cluster core and outside $Rb^{+/-}$ tumor suggests there may be different relationships between stem cell markers and pituitary tumor types.

The relationship between stem cell markers and pituitary adenomas is being actively investigated in human patients. We discuss two of the studies here. First, twelve human pituitary adenomas were subjected to cell sorting, and side population cells were reported to represent between 1.5–8% of the tumor cells.[44] This is higher than would be expected based on the rodent studies, but there were no normal human controls. We have observed that the genetic background influences the number of colony-forming units in postnatal pituitary glands from normal mice, and the proportion changes with age (unpublished). This suggests that future studies with human patients will require large numbers of samples from normal people and for each of the tumor types. In a second study of human patients, stem-like cells were found in both hormone-producing and hormone-null pituitary adenomas, and they were able to self-renew in pituisphere cultures. These pituitary adenoma-derived cells express both stem cell-associated markers and lineage-specific markers. When differentiated *in vitro*, they downregulate stem cell-associated genes and produce multiple pituitary hormones in response to hypothalamic-hormone stimulation. Finally, when transplanted into immune-compromised mice, these sphere-forming cells generate intracranial tumors that can be serially transplanted to recapitulate the human tumor.[45] These data are intriguing, and suggest a role of stem cells or progenitors in human adenomas. Future studies that explicitly demonstrate the role of stem cells in adenomas and an understanding of the mechanisms may lead to novel targets for therapeutic intervention. If this can be accomplished, the prognosis for patients with recurrent adenomas might improve.

CONCLUSION

In summary, using different strategies, several groups around the world have identified cell populations that meet many of the criteria to be called pituitary stem cells. In the future, we expect that there will be more clarity about the nature of the progenitors present in the embryonic, postnatal and adult pituitary gland, and a better understanding of the cell—cell signaling and transcriptional regulation that drives progenitors on particular paths of differentiation. These studies will provide an important foundation for finding new therapeutic approaches for pituitary deficiencies and tumors.

Acknowledgements

NIH HD30428, HD34283 (SAC), University of Michigan Organogenesis Postdoctoral Fellowship (MIPM), the Endocrine Society International Scholar Program (MIPM). We thank Cynthia L. Andoniadou and Juan Pedro Martinez Barbera for the stem cell differentiation protocol.

References

1. Carbajo-Perez E, Watanabe YG. Cellular proliferation in the anterior pituitary of the rat during the postnatal period. *Cell Tissue Res* 1990;**261**:333–8.
2. Landolt AM. Regeneration of the human pituitary. *J Neurosurg* 1973;**39**:35–41.
3. Borrelli E, et al. Transgenic mice with inducible dwarfism. *Nature* 1989;**339**:538–41.
4. Fu Q, et al. The adult pituitary shows stem/progenitor cell activation in response to injury and is capable of regeneration. *Endocrinology* 2012;**153**:3224–35.
5. Moore KA, Lemischka IR. Stem cells and their niches. *Science* 2006;**311**:1880–5.
6. Gage FH. Mammalian neural stem cells. *Science* 2000;**287**:1433–8.
7. Leri A, Kajstura J, Anversa P. Cardiac stem cells and mechanisms of myocardial regeneration. *Physiol Rev* 2005;**85**:1373–416.
8. Davis SW, Mortensen AH, Camper SA. Birthdating studies reshape models for pituitary gland cell specification. *Dev Biol* 2011;**352**:215–27.
9. Castinetti F, et al. Pituitary stem cell update and potential implications for treating hypopituitarism. *Endocr Rev* 2011;**32**:453–71.
10. Yoshimura F, et al. Differentiation of isolated chromophobes into acidophils or basophils when transplanted into the hypophysiotrophic area of hypothalamus. *Endocrinol Jpn* 1969;**16**:531–40.
11. Otsuka Y, et al. Effect of CRF on the morphological and functional differentiation of the cultured chromophobes isolated from rat anterior pituitaries. *Endocrinol Jpn* 1971;**18**:133–53.
12. Chen J, et al. The adult pituitary contains a cell population displaying stem/progenitor cell and early embryonic characteristics. *Endocrinology* 2005;**146**:3985–98.
13. Devnath S, Inoue K. An insight to pituitary folliculostellate cells. *J Neuroendocrinol* 2008;**20**:687–91.
14. Lepore DA, et al. Identification and enrichment of colony-forming cells from the adult murine pituitary. *Exp Cell Res* 2005;**308**:166–76.
15. Bauer K. Carnosine and homocarnosine, the forgotten, enigmatic peptides of the brain. *Neurochem Res* 2005;**30**:1339–45.
16. Lepore DA, et al. A role for angiotensin-converting enzyme in the characterization, enrichment, and proliferation potential of adult murine pituitary colony-forming cells. *Stem Cells* 2006;**24**:2382–90.

17. Lepore DA, et al. Survival and differentiation of pituitary colony-forming cells *in vivo*. *Stem Cells* 2007;**25**:1730−6.

18. Goodell MA, et al. Isolation and functional properties of murine hematopoietic stem cells that are replicating *in vivo*. *J Exp Med* 1996;**183**:1797−806.

19. Challen GA, Little MH. A side order of stem cells: the SP phenotype. *Stem Cells* 2006;**24**:3−12.

20. Chen J, et al. Pituitary progenitor cells tracked down by side population dissection. *Stem Cells* 2009;**27**:1182−95.

21. Takahashi K, Yamanaka S. Induction of pluripotent stem cells from mouse embryonic and adult fibroblast cultures by defined factors. *Cell* 2006;**126**:663−76.

22. Episkopou V. SOX2 functions in adult neural stem cells. *Trends Neurosci* 2005;**28**:219−21.

23. Seymour PA, et al. SOX9 is required for maintenance of the pancreatic progenitor cell pool. *Proc Natl Acad Sci USA* 2007;**104**:1865−70.

24. Poche RA, et al. Sox9 is expressed in mouse multipotent retinal progenitor cells and functions in Muller glial cell development. *J Comp Neurol* 2008;**510**:237−50.

25. Scott CE, et al. SOX9 induces and maintains neural stem cells. *Nat Neurosci* 2010;**13**:1181−9.

26. Sekido R. SRY: a transcriptional activator of mammalian testis determination. *Int J Biochem Cell Biol* 2010;**42**:417−20.

27. Wegner M, Stolt CC. From stem cells to neurons and glia: a Soxist's view of neural development. *Trends Neurosci* 2005;**28**:583−8.

28. Lee YH, Saint-Jeannet JP. Sox9 function in craniofacial development and disease. *Genesis* 2011;**49**:200−8.

29. Fauquier T, et al. SOX2-expressing progenitor cells generate all of the major cell types in the adult mouse pituitary gland. *Proc Natl Acad Sci USA* 2008;**105**:2907−12.

30. Rizzoti K, Akiyama H, Lovell-Badge R. Mobilized Adult Pituitary Stem Cells Contribute to Endocrine Regeneration in Response to Physiological Demand. *Cell Stem Cell* 2013;**89**(4):419−32.

31. Garcia-Lavandeira M, et al. A GRFa2/Prop1/stem (GPS) cell niche in the pituitary. *PLoS One* 2009;**4**: e4815.

32. Krylyshkina O, et al. Nestin-immunoreactive cells in rat pituitary are neither hormonal nor typical folliculo-stellate cells. *Endocrinology* 2005;**146**:2376−87.

33. Gleiberman AS, et al. Genetic approaches identify adult pituitary stem cells. *Proc Natl Acad Sci USA* 2008;**105**:6332−7.

34. Galichet C, Lovell-Badge R, Rizzoti K. Nestin-Cre mice are affected by hypopituitarism, which is not due to significant activity of the transgene in the pituitary gland. *PLoS One* 2010;**5**:e11443.

35. Suga H, et al. Self-formation of functional adenohypophysis in three-dimensional culture. *Nature* 2011;**480**:57−62.

36. Alatzoglou KS, Dattani MT. Genetic forms of hypopituitarism and their manifestation in the neonatal period. *Early Hum Dev* 2009;**85**:705−12.

37. Sheng HZ, et al. Specification of pituitary cell lineages by the LIM homeobox gene *Lhx3*. *Science* 1996;**272**:1004−7.

38. Raetzman LT, Ward R, Camper SA. Lhx4 and Prop1 are required for cell survival and expansion of the pituitary primordia. *Development* 2002;**129**:4229−39.

39. Gergics P, et al. LHX4 dysfunction causes pituitary hypoplasia by permitting ectopic expression of the cell cycle inhibitor p21. *Endocr Rev* 2012;**33** MON-493

40. Nasonkin IO, et al. Aged PROP1 deficient dwarf mice maintain ACTH production. *PLoS One* 2011;**6**:e28355.

41. Ward RD, et al. Cell proliferation and vascularization in mouse models of pituitary hormone deficiency. *Mol Endocrinol* 2006;**20**:1378−90.

42. Gaston-Massuet C, et al. Increased Wingless (Wnt) signaling in pituitary progenitor/stem cells gives rise to pituitary tumors in mice and humans. *Proc Natl Acad Sci USA* 2011;**108**:11482−7.

43. Andoniadou CL, et al. Sox2 Stem/Progenitor cells in the adult Mouse pituitary support organ homeostasis and have tumor-inducing potential. *Cell Stem Cell* 2013;**13**(4):433−45.

44. Vankelecom H, Gremeaux L. Stem cells in the pituitary gland: a burgeoning field. *Gen Comp Endocrinol* 2010;**166**:478−88.

45. Xu Q, et al. Isolation of tumour stem-like cells from benign tumours. *Br J Cancer* 2009;**101**:303−11.

The Na$^+$/I$^-$ Symporter (NIS) and Thyroid Hormone Biosynthesis

Juan Pablo Nicola and Nancy Carrasco

Yale University School of Medicine, New Haven, CT, USA

IODIDE METABOLISM

The thyroid hormones (THs) triiodothyronine (T$_3$) and tetraiodothyronine or thyroxine (T$_4$) are essential for the maturation of the central nervous system and lungs in the fetus and the newborn, and for intermediary metabolism in virtually all tissues throughout life. Iodine is an essential constituent of these hormones, the only ones in vertebrates that contain it. Thyroid hormones are biosynthesized in the thyroid follicular cells, so named because they form clearly defined microscopic structural units called follicles. Each follicle is a spheroidal structure whose wall is made up of a single layer of these epithelial cells, which surround an amorphous-looking extracellular substance called the colloid, located in the center (lumen) of the follicle. The thyroid follicular cells, also known as thyrocytes, are polarized, with their basolateral surface facing connective tissue and blood vessels, and their apical surface facing the colloid (Figure 5.1). THs are unique, not only because their biosynthesis occurs both intracellularly and extracellularly, but also because they do not resemble any other hormones chemically, and because they contain iodine as an essential constituent. Iodine is extremely scarce in the environment, and vertebrates, including humans, obtain it exclusively from their diet. Insufficient dietary iodide (I$^-$) intake, depending on its severity, causes hypothyroidism, goiter, stunted growth, retarded psychomotor development, and even irreversible mental retardation.[1] I$^-$ deficiency disorders (IDDs) are the leading preventable cause of mental retardation in the world, and were, therefore, slated by the World Health Organization for total global eradication by iodination of table salt. Although significant strides have been made in many regions, in 2011 there were still an estimated 1.88 billion people having insufficient iodine intake.[2]

A fundamental condition for TH biosynthesis is that I$^-$ be made available in sufficient amounts to the thyroid, for which the gland has developed a remarkably efficient and specialized system. That the thyroid actively accumulates I$^-$ has been known since 1896.[3] The thyroid concentrates I$^-$ >40-fold with

65

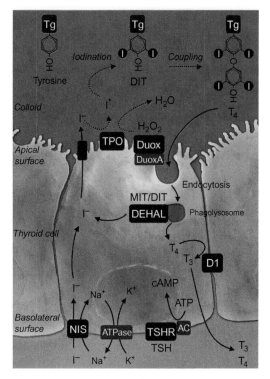

FIGURE 5.1 **Schematic representation of the thyroid hormone biosynthesis pathway.** The apical surface of thyroid epithelial cells is in contact with the extracellular colloid. Their basolateral surface, by contrast, faces the bloodstream. NIS, Na^+/I^- symporter; ATPase, Na^+/K^+ ATPase; TPO, thyroid peroxidase; Tg, thyroglobulin; Duox, dual oxidase; DuoxA, Duox maturation factor; DEHAL, iodotyrosine dehalogenase; D1, type 1 iodothyronine deiodinase; TSH, thyroid stimulating hormone; TSHR, TSH receptor; AC, adenylate cyclase; camp, cyclic adenosine monophosphate; MIT and DIT, mono- and diiodotyrosine; T_3, triiodothyronine; T_4, thyroxine. For further explanation, see text.

respect to the plasma under physiological conditions.[4–6] Moreover, the ability of the thyroid to concentrate I^- has long been used as a tool in the diagnosis and treatment of thyroid diseases, including cancer. However, the identity of the molecule responsible for active I^- transport in the thyroid was only elucidated in 1996 by Dai et al,[7] who isolated the cDNA that encodes the Na^+/I^- symporter

(NIS). Briefly, NIS is the key plasma membrane protein, located on the basolateral surface of thyrocytes, that mediates active I^- uptake in the thyroid, the first and one of the key steps in TH biosynthesis (Figure 5.1). NIS couples the inward *downhill* translocation of Na^+ to the inward *uphill* translocation of I^- (Figure 5.1).[8–11] The driving force for this process is the inwardly directed Na^+ gradient generated by the Na^+/K^+ ATPase.

Once I^- has reached the cytosol from the bloodstream via NIS, the next step in the hormonogenesis pathway is for I^- to efflux from the cytosol to the follicular lumen, a process that has been suggested to be passively mediated by a membrane transporter located on the apical surface of these cells (Figure 5.1). The identity of this transporter has yet to be unequivocally established.[12] Pendrin, a Cl^-/I^- transporter,[13] has been proposed to mediate I^- efflux. However, it is possible that I^- efflux is instead mediated by other Cl^- channels or transporters, such as Cl^-/H^+ antiporter 5 (ClC-5)[14] and the cystic fibrosis transmembrane conductance regulator (CFTR).[15,16] I^- efflux needs to be investigated further.

Thyroglobulin (Tg) is the most abundant protein in the thyroid. It is secreted by the thyrocytes into the follicular lumen, where it is the main components of the colloid. The human *Tg* gene, located in chromosome 8q24.2, consists of 48 exons and encodes a large protein of 2749 amino acids. In the colloid, Tg is present as a 660 kDa dimeric glycoprotein. Tg functions as the scaffold for TH synthesis and for the storage of the inactive form of THs and iodine.[17] Biallelic mutations in the *Tg* gene can also cause congenital hypothyroidism[18] – for instance, mutations that impair synthesis or folding. Most of these mutations in one way or another cause Tg to be retained in the endoplasmic reticulum (ER), preventing it from being secreted into the colloid.[19] Over 50 *Tg* mutations have been identified to date.

Originally identified as the major thyroid microsomal antigen targeted in autoimmune thyroid disease, thyroperoxidase (TPO), a single-pass transmembrane glycoprotein located on the apical membrane of thyrocytes, catalyzes the oxidation of I^-, iodination of tyrosines in Tg, and subsequent coupling of iodotyrosines to form iodothyronines (Figure 5.1).[20] Thus, the enzymatic activity of TPO is essential for thyroid hormonogenesis; inactivating mutations in the TPO gene cause thyroid dyshormonogenesis due to defective I^- organification.[21] The human *TPO* gene, located in chromosome 2p12, codes for a 933-residue 103-kDa protein.[22,23] The catalytic domain of TPO, located in the extracellular amino-terminal portion of the protein facing the thyroid follicular lumen, contains a prosthetic heme group.[24]

Hydrogen peroxide (H_2O_2) is a limiting step for TPO-mediated catalytic reactions and therefore for TH synthesis even when the supply of I^- is sufficient.[25] Acquired defects in H_2O_2 production have been identified as causes of congenital hypothyroidism due to deficient I^- organification.[26,27] De Deken *et al.* identified two nicotinamide adenine dinucleotide phosphate (NADPH) oxidases, which they named dual oxidases 1 and 2 (Duox1 and Duox2), as the catalytic enzymatic core for H_2O_2 generation in the thyroid using a strategy based on the functional similarities between H_2O_2 production in thyrocytes and leukocytes (Figure 5.1).[28] Both human Duox genes, closely located to each other on chromosome 15q15, encode for two proteins of 1550 amino acids displaying 83% sequence similarity. The Duox secondary structure model predicts a glycoprotein with seven transmembrane segments (TMSs), two EF-hand calcium-binding domains in the first intracellular loop, binding sites for NADPH and flavin adenine dinucleotide (FAD) in the cytoplasmic carboxy-terminal portion, and an extracellular amino-terminal domain with 43% homology with that of TPO. Duox protein expression is restricted to the apical membrane of thyrocytes, and the protein colocalizes with TPO. Therefore, the association of these two proteins at the plasma membrane, if it exists, would increase the efficiency of the H_2O_2 producer-consumer system.[29,30] Targeting of Duox enzymes to the apical plasma membrane requires the presence of a specific Duox maturation factor called DuoxA (Figure 5.1).[31] Two human DuoxA paralogs, named DuoxA1 and 2, have been identified in the thyroid. Both *DuoxA* genes are located on chromosome 15 in the Duox1/Duox2 intergenic region in a tail-to-tail orientation. DuoxA1 and 2 proteins are 58% identical, and are predicted to have five TMSs and a cytoplasmic carboxy-terminal region. Recent data indicate that Duox activators not only promote Duox maturation but are also parts of the H_2O_2 generating complex.[32]

Thus, as mentioned above, THs are the only hormones whose biosynthesis takes place both intracellularly, in the thyrocytes, and extracellularly, at the cell/colloid interface. In a complex sequential multistep reaction at the cell—colloid interface, which involves fusion of Tg-containing secretory vesicles, I^- released into the lumen is oxidized by TPO in the presence of H_2O_2 generated by Duox, and then it is covalently incorporated into selected tyrosyl residues within the Tg molecule (Figure 5.1).[20,25] This step, known as I^- organification, results in the formation of mono- and di-iodotyrosines (MIT and DIT, respectively).[20] However, even under I^- sufficiency, not all tyrosine residues in Tg are iodinated. The final step involves the coupling of two iodotyrosine residues that are near each other in the tertiary structure to produce THs (iodothyronines), with two DITs forming T_4 and one DIT and one MIT forming T_3 (Figure 5.1). The generation of the iodothyronines involves the formation of an ether bond between the iodophenol ring from a donor iodotyrosine and the hydroxyl group of an acceptor iodotyrosine. After the cleavage reaction, a dehydroalanine residue remains in the donor site.[33] Under conditions of I^- sufficiency, an average molecule of human Tg contains 6.5 residues of MIT, 4.8 of DIT, 2.3 of T_4, and 0.3 of T_3.[17] I^- deficiency decreases the ratios of DIT to MIT and T_4 to T3.[34] The mature hormone-containing Tg molecules are stored in the colloid. In response to TH demand, Tg-containing colloid enters the thyrocytes through micropinocytosis (Figure 5.1). The turnover of mature Tg varies according to thyroid status; in a euthyroid adult, turnover is about 1% per day. Internalized Tg molecules are probably subjected to a receptor-mediated sorting step for subsequent differential cellular handling of Tg molecules with different properties.[35] It is speculated that high-hormone content Tg molecules are directed to a lysosomal degradation pathway for the release of free THs and complete degradation of the protein, while Tg molecules with low-hormone content are either recycled back into the colloid or transported and released through the basolateral membrane (transcytosis). Although Tg transcytosis occurs only to a limited extent under physiological conditions, this process accounts for the presence of Tg in the plasma, a finding with high diagnostic value in thyroid cancer.[36] The importance of low-hormonogenic Tg transcytosis was uncovered by studying mice genetically deficient in megalin (megalin$^{-/-}$ mice). Lisi et al. reported that megalin$^{-/-}$ mice are hypothyroid and have a low serum Tg concentration.[37,38] These authors hypothesized that transcytosis blockage redirects low-hormonogenic Tg to lysosomes for degradation, and that, once there, low-hormonogenic Tg competes with high-hormonogenic Tg for lysosomal proteolytic enzymes, resulting in decreased TH release, and thus a hypothyroid phenotype.

THs released from the Tg backbone must exit the lysosomes and then the cell to reach the bloodstream. Although generally considered lipophilic, T_3 and T_4 are zwitterionic at physiological pH, and so may be unable to diffuse freely through cell membranes. Therefore, their exit from lysosomes and thyroid cells may involve membrane transporters, just like their entry into peripheral target cells.[39] A detailed characterization of TH transport across the lysosomal membrane and the basolateral membrane of thyroid cells remains to be carried out. Initial steps in this direction have been taken by Di Cosmo et al, who have provided evidence that monocarboxylate transporter 8 (MCT-8), which is located at the basolateral membrane of mouse thyrocytes, is involved in transporting THs out of these cells.[40]

Although T_4 is the main TH secreted, it is considered a prohormone and a circulating reservoir of T_3, which is the biologically active TH. Under I^- sufficiency, the daily secretion of THs is $\sim 100\ \mu g\ T_4$ and $\sim 15\ \mu g\ T_3$. Some of this T_3 starts out as T_4: about 10% of the T_4 initially produced is deiodinated intrathyroidally, converting it into T_3 (Figure 5.1). This process is catalyzed by two selenocysteine-containing transmembrane proteins: type I and type II iodothyronine deiodinase (D1 and D2).[41] Although these two enzymes have different subcellular localizations — D2 is in the ER membrane and D1 in the basolateral plasma membrane — they both have their catalytic domain in the cytoplasm.[42] The extent of this intrathyroidal deiodination changes significantly as a function of thyroid status, as TSH and T_3 upregulate transcriptional expression of human D1.[43]

Under I^- sufficiency, about 70% of the iodine present in Tg molecules is in the form of iodotyrosines. During the proteolysis of Tg, released iodotyrosines are deiodinated by iodotyrosine dehalogenase (DEHAL), and most of the I^- released returns to the intrathyroidal I^- pool (Figure 5.1), in what is an important I^- recycling mechanism. The nucleotide sequence of DEHAL was identified by the serial analysis of gene expression (SAGE) technique using mRNA from human thyroid tissue.[44] DEHAL is a transmembrane protein with a large extracellular amino-terminal region containing a conserved nitroreductase domain with a flavin mononucleotide (FMN)-binding prosthetic group, and a short carboxy-terminal cytoplasmic domain.[45]

DEHAL deiodinates both free iodotyrosines using NADPH and FMN as cofactors.[45] DEHAL is mainly located at the apical membrane of thyrocytes; therefore, the deiodination process may take place along with the proteolysis of Tg. The importance of DEHAL-catalyzed intrathyroidal I⁻ recycling was underscored by the identification of patients with congenital hypothyroidism due to biallelic mutations in the *DEHAL* gene.[46]

The biosynthesis of THs is regulated by thyroid stimulating hormone (TSH), which is released by the pituitary.[47] Upon activation of the basolaterally located TSH receptor (TSHR), a member of the Gαs protein-coupled receptor superfamily, TSH activates adenylyl cyclase, which increases the intercellular levels of cyclic adenosine monophosphate (cAMP). cAMP mediates most TSH-induced physiological processes in thyroid cells (Figure 5.1).[48,49]

MOLECULAR CHARACTERIZATION OF NIS

Given the physiological, diagnostic, and therapeutic significance of the I⁻ concentrating system of the thyroid, it is not surprising that many investigators sought to identify the relevant molecule for decades, but without success. A promising development came when poly-A⁺ RNA from the highly differentiated rat thyroid-derived cell line FRTL-5 was microinjected into *Xenopus laevis* oocytes, resulting in Na⁺-dependent and ClO₄⁻-sensitive I⁻ transport activity.[11] Dai *et al.* then finally isolated the cDNA encoding NIS by functional cloning in *X. laevis* oocytes using cDNA libraries from FRTL-5 cells.[7] The complete nucleotide sequence revealed an open reading frame of 1.854 nucleotides, which encodes a protein of 618 amino acids. Based on the rat NIS cDNA, Smanik *et al.* screened a human thyroid cDNA library to isolate a cDNA encoding human NIS, which exhibits 84% identity and

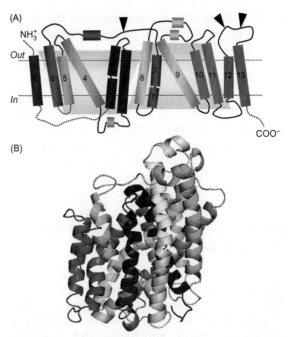

FIGURE 5.2 **Secondary and tertiary structure of NIS.** (A) Secondary structure. vSGLT-based NIS topology model showing the 13 transmembrane segments (TMSs) from the extracellular amino terminus to the intracellular carboxy terminus. TMSs are labeled with Arabic numerals. Gray trapezoids represent the inverted topology formed by TMSs 2–6 and TMSs 7–11. Black triangles mark N-linked glycosylation sites at Asn 225, 485, and 497. (B) Tertiary structure. Membrane plane of the 3-dimensional homology model of NIS based on the vSGLT X-ray structure.[57]

93% similarity to rat NIS.[50] Subsequently, Smanik *et al.* mapped the human NIS gene to chromosome 19p12-13.2.[51] The gene comprises 15 exons spanning 23.2 kb, and has an open reading frame of 1.929 nucleotides, encoding a protein of 643 amino acids.

According to the Online Mendelian Inheritance in Man, NIS belongs to the solute carrier family 5 (SLC5A5), which also includes — among many other proteins that mediate key physiological transport processes — all Na⁺/glucose cotransporters (SGLT).[52] Extensive biochemical analysis of NIS,[7,53–55] has led to the current 13-TMS secondary structure model for NIS (Figure 5.2A). This model has been confirmed by the determination of the X-ray structure of the *Vibrio parahaemolyticus* Na$^+$/galactose transporter (vSGLT), a bacterial homologue of the human SGLT1.[56] Remarkably, NIS shares significant identity (27%) and homology (58%) with vSGLT − almost as much as SGLT1 does (31% identity, 62% homology). This allowed Paroder-Belenitsky *et al.* to generate a 3D homology model of NIS,[57] comprising residues 50 to 476, using the X-ray structure of vSGLT as a template (Figure 5.2B).

Mechanism and Stoichiometry of NIS-Mediated Transport

Eskandari *et al.* elicited inward currents by adding I$^-$ to a perfusing solution containing Na$^+$ in *X. laevis* oocytes expressing NIS (Figure 5.3).[58] By simultaneously measuring radioactive tracer substrates and currents, these authors determined that NIS transports Na$^+$ and I$^-$ with a 2:1 stoichiometry. Similar inward currents were elicited using a wide range of anions, such as ClO$_3$$^-$, SCN$^-$, SeCN$^-$, NO$_3$$^-$, and Br$^-$, showing that NIS also transports these anions. Surprisingly, ClO$_4$$^-$, the best-known inhibitor of I$^-$ uptake, did not elicit currents.[58] Similarly, Yoshida *et al.* reported that ClO$_4$$^-$ did not induce current in Chinese hamster ovary cells expressing NIS.[59] As flux experiments could not be carried out with radioactive ^{36}ClO$_4$$^-$, the most likely interpretation of those results, at the time, seemed to be that ClO$_4$$^-$ was not transported by NIS but was rather a blocker, although the unlikely possibility could not be ruled out that ClO$_4$$^-$ was translocated by NIS with an electroneutral 1:1 Na$^+$/ClO$_4$$^-$ stoichiometry.

Dohán *et al.* later conclusively demonstrated that ClO$_4$$^-$ is, in fact, actively translocated by NIS − both *in vitro*, using a bioassay in a polarized bicameral model system, and *in vivo*, by administering ^{131}I$^-$ or ^{131}I$^-$/ClO$_4$$^-$ to lactating dams but not their pups and showing that

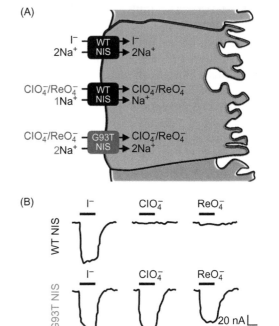

FIGURE 5.3 NIS stoichiometry. (A) Wild-type (WT) NIS mediates electrogenic I$^-$ transport (2 Na$^+$ per I$^-$) and electroneutral ClO$_4$$^-$ or ReO$_4$$^-$ transport (1 Na$^+$ per ClO$_4$$^-$ or ReO$_4$$^-$). Among other G93 NIS mutants, G93T NIS mediates electrogenic I$^-$ (not shown) and ClO$_4$$^-$ or ReO$_4$$^-$ transport (2 Na$^+$ per I$^-$, ClO$_4$$^-$, or ReO$_4$$^-$). (B) Current traces in *Xenopus laevis* oocytes expressing WT or G93T NIS. Downhill deflections represent an influx of positive charge into the oocyte, and hence NIS-mediated electrogenic transport of Na$^+$ and I$^-$ or ClO$_4$$^-$ or ReO$_4$$^-$ into the cell.

thyroidal I$^-$ uptake in both dams and pups was markedly lower in the ClO$_4$$^-$-treated animals than in the controls,[60] and that ClO$_4$$^-$ accumulated in the milk.[60] The kinetic parameters of NIS-mediated transport of a anion structurally similar to ClO$_4$$^-$, perrhenate (ReO$_4$$^-$), were analyzed, taking advantage of the availability of the radioisotope ^{186}ReO$_4$$^-$.[60] The initial rates of ReO$_4$$^-$ transport − as a function of extracellular Na$^+$ − yielded a hyperbolic curve indicative of an electroneutral stoichiometry, which stands in stark contrast to the electrogenic 2:1 Na$^+$/anion stoichiometry observed for several other NIS substrates. These authors thus uncovered that NIS translocates different substrates with different stoichiometries (Figure 5.3),[60] a property that had not been observed in any other transporter, and has not been since. That ClO_4^- is actively accumulated by NIS was later also reported by Tran et al.,[61] who used sensitive chromatography—electrospray ionization—tandem mass spectrometry, and by Cianchetta et al.,[62] who used a yellow fluorescent protein (YFP) variant as a genetic biosensor of intracellular ClO_4^- concentration.

The results discussed above indicating that ClO_4^- is a transported substrate and not just a NIS blocker are highly significant for public health. ClO_4^- is a contaminant in many drinking water sources in a number of regions of the United States. This has led to extensive debate concerning the health impact of exposure to ClO_4^-, which has already been demonstrated to increase serum TSH levels and decrease T_4 levels in women with a low I^- status.[63–66] That ClO_4^- is actively transported by NIS entails that the health effects of ClO_4^- exposure are more detrimental than previously supposed, not only for pregnant and nursing women with a low I^- status but also, and more worryingly, for their children, whose exposure to high ClO_4^- levels puts them at risk of impaired development, not only physically but also intellectually.

NIS Expression in Extrathyroidal Tissues

Several tissues, in addition to the thyroid, have also been reported to actively accumulate I^-, including the salivary and lacrimal glands, gastric mucosa, lactating breast, choroid plexus, and the cilliary body of the eye. Indeed, radioiodide accumulation in the stomach and salivary glands is routinely observed in radioiodide whole-body scintiscans.[67] Several reports of patients with a congenital I^- transport defect (ITD) who displayed no I^-

transport in the thyroid, salivary glands, and stomach suggested the existence of a protein common to the I^- transport processes in these various tissues. Even though different groups have detected NIS mRNA expression in various tissues by reverse transcription polymerase chain reaction (RT-PCR), the results on whether or not NIS is expressed in a particular tissue were inconsistent and sometimes conflicting.[68–71] Functional NIS expression has been demonstrated in extrathyroidal tissues previously known to actively accumulate I^-, such as salivary glands, stomach, and lactating breast,[72–74] as well as in the small intestine, kidney, and placenta.[75–80] Importantly, NIS cDNAs isolated from different tissues all have the same sequence.[70,81]

The functional significance of NIS expression is clear in some extrathyroidal tissues but unknown in others. TH synthesis in the fetus depends upon the supply of I^- from the maternal circulation through the placenta, where NIS protein expression has been mainly demonstrated in the apical membrane of trophoblasts.[77,78] Placental NIS expression has been reported to correlate with gestational age, suggesting that the I^- supply increases as the pregnancy progresses to meet an increased fetal requirement for TH synthesis.[82] An adequate supply of I^- is also essential for the newborn, who needs it to produce enough THs for proper development. The newborn obtains this iodide from maternal milk, into which it is translocated from the bloodstream by NIS in the lactating breast.[72] In this tissue, NIS is expressed at the basolateral membrane of ductal epithelial cells.

Because I^- is supplied exclusively by the diet, I^- absorption in the gastrointestinal tract is the first step in I^- metabolism. Studies involving ligation procedures of the gastrointestinal tract suggested that I^- might be absorbed in the small intestine.[83] Consistent with this notion, Nicola et al. reported functional NIS expression at the apical surface of the small intestinal epithelium as the central component involved in I⁻ absorption.[75] Conversely, in both gastric mucosa and salivary glands, NIS is expressed in the basolateral membrane. In these tissues, NIS transports I⁻ from the bloodstream into the epithelial cells, from which it is then released into the gastric juice and saliva, respectively. The functional significance of NIS-mediated I⁻ uptake in these two tissues is unknown, though it has been proposed that I⁻ acts as an antioxidant and antimicrobial agent in these tissues.[84,85] It seems plausible that NIS-mediated secretion of I⁻ into the saliva and gastric juice and NIS-mediated absorption of I⁻ in the small intestine are part of a systemic mechanism that recycles I⁻ and thereby conserves this essential anion.

IODIDE TRANSPORT DEFECT (ITD)

Underscoring the significance of NIS for thyroid function, inactivating NIS mutations have been identified as causes of congenital I⁻ transport defect (ITD) that results in congenital hypothyroidism.[86] In addition, the study of the ontogeny of various key thyroid proteins showed that NIS is the last to be expressed and that NIS activity is both the final step in thyroid differentiation and the final requirement for the onset of thyroid function in humans.[87] ITD is a rare autosomal recessive disorder clinically characterized, when untreated, by hypothyroidism and goiter of varying degrees, reduced I⁻ uptake, and a low I⁻ saliva-to-plasma ratio.[88,89] To date, 13 ITD-causing mutations in the NIS-gene coding region have been identified: V59E, G93R, R124H, Q267E, C272X, T354P, G395R, frame-shift 515X, Y531X, G543E, Δ143−323, Δ287−288, and Δ439−443. They are either nonsense, alternative splicing, frameshift, deletion, or missense mutations of the NIS gene (Figure 5.4). In addition, a mutation in the 5′ untranslated region of NIS, a C-to-T transition at nucleotide −54 (−54C > T) that

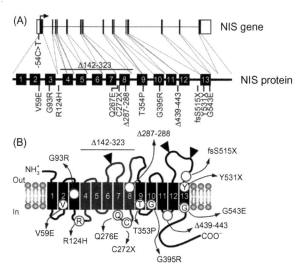

FIGURE 5.4 Schematic representation of the location of I⁻ transport defect (ITD)-causing NIS mutations. (A) Correlation of the structural organization of the human NIS gene with that of the human NIS protein. NIS coding exons are represented by black bars. The 5′ and 3′ untranslated regions coded by exons 1 and 15, respectively, are represented by white boxes. The 13 transmembrane segments (TMSs) in the NIS protein are represented by black boxes and labeled with Arabic numerals. NIS-coding exons are connected to the corresponding protein regions by gray lines. The 14 reported I⁻ transport defect (ITD)-causing NIS mutants are indicated in blue. In the NIS gene, nucleotide position +1 (black arrow in exon 1) corresponds to the adenine (A) of the ATG codon where NIS translation starts. Amino acids are indicated with the single-letter code. Δ, deletion; X, stop codon; fs, frame shift. (B) Secondary structure model showing NIS protein mutations identified in ITD patients. Black rectangles represent the 13 TMSs, which are numbered using Arabic numerals. Black triangles indicate N-linked glycosylation sites. Red circles indicate approximately where the mutated residues are located. The region of the protein lacking in the ITD mutant Δ142−323 NIS is shown in gray.

reduces NIS translation efficiency with a subsequent decrease in protein expression and function, has been reported (Figure 5.4).[90]

The study of ITD-causing NIS mutants has yielded important information on the molecular requirements for NIS function at the relevant positions.[57,91−98] The analysis of the

T354P substitution revealed that NIS function requires an OH group at the β-carbon at position 354,[94] highlighting the importance of other β-OH−containing residues in TMS 9. Thereafter, De la Vieja et al. showed that these residues are involved in Na^+ binding/translocation, and proposed a structural homology between the leucine transporter from Aquifex aeolicus (LeuT) and NIS, even though there is no primary sequence homology between them.[95] This prediction was confirmed when the elucidation of the structure of vSGLT,[56] which belongs to the same family as NIS, revealed that vSGLT and LeuT have the same fold.

Paroder-Belenitsky et al. have reported unexpected properties of NIS,[57] which they uncovered by the detailed molecular characterization of the ITD-causing G93R NIS mutant.[99] The lack of activity of the G93R NIS mutant was found not to be due to impaired trafficking to the plasma membrane or to the positive charge conferred by Arg in the middle of TMS 3. Instead, amino acid replacements at this position caused, for the first time in any NIS mutant, a significant change in the apparent affinity for I^-. Position 93 was shown to be critical for substrate specificity and affinity as, remarkably, a single amino acid substitution at this position switched the $Na^+:ReO_4^-$ or ClO_4^- stoichiometry from electroneutral to electrogenic (Figure 5.3). This notion was supported by the generation of currents upon addition of these substrates in electrophysiology experiments with oocytes. Strikingly, although G93E and Q do not transport I^-, they do transport ReO_4^- and ClO_4^-, and do so electrogenically, a property that may be relevant for gene transfer studies. Furthermore, the NIS 3D homology model suggested that, in going from an outwardly to an inwardly open conformation during the transport cycle, NIS uses G93 as a pivot.[57]

Recently, a thorough characterization of the ITD-causing NIS mutations R124H and Δ439−443 has demonstrated the importance of intramolecular interactions for proper NIS folding.[92,97] Although R124H NIS was initially reported to be located at the plasma membrane,[100] Paroder et al. found that, just like G543E NIS, R124H NIS is also incompletely glycosylated and retained intracellularly. As a result, R124H NIS is not targeted to the plasma membrane and therefore does not mediate any I^- transport in transfected cells. Strikingly, however, R124H is intrinsically active, as revealed by its ability to mediate I^- transport in membrane vesicles.[92] Gln is the only amino acid substitution at position 124, which is located in intracellular loop 2 (IL-2), that restored the targeting of NIS to the plasma membrane, and thus NIS activity. The intrinsic activity of R124H suggests a key structural role for the δ-amino group of R124, a highly conserved residue throughout the SLC5 family, in the cell surface targeting of NIS. An interaction between the δ-amino group of either R or Q124 and the thiol group of C440, located in IL-6, was also tested, showing that the interaction between IL-2 and IL-6 is critical for the local folding required for NIS plasma membrane trafficking.[92] Li et al. demonstrated that Δ439−443 NIS is intracellularly retained and intrinsically inactive.[97] Therefore, to study the defect caused by the deletion of residues 439−443, the authors engineered five consecutive Ala residues at positions 439−443, which partially recovered cell surface targeting and I^- transport. Strikingly, when only Asn was introduced at position 441 in this context, yielding NIS with the sequence 439−AANAA−443, the resulting NIS protein was fully targeted to the plasma membrane and exhibited I^- transport.[97] The authors proposed that the side chain of N441, a residue conserved throughout most of the SLC5 family, interacts with the main chain amino group of G444, capping the helix of TMS 12 and thus stabilizing the structure of the molecule.

A NOVEL MODE OF NIS REGULATION

The transcriptional and post-transcriptional regulation of NIS by TSH, I⁻, and Tg, among other factors, has been reviewed elsewhere, along with the pharmacological regulation of NIS expression.[101−103] Here, we will discuss a novel physiological mechanism of NIS regulation – namely, NIS regulation by another membrane protein, the K^+ channel KCNQ1/KCNE2. This channel is a complex comprising four large KCNQ1 subunits, which traverse the membrane six times each, and two small KCNE2 subunits, which traverse the membrane only once each.[104] KCNQ1/KCNE2 K^+ channels participate in the repolarization of the cardiac ventricles.[104] The regulation of NIS by this channel was uncovered when mice in which the *Kcne2* gene had been genetically deleted (Kcne2⁻/⁻ mice) presented not only cardiac symptoms (cardiomegaly, hypertrophy, and impaired contractility) but also several other, unexpected symptoms: alopecia, a 50% embryonic death rate, and growth retardation resulting in dwarfism. These findings led Roepke *et al.* to determine the thyroid status of these mice;[105] they were hypothyroid. Quantitative imaging of thyroidal I⁻ accumulation by positron-emission tomography (PET) revealed that the thyroids of Kcne2⁻/⁻ mice accumulate less I⁻ than those of wild-type mice (Figure 5.5). From this the authors concluded that KCNQ1/KCNE2 K^+ channels are required for thyroid hormone biosynthesis. Remarkably, all the aforementioned symptoms could be alleviated by administration of thyroid hormones, indicating that they were due to hypothyroidism. Strikingly, all these symptoms were also alleviated by having the pups fed by wild-type dams, the explanation being that wild-type dams provide their pups with an abundant supply of thyroid hormones through their milk.[105]

To uncover the mechanism that led to decreased I⁻ accumulation and hence hypothyroidism

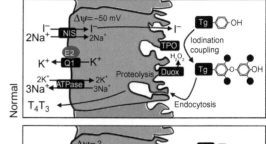
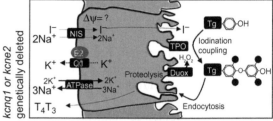
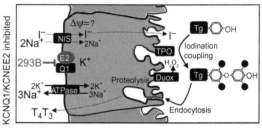

FIGURE 5.5 **Effect of genetic deletion or pharmacological inhibition of the KCNQ1/KCNE2 K^+ channel on thyroid hormone biosynthesis.** In the normal thyroid (upper panel), basolateral NIS-mediated I⁻ uptake supplies I⁻ for apical iodination of Tg and thyroid hormone biosynthesis. When the basolaterally located KCNQ1/KCNE2 K^+ channel is genetically deleted (middle panel) or pharmacologically inhibited (lower panel), thyroid hormone production is reduced as a result of a decrease in NIS-mediated I⁻ accumulation, not in Tg iodination.[106] The decrease in NIS activity may be due to an increase in membrane potential ($\Delta\Psi$) caused by the absence or inhibition of KCNQ1/KCNE2. E2, KCNE2; Q1, KCNQ1; 293B, KCNQ1/KCNE2 inhibitor (−)-[3R,4S]-chromanol 293B.

in Kcne2⁻/⁻ mice, Purtell *et al.* investigated which of the two steps in I⁻ accumulation had been affected – NIS-mediated I⁻ transport or I⁻ organification – by uncoupling the two processes.[106] Organification was intact; it was I⁻ transport that was decreased. The results of these *in vivo* experiments were recapitulated in

FRTL-5 cells, a line of thyroid cells that endogenously express both NIS and KCNQ1/KCNE2, by showing that the KCNQ1/KCNE2 inhibitor $(-)$-[3R,4S]-chromanol 293B decreases NIS-mediated I$^-$ transport (Figure 5.5). Complementary experiments demonstrated that COS-7 cells transport more I$^-$ when transfected with both NIS and KCNQ1/KCNE2 than when transfected with NIS only (COS-7 cells do not express either protein endogenously).[106]

It is worth noting that Frohlich et al. showed that targeted deletion of the Kcnq1 gene also causes hypothyroidism,[107] indicating that both subunits of KCNQ1/KCNE2 are required for adequate thyroid function.

The results reviewed in this section raise a series of interesting mechanistic and medically relevant questions. How does KCNQ1/KCNE2 regulate NIS function? One possibility is that the activity of this channel is crucial for maintaining the membrane potential in thyrocytes. If this is so, then alteration of KCNQ1/KCNE2 by either genetic or pharmacological means will decrease the membrane potential (Figure 5.5). Because this potential is part of the driving force for NIS-mediated I$^-$ uptake (which is electrogenic), such a decrease would impair NIS activity. Do NIS and KCNQ1/KCNE2 interact physically? Does NIS have an effect on the activity of the channel? Is NIS regulated by KCNQ1/KCNE2 in cells other than thyrocytes? Are there other channel−transporter interactions of physiological interest? Do the cardiac problems of patients with mutations in either KCNQ1 or KCNE2 subunits have an endocrine component? It is to be hoped that combining genetic, pharmacological, biochemical, and biophysical approaches will allow us to answer these intriguing questions.

NIS in Thyroid Cancer

NIS-mediated transport of radioiodide has been used for over 65 years in diagnostic scintigraphy and to target and destroy hyperfunctioning thyroid tissue, as in Graves' disease, and also I$^-$-transporting thyroid cancer remnants and their metastases (after thyroidectomy).[108,109] However, some thyroid tumors do not accumulate I$^-$ and are hence radioresistant, leading to poor prognoses: patients with radioiodide-sensitive metastatic thyroid cancer have a 10-year survival rate of $\sim 60\%$, whereas the figure is $\sim 10\%$ for patients whose metastases are radioresistant.[110] Therefore, understanding and overcoming the cause of reduced I$^-$ uptake in thyroid cancer will certainly have major implications for its treatment.[111]

Several studies have reported a decrease in NIS mRNA expression in thyroid cancer samples, based on conventional or quantitative RT-PCR.[112−118] These studies, however, did not address the levels of NIS protein expression or the subcellular localization of the protein. In fact, several groups have demonstrated that most thyroid tumors display higher levels of NIS expression than the histologically normal tissue surrounding them. Surprisingly, NIS was localized mainly intracellularly in most cases, strongly suggesting that the decrease in I$^-$ uptake in most thyroid carcinomas is due to impaired trafficking of NIS to, or retention of NIS at, the plasma membrane.[119−124] It should be emphasized that no mutations have been found in NIS obtained from papillary carcinomas, ruling out the possibility that a mutation might be impairing NIS function in these tumors.[125] The paradoxical co-occurrence of decreased I$^-$ transport and overexpression of NIS (which, however, is intracellularly retained) underscores the importance of uncovering the mechanisms involved in the biogenesis of NIS and its trafficking to and retention at the cell surface.

Since most thyroid cancers conserve a certain degree of differentiation, one logical therapeutic approach to this disease would be to redifferentiate the cells and reinduce or increase endogenous NIS function. In a clinical trial conducted to determine the ability of all-*trans*-retinoic acid to redifferentiate metastatic thyroid cancers and render them responsive to radioiodide therapy, all-*trans*-retinoic acid yielded only a modest clinical benefit:[126] 26% of the 50 patients showed a significant increase in radioiodide uptake, but only 16% displayed reduced tumor volume after radioiodide therapy.[126]

The oncogenes BRAF, RAS, and RET/PTC constitutively activate mitogen-activated protein kinase (MAPK) signaling and induce partial to complete loss of differentiation in thyroid cancers (Figure 5.6).[127,128] *In vitro* studies have found that the expression of any

of these three oncogenes downregulates NIS expression and impairs targeting of NIS to the plasma membrane, thereby reducing I^- uptake (Figure 5.6).[129–131] In patients, BRAF-positive papillary thyroid carcinoma correlated with scan-negative recurrences and defective NIS expression or cell surface targeting.[132] Transgenic mice engineered to express the oncogene BRAF^{V600E} in their thyroid follicular cells developed papillary thyroid tumors, and these tumors did not concentrate radioiodide.[133] $^{124}I^-$ PET imaging of the mice showed that the tumors recovered NIS-mediated I^- transport when BRAF kinase activity was blocked either with PLX4720 or, farther downstream, with the

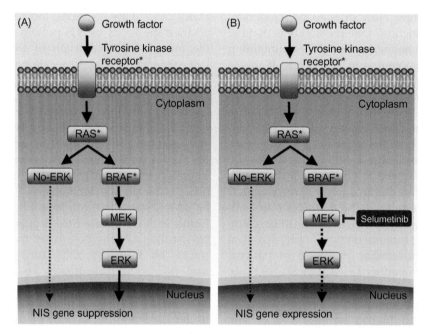

FIGURE 5.6   **Schematic representation of the effect of selumetinib on NIS expression in thyroid cancer.** Growth factors activate the RAS signaling pathway through tyrosine kinase receptors, leading to stimulation of the BRAF–MEK–ERK cascade, and also of ERK-independent pathways. The RAS pathway can be constitutively activated in thyroid tumors in any of two ways: through chromosomal rearrangement of RET/PTC, leading to constitutive activation of the tyrosine kinase receptor, or as a result of activating mutations in the genes coding for RAS or BRAF. (A) Constitutive activation of the RAS signaling pathway leads to thyroid cell dedifferentiation and loss of NIS expression, resulting in thyroid tumor radioresistance. (B) Pharmacological inhibition of MEK with selumetinib increases NIS expression, and hence radioiodide uptake, in patients with radioresistant thyroid tumors with activating mutations in RAS or BRAF, as reported by Ho *et al.*[134] Asterisks indicate constitutive activation.

MEK 1/2 kinase inhibitor selumetinib (Figure 5.6).[133] These exciting findings prompted a clinical study in patients with thyroid cancer metastases resistant to radioiodide. In this study, too, selumetinib induced I^- uptake, allowing radioiodide therapy to be used.[134] Selumetinib administration increased $^{124}I^-$ uptake in 12 of the 20 patients, and eight of these 12 were treated with $^{131}I^-$ because they had reached the dosimetry threshold (> 2000 cGy). Strikingly, of these eight patients, five had confirmed partial responses and three stable disease.[134] $^{131}I^-$ therapy following treatment with selumetinib was especially efficacious in patients whose tumors had a mutated *NRAS* gene (five out of five). These results suggest that this group of patients may benefit most from this approach. Understanding, at the molecular level, why tumors with *NRAS* mutations have increased levels of NIS expression and/or are particularly susceptible to $^{131}I^-$ treatment after MEK inhibition may help us devise novel therapies for thyroid cancer.

Considering the high efficacy and low rate of side effects of NIS-mediated radioiodide therapy in thyroid cancer, a whole new field has opened to develop strategies to apply this therapy in other cancers. To date, besides thyroid cancer, functional endogenous NIS expression has only been observed in breast cancer.[72] Although most tumors do not express endogenous NIS, ectopic NIS expression by gene transfer has paved the way for the development of new therapeutic strategies. Whether a tumor expresses functional NIS endogenously or ectopically, NIS function makes it possible to treat the tumor with radioiodide, just as in differentiated thyroid cancer. Since Shimura *et al.* successfully restored functional NIS expression to malignant transformed rat thyroid cells that did not show I^- transport activity rendering them sensitive to radioiodide treatment,[135] a large body of evidence has shown the feasibility of transferring, either constitutive or tissue-specific, functional NIS expression to several cancer cell lines, enabling these cells to concentrate radioiodide, thus making them susceptible to radioiodine therapy (discussed in recent reviews).[136,137]

CONCLUDING REMARKS

The biosynthesis of the THs is a highly complex mechanism — certainly far more complex than it was imagined to be when these hormones were first isolated. Although many mechanistic details of TH biosynthesis remain to be elucidated, major strides have been made in uncovering how this process occurs. In this chapter, although we cover all stages of the pathway, we concentrate on the protein that mediates the first step in TH biosynthesis — NIS — which has turned out to be an unending source of surprises, and hence far more fascinating than anyone could have expected. NIS transports different substrates with different stoichiometries, a property without precedent in any other transporter.[60] The stoichiometry of NIS-mediated ClO_4^- or ReO_4^- transport can be changed by a single amino acid substitution.[57] NIS, far from being restricted to the thyroid, has been found to be functionally expressed in numerous extrathyroidal tissues, including lactating breast, and in breast cancer and breast cancer metastases.[72] Among the mechanisms regulating NIS function are some that are highly novel, such as cross-talk with the K^+ channel KCNQ1/KCNE2.[105,106] Without a doubt, the more we investigate NIS, the more surprises of great physiological and pathophysiological significance it will yield, likely extending its clinical applications well beyond thyroid disease.

References

[1] Zimmermann MB. Iodine deficiency. *Endocr Rev* 2009;**30**:376–408.

[2] Andersson M, Karumbunathan V, Zimmermann MB. Global iodine status in 2011 and trends over the past decade. *J Nutr* 2012;**142**:744−50.

[3] Baumann E. Ueber das normale Vorkommen von Jod im Thierkörper. *Z Phys Chem* 1896;**21**:319−30.

[4] Wolff J, Maurey JR. Thyroidal iodide transport. II. Comparison with non-thyroid iodide-concentrating tissues. *Biochim Biophys Acta* 1961;**47**:467−74.

[5] Carrasco N. Iodide transport in the thyroid gland. *Biochim Biophys Acta* 1993;**1154**:65−82.

[6] Wolff J. Transport of iodide and other anions in the thyroid gland. *Physiol Rev* 1964;**44**:45−90.

[7] Dai G, Levy O, Carrasco N. Cloning and characterization of the thyroid iodide transporter. *Nature* 1996;**379**:458−60.

[8] Iff HW, Wilbrandt W. The dependency of iodine accumulation in thyroid slices on the ionic composition of the incubation medium and its influence by cardiac glycosides. *Biochim Biophys Acta* 1963;**78**:711−25.

[9] O'Neill B, Magnolato D, Semenza G. The electrogenic, Na^+-dependent I^- transport system in plasma membrane vesicles from thyroid glands. *Biochim Biophys Acta* 1987;**896**:263−74.

[10] Weiss SJ, Philp NJ, Grollman EF. Iodide transport in a continuous line of cultured cells from rat thyroid. *Endocrinology* 1984;**114**:1090−8.

[11] Vilijn F, Carrasco N. Expression of the thyroid sodium/iodide symporter in *Xenopus laevis* oocytes. *J Biol Chem* 1989;**264**:11901−3.

[12] Fong P. Thyroid iodide efflux: a team effort? *J Physiol* 2011;**589**:5929−39.

[13] Scott DA, Wang R, Kreman TM, Sheffield VC, Karniski LP. The Pendred syndrome gene encodes a chloride-iodide transport protein. *Nat Genet* 1999;**21**:440−3.

[14] van den Hove MF, Croizet-Berger K, Jouret F, et al. The loss of the chloride channel, ClC-5, delays apical iodide efflux and induces a euthyroid goiter in the mouse thyroid gland. *Endocrinology* 2006;**147**:1287−96.

[15] Li H, Ganta S, Fong P. Altered ion transport by thyroid epithelia from CFTR(−/−) pigs suggests mechanisms for hypothyroidism in cystic fibrosis. *Exp Physiol* 2010;**95**:1132−44.

[16] Devuyst O, Golstein PE, Sanches MV, et al. Expression of CFTR in human and bovine thyroid epithelium. *Am J Physiol* 1997;**272**:C1299−308.

[17] Deshpande V, Venkatesh SG. Thyroglobulin, the prothyroid hormone: chemistry, synthesis and degradation. *Biochim Biophys Acta* 1999;**1430**:157−78.

[18] Targovnik HM, Esperante SA, Rivolta CM. Genetics and phenomics of hypothyroidism and goiter due to thyroglobulin mutations. *Mol Cell Endocrinol* 2010;**322**:44−55.

[19] Vono-Toniolo J, Rivolta CM, Targovnik HM, Medeiros-Neto G, Kopp P. Naturally occurring mutations in the thyroglobulin gene. *Thyroid* 2005;**15**:1021−33.

[20] Ruf J, Carayon P. Structural and functional aspects of thyroid peroxidase. *Arch Biochem Biophys* 2006;**445**: 269−77.

[21] Ris-Stalpers C, Bikker H. Genetics and phenomics of hypothyroidism and goiter due to TPO mutations. *Mol Cell Endocrinol* 2010;**322**:38−43.

[22] Kimura S, Kotani T, McBride OW, et al. Human thyroid peroxidase: complete cDNA and protein sequence, chromosome mapping, and identification of two alternately spliced mRNAs. *Proc Natl Acad Sci USA* 1987;**84**:5555−9.

[23] Seto P, Hirayu H, Magnusson RP, et al. Isolation of a complementary DNA clone for thyroid microsomal antigen. Homology with the gene for thyroid peroxidase. *J Clin Invest* 1987;**80**:1205−8.

[24] Taurog A, Wall M. Proximal and distal histidines in thyroid peroxidase: relation to the alternatively spliced form, TPO-2. *Thyroid* 1998;**8**:185−91.

[25] Song Y, Driessens N, Costa M, et al. Roles of hydrogen peroxide in thyroid physiology and disease. *J Clin Endocrinol Metab* 2007;**92**:3764−73.

[26] Moreno JC, Bikker H, Kempers MJ, et al. Inactivating mutations in the gene for thyroid oxidase 2 (THOX2) and congenital hypothyroidism. *N Engl J Med* 2002;**347**:95−102.

[27] Zamproni I, Grasberger H, Cortinovis F, et al. Biallelic inactivation of the dual oxidase maturation factor 2 (DUOXA2) gene as a novel cause of congenital hypothyroidism. *J Clin Endocrinol Metab* 2008;**93**:605−10.

[28] De Deken X, Wang D, Many MC, et al. Cloning of two human thyroid cDNAs encoding new members of the NADPH oxidase family. *J Biol Chem* 2000;**275**:23227−33.

[29] Song Y, Ruf J, Lothaire P, et al. Association of duoxes with thyroid peroxidase and its regulation in thyrocytes. *J Clin Endocrinol Metab* 2010;**95**:375−82.

[30] Fortunato RS, Lima de Souza EC, Ameziane-el Hassani R, et al. Functional consequences of dual oxidase-thyroperoxidase interaction at the plasma membrane. *J Clin Endocrinol Metab* 2010;**95**:5403−11.

[31] Grasberger H, Refetoff S. Identification of the maturation factor for dual oxidase. Evolution of an eukaryotic operon equivalent. *J Biol Chem* 2006;**281**: 18269−72.

[32] Morand S, Ueyama T, Tsujibe S, Saito N, Korzeniowska A, Leto TL. Duox maturation factors form cell surface complexes with Duox affecting the specificity of reactive oxygen species generation. *FASEB J* 2009;**23**:1205−18.

[33] Dunn JT, Dunn AD. The importance of thyroglobulin structure for thyroid hormone biosynthesis. *Biochimie* 1999;**81**:505−9.

[34] World Health Organization. International Council for Control of Iodine Deficiency Disorders, and United Nations Children's Fund. Assessment of iodine deficiency disorders and monitoring their elimination: a guide for programme managers. 3rd ed. Geneva, Switzerland; 2007.

[35] Marino M, McCluskey RT. Role of thyroglobulin endocytic pathways in the control of thyroid hormone release. *Am J Physiol Cell Physiol* 2000;**279**:C1295−306.

[36] Spencer CA, Lopresti JS. Measuring thyroglobulin and thyroglobulin autoantibody in patients with differentiated thyroid cancer. *Nat Clin Pract Endocrinol Metab* 2008;**4**:223−33.

[37] Lisi S, Segnani C, Mattii L, et al. Thyroid dysfunction in megalin deficient mice. *Mol Cell Endocrinol* 2005;**236**:43−7.

[38] Lisi S, Pinchera A, McCluskey RT, et al. Preferential megalin-mediated transcytosis of low-hormonogenic thyroglobulin: a control mechanism for thyroid hormone release. *Proc Natl Acad Sci USA* 2003;**100**: 14858−63.

[39] Visser WE, Friesema EC, Visser TJ. Minireview: thyroid hormone transporters: the knowns and the unknowns. *Mol Endocrinol* 2011;**25**:1−14.

[40] Di Cosmo C, Liao XH, Dumitrescu AM, Philp NJ, Weiss RE, Refetoff S. Mice deficient in MCT8 reveal a mechanism regulating thyroid hormone secretion. *J Clin Invest* 2010;**120**:3377−88.

[41] Gereben B, Zavacki AM, Ribich S, et al. Cellular and molecular basis of deiodinase-regulated thyroid hormone signaling. *Endocr Rev* 2008;**29**:898−938.

[42] Baqui MM, Gereben B, Harney JW, Larsen PR, Bianco AC. Distinct subcellular localization of transiently expressed types 1 and 2 iodothyronine deiodinases as determined by immunofluorescence confocal microscopy. *Endocrinology* 2000;**141**:4309−12.

[43] Maia AL, Goemann IM, Meyer EL, Wajner SM. Deiodinases: the balance of thyroid hormone: type 1 iodothyronine deiodinase in human physiology and disease. *J Endocrinol* 2011;**209**:283−97.

[44] Moreno JC, Pauws E, van Kampen AH, Jedlickova M, de Vijlder JJ, Ris-Stalpers C. Cloning of tissue-specific genes using serial analysis of gene expression and a novel computational substraction approach. *Genomics* 2001;**75**:70−6.

[45] Gnidehou S, Caillou B, Talbot M, et al. Iodotyrosine dehalogenase 1 (DEHAL1) is a transmembrane protein involved in the recycling of iodide close to the thyroglobulin iodination site. *FASEB J* 2004;**18**: 1574−6.

[46] Moreno JC, Klootwijk W, van Toor H, et al. Mutations in the iodotyrosine deiodinase gene and hypothyroidism. *N Engl J Med* 2008;**358**:1811−8.

[47] Magner JA. Thyroid-stimulating hormone: biosynthesis, cell biology, and bioactivity. *Endocr Rev* 1990;**11**:354−85.

[48] Szkudlinski MW, Fremont V, Ronin C, Weintraub BD. Thyroid-stimulating hormone and thyroid-stimulating hormone receptor structure-function relationships. *Physiol Rev* 2002;**82**:473−502.

[49] Kleinau G, Neumann S, Gruters A, Krude H, Biebermann H. Novel insights on thyroid stimulating hormone receptor signal transduction. *Endocr Rev* 2013.

[50] Smanik PA, Liu Q, Furminger TL, et al. Cloning of the human sodium iodide symporter. *Biochem Biophys Res Commun* 1996;**226**:339−45.

[51] Smanik PA, Ryu KY, Theil KS, Mazzaferri EL, Jhiang SM. Expression, exon−intron organization, and chromosome mapping of the human sodium iodide symporter. *Endocrinology* 1997;**138**:3555−8.

[52] Wright EM, Loo DD, Hirayama BA. Biology of human sodium glucose transporters. *Physiol Rev* 2011;**91**:733−94.

[53] Levy O, Dai G, Riedel C, et al. Characterization of the thyroid Na$^+$/I$^-$ symporter with an anti-COOH terminus antibody. *Proc Natl Acad Sci U S A* 1997;**94** (11):5568−73.

[54] Levy O, De la Vieja A, Ginter CS, Riedel C, Dai G, Carrasco N. N-linked glycosylation of the thyroid Na$^+$/I$^-$ symporter (NIS). Implications for its secondary structure model. *J Biol Chem* 1998;**273**:22657−63.

[55] Paire A, Bernier-Valentin F, Selmi-Ruby S, Rousset B. Characterization of the rat thyroid iodide transporter using anti-peptide antibodies. Relationship between its expression and activity. *J Biol Chem* 1997;**272**:18245−9.

[56] Faham S, Watanabe A, Besserer GM, et al. The crystal structure of a sodium galactose transporter reveals mechanistic insights into Na$^+$/sugar symport. *Science* 2008;**321**:810−4.

[57] Paroder-Belenitsky M, Maestas MJ, Dohan O, et al. Mechanism of anion selectivity and stoichiometry of the Na$^+$/I$^-$ symporter (NIS). *Proc Natl Acad Sci USA* 2011;**108**:17933−8.

[58] Eskandari S, Loo DD, Dai G, Levy O, Wright EM, Carrasco N. Thyroid Na$^+$/I$^-$ symporter. Mechanism, stoichiometry, and specificity. *J Biol Chem* 1997;**272**: 27230−8.

[59] Yoshida A, Sasaki N, Mori A, et al. Different electrophysiological character of I$^-$, ClO$_4^-$, and SCN$^-$ in the transport by Na$^+$/I$^-$ symporter. *Biochem Biophys Res Commun* 1997;**231**:731−4.

[60] Dohan O, Portulano C, Basquin C, Reyna-Neyra A, Amzel LM, Carrasco N. The Na$^+$/I symporter (NIS) mediates electroneutral active transport of the environmental pollutant perchlorate. Proc Natl Acad Sci USA 2007;104:20250−5.

[61] Tran N, Valentin-Blasini L, Blount BC, et al. Thyroid-stimulating hormone increases active transport of perchlorate into thyroid cells. Am J Physiol Endocrinol Metab 2008;294:E802−6.

[62] Cianchetta S, di Bernardo J, Romeo G, Rhoden KJ. Perchlorate transport and inhibition of the sodium iodide symporter measured with the yellow fluorescent protein variant YFP-H148Q/I152L. Toxicol Appl Pharmacol 2010;243:372−80.

[63] Blount BC, Pirkle JL, Osterloh JD, Valentin-Blasini L, Caldwell KL. Urinary perchlorate and thyroid hormone levels in adolescent and adult men and women living in the United States. Environ Health Perspect 2006;114:1865−71.

[64] Pearce EN, Lazarus JH, Smyth PP, et al. Perchlorate and thiocyanate exposure and thyroid function in first-trimester pregnant women. J Clin Endocrinol Metab 2010;95:3207−15.

[65] Pearce EN, Spencer CA, Mestman JH, et al. Effect of environmental perchlorate on thyroid function in pregnant women from Cordoba, Argentina, and Los Angeles, California. Endocr Pract 2011;17:412−7.

[66] Leung AM, Braverman LE, He X, et al. Environmental perchlorate and thiocyanate exposures and infant serum thyroid function. Thyroid 2012;22:938−43.

[67] Bruno R, Giannasio P, Ronga G, et al. Sodium iodide symporter expression and radioiodine distribution in extrathyroidal tissues. J Endocrinol Invest 2004;27:1010−4.

[68] Ajjan RA, Kamaruddin NA, Crisp M, Watson PF, Ludgate M, Weetman AP. Regulation and tissue distribution of the human sodium iodide symporter gene. Clin Endocrinol (Oxf) 1998;49:517−23.

[69] Perron B, Rodriguez AM, Leblanc G, Pourcher T. Cloning of the mouse sodium iodide symporter and its expression in the mammary gland and other tissues. J Endocrinol 2001;170:185−96.

[70] Spitzweg C, Joba W, Eisenmenger W, Heufelder AE. Analysis of human sodium iodide symporter gene expression in extrathyroidal tissues and cloning of its complementary deoxyribonucleic acids from salivary gland, mammary gland, and gastric mucosa. J Clin Endocrinol Metab 1998;83:1746−51.

[71] Lacroix L, Mian C, Caillou B, et al. Na$^+$/I$^-$ symporter and Pendred syndrome gene and protein expressions in human extra-thyroidal tissues. Eur J Endocrinol 2001;144:297−302.

[72] Tazebay UH, Wapnir IL, Levy O, et al. The mammary gland iodide transporter is expressed during lactation and in breast cancer. Nat Med 2000;6:871−8.

[73] Altorjay A, Dohan O, Szilagyi A, Paroder M, Wapnir IL, Carrasco N. Expression of the Na$^+$/I$^-$ symporter (NIS) is markedly decreased or absent in gastric cancer and intestinal metaplastic mucosa of Barrett esophagus. BMC Cancer 2007;7:5.

[74] La Perle KM, Kim DC, Hall NC, et al. Modulation of sodium/iodide symporter expression in the salivary gland. Thyroid 2003;23:1029−36.

[75] Nicola JP, Basquin C, Portulano C, Reyna-Neyra A, Paroder M, Carrasco N. The Na$^+$/I$^-$ symporter mediates active iodide uptake in the intestine. Am J Physiol Cell Physiol 2009;296:C654−62.

[76] Nicola JP, Reyna-Neyra A, Carrasco N, Masini-Repiso AM. Dietary iodide controls its own absorption through post-transcriptional regulation of the intestinal Na$^+$/I$^-$ symporter. J Physiol 2012;590:6013−26.

[77] Mitchell AM, Manley SW, Morris JC, Powell KA, Bergert ER, Mortimer RH. Sodium iodide symporter (NIS) gene expression in human placenta. Placenta 2001;22:256−8.

[78] Di Cosmo C, Fanelli G, Tonacchera M, et al. The sodium-iodide symporter expression in placental tissue at different gestational age: an immunohistochemical study. Clin Endocrinol (Oxf) 2006;65:544−8.

[79] Donowitz M, Singh S, Salahuddin FF, et al. Proteome of murine jejunal brush border membrane vesicles. J Prot Res 2007;6:4068−79.

[80] Spitzweg C, Dutton CM, Castro MR, et al. Expression of the sodium iodide symporter in human kidney. Kidney Int 2001;59:1013−23.

[81] Kotani T, Ogata Y, Yamamoto I, et al. Characterization of gastric Na$^+$/I$^-$ symporter of the rat. Clin Immunol Immunopathol 1998;89:271−8.

[82] Li H, Patel J, Mortimer RH, Richard K. Ontogenic changes in human placental sodium iodide symporter expression. Placenta 2012;33:946−8.

[83] Josefsson M, Grunditz T, Ohlsson T, Ekblad E. Sodium/iodide-symporter: distribution in different mammals and role in entero-thyroid circulation of iodide. Acta Physiol Scand 2002;175:129−37.

[84] El Hassani RA, Benfares N, Caillou B, et al. Dual oxidase2 is expressed all along the digestive tract. Am J Physiol Gastrointest Liver Physiol; 288:G933−42.

[85] Geiszt M, Witta J, Baffi J, Lekstrom K, Leto TL. Dual oxidases represent novel hydrogen peroxide sources supporting mucosal surface host defense. FASEB J 2003;17:1502−4.

[86] Fujiwara H, Tatsumi K, Miki K, et al. Congenital hypothyroidism caused by a mutation in the Na$^+$/I$^-$ symporter. Nat Genet 1997;16:124−5.

[87] Szinnai G, Lacroix L, Carre A, et al. Sodium/iodide symporter (NIS) gene expression is the limiting step for the onset of thyroid function in the human fetus. *J Clin Endocrinol Metab* 2007;**92**:70–6.

[88] Stanbury JB, Chapman EM. Congenital hypothyroidism with goitre. Absence of an iodide-concentrating mechanism. *Lancet* 1960;**1**:1162–5.

[89] Wolff J. Congenital goiter with defective iodide transport. *Endocr Rev* 1983;**4**:240–54.

[90] Nicola JP, Nazar M, Serrano-Nascimento C, et al. Iodide transport defect: functional characterization of a novel mutation in the Na^+/I^- symporter 5′-untranslated region in a patient with congenital hypothyroidism. *J Clin Endocrinol Metab* 2011;**96**: E1100–7.

[91] Reed-Tsur MD, De la Vieja A, Ginter CS, Carrasco N. Molecular characterization of V59E NIS, a Na^+/I^- symporter (NIS) mutant that causes congenital I^- transport defect (ITD). *Endocrinology* 2008;**149**: 3077–84.

[92] Paroder V, Nicola JP, Ginter CS, Carrasco N. The iodide transport defect-causing mutation R124H: a delta-amino group at position 124 is critical for maturation and trafficking of the Na^+/I^- symporter (NIS). *J Cell Sci* 2013. Available from: http://dx.doi.org/10.1242/jcs.120246.

[93] De La Vieja A, Ginter CS, Carrasco N. The Q267E mutation in the sodium/iodide symporter (NIS) causes congenital iodide transport defect (ITD) by decreasing the NIS turnover number. *J Cell Sci* 2004;**117**:677–87.

[94] Levy O, Ginter CS, De la Vieja A, Levy D, Carrasco N. Identification of a structural requirement for thyroid Na^+/I^- symporter (NIS) function from analysis of a mutation that causes human congenital hypothyroidism. *FEBS Lett* 1998;**429**:36–40.

[95] De la Vieja A, Reed MD, Ginter CS, Carrasco N. Amino acid residues in transmembrane segment IX of the Na^+/I^- symporter play a role in its Na+ dependence and are critical for transport activity. *J Biol Chem* 2007;**282**:25290–8.

[96] Dohan O, Gavrielides MV, Ginter C, Amzel LM, Carrasco N. Na^+/I^- symporter activity requires a small and uncharged amino acid residue at position 395. *Mol Endocrinol* 2002;**16**:1893–902.

[97] Li W, Nicola JP, Amzel LM, Carrasco N. Asn441 plays a key role in folding and function of the Na^+/I^- symporter (NIS). *FASEB J* 2013;**27**: 3229–38.

[98] De la Vieja A, Ginter CS, Carrasco N. Molecular analysis of a congenital iodide transport defect: G543E impairs maturation and trafficking of the Na^+/I^- symporter. *Mol Endocrinol* 2005;**19**: 2847–58.

[99] Kosugi S, Inoue S, Matsuda A, Jhiang SM. Novel, missense and loss-of-function mutations in the sodium/iodide symporter gene causing iodide transport defect in three Japanese patients. *J Clin Endocrinol Metab* 1998;**83**:3373–6.

[100] Szinnai G, Kosugi S, Derrien C, et al. Extending the clinical heterogeneity of iodide transport defect (ITD): a novel mutation R124H of the sodium/iodide symporter gene and review of genotype–phenotype correlations in ITD. *J Clin Endocrinol Metab* 2006;**91**:1199–204.

[101] Kohn LD, Suzuki K, Nakazato M, Royaux I, Green ED. Effects of thyroglobulin and pendrin on iodide flux through the thyrocyte. *Trends Endocrinol Metab* 2001;**12**(1):10–6.

[102] Riesco-Eizaguirre G, Santisteban P. A perspective view of sodium iodide symporter research and its clinical implications. *Eur J Endocrinol* 2006;**155**: 495–512.

[103] Kogai T, Brent GA. The sodium iodide symporter (NIS): regulation and approaches to targeting for cancer therapeutics. *Pharmacol Ther* 2012;**135**:355–70.

[104] Jespersen T, Grunnet M, Olesen SP. The KCNQ1 potassium channel: from gene to physiological function. *Physiology* 2005;**20**:408–16.

[105] Roepke TK, King EC, Reyna-Neyra A, et al. Kcne2 deletion uncovers its crucial role in thyroid hormone biosynthesis. *Nat Med* 2009;**15**:1186–94.

[106] Purtell K, Paroder-Belenitsky M, Reyna-Neyra A, et al. The KCNQ1-KCNE2 K^+ channel is required for adequate thyroid I^- uptake. *FASEB J* 2012;**26**:3252–9.

[107] Frohlich H, Boini KM, Seebohm G, et al. Hypothyroidism of gene-targeted mice lacking Kcnq1. *Pflugers Archiv* 2011;**461**:45–52.

[108] Bonnema SJ, Hegedus L. Radioiodine therapy in benign thyroid diseases: effects, side effects, and factors affecting therapeutic outcome. *Endocr Rev* 2012;**33**:920–80.

[109] Mazzaferri EL, Kloos RT. Clinical review 128: Current approaches to primary therapy for papillary and follicular thyroid cancer. *J Clin Endocrinol Metab* 2001;**86**:1447–63.

[110] Durante C, Haddy N, Baudin E, et al. Long-term outcome of 444 patients with distant metastases from papillary and follicular thyroid carcinoma: benefits and limits of radioiodine therapy. *J Clin Endocrinol Metab* 2006;**91**:2892–9.

[111] Schlumberger M, Lacroix L, Russo D, Filetti S, Bidart JM. Defects in iodide metabolism in thyroid cancer and implications for the follow-up and treatment of patients. *Nat Clin Pract Endocrinol Metab* 2007;**3**: 260–9.

[112] Arturi F, Russo D, Giuffrida D, Schlumberger M, Filetti S. Sodium-iodide symporter (NIS) gene expression in lymph-node metastases of papillary thyroid carcinomas. *Eur J Endocrinol* 2000;**143**: 623–7.

[113] Arturi F, Russo D, Schlumberger M, et al. Iodide symporter gene expression in human thyroid tumors. *J Clin Endocrinol Metab* 1998;**83**:2493–6.

[114] Lazar V, Bidart JM, Caillou B, et al. Expression of the Na$^+$/I$^-$ symporter gene in human thyroid tumors: a comparison study with other thyroid-specific genes. *J Clin Endocrinol Metab* 1999;**84**:3228–34.

[115] Ryu KY, Senokozlieff ME, Smanik PA, et al. Development of reverse transcription-competitive polymerase chain reaction method to quantitate the expression levels of human sodium iodide symporter. *Thyroid* 1999;**9**:405–9.

[116] Tanaka K, Otsuki T, Sonoo H, et al. Semi-quantitative comparison of the differentiation markers and sodium iodide symporter messenger ribonucleic acids in papillary thyroid carcinomas using RT-PCR. *Eur J Endocrinol* 2000;**142**:340–6.

[117] Ringel MD, Anderson J, Souza SL, et al. Expression of the sodium iodide symporter and thyroglobulin genes are reduced in papillary thyroid cancer. *Mod Pathol* 2001;**14**:289–96.

[118] Ward LS, Santarosa PL, Granja F, da Assumpcao LV, Savoldi M, Goldman GH. Low expression of sodium iodide symporter identifies aggressive thyroid tumors. *Cancer Letters* 2003;**200**:85–91.

[119] Saito T, Endo T, Kawaguchi A, et al. Increased expression of the sodium/iodide symporter in papillary thyroid carcinomas. *J Clin Invest* 1998;**101**:1296–300.

[120] Dohan O, Baloch Z, Banrevi Z, Livolsi V, Carrasco N. Rapid communication: predominant intracellular overexpression of the Na$^+$/I$^-$ symporter (NIS) in a large sampling of thyroid cancer cases. *J Clin Endocrinol Metab* 2001;**86**:2697–700.

[121] Wapnir IL, van de Rijn M, Nowels K, et al. Immunohistochemical profile of the sodium/iodide symporter in thyroid, breast, and other carcinomas using high density tissue microarrays and conventional sections. *J Clin Endocrinol Metab* 2003;**88**: 1880–8.

[122] Tonacchera M, Viacava P, Agretti P, et al. Benign nonfunctioning thyroid adenomas are characterized by a defective targeting to cell membrane or a reduced expression of the sodium iodide symporter protein. *J Clin Endocrinol Metab* 2002;**87**:352–7.

[123] Liu YY, Morreau H, Kievit J, Romijn JA, Carrasco N, Smit JW. Combined immunostaining with galectin-3, fibronectin-1, CITED-1, Hector Battifora mesothelial-1, cytokeratin-19, peroxisome proliferator-activated receptor-{gamma}, and sodium/iodide symporter antibodies for the differential diagnosis of non-medullary thyroid carcinoma. *Eur J Endocrinol* 2008;**158**:375–84.

[124] Kollecker I, von Wasielewski R, Langner C, et al. Subcellular distribution of the sodium iodide symporter in benign and malignant thyroid tissues. *Thyroid* 2012;**22**:529–35.

[125] Russo D, Manole D, Arturi F, et al. Absence of sodium/iodide symporter gene mutations in differentiated human thyroid carcinomas. *Thyroid* 2001;**11**:37–9.

[126] Simon D, Korber C, Krausch M, et al. Clinical impact of retinoids in redifferentiation therapy of advanced thyroid cancer: final results of a pilot study. *Eur J Nucl Med Mol Imaging* 2002;**29**:775–82.

[127] Soares P, Trovisco V, Rocha AS, et al. BRAF mutations and RET/PTC rearrangements are alternative events in the etiopathogenesis of PTC. *Oncogene* 2003;**22**:4578–80.

[128] Kimura ET, Nikiforova MN, Zhu Z, Knauf JA, Nikiforov YE, Fagin JA. High prevalence of BRAF mutations in thyroid cancer: genetic evidence for constitutive activation of the RET/PTC-RAS-BRAF signaling pathway in papillary thyroid carcinoma. *Cancer Res* 2003;**63**:1454–7.

[129] Santoro M, Melillo RM, Grieco M, Berlingieri MT, Vecchio G, Fusco A. The TRK and RET tyrosine kinase oncogenes cooperate with ras in the neoplastic transformation of a rat thyroid epithelial cell line. *Cell Growth Differ* 1993;**4**:77–84.

[130] Knauf JA, Kuroda H, Basu S, Fagin JA. RET/PTC-induced dedifferentiation of thyroid cells is mediated through Y1062 signaling through SHC-RAS-MAP kinase. *Oncogene* 2003;**22**:4406–12.

[131] Mitsutake N, Knauf JA, Mitsutake S, Mesa Jr. C, Zhang L, Fagin JA. Conditional BRAFV600E expression induces DNA synthesis, apoptosis, dedifferentiation, and chromosomal instability in thyroid PCCL3 cells. *Cancer Res* 2005;**65**:2465–73.

[132] Riesco-Eizaguirre G, Gutierrez-Martinez P, Garcia-Cabezas MA, Nistal M, Santisteban P. The oncogene BRAF V600E is associated with a high risk of recurrence and less differentiated papillary thyroid carcinoma due to the impairment of Na$^+$/I$^-$ targeting to the membrane. *Endocr Relat Cancer* 2006;**13**: 257–69.

[133] Chakravarty D, Santos E, Ryder M, et al. Small-molecule MAPK inhibitors restore radioiodine incorporation in mouse thyroid cancers with conditional BRAF activation. *J Clin Invest* 2011;**121**: 4700–11.

[134] Ho AL, Grewal RK, Leboeuf R, et al. Selumetinib-enhanced radioiodine uptake in advanced thyroid cancer. *N Engl J Med* 2013;**368**:623–32.

[135] Shimura H, Haraguchi K, Miyazaki A, Endo T, Onaya T. Iodide uptake and experimental 131I therapy in transplanted undifferentiated thyroid cancer cells expressing the Na^+/I^- symporter gene. *Endocrinology* 1997;**138**:4493–6.

[136] Penheiter AR, Russell SJ, Carlson SK. The sodium iodide symporter (NIS) as an imaging reporter for gene, viral, and cell-based therapies. *Curr Gene Ther* 2012;**12**:33–47.

[137] Hingorani M, Spitzweg C, Vassaux G, et al. The biology of the sodium iodide symporter and its potential for targeted gene delivery. *Curr Cancer Drug Targets* 2010;**10**:242–67.

The Follicle-Stimulating Hormone Signaling Network in Sertoli Cells

Nathalie Gallay, Laurine Gagniac,
Florian Guillou and Pascale Crépieux

CNRS, Nouzilly, France, François Rabelais University, Tours, France

INTRODUCTION

The seminiferous tubule is a highly dynamic structure that undergoes permanent remodeling throughout life to enable spermatogenesis. It is characterized by intricate communications between intimately connected germ cells and somatic cells. These somatic cells are Sertoli cells, which undergo developmentally regulated processes and fine-tune germ cell mitosis and progression through meiosis. Their primary role is to nurture germ cells from the spermatogonia stage until they are released as elongated spermatids into the lumen of seminiferous tubules.[1] In the adulthood, the spermatogenic capacity, both quantitatively and qualitatively, is directly linked to the appropriate development of Sertoli cells. Thus, both the Sertoli cell population size,[2] as a function of Sertoli cell mitogenic potential, and Sertoli cell effective differentiation are crucial for sperm output and, hence, male reproductive ability.[3] Both processes are operational respectively round birth and at puberty, although pervading evidence also suggests that Sertoli cells proliferation might be relaunched in the adult.[4]

Postnatally, a mandatory endocrine controller is follicle-stimulating hormone (FSH), which sustains Sertoli cell mitotic activity in the neonate[5,6] and then drives cells to differentiate, in a concerted manner with other endocrine, paracrine or autocrine factors, such as thyroid hormones, basic fibroblast growth factor (bFGF), interleukin-1 beta (IL-1β), insulin-like growth factor 1 (IGF-1) or glial cell-derived neurotrophic factor (GDNF).[7] FSH is a dimeric glycoprotein hormone secreted by the pituitary and composed of an α and a β subunit.[8] Its importance in male reproduction has been firmly established experimentally by gene knockout of the FSH receptor (FSH-R)[9,10] in mouse. By the same token, a partial loss-of-function mutation in the FSH-R has been identified in man.[11]

Given the pleiotropic biological role of FSH, and to gain insight onto the dynamics that governs the switch between FSH-driven mitotic and differentiating cells, many studies have been dedicated to deciphering the signaling pathways induced by this hormone from birth to puberty. This is the topic of this chapter. The authors also summarize some of the

Cellular Endocrinology in Health and Disease.
DOI: http://dx.doi.org/10.1016/B978-0-12-408134-5.00006-8

studies done in rodents and in humans and propose new directions for research in line with recent findings in the field.

IS cAMP THE SOLE DRIVING FORCE TO FSH ACTION IN SERTOLI CELLS?

The FSH receptor (FSH-R) is a seven-helix transmembrane receptor, coupled to G proteins, that belongs to class A of G protein-coupled receptors (GPCR). This sub-class of

GPCR is characterized by a E/D-R-W/Y motif involved in coupling to G proteins. Coupling of the FSHR to the $G\alpha_s$/cyclic adenosine monophosphate (cAMP)/protein kinase A (PKA)/cAMP response element-binding protein (CREB) signaling pathway has been acknowledged as the sole effector mechanism of FSH for more than twenty years (Figure 6.1).[21] However, PKA is not the only kinase activated in response to a rise in cAMP, since phosphatidyl inositide-(3,4,5)-trisphosphate kinase (PI3K)-dependent signaling has also been reported to depend on the cAMP

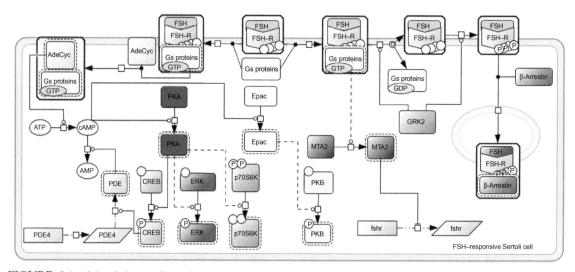

FIGURE 6.1 **Stimulation and regulation of the $G\alpha_s$/cAMP/PKA pathway, long considered as the sole signaling pathway triggered at the FSH-R.** Most events have been demonstrated in Sertoli cells. Others have been shown in immortalized cell lines, mainly HEK293 cells or in MSC-1 cells: agonist-induced phosphorylation of the FSH-R,[12–14] the role that GRKs play in these phosphorylation events,[15–17] β-arrestin-dependent FSH-R internalization.[14,18] The involvement of EPAC is deduced from the observation of cAMP-dependent, PKA-independent activation of PKB, reported in ref[19]. Upregulation of PDE4 expression is mediated by cAMP, but the involvement of CREB is still unclear. The Cell Designer program[20] has been used to formalize the topology of the biochemical reactions (as well as in Figures 6.2 to 6.7). For the sake of clarity, reversibility of the reactions is not figured. Complexes are surrounded by a box. Dashed lines indicate indirect reactions. The complete semantic is as follows:

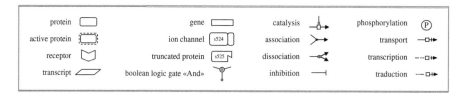

response, in Sertoli cells.[19] Particularly, protein kinase B (PKB) activation appears to be triggered independently of PKA, which suggests a putative role of exchange protein directly activated by cAMP (EPAC), although EPAC activity has not been measured yet in Sertoli cells.

Like other biomolecules, the intracellular concentration of second messengers results from the equilibrium between their synthesis and degradation rates. Hence, cAMP production results from its rate of synthesis by adenylate cyclase, and its rate of degradation by phosphodiesterases (PDE). Accordingly, in Sertoli cells, FSH regulates the expression of some PDE, mostly PDE4D1 and PDE4D2,[22] in a cAMP-dependent manner.[23,24] Another way to fine-tune the intracellular concentration of second messengers is to desensitize the FSH-R, which occurs within minutes following FSH exposure in *in vitro* primary cultures of Sertoli cells.[25,26] This phenomenon relies on the uncoupling of the ligand-bound receptor from its intracellular effectors. *In vitro* stimulation of Sertoli cells by FSH leads to the immediate recruitment at the plasma membrane of kinases which target GPCR, namely the GRKs (G-protein coupled receptor kinases).[25,26] β-arrestins are also recruited to the FSH-R, likely as the result of both agonist-induced conformational change and phosphorylation of the receptor. β-arrestins are adaptor proteins used by GPCR to launch internalization, by virtue of their ability to interact with components of the clathrin-coated pits.

This desensitization process is of clear physiological importance because, in the living organism, Sertoli cells are permanently exposed to the FSH, the FSH-R being localized at the basal pole of the cell, oriented towards the basal lamina and blood capillaries. Furthermore, another desensitization mechanism superimposes on the former one, in order to keep the cell from being overstimulated: it consists in the downregulation of the FSH-R mRNA, likely resulting from chromatin remodeling.[27] Accordingly, a protein belonging to the nucleosome remodeling and histone deacetylation (NuRD), the MTA2 (metastasis-associated) protein, has been recently found to recruit histone deacetylase at the *fshr* gene promoter to inhibit its transcription.[28]

cAMP-INDEPENDENT FSH SIGNALING IN SERTOLI CELLS

A dynamic computational model has predicted that inhibition of PKA does not compromise phosphatidyl-inositide (3,4,5)-trisphosphate (PIP3) production, in prepubertal rats,[29] indicating that cAMP-dependent and -independent signaling might co-exist in parallel. In turn, PKA and PI3K regulate downstream kinases, such as the ERK MAP kinases[30,31] or p70S6 kinase (p70S6K).[29,32] Strikingly, second messenger levels appear to be developmentally regulated, with the efficacy of FSH-induced cAMP production rising from birth to puberty,[30] whereas the sensitivity to FSH of PIP3 production decreases over time.[29] Consistently, PIP3-dependent molecular events, such as mTOR and PRAS-40 phosphorylation, appear to be involved in the mitogenic response to FSH, whereas AMPK and PTEN counteract this effect.[33,34] So far, the mechanisms whereby the FSH-R couples to PIP3 regulation is unclear, but could involve adaptor protein containing PH domain, PTB domain and leucine zipper motif 1 (APPL1). This adaptor protein has been identified in a two-hybrid screen done with the first and second intracellular loops of the FSH-R as bait. It is known to interact with the p110α catalytic subunit of PI3K in immortalized cell lines.[35,36] Since the interaction between the FSH-R and APPL1 is not altered by the presence of FSH, it could mediate the constitutive activation of the PI3K pathway that is observed in Sertoli cells from the prepubertal rat.[29]

In the early 2000s, a boost of discoveries have highlighted that FSH actually induces a complex network of molecular interactions within the cell, which largely encompasses G protein-dependent signaling. Coupling of the FSHR to the $G\alpha_s$/cAMP/PKA signaling pathway can now be viewed as one among several mechanisms contributing to the activation of the whole hormone-induced signaling network. Many diverse signaling pathways have been linked to activation of the FSH-R in granulosa cells[37] and in immortalized cell lines,[38] depending or not on G proteins. However, in Sertoli cells, promiscuity of G protein coupling has been demonstrated so far as the only way to promote cAMP-independent signaling. First, the FSH-R can couple not only to G_s but also to G_i, that appears to be involved in the postnatal mitotic response of Sertoli cell to FSH, via ERK MAP kinase-dependent signaling (Figure 6.2).[30] G_i coupling also occurs at a particular splicing variant of the FSH-R expressed as the hormone-binding ecto-domain and possessing a different carboxyl terminus. Intriguingly, this splicing creates a novel receptor with a single membrane-spanning domain.[40] Initially identified in granulosa cells,[41] it has also been observed in testicular membranes, hence it could presumably be expressed

at the Sertoli cell surface. Second, coupling of the FSH-R to G_h (also known as tissue transglutaminase), to mediate the Ca^{2+} influx via phosphatidylinositol-phospholipase C (PLC) $\delta1$ activation and inositol 1,4,5-triphosphate (IP3) accumulation has been unambiguously demonstrated by the use of peptide competition experiments (Figure 6.2).[39,42] These data add mechanistic insights onto the FSH-mediated Ca^{2+} response, which has been the matter of intense controversies for years.[43]

GENE REGULATION

Part of the FSH-induced signaling network ends up in transcriptional regulations, to promote developmentally appropriate biological responses (Figure 6.3). Accordingly, DNA microarray analyses of prepubertal rat Sertoli cells stimulated *in vitro* have demonstrated that FSH impacts on the steady-state level of many mRNA, as early as after 2 hours of hormone stimulation (Table 6.1).[44] Suppressing FSH by a 4-day passive immunization with neutralizing antibodies (Abs) led to the identification of hormone-regulated transcripts, among which some are well-known FSH-responsive genes in

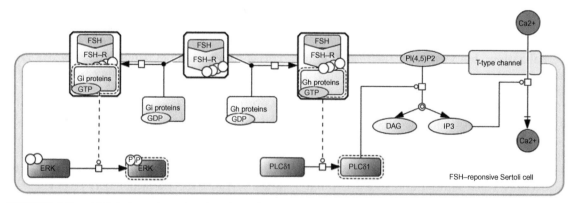

FIGURE 6.2 cAMP-independent FSH signaling. Shown are two pathways known to be activated via FSH-R coupling to G_h or to G_i, in Sertoli cells. In fact, ERK MAP kinases are activated through joint G_s- and G_i-dependent signaling. G_h- and G_i-dependent coupling of the FSH-R may not coexist, since G_i coupling has been reported in cells from neonate rats,[30] whereas G_h coupling has been observed in cells in prepubertal rats.[39]

Sertoli cells, such as StAR (steroidogenic acute regulatory), cathepsin L, insulin-like growth factor binding protein-3 (IGF-BP3), while others encode proteins involved in cell cycle and survival regulation, such as cyclin D1 or scavenger

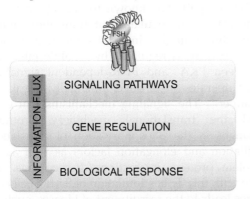

FIGURE 6.3 **Directed information flux from the FSH-R at the plasma membrane to biological responses integrating the dynamic properties of the FSH-dependent signaling network.** This flux is developmentally regulated in Sertoli cells.

receptor class B1. Genes of non-Sertoli cell[45] or genes involved in cell—cell communication at spermiation (extracellular matrix components, lysosomes implicated in residual body phagocytosis) were also identified as indirectly regulated by FSH.[47] Comparable results were obtained in *hpg* mice, a convenient model to analyze the effect of exogenously administrated FSH on gene regulation *in vivo*.[46,48] These mice lack circulating gonadotropins because of a natural mutation in the GnRH gene,[49] but they have retained responsiveness to gonadotropin treatments,[50] despite of an immature reproductive tract.

So far, all of these transcriptomic analyses have provided an atlas of genes regulated by FSH signaling either *in vitro* or *in vivo*, but they indicate neither the transcription factors that could recognize the gene regulatory regions nor the upstream signaling pathways that could be involved. Systemic analyses of the global FSH-induced network from the receptor to the target genes will upgrade our

TABLE 6.1 Summary of the main features from studies exploring the FSH-dependent transcriptome, either from *in vitro*[44] or from *in vivo*[45,46] experiments

Species	Model	FSH input	Microarray	Fold Change	Results	Reference
Rat 20 dpp	*In vitro* primary culture of Sertoli cells	*In vitro* stimulation 0, 2, 4, 8, 24 h	Affymetrix, 8799 probes	≧2	100 to 300 genes upregulated at each timepoint, with more genes upregulated than downregulated	44
Mouse	Whole testis of hypogonadal mouse	FSH injection 4, 8, 12, 24 h prior to sacrifice	Affymetrix, 12, 488 probes	≧2	436 genes differently regulated with FSH (4, 8 and 12 h vs. control)	46
Rat 18 dpp	Whole testis	Anti-FSH antibodies 4 days prior to sacrifice	Affymetrix, 8799 probes	1.5—6.4	30 genes upregulated and 30 downregulated	45
Adult rat	Seminiferous tubules	Anti-FSH antibodies 4 days prior to sacrifice	Affymetrix, 31, 099 probes	≧4	Depending on the stage of the seminiferous epithelium, cyclic repression and activation of transcripts	47
Mouse 12 dpp weeks	Whole testis of hypogonadal mouse	FSH injection 12, 24 or 72 h prior to sacrifice	Affymetrix, 14, 000 probes	≧2	Decreased level of genes encoding tight junction components upon prolonged exposure to FSH	48

understanding at the molecular level of FSH-induced biological responses. For example, PKA is known to target transcription factors of the CREB family. The importance of functional CREB in the whole testis has been demonstrated by injecting a non-phosphorylable version of the protein into the seminiferous tubules, which induced massive spermatocyte apoptosis.[51] However, when considering solely the response to FSH, the requirement of CREB in transcriptional regulation has to be reconsidered because not so many FSH-responsive genes include cAMP-responsive elements (CRE) in their promoter regions. For instance, PKA also regulates the activity of retinoic acid receptor α,[52] a well-known regulator of germ cell development in the testis.

Careful examination of the promoter regions will certainly highlight the involvement of unexpected transcriptional regulators. Similarly, kinases will probably appear to input at regulatory points where they are not expected. In line with this assumption, PKA has been discovered to phosphorylate histone H3 in the promoter regions of the c-fos, serum/glucocorticoid-inducible serine/threonine protein kinase (SGK) and α-inhibin genes.[53] Thus, PKA could be involved not only in transcription factor-specific regulations, but also in global chromatin remodeling.

The anabolic role of FSH in Sertoli cells is well established; hence, it is surprising that the whole FSH-induced proteome is not yet available. And strikingly, the pioneering works on the subject have identified only a limited number of neosynthesized proteins, which does not match the several hundreds of genes upregulated or downregulated in response to FSH, as indicated by transcriptomic analyses presented above. This suggests that the FSH proteome and transcriptome may not match and that important regulations have to take place in between. Previous reports have suggested that FSH could regulate the translation efficacy of preexisting mRNA. In agreement, recent evidence of FSH regulation at the level of mRNA translation has been gained in primary Sertoli cell cultures (Figure 6.4).[29,32,54] For example, FSH stimulates the mammalian target of rapamycin (mTOR)/p70 ribosomal S6 kinase (p70S6K) pathway, known to end up in phosphorylation of regulators of translation initiation. Interestingly, in FSH-stimulated Sertoli cells, p70S6K is regulated by a subtle interplay between PKA- and PI3K-dependent signaling that operates in a developmentally regulated manner.[29] The signaling mechanisms that control the translational machinery promote changes in the phosphorylation of translation initiation and elongation factors as well as in protein/protein interactions/dissociations. In line with this, FSH stimulation of Sertoli cell cultures leads to the recruitment of kinases, such as mTOR and p70S6K, to the 5′ methyl-guanosine mRNA moiety. Translation initiation factors are also inducibly regulated, such as eukaryotic initiation factor 4B (eIF4B), one of p70S6K targets, and eIF4G, the scaffolding protein for the translation pre-initiation complex. All these molecular rearrangements lead to an increased recruitment of mRNA to the polysomes. Importantly, this stimulatory effect of FSH on translation regards selective mRNA, such as the c-fos and vascular endothelial growth factor (VEGF) mRNA, but not all FSH target mRNA.[54] These data are in agreement with elegant analysis using the RiboTag technology, designed to determine the FSH-regulated, Sertoli cell-specific translatome in mouse.[56]

INTEGRATING miRNA INTO THE COMPLEXITY OF THE FSH SIGNALING NETWORK

The influence that FSH-induced signaling exerts on mRNA extends far beyond the regulation of mRNA transcription and translation. Notably, microRNAs (miRNAs) now appear as key players in the control of reproductive

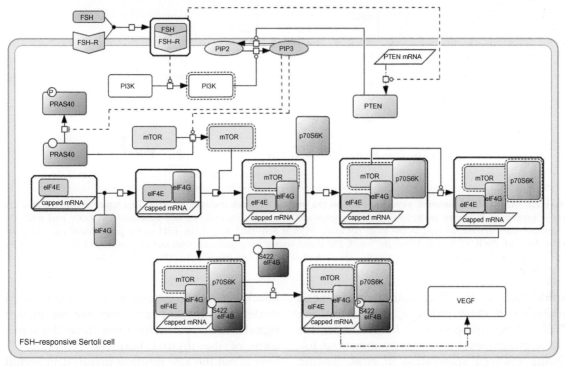

FIGURE 6.4 **Translational control regulated by FSH-induced signaling in Sertoli cells.** Adapted from ref[54]. FSH triggers cap-dependent translation by enhancing p70S6K, mTOR and eIF4G recruitment to the mRNA 5′ cap. 4E-BP1 dissociation from the cap has been observed in granulosa cells,[55] but not in Sertoli cells. PKA-dependent p70S6K activation by FSH in Sertoli cells has been reported elsewhere[32,54] and is not shown here. PRAS40 is a component of the mTORC1 complex.

function. Their biological action primarily consists in the destabilization of cytoplasmic mRNA,[57] but they can also regulate mRNA translation. As miRNAs constitute themselves a *bona fide* network, intertwined with cell signaling networks in the cell, it is worth considering the role that those miRNAs could potentially play in regulating FSH-induced signaling within their natural target cells in the gonad. In line with this, recent works have stressed that some miRNAs could regulate part of the FSH signaling network. Sertoli cell-selective knock-out in mouse of the gene encoding Dicer, an enzyme involved in miRNA processing, has unraveled the role of miRNAs in regulating the expression of genes essential for meiosis and spermiogenesis.[58] Consequently, Sertoli-cell

restricted Dicer knocked-out mice are infertile. A rat model where FSH and testosterone action was suppressed *in vivo* has been created to identify the miRNA network at spermiation.[59] This spermatogenesis stage is characterized by the release of elongated spermatids away from Sertoli cells following complex membrane rearrangements, and is particularly sensitive to hormone regulation. By probing a miRNA microarray, 163 miRNA appeared to be responsive to *in vitro* hormone treatment of Sertoli cells isolated from prepubertal hormone-suppressed rats. Four of the miRNAs that came out from this analysis were complementary to the PTEN mRNA, which the authors localized in the apical region of the cells, in the vicinity of mature spermatids. The hormonal input

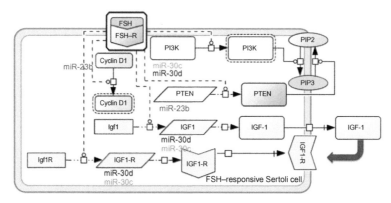

FIGURE 6.5   **miRNA network superimposed on the protein network induced by FSH.** This figure shows the miRNA whose expression is altered upon *in-vivo* FSH suppression, i.e., miR-23b, miR-30c and miR-30d.[59] MiR-30c and miR-30d target the IGF-1 system, as well as the PI3K pathway. PI3K is activated directly by FSH in the prepubertal rat, and indirectly, through the IGF1-R, in the neonate.[60] miR-23b could impair the post-transcriptional regulation of PTEN induced by FSH,[34] and could potentially target Cyclin D1. Whether FSH regulates Cyclin D1 at the transcriptional or post-transcriptional level is not known yet.

would lead to the degradation or synthesis inhibition of these miRNAs, then stabilizing PTEN at spermiogenesis.[59] Interestingly, the PTEN protein level is massively enhanced following FSH-cell stimulation *in vitro*, leading Sertoli cells to achieve terminal differentiation (Figure 6.5).[34] Since FSH enhances PTEN protein level within minutes, the mechanisms involved probably occur post-transcriptionally, an assumption consistent with the hormone-induced degradation of miRNA that prevents the accumulation of PTEN locally in Sertoli cells. In conclusion, miRNA networks might regulate the compartmentalization of FSH-signaling components and might control the kinetics of these biochemical reactions.

CROSS-TALK WITH OTHER FACTORS TO REGULATE CELL FATE

Sertoli cells represent a paradigmatic model of transition between a proliferative state and commitment to differentiation. This transition is controlled by FSH[61,62] and by thyroid hormones,[63] that both act antagonistically. Seminal knockout experiments in mouse have demonstrated that thyroid hormones are the master signal that arrests Sertoli growth, *via* the α isoform of the thyroid hormone receptor.[64–66] FSH is not the sole mitogen involved in Sertoli cell proliferation, and autocrine secretion of other factors also synergize, or at least complement FSH action. For example, GDNF[67] and IGF-I[68] stimulate Sertoli cell proliferation *in vitro*. Interestingly, the FSH-R- and the IGF-R-transduced signaling share the PI3K as a common effector to stimulate lactate production in the neonate (Figure 6.6),[69] whereas FSH acts independently of IGF-I before puberty.[76] Lactate production is a key nurturing function of Sertoli cells, which metabolize it from glucose that post-meiotic germ cells can then utilize as an energy source. The role of FSH in this process is first to accelerate the transport of glucose within Sertoli cells, presumably through GLUT-1,[77] and second to enhance the activity of lactate dehydrogenase, the enzyme that converts pyruvate to lactate at the end of the metabolic chain. *In vivo* ablation of the IGF system (both IGF-R and Ins-R knockout) have shown convincingly that it is required for FSH-mediated mitogenic action in the prepubertal

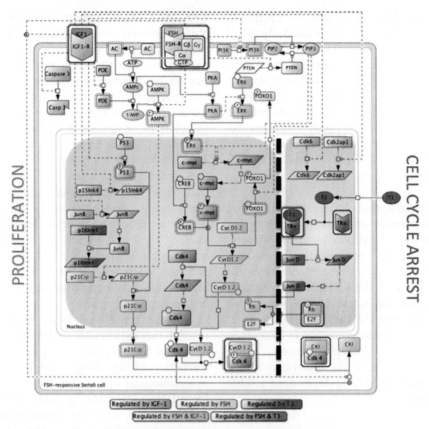

FIGURE 6.6 **Cross-talk between signaling pathways induced by FSH, testosterone, IGF1 and T3, to regulate Sertoli cell fate.** Both FSH and IGF-1 regulate phosphodiesterase expression and/or activity, but only the relationships with IGF-1 signaling are shown here. The impact of FSH on the cell cycle is from refs[29,30,69]. The references related to IGF-1 are refs[70,71] whereas the role of T3 on Sertoli cell cycle regulators has been studied in ref[64]. Considering the regulation by IGF-1 and T3, this network is essentially an extrapolation from transcriptomic data reported in refs[64,71]. Activation of Cdk4 by FSH has only been reported in granulosa cells.[72] Transcriptional regulation of p16Ink4 by JunB has been shown only in the skin.[73] The general mechanism of p21Cip1 regulation by p53 is described in ref[74]. Regulation of Cyclin D-dependent cell cycle progression is reviewed in ref[75]. Cyclin D2 upregulation depends on ERK activation by FSH, and c-myc could be a direct target of the kinase. In addition to p21Cip1, AMPK also regulates p27Kip1 and p19 Ink4d.[33] For convenience, nuclear proteins appear as translated in the nucleus.

mouse.[71] Based on previous observations, it can be assumed that FSH and IGF-I can both act on the PI3K[70] and ERK MAP kinase[30] pathways. In parallel, FSH, IGF-I and T3 directly impact on cell cycle regulators. For example, both FSH and IGF-I can stimulate the expression of the D1 and D2 cyclins,[30,71] instrumental in mediating progression through the G1 phase of the cell cycle (Figure 6.6). Furthermore, IGF-I

inhibits the transcription of two cycle inhibitors, namely the p15[Ink4] gene[71] and the p21[Cip] gene,[74] presumably via p53 dephosphorylation.[70] To counteract this action on the positive regulators of the cell cycle, T3 downregulates the expression of cyclin-dependent kinase 4,[64] an enzyme whose expression is upregulated by FSH in granulosa cells.[72] It is assumed that T3 inhibits molecular events occuring in early G1, and not later on, because neither the cyclins E, A and B nor Cdk2 are modulated by T3.[64] Summarizing the current data on the control of Sertoli cell commitment to differentiation indicates that it is mainly fined-tuned by FSH and T3. In support of this assumption, mice with a Sertoli cell-specific double knockouts of the IGF-R and the Ins-R exhibit only a maturation delay of Sertoli cells between the 5th and 21st days postpartum, which only leads to a shallow retardation of the first spermatogenic wave.[71]

A vast array of studies have established that FSH and testosterone are both mandatory for spermatogenesis to proceed appropriately. Interestingly, recent investigations on the respective role of FSH and testosterone by cell-specific knockout of their respective receptor have shown that the extent of the Sertoli cell population is primarily determined by FSH, and that FSH and testosterone act in concert to enable progression through meiosis.[78] For example, FSH stimulates the androgen receptor (AR) to mediate MTA-2-dependent FSH-R downregulation.[28]

OPENED QUESTIONS

FSH biological function shifts between 9 and 18 days postpartum, prior to the first wave of spermatogenesis in rodents, from sustaining Sertoli-cell proliferation to metabolic and architectural support for germ cells. Defining the molecular basis of this switch has been the matter of an intensive interest, but many possibilities still remain. One of them is the developmental change in the splicing variants of the FSH-R, switching from a cytokine receptor-like isoform in the neonate to the 7 transmembrane domain receptor isoform prior to puberty. This change would modify the coupling ability of the FSH-R, from a G_i-coupled receptor early in life to a G_s-coupled FSH-R later on. It would explain nicely G_i-mediated

ERK MAP kinase phosphorylation in neonates and increased responsiveness of the cAMP production prior puberty.[30,40] Initially identified in ovarian tissues,[41] this one-pass FSH-R has to be formally identified in Sertoli cells. Another appealing hypothesis would be a systemic change in the miRNA network, similar to what is observed at the maternal/zygotic transition.[79] This hypothesis is supported by the fact that, in the course of cell differentiation, miRNAs are now considered to consolidate the transcriptional gene expression program by repressing leaky transcripts that were specific of the anterior, less differentiated stage.[80] By these means they would support transcript cell specificity.[81,82] A genome-wide analysis of FSH-regulated miRNAs at different Sertoli cell developmental stages could probably address this option.

Selective miRNAs have been shown to restrict the expression of PTEN to the apical region of Sertoli cells,[59] suggesting that spatial restriction of signaling effectors on Sertoli cell morphological changes could correlate to differentiation. Notably, a body of recent data have pointed out the instrumental role of β-arrestins as direct interactors of the FSH-R, involved in the spatial and temporal organization of FSH-driven signaling events (Figure 6.7).[15,83,84] Beyond their ability to undergo clathrin-directed receptor internalization, β-arrestins are endowed with regulatory scaffolding properties generic to most, if not all, GPCRs.[87,88] For example, in HEK293 cells exogenously expressing the FSH-R, β-arrestins have been shown to regulate ERK MAP kinases with temporally distinct kinetics when compared to G proteins.[15] Sertoli cells being polarized, epithelioid cells, spatialization of the biochemical reactions is of paramount interest, and β-arrestins could be key players in the local activation of signaling molecules. For example, they have been found to differentially regulate PTEN activity as a function of the upstream signal, with important outcomes

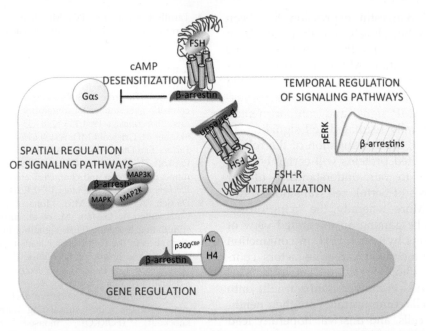

FIGURE 6.7 **Hypothesis on a putative role for β-arrestins in regulating parts of the FSH-induced signaling network.** Recruitment of β-arrestins at Sertoli cell plasma membrane in response to FSH has been reported in ref[26]. Otherwise, temporal regulation of MAP kinase pathways by β-arrestins has been observed by many authors, in response to most GPCRs. Regarding the FSH-R, β-arrestin-dependent MAPK activation has only been reported in HEK293 cells[15,83,84] transiently or stably expressing the FSH-R. FSH-dependent temporal regulation of ERK mediated by β-arrestins has been reported in ref[15]. ERK spatial regulation of MAP kinase pathways by β-arrestins have initially been reported in ref[85], in response to PAR2 activation, whereas a nuclear role of β-arrestins in histone H4 acetylation via CBP recruitment has been demonstrated in cells stimulated with an agonist of the δ-opioid receptor.[86]

on cell proliferation.[89] Even in the nucleus β-arrestins can potentially intervene, since they have been reported to target acetylation of histones *via* CBP recruitment, in response to agonist-activated δ-opioid receptor, another GPCR.[86] Whether FSH induces histone acetylation through β-arrestins still needs to be addressed. However, it is tempting to speculate that both β-arrestins and PKA-mediated transcriptional events might each contribute to different parts of the whole FSH-regulated gene network. The pattern of transcriptionally regulated genes through activation of PKA, on the one hand, and β-arrestins, on the other hand, might be dictated primarily by modifications of chromatin accessibility, and to a lesser extent by recognition of consensus binding

sites for specific transcription factors related to CREB.

In line with this, the authors' group has shown that when a low number of FSH-R is exposed at the cell surface, the signal is preferentially transduced via β-arrestins, at the expense of G proteins.[83] This is exactly what happens with the FSH-R A189V mutation, and explains nicely that men bearing this mutation have preserved fertility, despite loss of function effects.[11] Physiologically, the number of FSH-R varies during the seminiferous cycle, being the highest at stages XIII to I (spermatogonia proliferation, meiosis) and the lowest at stages VII and VIII (spermiation),[90] which supports a role of FSH in maintaining the germ stem cell niche and/or spermatogonia entry into meiosis. β-arrestin expression has been detected in rat Sertoli cells and they have been shown to be recruited to the plasma membrane upon FSH stimulation. Although it is not demonstrated yet in Sertoli cells, they are expected to promote FSH-R internalization, desensitization of the cAMP signal and compartmentalized or temporally regulated cell signaling. In the near future, the investigation of the role of these multifaceted proteins will certainly open new avenues to our understanding of the molecular basis of Sertoli cell transition to differentiation. More generally, the "icing on the cake" would be gaining an integrated view of the mechanisms by which FSH, in conjunction with other endocrine or paracrine factors, coordinates the whole signaling network. This knowledge will certainly provide insight onto how Sertoli cells communicate with their neighboring germ cells, and this will hopefully lead to the development of an innovative male contraceptive and unravel the bases of some forms of spermatogenetic failure.

Acknowledgements

We sincerely apologize to all the authors whose work could not be cited here, due to space limitations. We thank Olivier Galland for his kind help with the figures.

References

1. Griswold MD. The central role of Sertoli cells in spermatogenesis. *Semin Cell Dev Biol* 1998;**9**:411−6.
2. Orth JM, Gunsalus GL, Lamperti AA. Evidence from Sertoli cell-depleted rats indicates that spermatid number in adults depends on numbers of Sertoli cells produced during perinatal development. *Endocrinology* 1988;**122**:787−94.
3. Mruk DD, Cheng CY, Saunders PT. Sertoli-Sertoli and Sertoli-germ cell interactions and their significance in germ cell movement in the seminiferous epithelium during spermatogenesis. Germ cell-somatic cell interactions during spermatogenesis. *Endocr Rev* 2004;**25**:747−806.
4. Tarulli GA, Stanton PG, Meachem SJ. Is the adult Sertoli cell terminally differentiated? *Biol Reprod* 2012;**87**(13):1−11.
5. Orth JM. Proliferation of Sertoli cells in fetal and postnatal rats: a quantitative autoradiographic study. *Anat Rec* 1982;**203**:485−92.
6. Orth JM. The role of follicle-stimulating hormone in controlling Sertoli cell proliferation in testes of fetal rats. *Endocrinology* 1984;**125**:1248−55.
7. Skinner M, Griswold MD. *Sertoli Cell-Secreted Regulatory Factors*. San Diego: Elsevier Academic Press; 2005.
8. Papkoff H, Ekblad M. Ovine follicle stimulating hormone: preparation and characterization of its subunits. *Biochem Biophys Res Commun* 1970;**40**:614−21.
9. Dierich A, Sairam MR, Monaco L, Fimia GM, Gansmuller A, LeMeur M, et al. Impairing follicle-stimulating hormone (FSH) signaling in vivo: targeted disruption of the FSH receptor leads to aberrant gametogenesis and hormonal imbalance. *Proc Natl Acad Sci USA* 1998;**95**:13612−7.
10. Krishnamurthy H, Danilovich N, Morales CR, Sairam MR. Qualitative and quantitative decline in spermatogenesis of the follicle-stimulating hormone receptor knockout (FORKO) mouse. *Biol Reprod* 2000;**62**:1146−59.
11. Tapanainen JS, Aittomaki K, Min J, Vaskivuo T, Huhtaniemi IT. Men homozygous for an inactivating mutation of the follicle-stimulating hormone (FSH) receptor gene present variable suppression of spermatogenesis and fertility. *Nat Genet* 1997;**15**:205−6.
12. Quintana J, Hipkin RW, Sanchez-Yague J, Ascoli M. Follitropin (FSH) and a phorbol ester stimulate the phosphorylation of the FSH receptor in intact cells. *J Biol Chem* 1994;**269**:8772−9.
13. Nakamura K, Krupnick JG, Benovic JL, Ascoli M. Signaling and phosphorylation-impaired mutants of the rat follitropin receptor reveal an activation- and phosphorylation-independent but arrestin-dependent pathway for internalization. *J Biol Chem* 1998;**273**:24346−54.
14. Nakamura K, Hipkin RW, Ascoli M. The agonist-induced phosphorylation of the rat follitropin receptor maps to the first and third intracellular loops. *Mol Endocrinol* 1998;**12**:580−91.
15. Kara E, Crépieux P, Gauthier C, Martinat N, Piketty V, Guillou F, et al. A phosphorylation cluster of five serine and threonine residues in the C-terminus of the follicle-stimulating hormone receptor is important for desensitization but not for beta-arrestin-mediated ERK activation. *Mol Endocrinol* 2006;**20**:3014−26.
16. Lazari MF, Liu X, Nakamura K, Benovic JL, Ascoli M. Role of G protein-coupled receptor kinases on the

agonist-induced phosphorylation and internalization of the follitropin receptor. *Mol Endocrinol* 1999;**13**:866−78.

17. Troispoux C, Guillou F, Elalouf JM, Firsov D, Iacovelli L, De Blasi A, et al. Involvement of G-protein coupled receptor kinases and arrestins in desensitization to FSH action. *Mol Endocrinol* 1999;**9**:1599−614.

18. Krishnamurthy H, Galet C, Ascoli M. The association of arrestin-3 with the follitropin receptor depends on receptor activation and phosphorylation. *Mol Cell Endocrinol* 2003;**204**:127−40.

19. Meroni SB, Riera MF, Pellizzari EH, Cigorraga SB. Regulation of rat Sertoli cell function by FSH: possible role of phosphatidylinositol 3-kinase/protein kinase B pathway. *J Endocrinol* 2002;**174**:195−204.

20. Kitano H, Funahashi A, Matsuoka Y, Oda K. Using process diagrams for the graphical representation of biological networks. *Nat Biotechnol* 2005;**23**:961−6.

21. Dorrington JH, Roller NF, Fritz IB. Effects of follicle-stimulating hormone on cultures of Sertoli cell preparations. *Mol and Cell Endocrinol* 1975;**3**:57−70.

22. Vicini E, Conti M. Characterization of an intronic promoter of a cyclic adenosine 3′,5′-monophosphate (cAMP)-specific phosphodiesterase gene that confers hormone and cAMP inducibility. *Mol Endocrinol* 1997;**11**:839−50.

23. Conti M, Toscano MV, Petrelli L, Geremia R, Stefanini M. Regulation of follicle-stimulating hormone and dibutyryl adenosine 3′,5′-monophosphate of a phosphodiesterase isoenzyme of the Sertoli cell. *Endocrinology* 1982;**110**:1189−96.

24. Swinnen JV, Tsikalas KE, Conti M. Properties and hormonal regulation of two structurally related cAMP phosphodiesterases from the rat Sertoli cell. *J Biol Chem* 1991;**266**:18370−7.

25. Marion S, Kara E, Crépieux P, Piketty V, Martinat N, Guillou F, et al. G protein-coupled receptor kinase 2 and beta-arrestins are recruited to FSH receptor in stimulated rat primary Sertoli cells. *J Endocrinol* 2006;**190**:341−50.

26. Marion S, Robert F, Crépieux P, Martinat N, Troispoux C, Guillou F, et al. G protein-coupled receptor kinases and beta arrestins are relocalized and attenuate cyclic 3′,5′-adenosine monophosphate response to follicle-stimulating hormone in rat primary Sertoli cells. *Biol Reprod* 2002;**66**:70−6.

27. Griswold MD, Kim JS, Tribley WA. Mechanisms involved in the homologous down-regulation of transcription of the follicle-stimulating hormone receptor gene in Sertoli cells. *Mol Cell Endocrinol* 2001;**173**:95−107.

28. Zhang S, Li W, Zhu C, Wang X, Li Z, Zhang J, et al. Sertoli cell-specific expression of metastasis-associated protein 2 (MTA2) is required for transcriptional regulation of the follicle-stimulating hormone receptor (FSHR) gene during spermatogenesis. *J Biol Chem* 2012;**287**:40471−83.

29. Musnier A, Heitzler D, Boulo T, Tesseraud S, Durand G, Lécureuil C, et al. Developmental regulation of p70 S6 kinase by a G protein-coupled receptor dynamically modelized in primary cells. *Cell Mol Life Sci: CMLS* 2009;**66**:3487−503.

30. Crépieux P, Marion S, Martinat N, Fafeur V, Vern YL, Kerboeuf D, et al. The ERK-dependent signalling is stage-specifically modulated by FSH, during primary Sertoli cell maturation. *Oncogene* 2001;**20**:4696−709.

31. Crépieux P, Martinat N, Marion S, Guillou F, Reiter E. Cellular adhesion of primary Sertoli cells affects responsiveness of the extra-cellular signal-regulated kinases 1,2 to follicle-stimulating hormone but not to epidermal growth factor. *Arch Biochem Biophys* 2002;**399**:245−50.

32. Lécureuil C, Tesseraud S, Kara E, Martinat N, Sow A, Fontaine I, et al. Follicle-stimulating hormone activates p70 ribosomal protein S6 kinase by protein kinase A-mediated dephosphorylation of Thr 421/Ser 424 in primary Sertoli cells. *Mol Endocrinol* 2005;**19**:1812−20.

33. Riera MF, Regueira M, Galardo MN, Pellizzari EH, Meroni SB, Cigorraga SB. Signal transduction pathways in FSH regulation of rat Sertoli cell proliferation. *Am J Physiol Endocrinol Metab* 2012;**302**:E914−23.

34. Dupont J, Musnier A, Decourtye J, Boulo T, Lécureuil C, Guillou H, et al. FSH-stimulated PTEN activity accounts for the lack of FSH mitogenic effect in prepubertal rat Sertoli cells. *Mol Cell Endocrinol* 2010;**315**:271−6.

35. Nechamen CA, Thomas RM, Cohen BD, Acevedo G, Poulikakos PI, Testa JR, et al. Human follicle-stimulating hormone (FSH) receptor interacts with the adaptor protein APPL1 in HEK 293 cells: potential involvement of the PI3K pathway in FSH signaling. *Biol Reprod* 2004;**71**:629−36 Epub 2004 Apr 7.

36. Walker WH, Cheng J. FSH and testosterone signaling in Sertoli cells. *Reproduction* 2005;**130**:15−28.

37. Hunzicker-Dunn M, Maizels ET. FSH signaling pathways in immature granulosa cells that regulate target gene expression: branching out from protein kinase A. *Cellular Signalling* 2006;**18**:1351−9.

38. Gloaguen P, Crépieux P, Heitzler D, Poupon A, Reiter E. Mapping the follicle-stimulating hormone-induced signaling networks. *Front Endocrinol (Lausanne)* 2011;**2**:45.

39. Lin YF, Tseng MJ, Hsu HL, Wu YW, Lee YH, Tsai YH. A novel follicle-stimulating hormone-induced G alpha h/phospholipase C-delta1 signaling pathway

mediating rat sertoli cell Ca^{2+}-influx. *Mol Endocrinol* 2006;**20**:2514–27.

40. Babu PS, Krishnamurthy H, Chedrese PJ, Sairam MR. Activation of extracellular-regulated kinase pathways in ovarian granulosa cells by the novel growth factor type 1 follicle-stimulating hormone receptor. Role in hormone signaling and cell proliferation. *J Biol Chem* 2000;**275**:27615–26.

41. Babu PS, Jiang L, Sairam AM, Touyz RM, Sairam MR. Structural features and expression of an alternatively spliced growth factor type I receptor for follitropin signaling in the developing ovary. *Mol Cell Biol Res Commun* 1999;**2**:21–7.

42. Lai TH, Lin YF, Wu FC, Tsai YH. Follicle-stimulating hormone-induced Galphah/phospholipase C-delta1 signaling mediating a noncapacitative Ca^{2+} influx through T-type Ca^{2+} channels in rat sertoli cells. *Endocrinology* 2008;**149**:1031–7.

43. Rommerts FF, Lyng FM, von Ledebur E, Quinlan L, Jones GR, Warchol JB, et al. Calcium confusion--is the variability in calcium response by Sertoli cells to specific hormones meaningful or simply redundant? *J Endocrinol* 2000;**167**:1–5.

44. McLean DJ, Friel PJ, Pouchnik D, Griswold MD. Oligonucleotide microarray analysis of gene expression in follicle-stimulating hormone-treated rat Sertoli cells. *Mol Endocrinol* 2002;**16**:2780–92.

45. Meachem SJ, Ruwanpura SM, Ziolkowski J, Ague JM, Skinner MK, Loveland KL. Developmentally distinct in vivo effects of FSH on proliferation and apoptosis during testis maturation. *J Endocrinol* 2005;**186**:429–46.

46. Sadate-Ngatchou PI, Pouchnik DJ, Griswold MD. Follicle-stimulating hormone induced changes in gene expression of murine testis. *Mol Endocrinol* 2004;**18**:2805–16.

47. O'Donnell L, Pratis K, Wagenfeld A, Gottwald U, Muller J, Leder G, et al. Transcriptional profiling of the hormone-responsive stages of spermatogenesis reveals cell-, stage-, and hormone-specific events. *Endocrinology* 2009;**150**:5074–84.

48. Abel MH, Baban D, Lee S, Charlton HM, O'Shaughnessy PJ. Effects of FSH on testicular mRNA transcript levels in the hypogonadal mouse. *J Mol Endocrinol* 2009;**42**:291–303.

49. Mason AJ, Hayflick JS, Zoeller RT, Young III WS, Phillips HS, Nikolics K, et al. A deletion truncating the gonadotropin-releasing hormone gene is responsible for hypogonadism in the hpg mouse. *Science* 1986;**234**:1366–71.

50. Singh J, Handelsman DJ. Neonatal administration of FSH increases Sertoli cell numbers and spermatogenesis in gonadotropin-deficient (hpg) mice. *J Endocrinol* 1996;**151**:37–48.

51. Scobey M, Bertera S, Somers J, Watkins S, Zeleznik A, Walker W. Delivery of a cyclic adenosine 3′,5′-monophosphate response element-binding protein (creb) mutant to seminiferous tubules results in impaired spermatogenesis. *Endocrinology* 2001;**142**:948–54.

52. Santos NC, Kim KH. Activity of retinoic acid receptor-alpha is directly regulated at its protein kinase A sites in response to follicle-stimulating hormone signaling. *Endocrinology* 2010;**151**:2361–72.

53. Salvador LM, Park Y, Cottom J, Maizels ET, Jones JC, Schillace RV, et al. Follicle-stimulating hormone stimulates protein kinase A-mediated histone H3 phosphorylation and acetylation leading to select gene activation in ovarian granulosa cells. *J Biol Chem* 2001;**276**:40146–55.

54. Musnier A, León K, Morales J, Reiter E, Boulo T, Costache V, et al. mRNA-selective translation induced by FSH in primary Sertoli cells. *Mol Endocrinol* 2012;**26**:669–80.

55. Alam H, Maizels ET, Park Y, Ghaey S, Feiger ZJ, Chandel NS, et al. FSH activation of HIF-1 by the PI3-kinase/AKT/Rheb/mTOR pathway is necessary for induction of select protein markers of follicular differentiation. *J Biol Chem* 2004;**279**:19431–40.

56. Sanz E, Evanoff R, Quintana A, Evans E, Miller JA, Ko C, et al. RiboTag analysis of actively translated mRNAs in Sertoli and Leydig cells in vivo. *PLoS One* 2013;**8**:e66179.

57. Guo H, Ingolia NT, Weissman JS, Bartel DP. Mammalian microRNAs predominantly act to decrease target mRNA levels. *Nature* 2010;**466**:835–40.

58. Papaioannou MD, Pitetti JL, Ro S, Park C, Aubry F, Schaad O, et al. Sertoli cell Dicer is essential for spermatogenesis in mice. *Dev Biol* 2009;**326**:250–9.

59. Nicholls PK, Harrison CA, Walton KL, McLachlan RI, O'Donnell L, Stanton PG. Hormonal regulation of sertoli cell micro-RNAs at spermiation. *Endocrinology* 2011;**152**:1670–83.

60. Khan SA, Ndjountche L, Pratchard L, Spicer LJ, Davis JS. Follicle-stimulating hormone amplifies insulin-like growth factor I-mediated activation of AKT/protein kinase B signaling in immature rat Sertoli cells. *Endocrinology* 2002;**143**:2259–67.

61. Griswold M, Mably E, Fritz IB. Stimulation by follicle stimulating hormone and dibutyryl cyclic AMP of incorporation of 3H-thymidine into nuclear DNA of cultured Sertoli cell-enriched preparations from immature rats. *Curr Top Mol Endocrinol* 1975;**2**:413–20.

62. Meachem SJ, McLachlan RI, de Kretser DM, Robertson DM, Wreford NG. Neonatal exposure of rats to recombinant follicle stimulating hormone increases adult

Sertoli and spermatogenic cell numbers. *Biol Reprod* 1996;**54**:36–44.

63. Van Haaster LH, De Jong FH, Docter R, De Rooij DG. The effect of hypothyroidism on Sertoli cell proliferation and differentiation and hormone levels during testicular development in the rat. *Endocrinology* 1992;**131**:1574–6.

64. Fumel B, Guerquin MJ, Livera G, Staub C, Magistrini M, Gauthier C, et al. Thyroid hormone limits postnatal Sertoli cell proliferation in vivo by activation of its alpha1 isoform receptor (TRalpha1) present in these cells and by regulation of Cdk4/JunD/c-myc mRNA levels in mice. *Biol Reprod* 2012;**87**:1–9.

65. Holsberger DR, Kiesewetter SE, Cooke PS. Regulation of neonatal Sertoli cell development by thyroid hormone receptor alpha1. *Biol Reprod* 2005;**73**:396–403.

66. Quignodon L, Vincent S, Winter H, Samarut J, Flamant F. A point mutation in the activation function 2 domain of thyroid hormone receptor alpha1 expressed after CRE-mediated recombination partially recapitulates hypothyroidism. *Mol Endocrinol* 2007;**21**:2350–60.

67. Hu J, Shima H, Nakagawa H. Glial cell line-derived neurotropic factor stimulates sertoli cell proliferation in the early postnatal period of rat testis development. *Endocrinology* 1999;**140**:3416–21.

68. Borland K, Mita M, Oppenheimer CL, Blinderman LA, Massague J, Hall PF, et al. The actions of insulin-like growth factors I and II on cultured Sertoli cells. *Endocrinology* 1984;**114**(1):240–6.

69. Khan SA, Ndjountche L, Pratchard L, Spicer LJ, Davis JS. Follicle-stimulating hormone amplifies insulin-like growth factor I-mediated activation of AKT/protein kinase B signaling in immature rat Sertoli cells. *Endocrinology* 2002;**143**:2259–67.

70. Froment P, Vigier M, Negre D, Fontaine I, Beghelli J, Cosset FL, et al. Inactivation of the IGF-I receptor gene in primary Sertoli cells highlights the autocrine effects of IGF-I. *J Endocrinol* 2007;**194**:557–68.

71. Pitetti JL, Calvel P, Zimmermann C, Conne B, Papaioannou MD, Aubry F, et al. An essential role for insulin and IGF1 receptors in regulating sertoli cell proliferation, testis size, and FSH action in mice. *Mol Endocrinol* 2013;**27**:814–27.

72. Yang P, Roy SK. Follicle stimulating hormone-induced DNA synthesis in the granulosa cells of hamster preantral follicles involves activation of cyclin-dependent kinase-4 rather than cyclin d2 synthesis. *Biol Reprod* 2004;**70**:509–17.

73. Mark EB, Jonsson M, Asp J, Wennberg AM, Molne L, Lindahl A. Expression of genes involved in the regulation of p16 in psoriatic involved skin. *Arch Dermatol Res* 2006;**297**:459–67.

74. Marx J. How p53 suppresses cell growth. *Science* 1993;**262**:1644–5.

75. Buttitta LA, Edgar BA. Mechanisms controlling cell cycle exit upon terminal differentiation. *Curr Opin Cell Biol* 2007;**19**:697–704.

76. Meroni SB, Riera MF, Pellizzari EH, Galardo MN, Cigorraga SB. FSH activates phosphatidylinositol 3-kinase/protein kinase B signaling pathway in 20-day-old Sertoli cells independently of IGF-I. *J Endocrinol* 2004;**180**:257–65.

77. Galardo MN, Riera MF, Pellizzari EH, Chemes HE, Venara MC, Cigorraga SB, et al. Regulation of expression of Sertoli cell glucose transporters 1 and 3 by FSH, IL1 beta, and bFGF at two different time-points in pubertal development. *Cell Tissue Res* 2008;**334**:295–304.

78. Abel MH, Baker PJ, Charlton HM, Monteiro A, Verhoeven G, De Gendt K, et al. Spermatogenesis and sertoli cell activity in mice lacking sertoli cell receptors for follicle-stimulating hormone and androgen. *Endocrinology* 2008;**149**:3279–85.

79. Walser CB, Lipshitz HD. Transcript clearance during the maternal-to-zygotic transition. *Curr Opin Genet Dev* 2011;**21**:431–43.

80. Stark A, Brennecke J, Bushati N, Russell RB, Cohen SM. Animal MicroRNAs confer robustness to gene expression and have a significant impact on 3′UTR evolution. *Cell* 2005;**123**:1133–46.

81. Farh KK, Grimson A, Jan C, Lewis BP, Johnston WK, Lim LP, et al. The widespread impact of mammalian MicroRNAs on mRNA repression and evolution. *Science* 2005;**310**:1817–21.

82. Sood P, Krek A, Zavolan M, Macino G, Rajewsky N. Cell-type-specific signatures of microRNAs on target mRNA expression. *Proc Natl Acad Sci USA* 2006;**103**:2746–51.

83. Tranchant T, Durand G, Gauthier C, Crépieux P, Ulloa-Aguirre A, Royère D, et al. Preferential beta-arrestin signalling at low receptor density revealed by functional characterization of the human FSH receptor A189 V mutation. *Mol Cell Endocrinol* 2011;**331**:109–18.

84. Wehbi V, Tranchant T, Durand G, Musnier A, Decourtye J, Piketty V, et al. Partially deglycosylated equine LH preferentially activates beta-arrestin-dependent signaling at the follicle-stimulating hormone receptor. *Mol Endocrinol* 2010;**24**:561–73.

85. DeFea KA, Zalevsky J, Thoma MS, Dery O, Mullins RD, Bunnett NW. beta-arrestin-dependent endocytosis of proteinase-activated receptor 2 is required for intracellular targeting of activated ERK1/2. *J Cell Biol* 2000;**148**:1267–81.

86. Kang J, Shi Y, Xiang B, Qu B, Su W, Zhu M, et al. A nuclear function of beta-arrestin1 in GPCR signaling:

regulation of histone acetylation and gene transcription. *Cell* 2005;**123**:833–47.

87. Kovacs JJ, Hara MR, Davenport CL, Kim J, Lefkowitz RJ. Arrestin development: emerging roles for beta-arrestins in developmental signaling pathways. *Dev Cell* 2009;**17**:443–58.

88. Reiter E, Lefkowitz RJ. GRKs and beta-arrestins: roles in receptor silencing, trafficking and signaling. *Trends Endocrinol Metab* 2006;**17**:159–65.

89. Lima-Fernandes E, Enslen H, Camand E, Kotelevets L, Boularan C, Achour L, et al. Distinct functional outputs of PTEN signalling are controlled by dynamic association with beta-arrestins. *EMBO J* 2011;**30**:2557–68.

90. Simoni M, Gromoll J, Nieschlag E. The follicle-stimulating hormone receptor: biochemistry, molecular biology, physiology, and pathophysiology. *Endocr Rev* 1997;**18**:739–73.

Epigenetics of Pituitary Cell Growth and Survival

Toru Tateno, Karen Gomez-Hernandez and Shereen Ezzat

University of Toronto, Princess Margaret Hospital, and the Ontario Cancer Institute, Toronto, ON, Canada

DNA METHYLATION

DNA methylation plays an important role in chromatin organization and gene expression. It reflects an epigenetic code that is maintained and inherited throughout cellular division. It is generally regarded as a mechanism of gene repression. This process is maintained by DNA methyltransferases (DNMT) that catalyze the addition of a methyl group donated by S-adenosyl-methione to 5′ cytosines in the context of a CpG dinucleotide. Cytosines in CpG-poor regions are usually methylated whereas those in CpG-rich areas, often referred to as CpG islands, are typically hypomethylated. Hypomethylation allows an open chromatin structure that is usually situated in promoter regions of active genes; in contrast, methylation of CpG islands results in a restricted chromatin structure that is characteristic of silenced intergenic and intronic regions. It has been proposed that methylation silences gene expression by hindering the access of transcription factors to their binding sites. Additionally, gene silencing might be achieved by methyl-binding proteins that recruit chromatin-modifying factors, further contributing to a compacted inactive chromatin.

DNA Methylation in Pituitary Tumors

In a genome-wide study examining mouse adrenocorticotropic hormone (ACTH)-secreting pituitary tumor cells, AtT20 cells, knockdown of DNMT1 resulted in re-expression of the fully methylated and typically imprinted neuronatin (Nnat) gene.[1] Examination of the three principal members of the DNMT family in primary human pituitary samples, however, revealed little difference between DNMT1 and DNMT3a. In contrast, DNMT3b was expressed at significantly higher levels in pituitary tumors than in normal tissue. No specific cell-type differences were noted.[2] Interestingly, analysis of the human DNMT3b 5′ region failed to show differences in the degree of methylation despite the presence of a large CpG island, but showed evidence of histone modifications in the control of the DNMT3b gene.[2]

Cellular Endocrinology in Health and Disease.
DOI: http://dx.doi.org/10.1016/B978-0-12-408134-5.00007-X

The Importance of Chromatin Remodeling Through Histone Modifications

The assembly of DNA into a compact structure called chromatin is essential for packaging the genome into the relatively confined region of the nucleus. The nucleosome, the basic unit of chromatin, consists of approximately 200 bp of DNA coiled twice around an octamer, which is composed of dimers of the core histones H1−4 (Figure 7.1). Within the nucleus, the nucleosome is further compacted into a higher-order structure. As a result, the regulatory DNA elements of genes are inaccessible to regulatory proteins or transcription factors. This interferes with gene expression, resulting in gene silencing. Chromatin modifications occur mainly through covalent acetylation or methylation of histones. Acetylation through histone acetyltransferases (HATs) and methylation through histone methyltransferases (HMTs) are the most widely recognized dynamic processes (Figure 7.1). Acetylation allows unwinding of the local DNA structure and enabels RNA polymerase II to promote gene transcription. Acetylation of specific lysine (K) residues associated with histone 4 and methylation of K9 on H3 can be found at inactive gene loci. Dedicated histone demethyltransferases (HDMTs) are also being discovered including LSD1 (Figure 7.1 and Table 7.1), which is required for late cell-lineage determination and differentiation during pituitary organogenesis.[3] Instead, acetylation on K9 and K14 of H3, methylation on K4 of H3, or acetylation on K5 of H4 can be identified at gene loci that are either active or have the potential to become activated (Table 7.1). Histone deacetylase complex (HDAC) enzymes are classified into four main classes (Table 7.2). HDACs are now well recognized for their role in mediating steroid hormone repression.[4] Dedicated enzymes involved in these bidirectional modifications are being unmasked (Figure 7.1). They structure a set of histone marks associated with a state of gene expression.

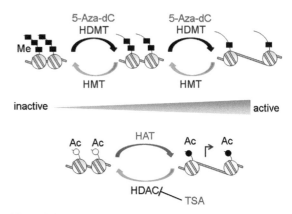

FIGURE 7.1　**Histone-tail modifications regulate DNA accessibility.** DNA is compacted into chromatin units of nucleosomes. Each unit consists of ~200 bp of DNA coiled twice (ribbons) around octamer dimers of the core histones H1−4 (circles). Chromatin modification occurs mainly through covalent methylation (upper) or acetylation (lower) of histone tails. Histone methylation (Me) is catalyzed by histone methyltransferase (HMT) enzymes that bring histone cores into a progressively more restricted structure. Histone demethyltransferase (HDMT) enzymes are a family of dedicated demethylases that reverse the trimethylation to dimethylation state, thus loosening the degree of histone dimer fomation. Similarly, histone acetylation (Ac) is mediated by histone acetyltransferase (HAT) enzymes, while histone deacetylation is mediated through at least four classes of histone deacetylase (HDAC) enzymes. These processes of HAT or HDAC are reversible and can be modified by specific chemical agents as indicated. Members of each enzyme family involved in these modifications and their targets are shown in Tables 7.1 and 7.2. 5-Aza-dC, 5-Aza-2′-deoxycytidine, DNA methyltransferase inhibitor; TSA; trichostatin-A, histone deacetylase inhibitor.

Epigenetic Mechanisms of Gene Silencing in Pituitary Tumors

Studies of human pituitary tumors identify altered epigenetic changes more frequently than intragenic mutations, loss of heterozygosity, or gene rearrangements.[5] For example, consistent with its well-recognized impact on pituitary neoplasia in genetically deficient mice,[6] the retinoblastoma (Rb) tumor suppressor gene is principally silenced through CpG

TABLE 7.1 Enzymes Involved in Histone Methyltransferase (HMT) and Histone Demethylase (HDMT) Activities

Histone Core	Histone Tail	HMT	HDMT
H3	Lys4	SET1 (MLL)	LSD1, JARID1
	Lys9	Suv39h, G9A ESET, RIZ	JMJD1A, JMJD2A, JMJD2C
	Lys27	G9A, EZH2	JMJD3, UTX
	Lys36	SET2	JMJD2A, JMJD2C, JHDM1A, JHDM1B
H4	Lys20	SET7	? (unclear)

TABLE 7.2 Characteristics of Acetyltransferase (HAT) and Histone De-acetyltransferase (HDAC)

Histone Core	Histone-Tail Residue	HAT	HDACa
H2A	Lys 5	P300, CBP	HDAC3
	Lys 14	Gcn5	
H2B	Lys 12	P300, CBP	
	Lys 15	P300, CBP	
H3	Lys 9	PCAF, Gcn5	SIRT1
	Lys 14	P300, CBP, PCAF, Gcn5	(Clr3)
	Lys 18	P300, CBP, PCAF, Gcn5	
	Lys 58	P300, CBP	SIRT1, 2
H4	Lys 5	P300, CBP, HBO1, HAT1	HDAC1, 3
	Lys 8	P300, CBP, HBO1, TIP60, Esa1	HDAC2
	Lys 12	P300, CBP, HBO1, TIP60, Esa1, HAT1	HDAC1, 2, 3
	Lys 16	TIP60, Esa1, Sas2, MYST1	SIRT1, 2, 3

A Class I: HDAC1, 2, 3, 8; Class II: HDAC4, 5, 6, 7A, 9, 10; Class III: SIRT1, 2, 3; Class IV: HDAC11.

island methylation in human pituitary tumor cells.[7–9] No inactivating mutations in the Rb promoter have been identified in pituitary tumors that are deficient of the Rb protein.[10] Similar to mice with the Rb mutation, mice lacking p27kip1 develop multi-organ neoplasia, including pituitary tumors,[11] and protein levels of this cyclin-dependent kinase inhibitor (CDKI) are reduced in human pituitary adenomas; however, the p27kip1 gene itself is not mutated.[12,13] Instead, the histone methyltransferase SET1/MLL-p27Kip1 complex (Table 7.1) is dysregulated in pituitary adenomas.[14] HDAC2-mediated H4 deacetylation of the POMC gene is defective in nearly 50% of corticotroph adenomas.[15] Expression of GADD45γ, a member of a growth arrest and DNA damage-inducible gene family, is reduced in pituitary tumors[16] through CpG island promoter methylation.[17] Similarly, the Ras-association domain family 1A gene (RASSF1A) is frequently inactivated by hyper-methylation of its promoter region in human pituitary tumors.[18] In addition, MEG3, a human homolog of the mouse maternally imprinted Gtl2 gene, is downregulated in human pituitary tumors through 5′ promoter hyper-methylation.[19] Pituitary tumor apoptosis gene (PTAG) is inactivated through its promoter hyper-methylation in human pituitary tumors. Conversely, induction of expression enhances the growth arresting effect of the dopamine agonist bromocriptine.[20] Expression of death-associated protein kinase (DAPK) is similarly reduced through methylation of the CpG islands contained within its promoter region in invasive human pituitary tumors.[21]

The Role of MicroRNAs in Pituitary Tumors

MicroRNAs (miRNAs) are small endogenous noncoding RNAs that play important

TABLE 7.3 Dysregulated MicroRNAs in Pituitary Tumors

MicroRNA	Expression	Target	Type of Pituitary Tumor	Reference
miR107	Upregulated	AIP	Somatotroph adenomas, non-functioning pituitary adenomas	22
miR-122, miR-493	Upregulated	LGALS3, RUNX2	Corticotroph carcinomas	23
miR-145, miR-21, miR141, let-7a, miR-15a, miR-16	Downregulated	MYC, KRAS, FOS, YES, FLI, CyclinD2, MAPKs	Corticotroph adenomas	24
miR-26a	Upregulated	PRKCD	Corticotroph adenomas	25,26
miR-15a, miR-16-1	Downregulated	RARS	Somatotroph adenomas, lactotroph adenomas	27
miR-26b	Upregulated	PTGS2, PTEN, HMGA1,	Somatotroph adenomas	25
miR-128	Downregulated	MAPK14, PLK2, BMI1	Somatotroph adenomas	25
miR-206, miR-516b, miR550	Upregulated	BMP4	Lactotroph adenomas	28
miR-671-5p	Downregulated	PDGFA		
let-7	Downregulated	HMGA2		29

roles in cell proliferation, differentiation, and apoptosis by regulating gene expression at the posttranscriptional level through direct cleavage of mRNA. In addition to their contribution to development and normal function, they can also act as tumor suppressor genes or oncogenes. The role of miRNAs depends on its target genes. An increasing repertoire of aberrant miRNA-expression profiling is currently unfolding in human pituitary tumors (Table 7.3). MicroR-107 overexpression in pituitary adenomas induced inhibition of aryl hydrocarbon receptor-interacting protein (AIP).[22] MicroR-122 and miR-493 were upregulated in corticotroph carcinomas compared with corticotroph adenomas.[23] Conversely, underexpression of miR-145, miR-21, miR-141, let-7a, miR-150, miR-15a, and miR-16 has been reported in ACTH-producing adenomas. Although miRNA expression did not correlate with tumor size in one study, lower miR-141 expression correlated with postoperative remission in patients with corticotroph adenomas.[24] MicroR-26a, which is overexpressed in human pituitary adenomas,[25] can control AtT20 corticotroph cell growth without involving caspase 3/7-mediated apoptosis.[26] Reduced expression of miR-15a and miR-16-1 has been linked to tumor size in growth hormone (GH)- and prolactin (PRL)-producing adenomas;[27] however, this was not the case in corticotroph adenomas.[24] The involvement of miRNAs, such as miR-26b and miR-128, regulates the PTEN−AKT pathway in GH-producing pituitary tumor formation.[25] MicroR-206, miR-516b, and miR-550 were significantly upregulated, while miR-671-5p was downregulated in prolactinomas treated with bromocriptine.[28] Platelet-derived growth factor alpha polypeptide (PDGFA) was upregulated, but bone morphogenetic protein 4 (BMP4) was downregulated.[28] In addition, a significant link between high-mobility group AT-hook2 (HMGA2) oncogene and let-7 miRNA has been shown in pituitary tumors.[29] In particular, reduced let-7 and high levels of HMGA2 correlated with the extent of tumor size, invasiveness, and the Ki-67 proliferation index.[29]

The Role of Ikaros in the Pituitary Gland

The factors involved in epigenetic silencing of tumor suppressor genes in pituitary tumors are still emerging. Of these, the hematopoietic stem cell chromatin remodeler Ikaros has received wide attention. Ikaros was originally described as a transcription factor that binds to regulatory sequences of genes expressed in cells of the lymphoid lineage.[30] The single-copy gene that encodes for this protein contains seven exons yielding at least eight isoforms by alternative splicing. The various isoforms can function as either transcriptional activators or repressors in a functionally diverse chromatin-remodeling network. Ikaros regulates hypothalamic neuroendocrine and pituitary cell population expansion during development by its transcriptional actions and chromatin-remodeling properties. It also plays an important role in the corticomelanotroph population expansion of the pituitary, where it binds and activates the pro-opiomelanocortin promoter.[31] Moreover, Ik serves a key developmental role in hypothalamic growth hormone releasing hormone (GHRH) neuronal development and an indirect role in pituitary somatotroph population expansion.[32] Interestingly, Ik is also expressed in pituitary cells of the Pit-1 lineage, where it mediates histone acetylation and chromatin remodeling in the selective regulation of pituitary GH and PRL gene expression. It does not bind directly to the proximal (−360) GH promoter, instead, it significantly abrogates the effect of the histone deacetylation inhibitor trichostatin-A on this promoter. Ikaros also selectively deacetylates histone 3 residues on the proximal GH promoter limiting access of the Pit-1 activator. In contrast, Ik acetylates histone 3 on the proximal PRL promoter and facilitates Pit-1 binding to this region in the same cells.[33] Human pituitary adenomas frequently express a dominant-negative Ik6 isoform that lacks the DNA-binding domain.[30] Forced expression of this dominant-negative form of Ik leads to histone 3 acetylation with activation of the anti-apoptotic Bcl-XL promoter.[33] Ikaros also plays an important role in cancer cell metabolism. Complementary DNA microarray analysis has revealed mediators of cholesterol uptake, including the low-density lipoprotein receptor (LDL-R) and the sterol-regulatory element binding protein 2, as targets of Ikaros action.[34] Ikaros regulates the LDL-R to alter lipid metabolism in pituitary corticotroph cells, expanding the repertoire of Ik actions to include regulation of the cholesterol-uptake metabolic pathway. The DNA-binding Ik isoform, Ik1, binds and directly activates the LDL-R promoter. It reduces methylation and increases acetylation of histone H3 lysine 9 at the LDL-R promoter. Consistent with these results, LDL-R −/− mice showed low circulating levels of ACTH phenocopying Ik −/− mice.[34] Furthermore, Ikaros is induced by hypoxia where it can modulate cell growth and survival by interacting with CtBP co-repressor in pituitary cells.[35]

FGF Signaling in Pituitary Development and Pituitary Tumorigenesis

FGF signaling play important roles in cell proliferation, differentiation, survival, migration, and tissue patterning during embryogenesis. There are currently 22 recognized members of the FGF ligand family. Their receptors are encoded by four independent genes, each giving rise to multiple isoforms.[36] The FGFR2 gene is alternatively spliced to generate FGFR2-IIIb, an isoform containing the second half of the third Ig-like domain encoded by exon 7, that binds FGF1, FGF3, FGF7, and FGF10 with high affinity.[37,38] In contrast, the FGFR2-IIIc isoform, in which the second half of the third Ig-like domain is encoded by exon 8, does not bind FGF7 or FGF10.[39,40] FGFR2-IIIb expression is restricted mainly to epithelial cells, whereas FGFR2-IIIc is more frequently detected in mesenchymal cells.[41] FGF signaling plays a critical role in pituitary development. Deletion of the FGFR2-IIIb isoform leads to failure of pituitary development.[42] Mid-gestational expression of a soluble dominant-negative FGFR results in severe pituitary dysgenesis.[43] Given this recognized role for FGFR2 in the development of the anterior pituitary gland[44] and the common mode of epigenetic silencing in human pituitary tumorigenesis,[45] cDNA microarray profiling of FGFR2-IIIb-expessing thyroid cells revealed that members of the cancer/testis-melanoma-associated antigen (MAGE-A) represent important targets.[46] These findings provided the basis for examining MAGE-A3 in pituitary tumors particularly where FGFR2-IIIb was reduced.

The Melanoma-Associated Antigens in Pituitary Tumors

The MAGE-A3 and -A6 genes are members of the MAGE-I family that includes the MAGE-A, -B, and -C subfamilies.[47] This family consists of a number of chromosome X-clustered genes, which are mainly expressed in testicular germ cells, placenta, and a variety of malignant tumors, resulting in their designation as cancer/testis antigens.[47-51] MAGE-A3 is highly expressed in pituitary tumors, and has been implicated in p53 downregulation through histone deacetylation.[52] In addition, MAGE-A3 downregulation results in p53 and p21 accumulation, which is consistent with its putative oncogenic functions. Although p53 is not intragenically mutated in pituitary adenomas, variable levels of p53 protein accumulation have been reported in these tumors.[53]

MAGE-A3 as a Target of FGF Signaling

MAGE-A3 is reciprocally expressed in tumors where FGFR2-IIIb is absent.[46] Its expression is intimately linked with the frequency and degree of promoter hypomethylation, particularly in tumors from female subjects. This is consistent with the induction of MAGE-A3 by estrogen through histone modification. This latter action inhibits the impact of FGFR2-IIIb, highlighting MAGE-A3 as a point of signaling integration between the opposing forces of estrogen and the FGFR. FGFR2-IIIb and FGFR2-IIIc isoforms are co-expressed in normal pituitary cells.[54] Pharmacologic methylation inhibition results in FGFR2 re-expression mainly of the FGFR2-IIIb isoform in pituitary cells,[55] and re-expression of FGFR2-IIIb results in potent tumor growth inhibition with enhanced apoptosis.[56] This effect has been ascribed to the ability of FGFR2-IIIb to sequester the FGFR substrate FRS2,[57] diverting it away from other competing FGFR signals.[46] Consistent with this model, FGFR2-IIIb signaling using a selective ligand FGF7 results in diminished pituitary tumor cell cycle progression.[55] This effect is recapitulated by downregulation of MAGE-A3, which results in p53 and p21 induction, providing functional evidence implicating MAGE-A3 as an important integrator of FGFR signaling in the pituitary.

MAGE-A3 Regulation Through Histone and DNA Modifications

The MAGE-A3 promoter is heavily methylated in normal human pituitary cells, resulting in lack of expression in this tissue. In contrast, in pituitary tumors the MAGE-A3 promoter is markedly hypomethylated.[58] These findings provide further evidence supporting epigenetic regulation through DNA methylation as a mechanism underlying the ectopic expression of putative oncogenes such as MAGE-A3 in human pituitary tumors. In addition, estrogen treatment, a potent stimulator of pituitary tumor progression[59,60] was associated with MAGE-A3 induction (Figure 7.2).

Histone Modification as Putative Therapeutic Targets in Pituitary Tumors

While FGFR2-IIIb plays a growth-inhibitory tumor-suppressive role,[56] MAGE-A3 is considered to have growth-promoting oncogenic functions.[52,61] These opposing forces rely on DNA methylation as well as histone modifications to interregulate each other. In addition, reduced expression of BMP-4 is associated with histone-tail modifications, not with CpG island methylation status, as enrichment for a modification associated with silent genes, H3K27me3, and depletion of a modification associated with active genes, H3K9Ac. In pituitary cell lines, reduced BMP-4 expression is also associated with similar histone-tail modifications and contemporaneous increase in CpG island methylation.[62] In addition, human pituitary adenomas showed increased acetylation of H3K9 compared to normal pituitary. The H3K9 acetylation increases along with tumor severity and atypical pituitary adenomas are more acetylated than typical pituitary adenomas. MIB-1 (Ki-67) overexpression is highly associated with increased acetylation of H3K9, as well as tumor severity. There is also a contribution of p53 expression to the altered global H3K9 acetylation pattern of pituitary adenomas.[63] Given the recognized importance of altered DNA methylation in pituitary tumors, the evidence reviewed here underscores the complex network of epigenetic changes. In some instances this control targets the balance of signals with opposing functions. It is quite likely that more examples of this nature will emerge, providing a newer understanding of pituitary tumorigenesis and their therapeutic targets.

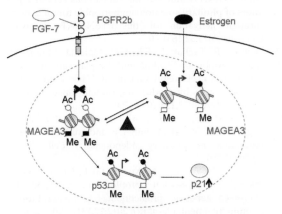

FIGURE 7.2 **The opposing actions of FGF7 via FGFR2-IIIb and estradiol on MAGE-A3.** The balance represents the converging actions of the two ligands on acetylation (Ac) and methylation modification of histone tails related to the MAGE-A3 promoter. Downregulation of MAGE-A3 supports enhanced acetylation and diminished methylation, resulting in p53 induction, protein accumulation and increased p21. Closed circles or squares represent enhanced modifications respectively; open circles or squares represent the opposing effect.

References

1. Dudley KJ, Revill K, Whitby P, Clayton RN, Farrell WE. Genome-wide analysis in a murine Dnmt1 knockdown model identifies epigenetically silenced genes in primary human pituitary tumors. *Mol Cancer Res* 2008;**6**:1567–74.
2. Zhu X, Mao X, Hurren R, Schimmer AD, Ezzat S, Asa SL, et al. Deoxyribonucleic acid methyltransferase 3B promotes epigenetic silencing through histone 3 chromatin modifications in pituitary cells. *J Clin Endocrinol Metab* 2008;**93**:3610–7.
3. Wang J, Scully K, Zhu X, Cai L, Zhang J, Prefontaine GG, et al. Opposing LSD1 complexes function in developmental gene activation and repression programmes. *Nature* 2007;**446**:882–7.
4. Sasaki S, Lesoon-Wood LA, Dey A, Kuwata T, Weintraub BD, Humphrey G, et al. Ligand-induced recruitment of a histone deacetylase in the negative-feedback regulation of the thyrotropin beta gene. *EMBO J* 1999;**18**:5389–98.
5. Ezzat S, Asa SL. Mechanisms of disease: the pathogenesis of pituitary tumors. *Nat Clin Pract Endocrinol Metab* 2006;**2**:220–30.
6. Jacks T, Fazeli A, Schmitt EM, Bronson RT, Goodell MA, Weinberg RA, et al. Effects of an Rb mutation in the mouse. *Nature* 1992;**359**:295–300.
7. Pei L, Melmed S, Scheithauer B, Kovacs K, Benedict WF, Prager D, et al. Frequent loss of heterozygosity at the retinoblastoma susceptibility gene (RB) locus in aggressive pituitary tumors: evidence for a chromosome 13 tumor suppressor gene other than RB. *Cancer Res* 1995;**55**:1613–6.

8. Woloschak M, Yu A, Xiao J, Post KD. Abundance and state of phosphorylation of the retinoblastoma gene product in human pituitary tumors. *Int J Cancer* 1996;**67**:16−9.

9. Simpson DJ, Hibberts NA, McNicol AM, Clayton RN, Farrell WE. Loss of pRb expression in pituitary adenomas is associated with methylation of the RB1 CpG island. *Cancer Res* 2000;**60**:1211−6.

10. Bates AS, Farrell WE, Bicknell EJ, McNicol AM, Talbot AJ, Broome JC, et al. Allelic deletion in pituitary adenomas reflects aggressive biological activity and has potential value as a prognostic marker. *J Clin Endocrinol Metab* 1997;**82**:818−24.

11. Nakayama K, Ishida N, Shirane M, Inomata A, Inoue T, Shishido N, et al. Mice lacking p27(Kip1) display increased body size, multiple organ hyperplasia, retinal dysplasia, and pituitary tumors. *Cell* 1996;**85**:707−20.

12. Bamberger CM, Fehn M, Bamberger AM, Ludecke DK, Beil FU, Saeger W, et al. Reduced expression levels of the cell-cycle inhibitor p27Kip1 in human pituitary adenomas. *Eur J Endocrinol* 1999;**140**:250−5.

13. Lidhar K, Korbonits M, Jordan S, Khalimova Z, Kaltsas G, Lu X, et al. Low expression of the cell cycle inhibitor p27Kip1 in normal corticotroph cells, corticotroph tumors, and malignant pituitary tumors. *J Clin Endocrinol Metab* 1999;**84**:3823−30.

14. Horiguchi K, Yamada M, Satoh T, Hashimoto K, Hirato J, Tosaka M, et al. Transcriptional activation of the mixed lineage leukemia-p27Kip1 pathway by a somatostatin analogue. *Clin Cancer Res* 2009;**15**:2620−9.

15. Bilodeau S, Vallette-Kasic S, Gauthier Y, Figarella-Branger D, Brue T, Berthelet F, et al. Role of Brg1 and HDAC2 in GR trans-repression of the pituitary POMC gene and misexpression in Cushing disease. *Genes Dev* 2006;**20**:2871−86.

16. Zhang X, Sun H, Danila DC, Johnson SR, Zhou Y, Swearingen B, et al. Loss of expression of GADD45 gamma, a growth inhibitory gene, in human pituitary adenomas: implications for tumorigenesis. *J Clin Endocrinol Metab* 2002;**87**:1262−7.

17. Bahar A, Bicknell JE, Simpson DJ, Clayton RN, Farrell WE. Loss of expression of the growth inhibitory gene GADD45gamma, in human pituitary adenomas, is associated with CpG island methylation. *Oncogene* 2004;**23**:936−44.

18. Qian ZR, Sano T, Yoshimoto K, Yamada S, Ishizuka A, Mizusawa N, et al. Inactivation of RASSF1A tumor suppressor gene by aberrant promoter hypermethylation in human pituitary adenomas. *Lab Invest* 2005;**85**:464−73.

19. Zhao J, Dahle D, Zhou Y, Zhang X, Klibanski A. Hypermethylation of the promoter region is associated with the loss of MEG3 gene expression in human pituitary tumors. *J Clin Endocrinol Metab* 2005;**90**:2179−86.

20. Bahar A, Simpson DJ, Cutty SJ, Bicknell JE, Hoban PR, Holley S, et al. Isolation and characterization of a novel pituitary tumor apoptosis gene. *Mol Endocrinol* 2004;**18**:1827−39.

21. Simpson DJ, Clayton RN, Farrell WE. Preferential loss of Death Associated Protein kinase expression in invasive pituitary tumours is associated with either CpG island methylation or homozygous deletion. *Oncogene* 2002;**21**:1217−24.

22. D'Angelo D, Palmieri D, Mussnich P, Roche M, Wierinckx A, Raverot G, et al. Altered microRNA expression profile in human pituitary GH adenomas: down-regulation of miRNA targeting HMGA1, HMGA2, and E2F1. *J Clin Endocrinol Metab* 2012;**97**: E1128−38.

23. Stilling G, Sun Z, Zhang S, Jin L, Righi A, Kovacs G, et al. MicroRNA expression in ACTH-producing pituitary tumors: up-regulation of microRNA-122 and -493 in pituitary carcinomas. *Endocrine* 2010;**38**:67−75.

24. Amaral FC, Torres N, Saggioro F, Neder L, Machado HR, Silva Jr WA, et al. MicroRNAs differentially expressed in ACTH-secreting pituitary tumors. *J Clin Endocrinol Metab* 2009;**94**:320−3.

25. Palumbo T, Faucz FR, Azevedo M, Xekouki P, Iliopoulos D, Stratakis CA, et al. Functional screen analysis reveals miR-26b and miR-128 as central regulators of pituitary somatomammotrophic tumor growth through activation of the PTEN-AKT pathway. *Oncogene* 2013;**32**:1651−9.

26. Gentilin E, Tagliati F, Filieri C, Mole D, Minoia M, Rosaria Ambrosio M, et al. miR-26a plays an important role in cell cycle regulation in ACTH-secreting pituitary adenomas by modulating protein kinase C-delta. *Endocrinology* 2013;**154**:1690−700.

27. Bottoni A, Piccin D, Tagliati F, Luchin A, Zatelli MC, degli Uberti EC, et al. miR-15a and miR-16-1 down-regulation in pituitary adenomas. *J Cell Physiol* 2005;**204**:280−5.

28. Wang C, Su Z, Sanai N, Xue X, Lu L, Chen Y, et al. microRNA expression profile and differentially-expressed genes in prolactinomas following bromocriptine treatment. *Oncol Rep* 2012;**27**:1312−20.

29. Qian ZR, Asa SL, Siomi H, Siomi MC, Yoshimoto K, Yamada S, et al. Overexpression of HMGA2 relates to reduction of the let-7 and its relationship to clinico-pathological features in pituitary adenomas. *Mod Pathol* 2009;**22**:431−41.

30. Ezzat S, Yu S, Asa SL. Ikaros isoforms in human pituitary tumors: distinct localization, histone acetylation, and activation of the 5′ fibroblast growth factor receptor-4 promoter. *Am J Pathol* 2003;**163**:1177−84.

31. Ezzat S, Mader R, Yu S, Ning T, Poussier P, Asa SL, et al. Ikaros integrates endocrine and immune system development. *J Clin Invest* 2005;**115**:1021−9.

32. Ezzat S, Mader R, Fischer S, Yu S, Ackerley C, Asa SL, et al. An essential role for the hematopoietic transcription factor Ikaros in hypothalamic-pituitary-mediated somatic growth. *Proc Natl Acad Sci USA* 2006;**103**:2214−9.

33. Ezzat S, Yu S, Asa SL. The zinc finger Ikaros transcription factor regulates pituitary growth hormone and prolactin gene expression through distinct effects on chromatin accessibility. *Mol Endocrinol* 2005;**19**:1004−11.

34. Loeper S, Asa SL, Ezzat S. Ikaros modulates cholesterol uptake: a link between tumor suppression and differentiation. *Cancer Res* 2008;**68**:3715−23.

35. Dorman K, Shen Z, Yang C, Ezzat S, Asa SL. CtBP1 interacts with Ikaros and modulates pituitary tumor cell survival and response to hypoxia. *Mol Endocrinol* 2012;**26**:447−57.

36. Itoh N, Ornitz DM. Evolution of the Fgf and Fgfr gene families. *Trends Genet* 2004;**20**:563−9.

37. Ornitz DM, Xu J, Colvin JS, McEwen DG, MacArthur CA, Coulier F, et al. Receptor specificity of the fibroblast growth factor family. *J Biol Chem* 1996;**271**:15292−7.

38. Luo Y, Ye S, Kan M, McKeehan WL. Structural specificity in a FGF7-affinity purified heparin octasaccharide required for formation of a complex with FGF7 and FGFR2IIIb. *J Cell Biochem* 2006;**97**:1241−58.

39. Thisse B, Thisse C, Weston JA. Novel FGF receptor (Z-FGFR4) is dynamically expressed in mesoderm and neurectoderm during early zebrafish embryogenesis. *Dev Dyn* 1995;**203**:377−91.

40. Thisse B, Thisse C. Functions and regulations of fibroblast growth factor signaling during embryonic development. *Dev Biol* 2005;**287**:390−402.

41. Baraniak AP, Lasda EL, Wagner EJ, Garcia-Blanco MA. A stem structure in fibroblast growth factor receptor 2 transcripts mediates cell-type-specific splicing by approximating intronic control elements. *Mol Cell Biol* 2003;**23**:9327−37.

42. De Moerlooze L, Spencer-Dene B, Revest JM, Hajihosseini M, Rosewell I, Dickson C, et al. An important role for the IIIb isoform of fibroblast growth factor receptor 2 (FGFR2) in mesenchymal-epithelial signalling during mouse organogenesis. *Development* 2000;**127**:483−92.

43. Celli G, LaRochelle WJ, Mackem S, Sharp R, Merlino G. Soluble dominant-negative receptor uncovers essential roles for fibroblast growth factors in multi-organ induction and patterning. *EMBO J* 1998;**17**:1642−55.

44. Asa SL, Ezzat S. The cytogenesis and pathogenesis of pituitary adenomas. *Endocr Rev* 1998;**19**:798−827.

45. Alexander JM. Tumor suppressor loss in pituitary tumors. *Brain Pathol* 2001;**11**:342−55.

46. Kondo T, Zhu X, Asa SL, Ezzat S. The cancer/testis antigen melanoma-associated antigen-A3/A6 is a novel target of fibroblast growth factor receptor 2-IIIb through histone H3 modifications in thyroid cancer. *Clin Cancer Res* 2007;**13**:4713−20.

47. Xiao J, Chen HS. Biological functions of melanoma-associated antigens. *World J Gastroenterol* 2004;**10**:1849−53.

48. van der Bruggen P, Traversari C, Chomez P, Lurquin C, De Plaen E, Van den Eynde B, et al. A gene encoding an antigen recognized by cytolytic T lymphocytes on a human melanoma. *Science* 1991;**254**:1643−7.

49. Weynants P, Lethe B, Brasseur F, Marchand M, Boon T. Expression of mage genes by non-small-cell lung carcinomas. *Int J Cancer* 1994;**56**:826−9.

50. Otte M, Zafrakas M, Riethdorf L, Pichlmeier U, Loning T, Janicke F, et al. MAGE-A gene expression pattern in primary breast cancer. *Cancer Res* 2001;**61**:6682−7.

51. Kim J, Reber HA, Hines OJ, Kazanjian KK, Tran A, Ye X, et al. The clinical significance of MAGEA3 expression in pancreatic cancer. *Int J Cancer* 2006;**118**:2269−75.

52. Monte M, Simonatto M, Peche LY, Bublik DR, Gobessi S, Pierotti MA, et al. MAGE-A tumor antigens target p53 transactivation function through histone deacetylase recruitment and confer resistance to chemotherapeutic agents. *Proc Natl Acad Sci USA* 2006;**103**:11160−5.

53. Thapar K, Scheithauer BW, Kovacs K, Pernicone PJ, Laws Jr. ER. p53 expression in pituitary adenomas and carcinomas: correlation with invasiveness and tumor growth fractions. *Neurosurgery* 1996;**38**:765−70.

54. Abbass SA, Asa SL, Ezzat S. Altered expression of fibroblast growth factor receptors in human pituitary adenomas. *J Clin Endocrinol Metab* 1997;**82**:1160−6.

55. Zhu X, Lee K, Asa SL, Ezzat S. Epigenetic silencing through DNA and histone methylation of fibroblast growth factor receptor 2 in neoplastic pituitary cells. *Am J Pathol* 2007;**170**:1618−28.

56. Kondo T, Zheng L, Liu W, Kurebayashi J, Asa SL, Ezzat S. Epigenetically controlled fibroblast growth factor receptor 2 signaling imposes on the RAS/BRAF/mitogen-activated protein kinase pathway to modulate thyroid cancer progression. *Cancer Res* 2007;**67**:5461−70.

57. Zhang Y, Wang H, Toratani S, Sato JD, Kan M, McKeehan WL, et al. Growth inhibition by keratinocyte growth factor receptor of human salivary adenocarcinoma cells through induction of differentiation and apoptosis. *Proc Natl Acad Sci USA* 2001;**98**:11336−40.

58. Zhu X, Asa SL, Ezzat S. Fibroblast growth factor receptor 2 and estrogen control the balance of histone 3

modifications targeting MAGE-A3 in pituitary neoplasia. *Clin Cancer Res* 2008;**14**:1984—96.

59. Lloyd RV, Cano M, Landefeld TD. The effects of estrogens on tumor growth and on prolactin and growth hormone mRNA expression in rat pituitary tissues. *Am J Pathol* 1988;**133**:397—406.

60. Asa SL, Ezzat S. The pathogenesis of pituitary tumours. *Nat Rev Cancer* 2002;**2**:836—49.

61. Yang B, O'Herrin S, Wu J, Reagan-Shaw S, Ma Y, Nihal M, et al. Select cancer testes antigens of the MAGE-A, -B, and -C families are expressed in mast cell lines and promote cell viability in vitro and in vivo. *J Invest Dermatol* 2007;**127**:267—75.

62. Yacqub-Usman K, Duong CV, Clayton RN, Farrell WE. Epigenomic silencing of the BMP-4 gene in pituitary adenomas: a potential target for epidrug-induced re-expression. *Endocrinology* 2012;**153**:3603—12.

63. Ebrahimi A, Schittenhelm J, Honegger J, Schluesener HJ. Histone acetylation patterns of typical and atypical pituitary adenomas indicate epigenetic shift of these tumours. *J Neuroendocrinol* 2011;**23**:525—30.

Epigenetic and Developmental Basis of Risk of Obesity and Metabolic Disease

Felicia M. Low*, Peter D. Gluckman* and Mark A. Hanson[†]

*University of Auckland, Auckland, New Zealand, [†]University of Southampton, Southampton, UK

GLOBAL BURDEN OF OBESITY AND METABOLIC DISEASE

The past few decades have witnessed an unrelenting rise in rates of metabolic diseases comprising a constellation of syndromes including obesity, type 2 diabetes (T2D), hypertension, dyslipidemia, and cardiovascular dysfunction. For example the prevalence of obesity in the US has been estimated to have more than doubled from 1980 to 2010, and now affects nearly 36% of adults,[1] with international data analysis demonstrating similar patterns occurring in all other parts of the world.[2] Fasting plasma glucose is rising,[3] and in 2011, 366 million people worldwide were afflicted by T2D, to which 4.6 million deaths could be attributed.[4] As a reflection of the transition to an obesogenic environment on a global level, particularly in industrialized countries and increasingly in developing countries, a recent comprehensive analysis showed that in the two decades since 1990, worldwide disease burden caused by overweight and obesity has now exceeded that attributable to undernutrition.[5]

In addition to the health, economic, and social burdens imposed by these secular trends, two particularly worrying aspects have emerged. The first is the decreasing age at which disease onset is occurring. While childhood obesity in several developed countries appears to have stabilized after years of increasing,[6] current worldwide trends predict that the prevalence could reach 9.1% by 2020, affecting 60 million children.[7] In 2008, 23% of US adolescents had pre-diabetes or T2D, up from 9% just 9 years prior, while 14% were pre-hypertensive or hypertensive.[8] Increasing rates of diabetes and hypertension are being documented in children and adolescents in China.[9,10] If left unaddressed, overweight and obesity generally persist through the life course; thus, earlier age of onset will exacerbate the already considerable costs associated with morbidity and mortality in adulthood.

A second issue is the intergenerational nature of metabolic disease where an obese/diabetic population is likely to have children at risk of developing a similar phenotype (described later). This situation where

Cellular Endocrinology in Health and Disease.
DOI: http://dx.doi.org/10.1016/B978-0-12-408134-5.00008-1

111

"diabesity begets diabesity" perpetuates the accompanying burdens in an intergenerational vicious cycle.[11,12]

Conventional interventions to tackle the obesity epidemic, such as public health education, financial (dis)incentives, and modifications to the built environment, have generally been aimed at two outcomes: dietary improvements and increased physical activity. However, these strategies share a common denominator – the simplistic assumption that correcting the imbalance in the energy intake:energy expenditure ratio through lifestyle choices in adulthood, often after disease onset has occurred, will address the root of the problem.[13] Yet, this disregards the extensive work over the past few decades, discussed below, which has established that early-life developmental factors play a critical role in determining an individual's susceptibility in an obesogenic environment towards a wide range of diseases including obesity and its comorbidities. The burgeoning field of molecular epigenetics is beginning to elucidate the mechanisms underlying the developmental basis of this disease risk. These considerations, together with the apparently inexorable rise in global prevalence of the metabolic syndrome, raise questions about the utility of current public health measures in which adult behavioral change tends to serve as the primary target.[14]

DEVELOPMENTAL ORIGINS OF HEALTH AND DISEASE (DOHaD)

Genetic Factors Provide an Incomplete Explanation for Common Complex Disease Risk

Completion of the mapping of the human genome in 2000 led to aspirations that identifying sites of functional mutations within the genomes of members of the population could pinpoint genetic determinants of various pathophysiological states. However, contrary to widespread predictions, genome-wide association studies (GWAS) have not been able to attribute a substantial genetic contribution to risk of disorders which have a non-Mendelian heritable component. Several genes have been identified at which polymorphisms have been linked to increased obesity risk, such as that at *FTO*, which modulates appetite control. However, combining genotypic data for 12 obesity-predisposing single nucleotide polymorphisms (SNPs) accounted for <1% of body mass index (BMI) variance and was found to contribute no more than 3% towards the predictive value for the risk of obesity.[15] Mathematical analyses using 20 SNPs to estimate a newborn's risk of becoming obese in childhood or adolescence could not improve predictive value beyond that provided by anthropomorphic or sociodemographic factors.[16] Furthermore, meta-analyses across multiple GWAS have implicated 44 independent loci significantly associated with T2D, yet when taken together they could only explain about 10% of the observed familial clustering in Europeans.[17] The so-called "missing heritability" observed for complex diseases led to suggestions that rare variants of larger effect size were not being detected by the genotyping arrays but could be revealed in larger sample sets. However, recent reports examining this hypothesis have found no evidence for such variants.[18] With fixed genomic variation no longer being a viable major explanation, it is necessary to consider the role of both cultural factors influencing lifestyle choice, and developmental factors that might provide an additional biological basis for this variation.

The Dawn of DOHaD

In 1934, a detailed cohort analysis study implicating a role for childhood conditions in later life mortality was published,[19] noted for its then-novel proposition that early life factors could have late-manifesting adverse effects on human health. Despite the initial interest generated, little work seeking to replicate or further probe this phenomenon was reported until the 1970s and 1980s, during which the literature become dotted with independent studies employing epidemiological and animal experimental approaches supporting the role of early life factors in the later development of metabolic disease (reviewed elsewhere).[20] Noteworthy examples include van Assche and colleagues' work on experimentally induced diabetes or intrauterine growth restriction in the pregnant rat leading to metabolic dysfunction in adult offspring,[21,22] and Freinkel's conceptual contribution of "fuel-mediated teratogenesis," which recognized the sensitivity of the developing fetus to maternal nutrition and the resultant long-term impact on metabolism, anthropometry and behavior.[23]

The premise that an individual's developmental history mediates later-life disease risk started to gain traction following the publication of a series of meticulous epidemiological analyses by Barker and colleagues, which showed an association between birth weight (an indicative marker of fetal nutrition) and the risk of death from coronary heart disease or cardiovascular disease in adulthood.[24–26] These observations were later extended to consideration of obesity[27] and T2D,[28] and have since been corroborated by a multitude of studies implicating fetal/infant life conditions in the etiopathogenesis of a broad range of other chronic diseases such as osteoporosis, depression, breast cancer, and asthma. This research field has been termed the "developmental origins of health and disease" (DOHaD), which today constitutes a major facet of our understanding of pathways to human health and disease.[29] Despite the strength of the epidemiological evidence, the DOHaD concept was initially met with skepticism. Some of this

arose because the initial observations were made on aged populations where the incidence of gestational obesity and diabetes was low, whereas in more recent studies it has become clearer that both lower and higher birth weight are associated with added disease risk.[30] The matter was further confused by the field's focus on low birth weight as part of the causal pathway,[31] and only more recently has it become clear that birth weight is merely a proxy for the conditions that a fetus has been exposed to, and that the pre- and postnatal risk pathways are in fact independent of birth weight itself.

Humans have evolved with an immense capacity for changing their physical (built) and social/cultural environments. It is evident that the modern-day nutritional environment, characterized by the readily availability of palatable food, high in calories, sugar, and glycemic index, is an evolutionary novelty in relation to the physiology and metabolism of humans as a species, which have been shaped through millennia of selective processes to cope with a hunter–gatherer lifestyle. It is generally accepted that obesity and its pathological consequences, such as T2D, have their origin in the evolutionary "mismatch" between our evolved physiology and the contemporary obesogenic environment and lifestyle.[32] Hence, it is argued that the rapid emergence of these diseases is a reflection of the recent substantial change in conditions, particularly of the nature of processed foods, in virtually all societies. However, it also needs to be acknowledged that there are substantial population and individual differences in the relationship between body composition and the lifetime risk of developing diabetes.[33] This argues for the consideration of obesity and its sequelae within a broader context, taking into account cultural, non-genomic trans-generational and developmental factors.

Much of the medical literature focuses on proximate causes of disease, invoking explanations at the genetic, cellular and physiological levels. While these reductionist approaches offer invaluable mechanistic insights into how disease is initiated and progresses, they need to be complemented by ultimate explanations that employ evolutionary insights to account for why individuals or populations are at risk of developing disease in the first place.[34] Ultimate explanations draw on evolutionary principles to offer an integrated framework of how evolutionary and developmental processes can influence phenotype and hence disease risk. In this context, it is important to consider pathways that have a potentially adaptive origin and those that do not. Here we are using the evolutionary rather than physiological meanings of "adaptive" and "adaptation" to refer to mechanisms and pathways that were selected because they provided advantage — that is, they increased reproductive success or (Darwinian) fitness — in the past. Given the changed environments in which modern humans exist, these previously advantageous pathways may have become disadvantageous in a modern environment.[35]

CONCEPTUAL FRAMEWORK

Developmental Plasticity and Predictive Adaptive Responses

Non-adaptive, overt disruption of an organism's developmental program may occur upon exposure to environmental cues of extreme severity, such as teratogens. In general, defenses have not evolved to meet very rare or novel environmental challenges. In contrast, the associations between early-life conditions and later-life disease risk reflect outcomes of developmentally plastic responses — the taxonomically widespread processes that confer an organism with the capacity to respond to external cues early in its life course by adjusting its phenotypic trajectory to match its environment.[36] Such potentially adaptive responses, which aptly illustrate how a single genotype can give rise to multiple phenotypes, may be invoked by a physiological range of factors such as altered maternal nutrition and maternal stress. It is important to note that they are normative processes that operate across the normal range of developmental experiences. There is now accumulating evidence that epigenetically mediated regulation of gene expression constitutes one of the mechanisms by which such plasticity is effected.[37,38]

Developmental plasticity represents an evolved strategy to maximize adaptive advantage in the face of either immediate or anticipated circumstances to sustain Darwinian fitness. The former, broadly classed as immediate adaptive responses, are generally prompted by more severe cues and involve trade-offs in the long term. For example, fetal exposure to an infected uterine environment may lead to early parturition to maximize its chance of survival, albeit at the cost of compromised health in infancy due to incomplete intrauterine growth. Predictive adaptive responses (PARs; sometimes called anticipatory responses), on the other hand, are prompted by cues of a more quotidian nature that provide information on the state of the anticipated postnatal environment. The organism is then able to capitalize on this information by altering its trajectory of development so as to be better matched to the environment predicted to prevail later.[39] Consequently, phenotypic effects and potential adaptive advantage derived from PARs show delayed manifestation in either infancy or childhood, as the biggest determinant of reproductive fitness in humans is survival to reproductive age.[40] Animal studies have provided substantial empirical support for the operation of PARs (reviewed elsewhere),[41] and we have recently presented direct evidence in humans that fetal responses can enhance fitness in later life. Within a cohort of survivors of marasmus and kwashiorkor — two forms of severe childhood malnutrition, the latter involving a poorer prognosis with respect to mortality and metabolic capacity to utilize tissue stores — those children developing marasmus had been born on average 333 g lighter.[42] This suggests that, as a result of fetal constraints, these individuals gained adaptive advantage by developing a metabolic or possibly metagenomic physiology that is better suited to coping with a nutritionally deficient postnatal environment, compared to counterparts who lacked such anticipatory responses and who were, therefore, more likely to develop the more threatening condition of kwashiorkor under conditions of severe malnutrition.

Developmental Mismatch

PARs are not an infallible strategy to enhance fitness; energetic costs and developmental constraints of maintaining the facility for plasticity generally limit its operation to a critical window early in the life course, beyond which the adjusted phenotypic trajectory becomes relatively invariant. Hence, PARs may potentially become maladaptive later in life should there be a disparity between the anticipated and actual environment, which may stem from cues being inaccurately transmitted or not reflecting the later-life environment. The outcome is a situation of developmental mismatch, where the adult phenotype has become incongruent with the environment, leading to elevated risk of disease.[43]

While this is clearly disadvantageous to the organism, it is essential to appreciate that anticipatory responses have evolved, and been retained, because of the overall fitness benefits conferred. The evolution and persistence of a trait is largely dependent on whether it promotes fitness up to at least the period of peak fecundity, so that gene transfer to the next generation can occur. Deleterious effects caused by inaccurate predictions generally become manifest in post-reproductive stages of life and are

inconsequential in evolutionary terms, as the major determinant of fitness in humans is pre-reproductive survival.[40] However, this has special impact on humans, as we are a long-lived species that has experienced marked extensions in longevity over the relatively recent past. Additionally, incorrectly anticipating a low-nutrient environment incurs a lower fitness cost than incorrectly predicting an environment of plenty; that is, the individual will be less disadvantaged if its physiology is set to meet under-nutrition but experiences overnutrition than in the reverse situation. Thus, formulation of PARs towards the former is likely to have been favored.[41]

This conceptual framework presents a heuristic approach to explain how traits increasing susceptibility to obesity and its comorbidities may be established early in life. Relative nutritional deprivation *in utero*, which could arise from many factors, such as maternal undernutrition, placental dysfunction, and/or maternal constraint (the evolved maternal and uteroplacental factors limiting fetal size to minimize risk of obstructed labor),[44] induces predictions of an energy-scarce postnatal environment in which a phenotype adapted to a low-nutrition plane will be better able to thrive. The resulting more metabolically conservative phenotype may be characterized by increases in appetite and sedentary behavior, and predilection toward dietary fat, as has been experimentally observed in the rat.[45,46] This, however, leaves the individual physiologically ill-equipped to cope with a postnatal environment of nutritional surfeit, in which degree of mismatch rises with level of obesogenicity. However even between these more extreme variations, slight differences in maternal nutritional status are sufficient to provide information to the fetus about its future environment. There is likely some inertia in the system such that the conceptus is not responding to every shift in maternal food intake, but rather to her overall nutritional status.[47]

The adverse effects of severe maternal undernutrition on offspring metabolic disease risk are clearly demonstrated in experimental animal models and human cohorts of famine survivors, described later. While these represent more extreme circumstances, there are many situations by which mismatch can arise on a population-wide level, discussed briefly here and extensively elsewhere.[41,48] For example the nutritional transition occurring in developing countries undergoing rapid economic growth,[49] and migration to more prosperous regions or countries, tend to involve pronounced environmental changes within a generation. Another example of relevance to the general unselected population is parity, which appears to be an important factor as maternal constraint operates in all pregnancies, but to a greater degree in primiparous women.[44] This has implications for populations experiencing declining birth rates and having a greater proportion of first-borns. Recent work has shown that prepubertal first-born children have reduced insulin sensitivity and greater blood pressure than subsequent siblings,[50] and they also have higher BMI than adults.[51] Furthermore, lower birth weight is linked to abdominal adiposity in adulthood,[52,53] showing that obesity risk can be modulated by physiological variations in fetal experience.

Maternal Obesity and Gestational Diabetes Mellitus

The discussion above has focused on normal and inadequate maternal nutrition. However, there is now increasing attention being paid to the effects of maternal adiposity and gestational diabetes mellitus (GDM) on the offspring's lifetime risks of diabetes and obesity. Pregnancy induces myriad physiological, endocrine and metabolic changes, such as a progressive increase in placental hormone-induced insulin resistance to enable glucose transfer to the fetus, and first- and second-trimester lipogenesis followed by third-trimester lipolysis to permit fetal fat accumulation. These features tend to be exaggerated with obesity and GDM — insulin resistance is greater and hyperlipidemia is more pronounced, with resulting increases in nutrient supply to the fetus.[54]

There is good evidence from clinical studies that maternal obesity is associated with increased risk of being born large independent of GDM,[55,56] of fetal hyperinsulinemia,[55] and of offspring obesity from the neonatal period through to childhood and at least young adulthood, with associations graded across the whole spectrum of maternal BMIs.[55,57,58] Offspring adiposity appears to mediate the negative impact of maternal BMI on cardiometabolic parameters such as blood pressure, insulin levels and lipid profile.[58] Higher gestational weight gain, of which obese women are at greater risk, is positively correlated with BMI in early adulthood independently of maternal obesity.[58,59]

The biological mechanisms by which GDM affects the fetus are well understood. High glucose exposure leads to fetal hyperinsulinemia, which in turn promotes excess adipogenesis; milder levels of maternal hyperglycemia produce increase in fetal lean body mass too. The increased number of fat cells presumably places the individual at greater risk in an adipogenic environment. The adverse clinical consequences of being exposed to hyperglycemic conditions *in utero* have been well documented. Importantly, such a relationship appears to be graded; very large multinational studies have indicated that a woman's glucose levels during pregnancy bear a continuous association with increased birth weight and cord blood C-peptide, a marker of fetal hyperinsulinemia.[60] An association between maternal glucose concentration and childhood obesity has also been described in a separate cohort after controlling for maternal pre-pregnancy BMI.[61] Even then, absence of excess offspring adiposity may still mask other symptoms of metabolic dysfunction such as poorer insulin sensitivity and β-cell function, suggesting developmental effects on the pancreas and other target tissues.[62]

Non-Adaptive Pathways to DOHaD

Although maternal under- and overnutrition are both risk factors for obesity and metabolic disease in adulthood, it should be stressed that the processes of developmental plasticity and the PAR model are unlikely to apply to the latter cue. Maternal obesity and GDM represent evolutionarily novel scenarios unlikely to have been commonly experienced throughout our evolutionary history, and, therefore, by definition would not have elicited evolved adaptive responses. Thus, increased vulnerability to disease is likely to be a result of non-adaptive pathways, although the specific proximate mechanisms remain unclear.

The origins of GDM can be accounted for by ultimate explanations. A degree of maternal insulin resistance is necessary in normal pregnancies for fetoplacental nutrient availability, especially in the later stages of gestation, during which fetal energy demands increase substantially. In addition, low-level intrauterine hyperglycemia leads to mild hyperinsulinemia, which in turn is likely to sustain fitness through its growth-promoting and adipogenic effects; neonatal fat mass is thought to assist thermogenesis and protect the brain, which has been estimated to account for more than 80% of a neonate's total energy requirements.[63,64] However, there is an absence of a saturation point for glucose transport across the placenta, suggesting that placental nutrient transfer has evolved to be biased towards protecting the growing fetus from undernutrition. This potentially leads to detrimental effects in today's obesogenic environment, where maternal hyperglycemia is becoming commonplace. Thus, maternal insulin resistance during pregnancy, despite having evolved as a normative physiological phenomenon, has become pathophysiological in the modern world.

EPIGENETICS AND DEVELOPMENT

The Advent of Epigenetics

The formulation of epigenetics as a concept dates back to the 1940s, when the term was first coined by Conrad Waddington, an eminent embryologist and geneticist, who described the "epigenetic landscape" as a metaphor for how gene–environment interactions during development could guide developmental pathways taken, and hence the phenotype generated.[65] This metaphorical landscape was used to highlight the probabilistic, rather than deterministic, relationship between genotype and phenotype. In now-classic experiments, Waddington showed that heat shock treatment of Drosophila pupae led to the occasional absence of crossvein wing patterns in adulthood; although phenotypically normal flies were used to continue breeding, some flies were found to display the crossveinless characteristic after about 14 generations of selection, despite no further exposure to heat treatment.[66] These observations demonstrated that environmental influences during development could have implications for the inheritance of phenotypic traits.

Waddington's foresight in his conceptualization of epigenetics is especially appreciable considering that it had taken place in a time that predates modern understanding of molecular genetics. Nevertheless, as mechanistic knowledge has accrued in the years since, epigenetics started to acquire various definitions, reflecting the variety of disciplines in which it plays a major role. For example developmental biologists may focus on gene-centric functional outcomes with less emphasis on transgenerational transfer, while evolutionary biologists have greater interest in developmentally induced phenotypes, non-genomic inheritance and the accompanying implications for evolutionary theory.[67] Indeed, the recognition that epigenetically induced phenotypes may be passed to subsequent generation(s) via gametic inheritance (see later) has prompted paradigmatic shifts in evolutionary biology thinking away from genetic determinism, as espoused by the Modern Synthesis, to more integrated models incorporating development, ecology and molecular biology.[68,69]

In the context of this chapter, we use the term "epigenetics" to refer to the interrelated molecular mechanisms that establish and maintain mitotically stable gene expression patterns independently of the underlying DNA sequence. This includes DNA methylation and histone protein modifications that induce structural changes in chromatin, and the activity of non-coding RNA strands that regulate gene expression at the transcriptional and post-translational level or through chromatin remodeling. "Mitotic stability" is used to avoid the ambiguity of discipline-dependent definitions of "heritable."

Epigenetic Mechanisms

DNA Methylation

DNA methylation refers to the covalent addition of a methyl group to specific sites on DNA. In eukaryotes, this usually occurs at the cytosine residue of a CpG dinucleotide, giving rise to 5-methylcytosine (5mC). CpG dinucleotides occur less frequently than would be predicted by chance throughout the mammalian genome, owing to the propensity for 5mC to spontaneously or enzymatically undergo deamination into thymine;[70] this conversion into a functional base in somatic cells is a risk factor for oncogenesis, while that in the germline may give rise to pathological mutations. They tend to cluster in roughly 1-kb long sequences known as CpG islands that are often located at transcription start sites and regulatory regions. In somatic cells, CpG islands are generally only methylated at imprinted genes, which display parent-of-origin dependent expression.

Due to the palindromic nature of CpG sites in double-stranded DNA, hemimethylated DNA can be recognized by methyltransferases that are directed to add a methyl group to the nascent daughter strand, thus preserving the appropriate methylation pattern during cell division. The enzymes DNMT1, DNMT3a and DNMT3b — all of which are developmentally indispensable — are responsible for the establishment and maintenance of methylation patterns. The transcriptional outcome of DNA methylation is highly context-dependent. Previously, it had been associated primarily with gene repression due to observations that 5mC interferes with binding of transcription factors and other related components, and that methyl-binding proteins are recruited, which in turn attract other repressor complexes that inhibit transcription. However, this is now known to apply to methylation at gene promoter regions; intragenic methylation, instead, is associated with gene activation, although this is dependent on whether the CpG site is located within a CpG island.[71] Instances of non-CpG DNA methylation are increasingly being identified, such as in human adult brains[71] and in human embryonic stem cells, in which almost one-quarter of all methylation is not in a CpG context.[72] It is dynamically accumulated and then lost during human male germ cell development,[73] but its function remains unclear at present.

5mC has long been regarded as the fifth DNA base, and in more recent years has been joined by a putative sixth base — its oxidized form, 5-hydroxymethylcytosine (5hmC) — following its identification in the mammalian brain and in mouse embryonic stem cells.[74,75] The oxidation of 5mC into 5hmC is catalyzed by the ten-eleven translocation 1 (TET) enzyme family, and it appears that this conversion comprises part of the molecular pathway by which DNA methylation is reversed, at least in the germline.[76] Indeed, it was recently reported that DNMT3a and 3b (but not DNMT1) are dehydroxymethylases that convert 5hmC to 5mC.[77] The complete mechanisms are still being elucidated, with another layer of complexity added by the identification of further intermediary oxidation products of 5mC such as 5-formylcytosine and 5-carboxylcytosine in mouse embryonic stem cells. Intriguingly 5hmC is enriched in the brain and its levels correlate with cerebellum development in humans;[78] the potential functional significance of this presents an exciting research avenue.

Histone Modifications

Histone modifications refer to the addition and removal of various covalent moieties to specific amino acid residues on histone protein complexes, which are encircled by DNA strands to form nucleosomes. These modifications, which include acetylation, methylation and phosphorylation, effect structural remodeling of chromatin — condensation to heterochromatin is associated with gene repression, and unwinding to euchromatin with gene activation. While acetylation generally reduces compactness and promotes gene expression by interfering with protein—DNA interactions, the effect of each modification can be highly residue-specific; for example lysine methylation at the histone H3 subunit leads to transcriptional permissiveness or silence, depending on which residue is modified.

A widely proposed histone code has been thought to underlie how the myriad modifications function in concert to regulate chromatin structure and hence gene expression, although recent work suggests that chromatin interaction with genetic factors may provide a more relevant framework.[79] Unlike DNA methylation, how histone modifications are mitotically transmitted following DNA replication remains unclear. Recent evidence points towards the polycomb repressive complex 2, a gene-silencing protein complex with H3K27 methyltransferase activity, reading and maintaining the extant chromatin state post-cell division.[80]

Noncoding RNA

A major finding from the ongoing Encyclopedia of DNA Elements (ENCODE) project documenting functional elements of the human genome has been that much of the genome possesses some form of biochemical function, confounding previous beliefs that non-protein coding sequences should be relegated to "junk DNA" status.[81] Indeed it is known that vast numbers of non protein-coding RNA (ncRNAs) are transcribed in tissue- and cell-specific patterns to dynamically regulate cell differentiation and development. ncRNAs fall into subclasses based on length, characteristics and function including microRNAs, piwi-interacting RNAs, small interfering RNAs, and long non-coding RNAs.[82] For example, the 20—24 nucleotide-long microRNAs — of which 2578 mature types have been identified to date in humans[83] — destabilize cognate mRNA to reduce protein expression, while the >200 nucleotide-long lncRNAs employ a broader range of mechanisms to partake in various gene-regulatory mechanisms such as transcription, translation, and chromatin modification. They have been shown to interact with chromatin-modifying complexes and transcription factors in human embryonic stem cells to guide neurogenesis.[84]

The capacity of RNA to undergo enzymatically directed editing, particularly adenosine to inosine, has been proposed to confer a degree of plasticity on the extensive regulatory network of ncRNAs. Taken together with the abundance and specific expression of ncRNA in the brain, and with the marked expansion of RNA editing in the human lineage, it has been argued that evolution of cognitive function may be at least partially underpinned by RNA-mediated mechanisms, which act as a conduit between the environment and the epigenome to effect physiological adaptation and brain function.[85]

A common postulate for the evolutionary basis of epigenetic mechanisms is that they first arose as defense measures to provide genomic protection against the deleterious effects of transposable element insertion.[86] It appears that these mechanisms, whilst remaining highly evolutionarily conserved, then became co-opted for other functions, such as gene dosage control during prokaryotic fusion and tissue differentiation in metazoan development. This may have then further extended to the processes of genomic imprinting in eutherian mammals and marsupials, and developmental plasticity in metazoans. The discussion in this chapter focuses primarily on how epigenetic mechanisms contribute to effecting developmental plasticity to generate a continuous range of phenotypes arising from developmental cues, known as "norms of reaction."[87]

Early-Life Nutritional Variation and Epigenetics

Animal models in which the mother is nutritionally or hormonally manipulated during pregnancy have provided a strong evidence base that epigenetic change in offspring contributes to adverse health in later life. In a well-established rat model in which dams are fed a protein-restricted diet *in utero*, offspring become prone to hypertension and present with disrupted lipid metabolism and vascular endothelial function as adults. Epigenetic investigations in offspring liver tissue have revealed hypomethylation at the promoters of *Ppara*,

encoding a regulator of fatty acid oxidation, and *GR*, encoding a key regulator of gluconeogenesis, and these changes were accompanied by increases in gene expression.[88] Supplementation of the maternal low-protein diet with folic acid, previously shown to reverse the induced phenotype, abolished the methylation and gene expression alterations, supporting the candidacy of epigenetic pathways as mechanisms underlying differential adult phenotypes.[88] Multiple repressive histone modifications such as decreased acetylation and increased methylation at H3K9 have also been observed at the promoter region of hepatic *Cyp7a1*, encoding a key cholesterol catabolizer.[89] The resultant reduction in Cyp7a1 protein levels was coincident with increased circulating cholesterol in adulthood.

In another rat model of maternal undernutrition, where food intake is limited to 30% of that consumed *ad libitum* by controls, offspring develop several characteristics typifying the metabolic syndrome and an energy-conserving phenotype in adulthood, including obesity, insulin and leptin resistance, hyperphagia and reduced locomotion.[45,46] Augmented methylation of *Ppara* concomitant with its decreased gene expression were observed in hepatic tissue in these offspring.[90] In further support of the epigenetic contribution to the developmentally induced phenotype, administration of leptin to neonatal pups prevented the onset of metabolic dysfunction and normalized *Ppara* methylation and expression levels to that of control pups.[90]

Changes in epigenetic marks have been observed even when maternal food intake is mildly reduced. Hepatic tissue from offspring of baboons fed 70% of control *ad libitum* diet during pregnancy showed hypomethylation at the promoter of *PCK1*, coding for a critical component of the gluconeogenesis pathway.[91] Epigenetic effects were induced in the absence of birth weight differences, while subsequent work linked the maternal manipulation to increased insulin resistance in juvenile offspring.[92] This demonstrates the malleability of the fetal developmental program to moderate environmental cues, and that the absence of an overt phenotype at birth cannot preclude possible adverse later health outcomes.

There is growing evidence of relationships between early-life events and epigenetic changes in humans, some of which has been derived from cohorts of individuals gestationally exposed to the Dutch Hunger Winter wartime famine of 1944–45. For example, exposure during early gestation has been linked to increased vulnerability to obesity, lipid metabolism dysfunction, and coronary heart disease at age 50 or 58.[93] Remarkably, these individuals also harbored differential methylation at genes associated with growth, development, and metabolic disease, such as the imprinted *IGF2*, compared to unexposed siblings,[94,95] underscoring the long-term stability of molecular marks arising from brief environmental cues in early life. Although effect sizes were small, the detection in these studies of persistent epigenetic changes via candidate gene (rather than genome-wide) approaches is noteworthy, particularly because in humans, smaller effect sizes may be obscured by variations in postnatal environmental conditions, by limitations of peripheral blood as a target tissue, or by epigenetic drift in older individuals. These issues may be reflected in a recent study of famine survivor offspring that did not detect methylation differences at four genes (examined on the basis of prior animal work implicating their amenability to epigenetic regulation by maternal diet), although some associations between methylation status and markers of metabolic disease and adult lifestyle were found.[96]

In contrast to the extreme circumstance of famine, it has recently been demonstrated in two large-scale independent cohorts of children born from normal pregnancies that epigenetic state at birth can both reflect relatively minor variations in maternal pregnancy diet and be predictive of a clinically relevant trait later in life.[97] This study found an inverse correlation

between carbohydrate intake in early pregnancy and methylation at specific CpG sites in the promoter of *RXRA*, the gene product of which regulates energy balance and glucose homeostasis. Furthermore, *RXRA* methylation was positively correlated with adiposity in children measured 6 or 9 years later. The degree of variance in body composition explained by this association was much higher than that found in genomic studies discussed earlier. These associations are especially important as they were found in uncomplicated pregnancies of a healthy population, suggesting that there is a high level of sensitivity toward developmental programming of obesity and metabolic risk via epigenetic mechanisms in essentially all pregnancies, and that environmental cues need not be severe to elicit such responses.

A number of small studies have since also assessed how other subtle changes in prenatal nutritional exposures in normal pregnancies can modulate epigenetic state at birth. For example, in cord blood, folic acid supplementation was associated with hypomethylation at the *H19* differentially methylated region (DMR), which regulates *IGF2* expression;[98] choline supplementation with hypomethylation at the cortisol-regulating genes *CRH* and *NR3C1*;[99] and a multi-micronutrient supplement with sex-specific effects on the *IGF2R* and *GTL2* DMRs.[100] The possible phenotypic and functional outcomes of epigenomic alterations observed in these studies await longitudinal investigations with larger cohort sizes, but they underscore the operation of epigenetically mediated, developmentally plastic responses under normal conditions.

Maternal Obesity/GDM and Developmental Epigenetic Dysregulation

Growing evidence from animal studies suggests that early overnourishment may induce epigenetic alterations at physiologically relevant genes in association with later-life sequelae. Rats subjected to neonatal overfeeding through artificial reduction of litter size become obese, hyperglycemic, hyperinsulinemic and hyperleptinemic. An examination of hypothalamic tissue in these animals has shown elevated methylation levels at two sites of anorexigenic neurohormone-encoding *Pomc*, this being a prerequisite for leptin and insulin to exert regulatory effects on *Pomc* expression,[101] as well as at the insulin-receptor promoter, which was positively correlated with glucose levels.[102]

A well-characterized nonhuman primate model of maternal high-fat diet exposure has provided insights into physiological pathways affected and attendant epigenetic alterations.[103] Fetal Japanese macaque livers displayed hyperacetylation at H3K14, which was particularly prevalent at the promoter region of *Npas*, a peripheral circadian regulator.[104] This histone modification was accompanied by decreased expression of SIRT1, a lysine deacetylase with broad roles including regulation of cellular metabolism, and altered expression of its downstream targets.[105] H3K18 acetylation at the thyroid hormone receptor β promoter together with differential recruitment of coactivators and corepressors was also observed.[106]

Molecular support for the effects of early-life overnutrition on epigenetic regulation in humans, while limited, is growing. There is some support for an epigenetic basis for how maternal preconceptional body composition influences offspring metabolic profile – a study of pregnant women showed that pre-pregnancy BMI was moderately correlated with promoter methylation of offspring umbilical cord *PPARGC1A*, the gene product of which regulates energy metabolism.[107] As diabetic women were excluded, and the relationship persisted throughout the normal range of BMIs, it can be inferred that developmental influences may still be exerted in unexceptional pregnancies. This graded effect has been mirrored in

another study, which found that maternal glycemic status was negatively correlated with fetal-side placental adiponectin gene methylation through the entire range of glucose levels.[108] In related work on women with impaired glucose tolerance but who did not necessarily meet international guideline thresholds for GDM diagnosis, maternal- and fetal-side placental leptin gene methylation were divergently correlated with glycemia throughout this upper range of glucose levels, suggesting that even mildly raised glycemic status may have physiological implications.[109] It was recently shown that cord blood and placental tissues in offspring exposed to GDM display hypomethylation at *MEST*, a maternally imprinted gene known to be upregulated in obese human adipose tissue and by early overnutrition in mice.[110] This pattern was also observed in blood from morbidly obese adults compared to lean controls, supporting the notion that *MEST* epigenetic dysregulation in early life may have lifelong consequences for metabolic disease risk.[110]

THE PATERNAL ROLE

Although research on the early-life determinants of obesity and metabolic disease has been dominated by investigations of maternal effects, it is now gradually emerging that the paternal state at conception may also play a role in determining offspring metabolic disease risk. A recent small clinical study has linked increased paternal BMI with lower methylation levels (mean relative difference of 3.5%) of the *IGF2* DMR in offspring umbilical cord blood.[111] No significant differences were found when maternal obesity was analyzed as the independent variable. It is surprising that paternal obesity in this study and maternal undernutrition in the Dutch famine cohort[94] are associated with a similar aberrant epigenetic state of similar effect size in offspring.

Clearly, follow-on metabolic phenotyping of the offspring is warranted to yield clues on possible functional significance.

Although this study hints that excessive paternal body size may interfere with appropriate establishment of genomic imprinting in germ cells, there are many confounding factors that encumber efforts to determine the precise nature of parental influence on progeny in humans. These include our outbred nature, maternal and grandmaternal effects, parental behavior and cultural influences. These factors are better controlled in rodent studies, in which paternal contribution is solely restricted to sperm. Mice whose sires were subjected to a protein-restricted diet showed numerous transcriptomic changes in hepatic genes with roles in lipid and cholesterol metabolism, as well as increased methylation at a locus upstream of *Ppara* that encodes its putative enhancer.[112] Female offspring of male rats fed a high-fat diet developed impaired insulin secretion, progressive decreases in glucose tolerance and disrupted β-cell histology, in addition to widespread islet gene expression changes and hypomethylation at a proposed regulatory region of *Il13ra2*, encoding a component of the Jak-Stat signal transduction pathway.[113]

There is increasing evidence that such male-line mediated phenotypic inheritance is associated with epigenetic changes in paternal[114,115] and offspring[116,117] sperm, which is counter to the widely held view that methylation patterns are globally erased during developmental epigenetic reprogramming.[118] Murine studies have shown that specific methylation marks in primordial germ cells and zygotes can survive past this phase.[76,119] In humans, a subset of nucleosomes at gene regulatory sites with potentially significant roles in development are retained during spermiogenesis in humans.[120,121] Furthermore, small ncRNAs — previously implicated in mouse models of transgenerational inheritance[122] — have been detected in human spermatozoa and may be involved in early embryo development.[123] Cumulatively, these findings provide proof of concept that epigenetic marks in the mammalian paternal germline may be directly inherited; this may be one of the pathways by which genomic fixation of heritable, developmentally induced epigenetic changes, such as that previewed by Waddington's experiments with fruit flies described earlier, are achieved.

There are obvious public health ramifications should the paternal effects seen in animal studies be found to operate in humans. However the human data to date are limited and essentially inferential (reviewed elsewhere).[37] Transmeiotic inheritance through the germline represents a pathway through which induced phenotypes can be directly inherited,[37] and this line of research could, therefore, yield molecular insights into phenotypic driven evolutionary processes.[38]

EPIGENETIC DYSREGULATION IN LATER-LIFE METABOLIC DISEASE

Multiple studies have revealed associations between global (non-gene specific) DNA methylation in peripheral blood from adults and risk of cardiovascular and metabolic disorders. For example, lower global methylation as measured by methylation at long interspersed nucleotide elements (LINE-1) repetitive elements was linked to increased subsequent risk of cardiovascular disease in elderly men.[124] However, another report has found that higher LINE-1 methylation among adults is associated with a poorer glycemic and lipid profile,[125] and a study using monozygotic twin sets discordant for T2D to account for genetic factors found that higher global methylation (measured in Alu repeats, the most abundant family of repetitive sequences) was associated with a greater degree of insulin resistance.[126]

These studies suggest that epigenetic dysregulation on a global level is associated with disease state, but beyond that they do not provide insight into how the different methylation patterns affect or are affected by specific functions mediating increased disease risk. It is important to use approaches that have the granularity to explore regulation at specific sites within the epigenome. Indeed, it is now clear that there is exquisite regulation at the level of individual epigenetic marks, and it is these that must be the focus of clinical investigation. To this end, studies focusing on candidate genes have identified specific epigenotype–phenotype associations, while genome-wide massive parallel sequencing technologies have enabled unbiased approaches to cataloguing large-scale transcriptional changes. Moreover although blood samples are a convenient tissue source, possible tissue specificity of epigenetic changes can preclude firm conclusions so other target tissues known to be involved in disease pathology may be more elucidatory.

Type 2 Diabetes

Profiling of the DNA methylation landscape in pancreatic islets from type 2 diabetic individuals has revealed differential methylation for 254 of 14,475 genes analyzed compared to nondiabetics.[127] The methylation difference for many of these genes was around 20–30%. Notably, methylation changes were not detected in blood, demonstrating the tissue specificity of epigenetic discordance. In contrast, a genome-wide exploratory analysis of CpG sites in skeletal muscle and subcutaneous adipose tissues from monozygotic twins discordant for T2D uncovered small differences in gene promoter methylation;[128] this may reflect the choice of tissue (e.g., visceral fat may be more informative, or the pancreas may be the primary site of pathological changes) or methodological limitations of a small sample size. A larger study performing genome-wide analyses in blood cells has identified

hypomethylation at a non-promoter CpG site of the obesity-associated gene FTO as a T2D risk factor; having this mark at age 30 was predictive of later development of impaired glucose metabolism.[129]

The pathophysiology of diabetes appears to be associated with aberrant methylation of several other physiologically relevant genes. For example, based on prior observations that expression of PPARGC1A is decreased in skeletal muscle of type 2 diabetic patients, Ling et al. investigated and detected hypermethylation of its promoter and gene underexpression in pancreatic islets in diabetic patients.[130] The positive correlation between PPARGC1A expression and glucose-stimulated insulin secretion in human islets lends support to the epigenetic regulation of the latter. Indeed in normal human pancreatic β-cells, the insulin gene promoter is completely demethylated at CpG sites, which are otherwise methylated in non-insulin expressing cells; methylation promotes interaction with the methylated DNA binding protein MeCP2, thus facilitating gene silencing.[131]

Subsequent studies in diabetic islets have demonstrated higher methylation levels at four CpG sites within the insulin promoter, in conjunction with downregulated insulin gene expression.[132] Elevated levels of glycated hemoglobin (HbA1c; a marker of hyperglycemia) were further correlated with insulin promoter methylation, and in vitro studies with clonal rat β-cells suggest that hyperglycemia is a causal factor for insulin promoter hypermethylation.[132] This group has further demonstrated similar relationships for PDX-1, which encodes a pancreatic homeobox transcription factor responsible for β-cell function and with a well defined role in the development of diabetes. Compared to controls, PDX-1 methylation is increased and its expression decreased in human diabetic islets.[133] Hypermethylation of Pdx-1 in clonal rat β-cells cultured in high glucose medium, together with associations of lower gene expression with higher HbA1c levels in humans, implicate exposure to hyperglycemia as a basis for epigenetic dysregulation.

There is good evidence that ncRNAs are involved in pancreatic β-cell function.[134] MicroRNA precursors are cleaved by the endo-nuclease Dicer during processing to maturity, and mice lacking the pancreas-specific Dicer1 protein present with progressive hyperglyce-mia culminating in adult-onset diabetes, underpinned by decreases in insulin gene expression and insulin secretion.[135] Cell cul-ture work from islets of this mouse model showed that a specific set of microRNAs is responsible for maintaining insulin gene acti-vation and expression, in part mediated through reducing the expression of insulin transcriptional repressors.[136] The pancreatic islet-specific microRNA-375, which has *in vitro* insulin-secretion regulating properties, has been reported to be highly upregulated in dia-betic human pancreatic islets compared to nondiabetic controls, with increased levels associated with greater islet amyloid forma-tion, a histological marker of β-cell failure.[137]

Type 1 and type 2 diabetes can lead to microvascular and macrovascular complica-tions including nephropathy, retinopathy, ath-erosclerosis and stroke. However, persistence of diabetes-related pathologies despite achieve-ment of normoglycemia can occur, and is referred to as metabolic memory or glycemic memory. The molecular basis for the legacy effects of prior hyperglycemic insult in the establishment and progression of vascular complications has long stymied diabetes researchers, but in recent years has begun to be partially explained by epigenetic changes. El-Osta *et al.* showed that short-term incuba-tion of human aortic endothelial cells with high glucose induced persistently elevated pro-atherogenic gene expression despite nor-malization of glucose levels,[138] likely due to chromatin remodeling effected by recruitment of the H3K4 methyltransferase Set7 and monomethylation at H3K4. Set7 was shown to translocate to the nucleus, where it also activated proinflammatory genes in H3K4 methylation-dependent and -independent manners.[139] Genome-wide analysis of these cells has also revealed numerous instances of hyperglycemia-induced hyperacetylation at histone H3K9/K14 within the gene promoters.[140] Hyperacetylation was associated with increased gene expression, while hypoacetylated sequences tended to be colocalized with methylated CpG sites.

Obesity

Genome-wide analysis of peripheral blood DNA from a small cohort of obese and lean adolescents has identified higher methylation at a CpG site in *UBASH3A* and lower methyla-tion at a CpG site in *TRIM3* in obese indivi-duals; these observations were upheld in a replication cohort of young adults.[141] Both gene products are implicated in immune func-tion — UBASH3A is a T-cell signaling regula-tor and TRIM3 a multifunctional protein with roles in regulating immunity — which could be relevant to the involvement of inflammatory processes in the pathophysiology of obesity.[142]

Hypermethylation at specific CpG sites of the *POMC* gene, which encodes a protein cru-cial to energy homeostasis, has also been observed in blood cells from obese children compared to their normal weight counter-parts.[143] Based on these results, the authors revisited longitudinal data from a separate cohort of obese adolescents whose DNA was collected before and after disease onset. All individuals in a subset who demonstrated *POMC* hypermethylation at adolescence had the same marker before development of obe-sity; as a group, methylation levels did not change between the two time points.[143] This was offered as support that differential meth-ylation is established prior to and not as a con-sequence of disease state, although such an interpretation necessitates caution due to the small sample size and the difficulties associated with disentangling the issue of causality.

A study examining genome-wide methylation changes induced by a five-day high-fat, hypercaloric diet in skeletal muscle of healthy young adults revealed expansive DNA methylation changes of modest magnitude.[144] In related work comparing low and normal birth weight young adults, the authors found that *PPARGC1A* promoter methylation is constitutively elevated in skeletal muscle of those who were born small, and that the program of overfeeding induced reversible hypermethylation only among those of normal birth weight.[145] The birth weight-dependent differential response in methylation suggests that developmental factors can influence later-life epigenetic regulatory pathways that cope with changes in the nutritional environment. It should be noted that both studies were unable to detect significant correlations between methylation and expression levels, which may reflect a buffering effect wherein the highly sensitive and responsive epigenetic marks do not evoke an immediate functional impact in response to a transient challenge, at least under these experimental conditions.

In recent years, several studies examining blood or adipose tissue in obese adults have reported finding that baseline epigenetic status at specific genes may underlie inter-individual variation in response to weight loss regimes, and that caloric restriction can also alter methylation patterns.[146–149] Many of the genes identified have known involvement in physiologically relevant processes such as gluconeogenesis, fat metabolism, and inflammation. These studies could have translational value for identification of predictive epigenetic biomarkers of weight loss success, and enable greater understanding of the molecular and physiological consequences of energy deprivation in later life.

DEVELOPMENTAL EPIGENOMICS AND HUMAN HEALTH

There are several methodological and interpretative caveats that need to be borne in mind when undertaking developmental epigenomic research. Differentiating between epigenetic changes in the pathway to disease causation and those secondary to the effects of disease itself should be informative in pinpointing specific molecular targets for intervention. However the most obvious approach to determining the direction of causality would require longitudinal studies involving large numbers of participants, in whom various target tissues other than blood are periodically subject to epigenomic analysis over the life course at least until disease onset, generating vast data sets that will require appropriate bioinformatic analyses. The major logistical and ethical challenges posed are obvious, and the absence of conclusive evidence for causality in the literature has been noted.[150] Indeed, it is telling that the cause-or-consequence debate still remains in the older and more intensively studied field of cancer epigenetics. Establishing a directional link would also require that DNA methylation differences are concordant with corresponding changes in gene expression, which may be context-specific rather than basal, and/or with translation into the gene product. That this cannot be left to assumption has been noted in a number of human studies.[144,145] Nevertheless, there is precedent for methylation profiling to have greater utility than gene expression in informing the clinical management of breast cancer[151] and in identifying individuals who are more responsive to dietary induced weight loss.[148,152] Alternative approaches such as *in vitro* studies may be necessary. For example *in vitro* incubation of preadipocytes with altered nutrient levels did not affect differentiation but did perturb mature adipocyte functionality in association with epigenetic changes in relevant genes (Ngo, Gluckman, Sheppard *et al.*, submitted).

Based on the development of several promising epigenetically based cancer treatments, epigenetic biology holds latent potential as an invaluable platform from which clinical therapeutic applications for the more complex metabolic diseases can be devised. Several translational outcomes may be envisaged: the prognostic identification of at-risk individuals by discordance in epigenetic marks associated with later-life metabolic disease, thus enabling preventative interventions such as the nutritional management of fetuses and infants; the development of diagnostic epigenetic biomarkers so disease onset can be determined at the prodromal stage, before becoming fully manifest; the development of epigenetic drugs for curative purposes; and the identification of individuals likely to respond to treatment programs by baseline epigenetic profiling. The feasibility of epigenetic stratification of neonates by susceptibility to obesity and metabolic disease has been shown,[97] and there is substantive evidence from animal studies that dietary or pharmacological interventions administered early in life may attenuate or even abrogate the metabolic programming effects of adverse developmental cues (reviewed in ref[153]).

Rapid technological advances in epigenetic research have made the study of the molecular underpinnings of developmental pathways to metabolic disease risk increasingly tractable. The recent scientific insights relating prenatal and early postnatal environmental influences to the later risks of obesity and metabolic compromise can no longer be overlooked. Indeed they appear to have a far greater role than has been generally considered,[97] which strongly argues for the incorporation of early-life preventative measures in the formulation of public health measures for maximal clinical and public health value.

Acknowledgements

Peter D. Gluckman and Felicia M. Low are supported by Gravida: National Centre for Growth and Development; Mark A. Hanson is supported by the British Heart Foundation.

References

1. Flegal KM, Carroll MD, Kit BK, Ogden CL. Prevalence of obesity and trends in the distribution of body mass index among us adults, 1999–2010. *JAMA* 2012;**307**:491–7.
2. Malik VS, Willett WC, Hu FB. Global obesity: trends, risk factors and policy implications. *Nat Rev Endocrinol* 2012;**9**:13–27.
3. Danaei G, Finucane MM, Lu Y, Signh GM, Cowan MJ, Paciorek CJ, et al. National, regional, and global trends in fasting plasma glucose and diabetes prevalence since 1980: systematic analysis of health examination surveys and epidemiological studies with 370 country-years and 2·7 million participants. *Lancet* 2011;**378**:31–40.
4. International Diabetes Federation. *IDF diabetes atlas*. 5th ed. Brussels: International Diabetes Federation; 2011.
5. Lim SS, Vos T, Flaxman AD, Danaei G, Shibuva K, Adair-Rohani H, et al. A comparative risk assessment of burden of disease and injury attributable to 67 risk factors and risk factor clusters in 21 regions, 1990–2010: a systematic analysis for the Global Burden of Disease Study 2010. *Lancet* 2012;**380**:2224–60.
6. Lakshman R, Elks CE, Ong KK. Childhood Obesity. *Circulation* 2012;**126**:1770–9.
7. de Onis M, Blössner M, Borghi E. Global prevalence and trends of overweight and obesity among preschool children. *Am J Clin Nutr* 2010;**92**:1257–64.
8. May AL, Kuklina EV, Yoon PW. Prevalence of cardiovascular disease risk factors among US adolescents, 1999–2008. *Pediatrics* 2012;**129**:1035–41.
9. Zhang Y-X, Zhao J-S, Sun G-Z, Lin M, Chu Z-H. Prevalent trends in relatively high blood pressure among children and adolescents in Shandong, China. *Ann Hum Biol* 2012;**39**:259–63.
10. Yan S, Li J, Li S, et al. The expanding burden of cardiometabolic risk in China: the China Health and Nutrition Survey. *Obes Rev* 2012;**13**:810–21.
11. Poston L. Intergenerational transmission of insulin resistance and type 2 diabetes. *Prog Biophys Mol Biol* 2011;**106**:315–22.
12. Cnattingius S, Villamor E, Lagerros YT, Wikstrom AK, Granath F. High birth weight and obesity—a vicious circle across generations. *Int J Obes* 2012;**36**:1320–4.

13. Beaglehole R, Bonita R, Horton R, Adams C, Alleyne G, Asaria P, et al. Priority actions for the non-communicable disease crisis. *Lancet* 2011;**377**:1438–47.

14. Gluckman PD, Hanson M, Zimmet P, Forrester T. Losing the war against obesity: the need for a developmental perspective. *Science Transl Med* 2011;**3**:93cm19.

15. Li S, Zhao JH, Ja Luan, Luben RN, Rodwell SA, Khaw KT, et al. Cumulative effects and predictive value of common obesity-susceptibility variants identified by genome-wide association studies. *Am J Clin Nutr* 2010;**91**:184–90.

16. Morandi A, Meyre D, Lobbens S, Kleinman K, Kaakinen M, Rifas-Shiman SL, et al. Estimation of newborn risk for child or adolescent obesity: lessons from longitudinal birth cohorts. *PLoS One* 2012;**7**:e49919.

17. Wheeler E, Barroso I. Genome-wide association studies and type 2 diabetes. *Brief Funct Genomics* 2011;**10**:52–60.

18. Kaiser J. Genetic influences on disease remain hidden. *Science* 2012;**338**:1016–7.

19. Kermack W, McKendrick A, McKinlay P. Death rates in Great Britain and Sweden: some general regularities and their significance. *Lancet* 1934;**223**:698–703.

20. Gluckman PD, Hanson MA, Buklijas T. A conceptual framework for the Developmental Origins of Health and Disease. *J Dev Orig Health Dis* 2010;**1**:6–18.

21. Aerts L, Van Assche FA. Is gestational diabetes an acquired condition? *J Devel Physiol* 1979;**1**:219–25.

22. De Prins FA, Van Assche FA. Intrauterine growth retardation and development of endocrine pancreas in the experimental rat. *Biol Neonat* 1982;**41**:16–21.

23. Freinkel N. Banting Lecture 1980. Of pregnancy and progeny. *Diabetes* 1980;**29**:1023–35.

24. Barker DJP, Winter PD, Osmond C, Margetts B, Simmonds SJ. Weight in infancy and death from ischaemic heart disease. *Lancet* 1989;**2**:577–80.

25. Barker DJP, Osmond C, Simmonds SJ, Weild GA. The relation of small head circumference and thinness at birth to death from cardiovascular disease in adult life. *Br Med J* 1993;**306**:422–6.

26. Osmond C, Barker DJP, Winter PD, Fall CHD, Simmonds SJ. Early growth and death from cardiovascular disease in women. *Br Med J* 1993;**307**:1519–24.

27. Kensara OA, Wootton SA, Phillips DI, Patel M, Jackson AA, Elia M, et al. Fetal programming of body composition: relation between birth weight and body composition measured with dual-energy X-ray absorptiometry and anthropometric methods in older Englishmen. *Am J Clin Nutr* 2005;**82**:980–7.

28. Eriksson JG, Forsen T, Tuomilehto J, Jaddoe VWV, Osmond C. Barker DJP. Effects of size at birth and childhood growth on the insulin resistance syndrome in elderly individuals. *Diabetologia* 2002;**45**:342–8.

29. Gluckman PD, Hanson MA, Cooper C, Thornburg KL. Effect of in utero and early-life conditions on adult health and disease. *N Engl J Med* 2008;**359**:61–73.

30. Gluckman PD, Hanson MA. *The fetal matrix: Evolution, development, and disease*. Cambridge: Cambridge University Press; 2005.

31. Hales CN, Barker DJ. Type 2 (non-insulin-dependent) diabetes mellitus: the thrifty phenotype hypothesis. *Diabetologia* 1992;**35**:595–601.

32. Gluckman P, Hanson M. *Mismatch: Why our world no longer fits our bodies*. Oxford: Oxford University Press; 2006.

33. Chiu M, Austin PC, Manuel DG, Shah BR, Tu JV. Deriving ethnic-specific BMI cutoff points for assessing diabetes risk. *Diabetes Care* 2011;**34**:1741–8.

34. Gluckman PD, Low FM, Buklijas T, Hanson MA, Beedle AS. How evolutionary principles improve the understanding of human health and disease. *Evol Appl* 2011;**4**:249–63.

35. Gluckman PD, Beedle AS, Hanson MA. *Principles of evolutionary medicine*. Oxford: Oxford University Press; 2009.

36. Low FM, Gluckman PD, Hanson MA. Developmental plasticity, epigenetics and human health. *Evol Biol* 2012;**39**:650–65.

37. Low FM, Gluckman PD, Hanson MA. Developmental plasticity and epigenetic mechanisms underpinning metabolic and cardiovascular diseases. *Epigenomics* 2011;**3**:279–94.

38. Bateson P, Gluckman P. *Plasticity, robustness, development and evolution*. Cambridge: Cambridge University Press; 2011.

39. Gluckman PD, Hanson MA, Spencer HG. Predictive adaptive responses and human evolution. *Trends Ecol Evol* 2005;**20**:527–33.

40. Jones JH. The force of selection on the human life cycle. *Evol Hum Behav* 2009;**30**:305–14.

41. Low FM, Gluckman PD, Hanson MA. Evolutionary and developmental origins of obesity and associated chronic diseases. In: Muehlenbein MP, ed. *Encyclopedia of Human Biology*. 3rd ed. Oxford: Elsevier; in press. Available from http://dx.doi.org/10.1016/B978-0-12-226980-6.00555-0.

42. Forrester TE, Badaloo AV, Boyne MS, Osmond C, Thompson D, Green C, et al. Prenatal factors contribute to emergence of kwashiorkor or marasmus in response to severe undernutrition: evidence for the predictive adaptation model. *PLoS One* 2012;**7**:e35907.

43. Gluckman PD, Hanson MA, Beedle AS. Early life events and their consequences for later disease: a life history and evolutionary perspective. *Am J Hum Biol* 2007;**19**:1–19.

44. Gluckman PD, Hanson MA. Maternal constraint of fetal growth and its consequences. *Semin Fet Neonat Med* 2004;**9**:419–25.

45. Vickers MH, Breier BH, Cutfield WS, Hofman PL, Gluckman PD. Fetal origins of hyperphagia, obesity, and hypertension and postnatal amplification by hypercaloric nutrition. *Am J Physiol* 2000;**279**:E83−7.

46. Vickers MH, Breier BH, McCarthy D, Gluckman PD. Sedentary behavior during postnatal life is determined by the prenatal environment and exacerbated by postnatal hypercaloric nutrition. *Am J Physiol* 2003;**285**:R271−3.

47. Kuzawa CW. Fetal origins of developmental plasticity: are fetal cues reliable predictors of future nutritional environments? *Am J Hum Biol* 2005;**17**:5−21.

48. Gluckman PD, Low FM, Hanson MA. Developmental epigenomics and metabolic disease. In: Jirtle RL, Tyson FL, editors. *Environmental epigenomics in health and disease: Epigenetics and disease origins*. Heidelberg: Springer-Verlag; 2013. Available from http://dx.doi.org/10.1007/978-3-642-23380-7_2.

49. Popkin BM, Adair LS, Ng SW. Global nutrition transition and the pandemic of obesity in developing countries. *Nutr Rev* 2012;**70**:3−21.

50. Ayyavoo A, Savage T, Derraik JGB, Hofman PL, Cutfield WS. First-born children have reduced insulin sensitivity and higher daytime blood pressure compared to later-born children. *J Clin Endocrinol Metab* 2013;**98**:1248−53.

51. Reynolds RM, Osmond C, Phillips DIW, Godfrey KM. Maternal BMI, parity, and pregnancy weight gain: influences on offspring adiposity in young adulthood. *J Clin Endocrinol Metab* 2010;**95**:5365−9.

52. Laitinen J, Pietilainen K, Wadsworth M, Sovio U, Järvelin MR. Predictors of abdominal obesity among 31-y-old men and women born in Northern Finland in 1966. *Eur J Clin Nutr* 2004;**58**:180−90.

53. Tian J-Y, Cheng Q, Song X-M, Li G, Jiang G, Gu Y, et al. Birth weight and risk of type 2 diabetes, abdominal obesity and hypertension among Chinese adults. *Eur J Endocrinol* 2006;**155**:601−7.

54. O'Reilly JR, Reynolds RM. The risk of maternal obesity to the long-term health of the offspring. *Clin Endocrinol* 2013;**78**:9−16.

55. Hapo Study Cooperative Research Group. Hyperglycaemia and Adverse Pregnancy Outcome (HAPO) Study: associations with maternal body mass index. *BJOG* 2010;**117**:575−84.

56. Black MH, Sacks DA, Xiang AH, Lawrence JM. The relative contribution of prepregnancy overweight and obesity, gestational weight gain, and IADPSG-defined gestational diabetes mellitus to fetal overgrowth. *Diabetes Care* 2012.

57. Modi N, Murgasova D, Ruager-Martin R, et al. The influence of maternal body mass index on infant adiposity and hepatic lipid content. *Pediatr Res* 2011;**70**:287−91.

58. Hochner H, Friedlander Y, Calderon-Margalit R, Meiner FV, Sagy Y, Avgil-Tsadok M, et al. Associations of maternal pre-pregnancy body mass index and gestational weight gain with adult offspring cardio-metabolic risk factors: the Jerusalem Perinatal Family follow-up study. *Circulation* 2012;**125**(11):1381−9.

59. Mamun AA, O'Callaghan M, Callaway L, Williams G, Najman J, Lawlor DA. Associations of gestational weight gain with offspring body mass index and blood pressure at 21 years of age: evidence from a birth cohort study. *Circulation* 2009;**119**:1720−7.

60. The HAPO Study Cooperative Research Group. Hyperglycemia and Adverse Pregnancy Outcomes. *New Engl J Med* 2008;**358**:1991−2002.

61. Deierlein AL, Siega-Riz AM, Chantala K, Herring AH. The association between maternal glucose concentration and child BMI at age 3 years. *Diabetes Care* 2011;**34**:480−4.

62. Bush NC, Chandler-Laney PC, Rouse DJ, Granger WM, Oster RA, Gower BA. Higher maternal gestational glucose concentration is associated with lower offspring insulin sensitivity and altered β-cell function. *J Clin Endocrinol Metab* 2011;**96**:E803−9.

63. Kuzawa CW. Beyond feast−famine: Brain evolution, human life history, and the metabolic syndrome. In: Muehlenbein MP, editor. *Human evolutionary biology*. Cambridge: Cambridge University Press 2010; p. 518−27.

64. Holliday MA. Body composition and energy needs during growth. In: Falkner F, Tanner JM, editors. *Human growth: A comprehensive treatise*. New York: Plenum 1986; p. 101−17.

65. Waddington CH. *Organisers and genes*. Cambridge: Cambridge University Press; 1940.

66. Waddington CH. Genetic assimilation of an acquired character. *Evolution* 1953;**7**:118−26.

67. Ho DH, Burggren WW. Epigenetics and transgenerational transfer: a physiological perspective. *J Exp Biol* 2010;**213**:3−16.

68. Bonduriansky R. Rethinking heredity, again. *Trends Ecol Evol* 2012;**27**:330−6.

69. Jablonka E. Epigenetic variations in heredity and evolution. *Clin Pharmacol Ther* 2012;**92**:683−8.

70. Pfeifer GP. Mutagenesis at methylated CpG sequences. *Curr Top Microbiol Immunol* 2006;**301**:259−81.

71. Varley KE, Gertz J, Bowling KM, Parker SL, Reddy TE, Pauli-Behn F, et al. Dynamic DNA methylation across diverse human cell lines and tissues. *Genome Res* 2013;**23**:555−67.

72. Lister R, Pelizzola M, Dowen RH, Hawkins RD, Hon G, Tonti-Filippini J, et al. Human DNA methylomes at base resolution show widespread epigenomic differences. *Nature* 2009;**462**:315−22.

73. Ichiyanagi T, Ichiyanagi K, Miyake M, Sasaki H. Accumulation and loss of asymmetric non-CpG methylation during male germ-cell development. *Nucleic Acids Res* 2013;**41**:738−45.

74. Kriaucionis S, Heintz N. The nuclear DNA base 5-hydroxymethylcytosine is present in Purkinje neurons and the brain. *Science* 2009;**324**:929–30.

75. Tahiliani M, Koh KP, Shen YH, Pastor WA, Bandukwala H, Brudno Y, et al. Conversion of 5-methylcytosine to 5-hydroxymethylcytosine in mammalian DNA by MLL partner TET1. *Science* 2009;**324**:930–5.

76. Hackett JA, Sengupta R, Zylicz JJ, Murakami K, Lee C, Down TA, Surani MA. Germline DNA demethylation dynamics and imprint erasure through 5-hydroxymethylcytosine. *Science* 2013;**339**:448–52.

77. Chen C-C, Wang K-Y, Shen C-KJ. The mammalian *de novo* DNA methyltransferases DNMT3A and DNMT3B are also DNA 5-hydroxymethylcytosine dehydroxymethylases. *J Biol Chem* 2012;**287**:33116–21.

78. Wang T, Pan Q, Lin L, et al. Genome-wide DNA hydroxymethylation changes are associated with neurodevelopmental genes in the developing human cerebellum. *Hum Mol Genet* 2012;**15**:5500–10.

79. Chen M, Licon K, Otsuka R, Pillus L, Ideker T. Decoupling epigenetic and genetic effects through systematic analysis of gene position. *Cell Rep* 2013;**3**:128–37.

80. Pirrotta V. How to read the chromatin past. *Science* 2012;**337**:919–20.

81. The ENCODE Project Consortium. An integrated encyclopedia of DNA elements in the human genome. *Nature* 2012;**489**:57–74.

82. Kaikkonen MU, Lam MTY, Glass CK. Non-coding RNAs as regulators of gene expression and epigenetics. *Cardiovasc Res* 2011;**90**:430–40.

83. miRBase. Browse miRBase by species. <http://www.mirbase.org/cgi-bin/browse.pl?org=hsa>; [accessed 06.01.14].

84. Ng S-Y, Johnson R, Stanton LW. Human long non-coding RNAs promote pluripotency and neuronal differentiation by association with chromatin modifiers and transcription factors. *EMBO J* 2012;**31**:522–33.

85. Barry G, Mattick JS. The role of regulatory RNA in cognitive evolution. *Trends Cogn Sci* 2012;**16**:497–503.

86. Slotkin RK, Martienssen R. Transposable elements and the epigenetic regulation of the genome. *Nat Rev Genet* 2007;**8**:272–85.

87. Schlichting CD, Pigliucci M. *Phenotypic evolution: A reaction norm perspective.* Sunderland: Sinauer Associates; 1998.

88. Lillycrop KA, Phillips ES, Jackson AA, Hanson MA, Burdge GC. Dietary protein restriction of pregnant rats induces and folic acid supplementation prevents epigenetic modification of hepatic gene expression in the offspring. *J Nutr* 2005;**135**:1382–6.

89. Sohi G, Marchand K, Revesz A, Arany E, Hardy DB. Maternal protein restriction elevates cholesterol in adult rat offspring due to repressive changes in histone modifications at the *cholesterol 7α-hydroxylase* promoter. *Mol Endocrinol* 2011;**25**:785–98.

90. Gluckman PD, Lillycrop KA, Vickers MH, Pleasants AB, Phillips ES, Beedle AS, et al. Metabolic plasticity during mammalian development is directionally dependent on early nutritional status. *Proc Natl Acad Sci USA* 2007;**104**:12796–800.

91. Nijland MJ, Mitsuya K, Li C, Ford SP, McDonald TJ, Nathanielsz PW, Cox LA. Epigenetic modification of fetal baboon hepatic phosphoenolpyruvate carboxykinase following exposure to moderately reduced nutrient availability. *J Physiol* 2010;**588**:1349–59.

92. Choi J, Li C, McDonald TJ, Comuzzie A, Mattern V, Nathanielsz PW. Emergence of insulin resistance in juvenile baboon offspring of mothers exposed to moderate maternal nutrient reduction. *Am J Physiol Regul Integr Comp Physiol* 2011;**301**:R757–62.

93. Roseboom T, de Rooij S, Painter R. The Dutch famine and its long-term consequences for adult health. *Early Hum Devel* 2006;**82**:485–91.

94. Heijmans BT, Tobi EW, Stein AD, Putter H, Blauw GJ, Susser ES, et al. Persistent epigenetic differences associated with prenatal exposure to famine in humans. *Proc Natl Acad Sci USA* 2008;**105**:17046–9.

95. Tobi EW, Lumey LH, Talens RP, et al. DNA methylation differences after exposure to prenatal famine are common and timing- and sex-specific. *Hum Mol Genet* 2009;**18**:4046–53.

96. Veenendaal MV, Costello PM, Lillycrop KA, et al. Prenatal famine exposure, health in later life and promoter methylation of four candidate genes. *J Dev Orig Health Dis* 2012;**3**:450–7.

97. Godfrey KM, Sheppard A, Gluckman PD, Lillycrop KA, Burdge GC, McLean C, et al. Epigenetic gene promoter methylation at birth is associated with child's later adiposity. *Diabetes* 2011;**60**:1528–34.

98. Hoyo C, Murtha AP, Schildkraut JM, et al. Methylation variation at *IGF2* differentially methylated regions and maternal folic acid use before and during pregnancy. *Epigenetics* 2011;**6**:928–36.

99. Jiang X, Yan J, West AA, et al. Maternal choline intake alters the epigenetic state of fetal cortisol-regulating genes in humans. *FASEB J* 2012;**26**:3563–74.

100. Cooper WN, Khulan B, Owens S, et al. DNA methylation profiling at imprinted loci after periconceptional micronutrient supplementation in humans: results of a pilot randomized controlled trial. *FASEB J* 2012;**26**:1782–90.

101. Plagemann A, Harder T, Brunn M, et al. Hypothalamic proopiomelanocortin promoter methylation becomes altered by early overfeeding: an epigenetic model of obesity and the metabolic syndrome. *J Physiol* 2009;**587**:4963–76.

102. Plagemann A, Roepke K, Harder T, et al. Epigenetic malprogramming of the insulin receptor promoter due to developmental overfeeding. *J Perinat Med* 2010;**38**:393–400.

103. Aagaard-Tillery KM, Grove K, Bishop J, et al. Developmental origins of disease and determinants of chromatin structure: maternal diet modifies the primate fetal epigenome. *J Mol Endocrinol* 2008;**41**:91–102.

104. Suter M, Bocock P, Showalter L, et al. Epigenomics: maternal high-fat diet exposure *in utero* disrupts peripheral circadian gene expression in nonhuman primates. *FASEB J* 2011;**25**:714–26.

105. Suter MA, Chen A, Burdine MS, et al. A maternal high-fat diet modulates fetal SIRT1 histone and protein deacetylase activity in nonhuman primates. *FASEB J* 2012;**26**:5106–14.

106. Suter MA, Sangi-Haghpeykar H, Showalter L, et al. Maternal high-fat diet modulates the fetal thyroid axis and thyroid gene expression in a nonhuman primate model. *Mol Endocrinol* 2012;**26**:2071–80.

107. Gemma C, Sookoian S, Alvarinas J, García SI, Quintana L, Kanevsky D, et al. Maternal pregestational BMI is associated with methylation of the *PPARGC1A* promoter in newborns. *Obesity* 2009;**17**:1032–9.

108. Bouchard L, Hivert M-F, Guay S-P, St-Pierre J, Perron P, Brisson D. Placental adiponectin gene DNA methylation levels are associated with mothers' blood glucose concentration. *Diabetes* 2012;**61**:1272–80.

109. Bouchard L, Thibault S, Guay S-P, et al. Leptin gene epigenetic adaptation to impaired glucose metabolism during pregnancy. *Diabetes Care* 2010;**33**:2436–41.

110. El Hajj N, Pliushch G, Schneider E, et al. Metabolic programming of *MEST* DNA methylation by intrauterine exposure to gestational diabetes mellitus. *Diabetes* 2012;**62**:1320–8.

111. Soubry A, Schildkraut J, Murtha A, et al. Paternal obesity is associated with *IGF2* hypomethylation in newborns: results from a Newborn Epigenetics Study (NEST) cohort. *BMC Med* 2013;**11**:29.

112. Carone BR, Fauquier L, Habib N, et al. Paternally induced transgenerational environmental reprogramming of metabolic gene expression in mammals. *Cell* 2010;**143**:1084–96.

113. Ng S-F, Lin RCY, Laybutt DR, Barres R, Owens JA, Morris MJ. Chronic high-fat diet in fathers programs β-cell dysfunction in female rat offspring. *Nature* 2010;**467**:963–6.

114. Fullston T, Ohlsson Teague EMC, Palmer NO, DeBlasio MJ, Mitchell M, Corbett M, et al. Paternal obesity initiates metabolic disturbances in two generations of mice with incomplete penetrance to the F2 generation and alters the transcriptional profile of testis and sperm microRNA content. *FASEB J* 2013;**27**:4226–43.

115. Vassoler FM, White SL, Schmidt HD, Sadri-Vakili G, Pierce RC. Epigenetic inheritance of a cocaine-resistance phenotype. *Nat Neurosci* 2013;**16**:42–7.

116. Guerrero-Bosagna C, Settles M, Lucker B, Skinner MK. Epigenetic transgenerational actions of vinclozolin on promoter regions of the sperm epigenome. *PLoS One* 2010;**5**:e13100.

117. Ding G-L, Wang F-F, Shu J, Tian S, Jiang Y, Zhang D, et al. Transgenerational glucose intolerance with *Igf2/H19* epigenetic alterations in mouse islet induced by intrauterine hyperglycemia. *Diabetes* 2012;**61**:1133–42.

118. Cedar H, Bergman Y. Programming of DNA Methylation Patterns. *Annu Rev Biochem* 2012;**81**:97–117.

119. Borgel J, Guibert S, Li Y, et al. Targets and dynamics of promoter DNA methylation during early mouse development. *Nat Genet* 2010;**42**:1093–100.

120. Arpanahi A, Brinkworth M, Iles D, Krawetz SA, Paradowska A, Platts AE, et al. Endonuclease-sensitive regions of human spermatozoal chromatin are highly enriched in promoter and CTCF binding sequences. *Genome Res* 2009;**19**:1338–49.

121. Hammoud SS, Nix DA, Zhang H, Purwar J, Carrell DT, Cairns BR. Distinctive chromatin in human sperm packages genes for embryo development. *Nature* 2009;**460**:473–8.

122. Rassoulzadegan M, Grandjean V, Gounon P, Vincent S, Gillot I, Cuzin F. RNA-mediated non-Mendelian inheritance of an epigenetic change in the mouse. *Nature* 2006;**441**:469–74.

123. Krawetz SA, Kruger A, Lalancette C, et al. A survey of small RNAs in human sperm. *Hum Reprod* 2011;**26**:3401–12.

124. Baccarelli A, Wright R, Bollati V, et al. Ischemic heart disease and stroke in relation to blood DNA methylation. *Epidemiology* 2010;**21**:819–28.

125. Pearce MS, McConnell JC, Potter C, et al. Global LINE-1 DNA methylation is associated with blood glycaemic and lipid profiles. *Int J Epidemiol* 2012;**41**:210–7.

126. Zhao J, Goldberg J, Bremner JD, Vaccarino V. Global DNA methylation is associated with insulin resistance: a monozygotic twin study. *Diabetes* 2012;**61**:542–6.

127. Volkmar M, Dedeurwaerder S, Cunha DA, et al. DNA methylation profiling identifies epigenetic dysregulation in pancreatic islets from type 2 diabetic patients. *EMBO J* 2012;**31**:1405–26.

128. Ribel-Madsen R, Fraga MF, Jacobsen S, Bork-Jensen J, Lara E, Calvanese V, et al. Genome-wide analysis of DNA methylation differences in muscle and fat from monozygotic twins discordant for type 2 diabetes. *PLoS One* 2012;**7**:e51302.

129. Toperoff G, Aran D, Kark JD, Rosenberg M, Dubnikov T, Nissan B, et al. Genome-wide survey reveals predisposing diabetes type 2-related DNA methylation variations in human peripheral blood. *Hum Mol Genet* 2011;**21**:371–83.

130. Ling C, Del Guerra S, Lupi R, Rönn T, Granhall C, Luthman H, et al. Epigenetic regulation of *PPARGC1A* in human type 2 diabetic islets and effect on insulin secretion. *Diabetologia* 2008;**51**:615–22.

131. Kuroda A, Rauch TA, Todorov I, Ku HT, Al-Abdullah IH, Kandeel F, et al. Insulin gene expression is regulated by DNA methylation. *PLoS One* 2009;**4**:e6953.

132. Yang B, Dayeh T, Kirkpatrick C, et al. Insulin promoter DNA methylation correlates negatively with insulin gene expression and positively with HbA(1c) levels in human pancreatic islets. *Diabetologia* 2011;**54**:360–7.

133. Yang BT, Dayeh TA, Volkov PA, et al. Increased DNA methylation and decreased expression of *PDX-1* in pancreatic islets from patients with type 2 diabetes. *Mol Endocrinol* 2012;**26**:1203–12.

134. Guay C, Jacovetti C, Nesca V, Motterle A, Tugay K, Regazzi R. Emerging roles of non-coding RNAs in pancreatic β-cell function and dysfunction. *Diabetes Obes Metab* 2012;**14**:12–21.

135. Kalis M, Bolmeson C, Esguerra JLS, Gupta S, Edlund A, Tormo-Badia N, et al. Beta-cell specific deletion of *Dicer1* leads to defective insulin secretion and diabetes mellitus. *PLoS One* 2011;**6**:e29166.

136. Melkman-Zehavi T, Oren R, Kredo-Russo S, et al. miRNAs control insulin content in pancreatic [beta]-cells via downregulation of transcriptional repressors. *EMBO J* 2011;**30**:835–45.

137. Zhao H, Guan J, Lee HM, et al. Up-regulated pancreatic tissue microRNA-375 associates with human type 2 diabetes through beta-cell deficit and islet amyloid deposition. *Pancreas* 2010;**39**:843–6.

138. El-Osta A, Brasacchio D, Yao D, et al. Transient high glucose causes persistent epigenetic changes and altered gene expression during subsequent normoglycemia. *J Exp Med* 2008;**205**:2409–17.

139. Okabe J, Orlowski C, Balcerczyk A, et al. Distinguishing hyperglycemic changes by Set7 in vascular endothelial cells. *Circ Res* 2012;**110**:1067–76.

140. Pirola L, Balcerczyk A, Tothill RW, et al. Genome-wide analysis distinguishes hyperglycemia regulated epigenetic signatures of primary vascular cells. *Genome Res* 2011;**21**:1601–15.

141. Wang X, Zhu H, Snieder H, et al. Obesity related methylation changes in DNA of peripheral blood leukocytes. *BMC Med* 2010;**8**:87.

142. Karalis KP, Giannogonas P, Kodela E, Koutmani Y, Zoumakis M, Teli T. Mechanisms of obesity and related pathology: linking immune responses to metabolic stress. *FEBS J* 2009;**276**:5747–54.

143. Kuehnen P, Mischke M, Wiegand S, Sers C, Horsthemke B, Lau S, et al. An Alu element–associated hypermethylation variant of the *POMC* gene is associated with childhood obesity. *PLoS Genet* 2012;**8**:e1002543.

144. Jacobsen SC, Brøns C, Bork-Jensen J, et al. Effects of short-term high-fat overfeeding on genome-wide DNA methylation in the skeletal muscle of healthy young men. *Diabetologia* 2012;**55**:3341–9.

145. Brøns C, Jacobsen S, Nilsson E, et al. Deoxyribonucleic acid methylation and gene expression of PPARGC1A in human muscle is influenced by high-fat overfeeding in a birth-weight-dependent manner. *J Clin Endocrinol Metab* 2010;**95**:3048–56.

146. Bouchard L, Rabasa-Lhoret R, Faraj M, et al. Differential epigenomic and transcriptomic responses in subcutaneous adipose tissue between low and high responders to caloric restriction. *Am J Clin Nutr* 2010;**91**:309–20.

147. Cordero P, Campion J, Milagro F, et al. Leptin and TNF-alpha promoter methylation levels measured by MSP could predict the response to a low-calorie diet. *J Physiol Biochem* 2011;**67**:463–70.

148. Milagro FI, Campión J, Cordero P, et al. A dual epigenomic approach for the search of obesity biomarkers: DNA methylation in relation to diet-induced weight loss. *FASEB J* 2011;**25**:1378–89.

149. Moleres A, Campión J, Milagro F, et al. Differential DNA methylation patterns between high and low responders to a weight loss intervention in overweight or obese adolescents: the EVASYON study. *FASEB J* 2013.

150. Rakyan VK, Down TA, Balding DJ, Beck S. Epigenome-wide association studies for common human diseases. *Nat Rev Genet* 2011;**12**:529–41.

151. Dedeurwaerder S, Desmedt C, Calonne E, et al. DNA methylation profiling reveals a predominant immune component in breast cancers. *EMBO Mol Med* 2011;**3**:726–41.

152. Mutch DM, Temanni MR, Henegar C, Combes F, Pelloux V, Holst C, et al. Adipose gene expression prior to weight loss can differentiate and weakly predict dietary responders. *PLoS One* 2007;**2**:e1344.

153. Vickers MH, Sloboda DM. Strategies for reversing the effects of metabolic disorders induced as a consequence of developmental programming. *Front Physiol* 2012;**3**:242.

Unraveling the Mechanism of Action of the GnRH Pulse Generator: A Possible Role for Kisspeptin/Neurokinin B/Dynorphin (KNDy) Neurons

Robert L. Goodman*, Lique M. Coolen[†] and Michael N. Lehman[†]

*West Virginia University, Morgantown, WV, USA,
[†]University of Mississippi Medical Center, Jackson, MS, USA

INTRODUCTION

Classic work on neuroendocrine control of gonadal function led to the concept of two types of gonadotropin secretion: (1) basal secretion that is necessary for steroid synthesis and release, which occurs in males and throughout most of the ovarian cycle of females; and (2) surge secretion that induces ovulation and only occurs in females late in the follicular phase. Basal secretion was often referred to as "tonic" secretion because it was thought to be released "more or less steadily"[1] in contrast to the cyclic release of the preovulatory gonadotropin surge.

The modern era in reproductive neuroendocrinology began with the advent of radioimmunoassays in the late 1960s, and the ability to monitor luteinizing hormone (LH) in frequently-collected blood samples soon led to the discovery that tonic LH secretion was not the steady, continuous type of release originally envisioned, but instead occurred episodically. The first report (in 1970) of episodic LH secretion was based on work in ovariectomized (OVX) monkeys,[2] and these observations were soon extended to include a large number of species and endocrine conditions.[3] One obvious inference from these data is that gonadotropin-releasing hormone (GnRH) release into the hypophysial portal circulation also occurs in an episodic fashion and this was later confirmed by GnRH measurements.[4–6] Subsequent work demonstrated that the episodic pattern of GnRH release was required for normal secretion of gonadotropins from the pituitary.[7] These observations reinforced the importance of pulsatile GnRH secretion and led to the development of long-acting GnRH

Cellular Endocrinology in Health and Disease.
DOI: http://dx.doi.org/10.1016/B978-0-12-408134-5.00009-3

analogs that are used extensively in clinical medicine today.

Given the central role of episodic GnRH release to reproductive function, it is not surprising that the underlying neural mechanisms are of considerable interest. Before reviewing attempts to identify these mechanisms, it may be useful to provide a detailed description of episodic GnRH secretion. The best available description comes from sheep, because the hypophysial portal circulation can be sampled at frequent intervals for extended periods of time in the absence of anesthesia in this species. In ewes, GnRH pulses occur as square waves, with very rapid onsets (within 1 min) from undectable to near maximal concentrations; GnRH release then remains elevated for an average of 5 min (range 2−8 min), before declining over about 3 min back to undetectable levels.[8] Based on this pattern, it is clear that the secretory activity of the GnRH neurons responsible for tonic release must be highly synchronized, but the mechanisms responsible for this synchronization are still under active investigation. In this review, we will: (1) briefly review earlier work on the possible neural mechanisms responsible for GnRH pulses; (2) describe the recent development of a new model for the GnRH pulse generator; (3) discuss experiments that have tested this model and propose modifications of it based on these data; and (4) close with a consideration of unresolved questions.

EARLY WORK ON THE GnRH PULSE GENERATOR

Anatomical Studies

The first approach to identifying the neural substrate necessary for episodic GnRH release was to use classic deafferentation techniques, and these pointed to the medial basal hypothalamus (MBH). Thus knife cuts that partially, or

completely, disrupt input to the MBH have little or no effect on tonic gonadotropin secretion in intact or OVX rats,[9−11] sheep,[12] monkeys,[13] or guinea pigs.[14] In some cases, complete cuts disrupted LH secretion in rats, but this may have been due to damage of tissue within the island because it was not always seen with identical cuts.[9] The conclusion that the MBH could produce normal episodic GnRH release is consistent with the location of GnRH cell bodies in primates,[15] sheep,[16] and guinea pigs,[17] but is apparently at odds with the general consensus that no GnRH cell bodies are found in the MBH of rats.[18] On the other hand, a few GnRH cell bodies have been observed in the MBH of rats[19,20] and these may be sufficient for normal GnRH secretion (see below). Alternatively, a few GnRH fibers in a subchiasmatic projection from the preoptic area to the median eminence that exists in the rat may have been spared by the knife cuts.[21]

There is less information available on the areas within the MBH necessary for tonic GnRH secretion, although most data indicate the arcuate nucleus (ARC) is a key structure. Lesion studies point to the anterior portion of the ARC as critical for episodic LH secretion in rats,[10,11] and the posterior portion of this nucleus in monkeys.[22] There is one report in ewes that knife cuts through the anterior ARC disrupted pulsatile LH secretion,[12] but this may have reflected damage posterior to the cut, because more rostral cuts that spared the ARC produced a similar suppression in tonic LH concentrations. Moreover, like many of these early knife-cut studies, this report was focused primarily on the effects of deafferentation on the ability of estradiol to induce an LH surge.

Electrical Activity

Perhaps one of the most remarkable observations in reproductive neuroendocrinology is the high correlation of bursts in multiunit electrical activity (MUA) with episodic LH secretion. This correlation was first reported in ewes[23] although the amplitude of changes was not robust and not every increase in MUA was associated with an LH pulse in that study. Subsequent work found essentially a one-to-one correlation between bursts of MUA and LH pulses in monkeys,[24] rats,[25] and goats.[26] In OVX animals, MUA (spikes/s) activity increased approximately 2–3 fold, and lasted from 1–5 min in rats, 3–4 min in goats, and 8–16 min in monkeys. Recording electrodes tended to be concentrated in the ARC or median eminence,[25,27] but positive signals were sometimes recorded some distance from these areas.[27] This phenomenon has been extensively studied in monkeys[28] and goats,[29] but it has been difficult to identify the neural source of these signals because of the large size of the recording electrode and the presence of inactive sites adjacent to sites where positive signals were recorded.[25,27]

Because GnRH neurons are scattered over a large volume of tissue, heroic efforts were required to monitor activity from individual cells until the creation of GnRH-green fluorescent protein (GFP) transgenic mice. The ability to identify individual GnRH neurons in slice preparations has led to a wealth of information on the electrical activity and membrane channels in these neurons.[18,30] However, single cell recordings "have yet to shed significant light on the issue of inherent pulsatility"[18] and we are unaware of any reports of synchronous firing of two individual GnRH neurons in situ. These negative results most likely reflect three factors: (1) it is technically difficult to record from a single neuron for the prolonged period (e.g., 40–60 min) necessary to identify activity that correlates with pulsatile GnRH release; (2) GnRH neurons are heterogeneous and exhibit a variety of different firing patterns; and (3) only a small percentage of them may be involved in episodic GnRH release; it has been estimated that as few as 70–80 GnRH

neurons are sufficient for normal tonic LH secretion in mice.[31] Thus one has to look to other approaches to understand the neural basis of the GnRH pulse generator.

Pulsatility from GnRH Neurons

The first major breakthrough on possible mechanisms responsible for pulse generation came from work on immortalized GnRH cells created by coupling the GnRH promoter with an oncogene.[33] The resulting GT1 and Gn cell lines can release GnRH in an episodic pattern,[34,35] although this capability develops over time in culture,[36] and is not seen in all cultures.[37] It was therefore concluded that episodic secretion is an inherent characteristic of GnRH neurons, and that the GnRH pulse generator may not need other neural input in vivo. Further work demonstrated that these cells produce high-frequency (every few seconds) bursts of action potentials and oscillations of intracellular calcium concentrations.[38] In cultures, much slower frequency (similar to that of endogenous GnRH pulse frequency) bursts in action potentials have been recorded from individual GT1 cells and these bursts are often synchronous across several neurons.[39] It has thus been suggested that a high-frequency generator endogenous to individual GnRH cells serves as a fundamental unit of activity that is converted into the low-frequency oscillation needed for GnRH pulses by the network properties of the whole population.[37] However, the mechanisms by which the network produces this transformation remain largely unknown.

While work with GT1 cell lines have provided considerable insight into the membrane channels, intracellular signaling mechanisms, and possible sources of GnRH pulses, extrapolation of these data to the normal animal remains problematic for several reasons. First, these are transformed cells that originated from immature GnRH neurons. They are thus

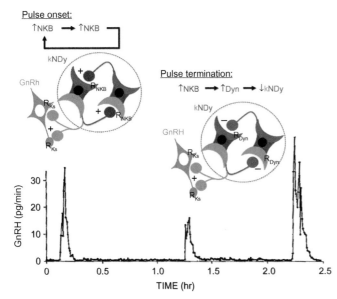

Pulse onset:

Pulse termination:

FIGURE 9.1 **Model for actions of KNDy peptides in initiation and termination of each GnRH pulse.** Each GnRH pulse is initiated by NKB (purple) acting within the KNDy network (within dashed oval), which stimulates kisspeptin (green) release to drive GnRH (blue) secretion. GnRH release is then terminated by dynorphin (red) release from KNDy neurons acting directly on KNDy neurons. Note that the color in each terminal indicates the biologically active transmitter (possibly due to selective expression of postsynaptic receptors) not selective transport of that peptide to the terminal. R_{dyn}, κ-opioid receptor; R_{Ks}, Kiss1r; R_{NKB}, NK3R. *Redrawn from Lehman* et al.[32]

likely to differ markedly from adult GnRH neurons,[40] and indeed GT1 cells have many characteristics (e.g., produce nitric oxide[41] and kisspeptin[42] and respond to β-adrenergic agonists[43]) not seen in normal GnRH neurons. There is also a remarkable quantitative difference: as noted above, GnRH pulses can be produced in normal animals by less than 100 GnRH neurons, but a typical GT1 preparation that secretes pulsatile GnRH requires approximately 10^6 cells.

A more physiological preparation that addresses some of the limitations of GT1 and Gn cells is the culture of embryonic GnRH neurons from olfactory placodes, which, after several days in culture, are capable of producing episodic GnRH patterns.[44] This approach has the advantage that the GnRH neurons have not been transformed and that relative low cell numbers are involved, although they are immature, not adult, cells. Moreover, much of this work has been done using primate cells,[45] which presumably closely resemble human GnRH neurons. Interestingly, individual cells showed relatively high-frequency (every 8 min) oscillations in calcium concentrations, but

approximately once an hour these oscillations were synchronized among most cells.[46] These data are consistent with the critical role for a GnRH cell network in pulse generation, but the mechanism of this synchronization is unknown. It should also be noted that these cultures do contain non-GnRH neurons[47] and some of these appear to participate in synchronization of these calcium oscillations.[48]

Although these *in vitro* studies clearly demonstrate that GnRH neurons can function episodically either alone (GT1 cells) or with some other elements (embryonic olfactory placodes), it has never been clear how activity of the widely-dispersed adult GnRH population *in situ* could be synchronized to produce episodic GnRH secretion. Moreover, the discovery that mutations, first in kisspeptin receptors[49,50] and subsequently in neurokinin B (NKB) signaling,[51] completely block tonic GnRH release in humans indicates that other neural elements are normally required for episodic GnRH secretion. Interestingly, the identification of the roles of kisspeptin and NKB to human fertility also played an important part in the development of a new model for the GnRH pulse generator (Figure 9.1).

A NEW MODEL FOR THE NEURAL ELEMENTS OF THE GnRH PULSE GENERATOR

Development of the Model

Retrospectively, it is now clear that the first key observation that led to the current model was the discovery that kisspeptin is critical for normal GnRH secretion in humans and mice.[49,50] This triggered an explosion of studies on the role of kisspeptin in control of GnRH secretion,[52–54] many of which supported the hypothesis that the negative feedback actions of estradiol were mediated by inhibition of the activity of kisspeptin neurons in the ARC. One implication of this hypothesis is that kisspeptin plays an important role in driving episodic GnRH secretion, and this is supported by the ability of kisspeptin receptor (Kiss1r) antagonists to inhibit GnRH neural activity in mice and block episodic LH secretion in OVX ewes.[55]

While much of the attention of neuroendocrinologists was focused on kisspeptin, several other important pieces of this puzzle were being assembled from studies focused on a set of neurons in the ARC that contain NKB and dynorphin. Simultaneous work in sheep[56] and rats[57] observed that all NKB neurons in this region contain dynorphin and that many (in rats) or all (in sheep) dynorphin neurons contain NKB. Both groups also reported evidence for extensive reciprocal innervation within this population (Figure 9.2A–B); for example, over 90% of NKB/dynorphin-immunoreactive (ir) cell bodies in sheep receive close contacts that also contain these two peptides.[56] Similar reciprocal innervation has since been reported in goats,[61] and tract tracing studies have demonstrated that it includes bilateral connections between ARC nuclei on each side of the third ventricle.[61,62] These connections probably reflect fibers that pass under the ventricle in the internal zone of the median eminence that have been observed in a number of species

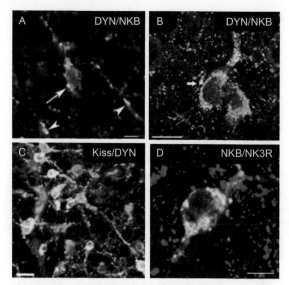

FIGURE 9.2 **Confocal images of ARC KNDy neurons illustrating characteristics that provide the anatomical foundation for synchronous firing.** Panels A and B: Reciprocal connections (arrows) among this population in sheep and rats, respectively. Both images are composites of two individual channels that identified dynorphin-immunoreactive (ir) (red in sheep/green in rats) and NKB-ir (green in sheep/red in rats) cells. Panel C is a similar composite showing virtually complete co-localization of kisspeptin-ir (red) and dynorphin-ir (green) in ewes. The composite in Panel D illustrates presence of NK3R-ir (red) in a NKB-containing (green) ARC neuron in sheep. Reprinted from: Panel A, Foradori et al.;[56] Panel B, Lehman et al.;[58] Panel C, Goodman et al.;[59] Panel D, Amstalden et al.[60] Magnification bars: 10 μm in Panels A, B, and D; 20 μm in Panel C.

including primates.[63] Subsequent work indicated that many of these NKB-ir neurons also contained NK3R, the receptor for NKB, in rats[64] and sheep (Figure 9.2D).[60] Thus, this subpopulation of neurons had important anatomical attributes necessary to function as an autoregulatory neural network.

The description of these NKB/dynorphin-ir neurons attracted little attention at the time because important neuroendocrine roles for these two peptides had not been described in many species (although there was strong evidence that dynorphin mediated progesterone negative feedback in ewes[65]). However, the final two pieces of the puzzle soon rectified this lack of interest. The first was the authors' report that these neurons also contain kisspeptin, and *vice versa*, so there is a single population of neurons in the ovine ARC containing all three peptides (Figure 9.2C).[59] These cells, which have become known as KNDy (for kisspeptin, NKB, and dynorphin) neurons[66] have since been described in rats,[57,67] mice,[68] goats,[69] and women.[70] The final major piece of the puzzle was the report that loss-of-function mutations in the genes for NKB or its receptor caused infertility in humans.[51] Thus both NKB and kisspeptin are necessary for normal secretion of GnRH in humans, although the loss of NKB signaling apparently can be overcome in some individuals.[71]

These data, together with work in goats indicating that the bursts of MUA that have long been considered the electrophysiological manifestation of the pulse generator were recorded from the vicinity of KNDy neurons,[72] led four groups, working largely independently of each other, to propose in late 2009/ early 2010 that KNDy neurons represented the long-sought GnRH pulse generator.[32,68,73,74]

Role of KNDy Peptides

All four groups proposed the same roles for kisspeptin, NKB, and dynorphin in their models (Figure 9.1). First, because GnRH neurons contain Kiss1r in several species,[75–77] while KNDy neurons apparently do not, at least in sheep,[78] kisspeptin was proposed to be the output from KNDy neurons that drove GnRH pulses. This hypothesis was also consistent with earlier data that GnRH neurons do not contain the κ-opioid receptor that mediates dynorphin action,[79] and with more recent reports that few,[64] or no,[60] GnRH neurons contain NK3R. Evidence for episodic kisspeptin release in monkeys[80,81] and that Kiss1r antagonists block pulsatile LH secretion[55] also

pointed to this role for kisspeptin. Moreover, exogenous kisspeptin failed to alter MUA in the ARC of goats[72] and rats,[82] so it was thought that kisspeptin had no effects on the pulse generator. Second, based on the anatomical characteristics of KNDy neurons described above, it was hypothesized that a small increase in NKB would act on these neurons to initiate a positive feedback loop within their network at the start of a GnRH pulse. A positive feedback loop is ideal for producing a dramatic increase in kisspeptin, and thus GnRH secretion, that occurs at the beginning of each pulse (Figure 9.1). The final component of the model was that dynorphin release would be stimulated from these same neurons by NKB and, after a short delay, it would inhibit activity of the KNDy network and thus terminate GnRH release at the end of a pulse. There was little direct support for the proposed role of dynorphin, but earlier work demonstrated that iv administration of naloxone, an antagonist for κ- and μ-opioid receptors, increased the amplitude and prolonged GnRH pulses in ewes.[83] This proposal generated considerable interest, and a fair amount of controversy, and has led to a variety of studies to test its viability. We will next consider evidence from sheep and goats testing this model, and then compare these data to similar studies in rodents.

TEST OF THE MODEL: DATA FROM SHEEP AND GOATS

In general, two types of experimental approaches have been used to test the working model in these animals: (1) Wakabayashi and colleagues have examined the effects of intracerebroventricular (icv) administration of receptor agonists and antagonists on MUA in the ARC of OVX goats; and (2) the authors have monitored episodic LH secretion before and during the local administration of similar drugs directly to the ARC of OVX ewes.

Does NKB Initiate each GnRH Pulse?

Three studies have examined the effects of NKB in the ARC of sheep and goats. The first of these reported that icv administration of NKB increased the frequency of LH pulses and MUA in the ARC of goats, while simultaneously inhibiting mean LH concentrations.[69] That the increase in MUA reflected stimulation of KNDy neurons is supported by the recent report that icv infusion of NKB in ovary-intact anestrous ewes increased Fos expression in KNDy neurons.[84] Using local microimplants of NKB in the ARC, we have confirmed that this tachykinin can act in this region to increase the frequency of LH pulses in OVX ewes.[85] Interestingly, unlike the effects of icv NKB in goats, NKB had no inhibitory effects on LH secretion in this study. Thus it appears that NKB acts in the ARC to stimulate the GnRH pulse generator, but elsewhere in the hypothalamus to inhibit GnRH secretion.

While NK3R agonists demonstrate possible effects of NKB, they provide no information on the role of endogenous NKB. Therefore, the authors repeated the previous experiment with local NKB treatment except they used microimplants containing SB222200 and observed that this NK3R antagonist disrupted episodic LH secretion when placed in the ARC of OVX ewes (Figure 9.3).[85] The duration of this effect was variable, ranging from 84 min to at least 240 min, but was evident in all ewes. These results thus demonstrate that the actions of endogenous NKB within the ARC are essential for normal episodic LH secretion in ewes, and are consistent with the proposed role for NKB in pulse generation.

Does Dynorphin Terminate each Pulse?

Two experiments using the specific κ-opioid receptor antagonist, nor-binatorphimine (BNI), provide strong evidence for inhibitory effects of dynorphin on the pulse generator. Specifically, icv administration of BNI increased bursts of MUA in goats[69] and the same antagonist

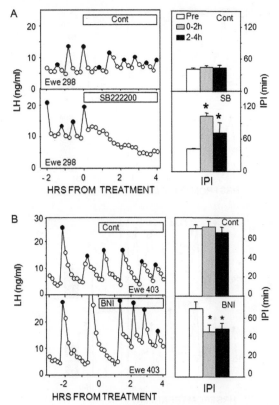

FIGURE 9.3 **Effects of local administration of receptor antagonists to NK3R (Panel A) and to the κ-opioid receptor (Panel B) into the ARC of OVX ewes.** LH pulse patterns in the same animal receiving control (empty) or antagonist-filled microimplants (bars) are illustrated on the left. Solid circle depict peaks of identified LH pulses. Bars on the right present mean (± SEM) interpulse intervals (IPI) for the period before insertion of microimplants (open bars) and 0–2 h (shaded bars) and 2–4 h (black bars) after insertion of microimplants. *$p < 0.05$ vs. pretreatment values. *Redrawn from Goodman* et al.[85]

significantly increased LH pulse frequency when given into the ARC of OVX ewes (Figure 9.3).[85] Thus endogenous dynorphin acts in the ARC to hold MUA and GnRH pulse frequency in check, but whether this reflects a direct action of dynorphin on KNDy neurons is unclear because there is no data available on cellular expression of κ-opioid receptors in sheep or goats at this time. The effects of naloxone on

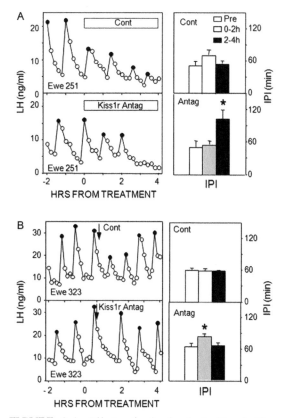

FIGURE 9.4 **Effects of microimplants (Panel A) or microinjections (Panel B) of a Kiss1r antagonist into the ARC of OVX ewes.** Left: LH pulse patterns in the same animal receiving control or antagonist treatments. Solid circle depict peaks of identified LH pulses. Bars on the right present mean (± SEM) interpulse intervals (IPI) for the period before insertion of microimplants or microinjections (open bars) and 0–2 h (shaded bars) and 2–4 h (black bars) after insertion of microimplants or microinjections. *p < 0.05 vs. pretreatment values. *Redrawn from Goodman et al.*[85]

Does Kisspeptin Affect the Pulse Generator or Simply Drive GnRH Release?

There is only limited data available to address this question in sheep and goats. As noted above, the ability of a Kiss1r antagonist to inhibit episodic LH secretion in OVX sheep[55] and the expression of Kiss1r in ovine GnRH neurons[77] are consistent with the proposed role of kisspeptin in driving GnRH secretion. Moreover, ovine KNDy neurons do not contain Kiss1r[78] and kisspeptin administration to OVX goats increased LH secretion but had no effect on the pattern of MUA in the ARC.[72]

Because of these data we were surprised to recently observe that local administration of the Kiss1r antagonist, peptide 271, to the ARC inhibited LH pulse frequency.[85] This inhibition was evident with both microimplants of this antagonist and bilateral microinjection of 2 nmole of antagonist (Figure 9.4). This decrease in LH pulse frequency is unlikely to reflect diffusion of the antagonist to nearby GnRH nerve terminals because no specific inhibition of LH pulse amplitude was observed in these experiments or other tests of lower doses of peptide 271.[85] It thus appears that endogenous kisspeptin does play a role in the ARC in stimulating the pulse generator in sheep, in addition to its accepted role as the primary output of KNDy neurons.

GnRH pulse patterns in ewes[83] suggest that inhibition by endogenous dynorphin is evident shortly after pulse onset and thus indicate that there is a slight delay between the actions of NKB and dynorphin on KNDy neurons. However, the mechanism responsible for this delay, and whether it occurs at a presynaptic and/or postsynaptic level, is another important unresolved question.

TEST OF THE MODEL: DATA FROM RODENTS

Work in rats and mice have used a variety of approaches to test the interactions of NKB, dynorphin, and kisspeptin in the control of LH secretion, including genetic manipulations in mice, monitoring of MUA in rats, and assessment of the effects of receptor agonists and antagonists in both species.

Does NKB Initiate each GnRH Pulse?

There is now considerable evidence that NKB most likely stimulates GnRH secretion via kisspeptin release from KNDy neurons, rather than directly, in mice and rats, although most of these studies have used the NK3R agonist, senktide, rather than the endogenous transmitter. However, this story is more complicated because this agonist can have stimulatory and inhibitory actions in rodents; in general, senktide is stimulatory when endogenous LH secretion is low and inhibitory when LH concentrations are elevated.[86–89] The stimulatory effects are clearly dependent on kisspeptin because they are not evident in Kiss1r knockout mice[90] and can be blocked by a Kiss1r antagonist in rats.[88] Moreover, this probably reflects kisspeptin release from KNDy neurons because senktide significantly increases Fos expression in this population *in vivo*[87] and NKB stimulates electrical activity in them in slice preparations.[91–93]

Although these data provide strong evidence that activation of NK3R receptors in KNDy neurons induces kisspeptin release and thereby increases GnRH secretion, the relevance of these observations to normal episodic GnRH secretion is debatable. Specifically, at this time it is unclear whether endogenous NKB is critical to pulse generation, because NK3R antagonists have no effect on episodic LH secretion in OVX rats (Figure 9.5).[94,95] It should be noted that blockade of all three tachykinin receptors (NK1R, NK2R, and NK3R) did inhibit LH pulse in these animals.[95] Moreover, blockade of all three receptors was also required to prevent the stimulatory effect of NKB on electrical activity of KNDy neurons, and activation of NK1R or NK2R with substance P or neurokinin A, respectively, had effects similar to those of NKB on these neurons.[93] These data raise the possibilities that either NKB acts via multiple tachykinin receptors to stimulate GnRH secretion or that the other tachykinins can substitute for NKB in maintaining GnRH pulses. At this time there is insufficient data to choose between these possibilities, but this redundancy may explain the relative modest effects of NK3R knockouts in mice.[96]

Does Dynorphin Terminate each Pulse?

There is little experimental support for the hypothesis that dynorphin plays a critical role in termination of GnRH pulses in rodents, although activation of NK3R in the ARC, and probably in KNDy neurons, appears to inhibit GnRH via dynorphin.[94] As noted above, senktide administration to OVX rats and mice inhibits tonic LH secretion;[86,87] in rats this effect reflects an action in the ARC to inhibit LH pulse frequency (Figure 9.5) that correlates with inhibition of MUA in this region.[89] Moreover, this inhibitory action of senktide is independent of kisspeptin, but can be blocked by the κ-opioid receptor antagonist, BNI (Figure 9.5).[94] Although apparently few murine KNDy neurons contain κ-opioid receptors,[68,91,92] activation of these receptors with exogenous dynorphin inhibits the electrical activity of these neurons in slice preparations.[92,93] These data all support the proposed role for dynorphin in pulse termination, yet BNI has no effect on MUA or episodic LH secretion in OVX rats (Figure 9.5),[94] or single-cell electrical activity in mice.[92,93] Thus, it appears that all the machinery is in place for termination of GnRH pulses by dynorphin, but at least in OVX rats, and probably mice, there is little or no endogenous dynorphin release to produce this effect. Of course, these data do not rule out the possibility that redundant systems have developed for pulse termination in rodents so that blockade of dynorphin is insufficient by itself to effect episodic LH secretion.

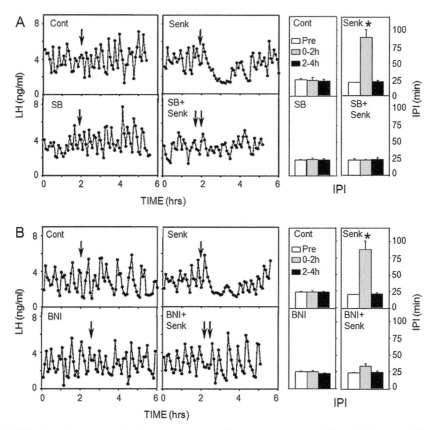

FIGURE 9.5   **Microinjections (arrows) of receptor antagonists to NK3R (Panel A) and the κ-opioid receptor (Panel B) into the ARC have no effect on LH pulses in OVX rats.** Individual LH pulse patterns in rats receiving control or antagonist treatments are illustrated in right panels. LH pulse patterns of positive-controls illustrating that the receptor antagonists act in the ARC to block the inhibitory effects of the NK3R agonist, senktide, are shown in middle panels (double arrows indicate antagonists given before senktide). Bars on the right depict mean (± SEM) interpulse intervals (IPI) before and 0–2 h and 2–4 h after microinjections as in Figures 9.3 and 9.4. *$p < 0.05$ vs. pretreatment values. *Redrawn from Grachev et al.*[94]

Does Kisspeptin Affect the Pulse Generator or Simply Drive GnRH Release?

In contrast to the previous two KNDy peptides, there appear to be close parallels between the roles of kisspeptin in rodents and ruminants. Thus, several lines of evidence indicate that kisspeptin is critical for GnRH pulses in rodents, including: (1) the effects of genetic deletion of Kiss1r;[97] (2) expression of Kiss1r in GnRH neurons;[75,76] (3) stimulation of electrical activity of GnRH neurons in slices;[76,98] and (4) inhibitory effects of Kiss1r antagonists on endogenous GnRH neural activity *in vitro* and LH secretion *in vivo*.[55] Moreover, administration of kisspeptin increased LH secretion in OVX rats without altering MUA[82] just like in goats, but administration of a Kiss1r antagonist to the ARC inhibited LH pulse frequency[99] just like in ewes. Thus, it appears that endogenous release of kisspeptin in the ARC is essential for function of the GnRH pulse generator in both rats and sheep. Interestingly, there is also evidence in humans that kisspeptin can reset the GnRH pulse generator[100] so this role for kisspeptin may be conserved across several species. Whether this reflects a direct action on KNDy neurons or an indirect effect via other neurons is unclear because the cellular distribution of Kiss1r in the rodent ARC has not yet been described. However, recent evidence that kisspeptin has no effect on the electrical activity of murine KNDy neurons[93] supports the latter.

CONCLUSIONS AND IMPORTANT UNRESOLVED ISSUES

Modifications in the Original Model

The experiments designed to test the role of KNDy neurons in generation of GnRH pulses in sheep and goats have provided strong support for the proposed roles for NKB and dynorphin. There is also good evidence that kisspeptin is the output signal from KNDy neurons, but it now appears likely that kisspeptin also has effects on the neural circuitry responsible for pulse generation in the ARC of sheep. We thus propose a modified model for the pulse generator in sheep and goats in which endogenous kisspeptin plays a role in stimulating the activity of the KNDy neural network. However, this action of kisspeptin must be indirect because, although Kiss1r-containing neurons are evident in the ovine ARC, KNDy neurons do not contain this receptor.[78] Consequently, we have added a kisspeptin-responsive interneuron to the model (Figure 9.6A) and propose that these neurons are stimulated by kisspeptin released from KNDy neurons and act in concert with NKB to stimulate release of kisspeptin onto GnRH neurons.

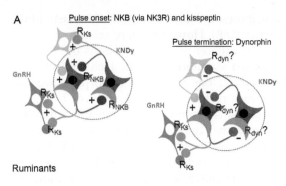

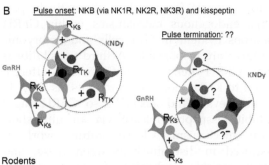

FIGURE 9.6 **Modified model for control of KNDy neural activity proposed to drive episodic GnRH secretion in ruminants (Panel A) and rodents (Panel B).** In ruminants, each GnRH pulse is initiated by NKB (purple) acting within the KNDy network (within dashed oval), which stimulates kisspeptin (green) release to drive GnRH (blue) secretion and activate unidentified Kiss1r-containing ARC neurons (grey) that reinforces the stimulatory actions of NKB on KNDy neurons. GnRH release is then terminated by dynorphin (red) release from KNDy neurons acting either directly on KNDy neurons and/or the unidentified Kiss1r-containing neurons. In rodents, NKB initiates each pulse, acting via multiple tachykinin receptors, and its actions are facilitated by kisspeptin release as in ruminants. The mechanisms responsible for pulse termination in rodents are unknown at this time. R_{dyn}, κ-opioid receptor; R_{Ks}, Kiss1r; R_{NKB}, NK3R; R_{TK}, tachykinin receptors (NK1R, NK2R, NK3R). Note that the color in each terminal indicates the biologically active transmitter (possibly due to selective expression of post-synaptic receptors) and does not reflect selective transport of that peptide to the terminal.

The model that has emerged from studies in rodents is similar to that in sheep and goats in two respects: (1) NKB likely initiates each GnRH pulse and (2) endogenous kisspeptin contributes to the effects of this tachykinin (Figure 9.6B). However, there are also two significant differences: (1) the actions of NKB are mediated via more than one tachykinin receptor and (2) the actions of dynorphin are not required for termination of a pulse. Thus, the mechanisms responsible for pulse termination in rodents remain completely unknown. Perhaps, as in the case of the tachykinins, there are multiple neurotransmitters/receptors in this circuitry that can act as redundant "stop" signals for a pulse. Alternatively perhaps endogenous mechanisms within GnRH neurons evident in GT1 cells and embryonic cultures are responsible. A third possibility is that GnRH neurons may provide afferent input to KNDy neurons in rodents, as they do in monkeys[101] and sheep.[85] If so, GnRH might inhibit kisspeptin release via an ultra-short loop mechanism. Clearly future work is required to distinguish between these, and other, possible mechanisms for termination of GnRH secretion at the end of each pulse in rodents.

Important Unresolved Issues

In the course of this review we have touched on several questions raised by the KNDy model for pulse generation that have not yet been answered. These include: (1) the site(s) where dynorphin acts to terminate each pulse in sheep and goats; (2) the mechanisms responsible for the lag of a few minutes between initiation of a pulse by NKB and its termination by dynorphin in sheep and goats; (3) the identity of kisspeptin-responsive neurons that contribute to pulse generation in all these species; and (4) the mechanisms that terminate GnRH secretion at the end of a pulse in rodents. In addition to these specific questions there are three more general questions that are worthy of some brief consideration.

Does Episodic Kisspeptin Release Drive GnRH Pulses?

One important underlying assumption of the KNDy model for pulse generation is that each GnRH pulse is driven by a brief increase in kisspeptin release from KNDy terminals on GnRH neurons. Yet the evidence for this assumption remains somewhat ambiguous. Specifically, it is unclear whether kisspeptin drives each pulse or is simply permissive for the function of a pulse generator inherent in the GnRH neural network. Kisspeptin is released in an episodic pattern in monkeys,[80,81] and data from monkeys,[102–104] rats,[105] and sheep[106] indicate that continuous elevation of kisspeptin is unable to maintain sustained elevations in LH secretion. In humans some studies have reported a decreased responsiveness to continuous kisspeptin stimulation,[107,108] but this was not always seen,[109] possibly because different kisspeptin preparations were used. Moreover, kisspeptin infusion into patients with low, or absent, episodic LH secretion due to defects in NKB signaling appeared to restore LH pulse patterns.[110] Although kisspeptin may have simply augmented the amplitude of undetectable LH pulses in this study, these results raise the possibility that kisspeptin may be permissive for GnRH pulses in humans. Thus, although most data support the assumption that kisspeptin pulses drive GnRH pulses, further work is clearly needed to rigorously test this aspect of the KNDy hypothesis.

Where Do KNDy Neurons Act to Stimulate GnRH Secretion?

Regardless of whether kisspeptin drives GnRH pulses or is permissive for episodic release, this peptide most likely acts directly on GnRH neurons to produce its effects. However, whether this action occurs at GnRH cell bodies or terminals remains an open question. There is strong evidence that KNDy neurons project to GnRH cell bodies in the

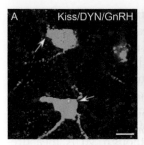
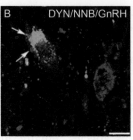

FIGURE 9.7   **Confocal images illustrating that KNDy neurons provide direct input to GnRH neurons in ewes.** Panel A: A fiber containing both kisspeptin (red) and dynorphin (blue) contacts (arrows) two GnRH neurons (green) in the POA. Panel B: MBH GnRH neuron (green) with two close contacts (arrows) containing both dynorphin (red) and NKB (blue); note nearby KNDy neuron (magenta) also contains both NKB and dynorphin. Magnification bar: 10 μm. *Panel A modified from Merkley et al.,[111] (cover illustration for issue); Panel B reprinted from Lehman et al.[58]*

preoptic area (POA), anterior hypothalamus, and MBH in sheep (Figure 9.7),[58] and to GnRH terminals in the median eminence in sheep[56,78] and goats.[112] It is, thus, likely that kisspeptin acts at both cell bodies and terminals in these species. Moreover, GnRH perikarya in the MBH have been implicated as the source of GnRH release during episodic LH secretion based on selective expression of Fos in this sub-population.[113] This hypothesis is supported by earlier deafferentation studies described above and by evidence that a greater percentage of this sub-population of GnRH cells receives inputs from KNDy neurons than other GnRH neurons.[65,114] However, it is also clear that kisspeptin can act in the ovine median eminence, presumably at GnRH terminals, to increase GnRH release.[78]

In rodents, there is similarly strong evidence for KNDy projections to the median eminence,[57,62,67] but the data supporting inputs to GnRH cell bodies is less clear. Specifically, only a very small percentage (2–5%) of kisspeptin-ir close contacts onto GnRH neurons[115] or nearby fibers[67] also contained NKB in mice[115] and rats.[67] It should also be noted

that most KNDy fibers in the median eminence are found in the internal zone where they contact GnRH fibers,[116] although a few are usually also found in the external zone.[57,62,67] Moreover, kisspeptin neurons do not take up the retrograde tracer, Fluorogold, when it is given intravenously, indicating that their axon terminals do not reach capillaries in the external zone of the median eminence.[64,116,117] In addition, the axo-axonic contacts observed between KNDy and GnRH fibers in the median eminence of ruminants and rodents are devoid of synaptic specializations, and Kiss1r protein has yet to be demonstrated in GnRH terminals in the median eminence in any species. Nevertheless, kisspeptin can stimulate GnRH from the median eminence of mice[118] and rats[116] indicating a potential site of action on GnRH fibers or terminals in this region.

Is the KNDy Model for Pulse Generation Applicable to Humans?

This issue is obviously difficult to address because of limitations in experimental work in humans. The critical roles of both kisspeptin and NKB to fertility in humans are consistent with this model, but the presence of this population in men has been questioned because of low numbers of dynorphin-immunoreactive cells in men.[119] In contrast, *in situ* hybridization for preprodynorphin mRNA detected over 100 cells capable of synthesizing dynorphin in normal adult women.[120] Moreover, because structural changes in these putative dynorphin neurons evident in postmenopausal women[120] closely paralleled those seen in NKB[121] and kisspeptin[122] neurons, these workers proposed that that KNDy neurons do exist in women.[73] Two factors, one methodological and the other biological, probably account for these apparent marked differences between men and women. First, low levels of dynorphin-ir may have occurred in men because of loss of antigen during the postmortem period. This possibility is consistent with the observation that only 30% of the NKB-ir neurons in men also contained kisspeptin;[119] in comparison, all NKB-positive neurons in the male monkey ARC were dual-labeled for kisspeptin.[63] Second, there is a marked sexual dimorphism in kisspeptin expression, with men having fewer kisspeptin-ir somata and fibers than women.[123] If this sexual dimorphism extends to dynorphin expression in the ARC, as it does in sheep,[66] one would expect low numbers of dynorphin-ir cells in men.

Thus, KNDy neurons most likely exist in women, but their presence in men is still under debate. There is even less evidence as to whether they have a role in episodic GnRH secretion, but the activity of KNDy neurons appears to increase after menopause when GnRH pulse generator activity would be maximal.[70] It is also interesting to note that NKB release from KNDy neurons has been proposed to induce hot flushes, based on the effects of senktide on temperature regulation in the POA[124] and the temporal correlation of LH pulses and hot flushes in postmenopausal women.[125] If this hypothesis proves to be correct it would provide strong circumstantial evidence that KNDy neurons are important for pulse generation in women.

More generally it is important to keep in mind that different mechanisms may have evolved for the production of GnRH pulses, so there may be significant species differences in the hypothalamic GnRH pulse generator. At first blush, one might expect something as important to reproductive function as episodic GnRH secretion to be driven by the same basic mechanisms across species. But this reasoning also applies to the preovulatory GnRH surge, which is just as critical to reproduction but is driven by markedly different neural systems in different species. Thus, care must be taken in extrapolating data from one species to another, unless that data is directly related to episodic

LH secretion. In this regard, a number of studies have directly addressed the role of KNDy neurons on episodic LH secretion in rats, sheep, and goats. On the other hand, although we now have a wealth of information on the control of KNDy neurons in mice, its relevance to episodic LH secretion remains somewhat speculative until direct study of LH pulses are done in this species. Finally, although MUA measurements and individual LH pulses are reliable surrogates for GnRH pulses, truly understanding the role of KNDy neurons in pulse generation may require determining the effects of disrupting signaling of individual KNDy peptides on GnRH pulse shape and secretion between pulses.

In conclusion, critical genetic studies in humans that identified the key role for kisspeptin and NKB in fertility have triggered a new wave of research in reproductive neuroendocrinology, some of which led to the KNDy model for GnRH pulse generator. Subsequent work has largely supported this model in sheep and goats, but indicated that its functioning, and particularly the mechanism of pulse termination, may differ in rodents.

Acknowledgements

We thank our many students and colleagues who are responsible for much of the work described in this chapter (see original citations) and whose intellectual input has been invaluable in the development of the concepts discussed. This work was supported by grants from NIH (R01-HD039916, RO1-HD017864).

References

1. Everett JW, Sawyer CH, Markee JE. A neurogenic timing factor in control of the ovulatory discharge of luteinizing hormone in the cyclic rat. *Endocrinology* 1949;**44**:234−50.
2. Dierschke DJ, Bhattacharya AN, Atkinson LE, Knobil E. Circhoral oscillations of plasma LH levels in the ovariectomized rhesus monkey. *Endocrinology* 1970;**87**:850−3.

3. Karsch FJ. Central actions of ovarian steroids in the feedback regulation of pulsatile secretion of luteinizing hormone. *Ann Rev Physiol* 1987;**49**:365–82.

4. Clarke IJ, Cummins JT. The temporal relationship between gonadotropin releasing hormone (GnRH) and luteinizing hormone (LH) secretion in ovariectomized ewes. *Endocrinology* 1982;**111**:1737–9.

5. Levine JE, Pau KY, Ramirez VD, Jackson GL. Simultaneous measurement of luteinizing hormone-releasing hormone and luteinizing hormone release in unanesthetized, ovariectomized sheep. *Endocrinology* 1982;**111**:1449–55.

6. Gearing M, Terasawa E. Luteinizing hormone releasing hormone (LHRH) neuroterminals mapped using the push-pull perfusion method in the rhesus monkey. *Brain Res Bull* 1988;**21**:117–21.

7. Belchetz PE, Plant TM, Nakai Y, Keogh EJ, Knobil E. Hypophysial responses to continuous and intermittent delivery of hypopthalamic gonadotropin-releasing hormone. *Science* 1978;**202**:631–3.

8. Moenter SM, Brand RM, Midgley AR, Karsch FJ. Dynamics of gonadotropin-releasing hormone release during a pulse. *Endocrinology* 1992;**130**:503–10.

9. Blake CA, Sawyer CH. Effects of hypothalamic deafferentation on the pulsatile rhythm in plasma concentrations of luteinizing hormone in ovariectomized rats. *Endocrinology* 1974;**94**:730–6.

10. Soper BD, Weick RF. Hypothalamic and extrahypothalamic mediation of pulsatile discharges of luteinizing hormone in the ovariectomized rat. *Endocrinology* 1980;**106**:348–55.

11. Ohkura S, Tsukamura H, Maeda K. Effects of various types of hypothalamic deafferentation on luteinizing hormone pulses in ovariectomized rats. *J Neuroendocrinol* 1991;**3**:503–8.

12. Jackson GL, Kuehl D, McDowell K, Zaleski A. Effect of hypothalamic deafferentation on secretion of luteinizing hormone in the ewe. *Biol Reprod* 1978;**18**:808–19.

13. Krey LC, Butler WR, Knobil E. Surgical disconnection of the medial basal hypothalamus and pituitary function in the rhesus monkey. I. Gonadotropin secretion. *Endocrinology* 1975;**96**:1073–87.

14. King JC, Ronsheim P, Liu E, Powers L, Slonimski M, Rubin BS. Fos expression in luteinizing hormone-releasing hormone neurons of guinea pigs, with knife cuts separating the preoptic area and the hypothalamus, demonstrating luteinizing hormone surges. *Biol Reprod* 1998;**58**:323–9.

15. Silverman AJ, Antunes JL, Abrams GM, et al. The luteinizing hormone-releasing hormone pathways in rhesus (Macaca mulatta) and pigtailed (Macaca nemestrina) monkeys: new observations on thick, unembedded sections. *J Comp Neurol* 1982;**211**:309–17.

16. Lehman MN, Robinson JE, Karsch FJ, Silverman AJ. Immunocytochemical localization of luteinizing hormone-releasing hormone (LHRH) pathways in the sheep brain during anestrus and the mid-luteal phase of the estrous cycle. *J Comp Neurol* 1986;**244**:19–35.

17. Krey LC, Silverman AJ. The luteinizing hormone-releasing hormone (LH-RH) neuronal networks of the guinea pig brain. II. The regulation on gonadotropin secretion and the origin of terminals in the median eminence. *Brain Res* 1978;**157**:247–55.

18. Herbison AE. Physiology of the gonadotropin-releasing hormone neuronal network. In: Neill JD, editor. *Knobil and Neill's physiology of reproduction*. 3rd ed. Amsterdam: Elsevier; 2006. p. 1415–82..

19. Ronnekleiv OK, Kelly MJ. Luteinizing hormone-releasing hormone neuronal system during the estrous cycle of the female rat: effects of surgically induced persistent estrus. *Neuroendocrinology* 1986;**43**:564–76.

20. Kimura F, Nishihara M, Hiruma H, Funabashi T. Naloxone increases the frequency of the electrical activity of luteinizing hormone-releasing hormone pulse generator in long-term ovariectomized rats. *Neuroendocrinology* 1991;**53**:97–102.

21. Hoffman GE, Gibbs FP. LHRH pathways in rat brain: 'deafferentation' spares a sub-chiasmatic LHRH projection to the median eminence. *Neuroscience* 1982;**7**:1979–93.

22. Plant TM, Krey LC, Moossy J, McCormack JT, Hess DL, Knobil E. The arcuate nucleus and the control of gonadotropin and prolactin secretion in the female rhesus monkey (Macaca mulatta). *Endocrinology* 1978;**102**:52–62.

23. Thiery JC, Pelletier J. Multiunit activity in the anterior median eminence and adjacent areas of the hypothalamus of the ewe in relation to LH secretion. *Neuroendocrinology* 1981;**32**:217–24.

24. Wilson RC, Kesner JS, Kaufman JM, Uemura T, Akema T, Knobil E. Central electrophysiologic correlates of pulsatile luteinizing hormone secretion in the rhesus monkey. *Neuroendocrinology* 1984;**39**:256–60.

25. Kawakami M, Uemura T, Hayashi R. Electrophysiological correlates of pulsatile gonadotropin release in rats. *Neuroendocrinology* 1982;**35**:63–7.

26. Mori Y, Nishihara M, Tanaka T, et al. Chronic recording of electrophysiological manifestation of the hypothalamic gonadotropin-releasing hormone pulse generator activity in the goat. *Neuroendocrinology* 1991;**53**:392–5.

27. Silverman AJ, Wilson R, Kesner JS, Knobil E. Hypothalamic localization of multiunit electrical activity associated with pulsatile LH release in the rhesus monkey. *Neuroendocrinology* 1986;**44**:168–71.

28. O'Byrne KT, Knobil E. Electrophysiological approaches to gonadotrophin releasing hormone pulse generator activity in the rhesus monkey. *Human reproduction* 1993;8(**Suppl 2**):37−40.

29. Okamura H, Murata K, Sakamoto K, et al. Male effect pheromone tickles the gonadotrophin-releasing hormone pulse generator. *Journal of neuroendocrinology* 2010;**22**:825−32.

30. Moenter SM. Identified GnRH neuron electrophysiology: a decade of study. *Brain Res* 2010;**1364**:10−24.

31. Herbison AE, Porteous R, Pape JR, Mora JM, Hurst PR. Gonadotropin-releasing hormone neuron requirements for puberty, ovulation, and fertility. *Endocrinology* 2008;**149**:597−604.

32. Lehman MN, Coolen LM, Goodman RL. Minireview: kisspeptin/neurokinin B/dynorphin (KNDy) cells of the arcuate nucleus: a central node in the control of gonadotropin-releasing hormone secretion. *Endocrinology* 2010;**151**:3479−89.

33. Mellon PL, Windle JJ, Goldsmith PC, Padula CA, Roberts JL, Weiner RI. Immortalization of hypothalamic GnRH neurons by genetically targeted tumorigenesis. *Neuron* 1990;**5**:1−10.

34. Wetsel WC, Valenca MM, Merchenthaler I, et al. Intrinsic pulsatile secretory activity of immortalized luteinizing hormone-releasing hormone-secreting neurons. *Proc Natl Acad Sci USA* 1992;**89**:4149−53.

35. Martinez de la Escalera G, Choi AL, Weiner RI. Generation and synchronization of gonadotropin-releasing hormone (GnRH) pulses: intrinsic properties of the GT1-1 GnRH neuronal cell line. *Proc Natl Acad Sci USA* 1992;**89**:1852−5.

36. Vazquez-Martinez R, Shorte SL, Boockfor FR, Frawley LS. Synchronized exocytotic bursts from gonadotropin-releasing hormone-expressing cells: dual control by intrinsic cellular pulsatility and gap junctional communication. *Endocrinology* 2001;**142**:2095−101.

37. Moenter SM, DeFazio AR, Pitts GR, Nunemaker CS. Mechanisms underlying episodic gonadotropin-releasing hormone secretion. *Front Neuroendocrin* 2003;**24**:79−93.

38. Charles AC, Hales TG. Mechanisms of spontaneous calcium oscillations and action potentials in immortalized hypothalamic (GT1-7) neurons. *J Neurophysiol* 1995;**73**:56−64.

39. Nunemaker CS, DeFazio RA, Geusz ME, Herzog ED, Pitts GR, Moenter SM. Long-term recordings of networks of immortalized GnRH neurons reveal episodic patterns of electrical activity. *J Neurophysiol* 2001;**86**:86−93.

40. Selmanoff M. Commentary on the use of immortalized neuroendocrine cell lines for physiological research. *Endocrine* 1997;**6**:1−3.

41. Mahachoklertwattana P, Black SM, Kaplan SL, Bristow JD, Grumbach MM. Nitric oxide synthesized by gonadotropin-releasing hormone neurons is a mediator of N-methyl-D-aspartate (NMDA)-induced GnRH secretion. *Endocrinology* 1994;**135**:1709−12.

42. Quaynor S, Hu L, Leung PK, Feng H, Mores N, Krsmanovic LZ, et al. Expression of a functional g protein-coupled receptor 54-kisspeptin autoregulatory system in hypothalamic gonadotropin-releasing hormone neurons. *Mol Endocrinol* 2007;**21**:3062−70.

43. Martinez de la Escalera G, Choi AL, Weiner RI. Beta 1-adrenergic regulation of the GT1 gonadotropin-releasing hormone (GnRH) neuronal cell lines: stimulation of GnRH release via receptors positively coupled to adenylate cyclase. *Endocrinology* 1992;**131**:1397−402.

44. Terasawa E, Keen KL, Mogi K, Claude P. Pulsatile release of luteinizing hormone-releasing hormone (LHRH) in cultured LHRH neurons derived from the embryonic olfactory placode of the rhesus monkey. *Endocrinology* 1999;**140**:1432−41.

45. Terasawa E. Cellular mechanism of pulsatile LHRH release. *Gen Comp Endocrinol* 1998;**112**:283−95.

46. Terasawa E, Schanhofer WK, Keen KL, Luchansky L. Intracellular Ca^{2+} oscillations in luteinizing hormone-releasing hormone neurons derived from the embryonic olfactory placode of the rhesus monkey. *J Neurosci* 1999;**19**:5898−909.

47. Richter TA, Keen KL, Terasawa E. Synchronization of Ca^{2+} oscillations among primate LHRH neurons and nonneuronal cells in vitro. *J Neurophysiol* 2002;**88**:1559−67.

48. Terasawa E, Richter TA, Keen KL. A role for non-neuronal cells in synchronization of intracellular calcium oscillations in primate LHRH neurons. *Prog Brain Res* 2002;**141**:283−91.

49. Seminara SB, Messager S, Chatzidaki EE, et al. The GPR54 gene as a regulator of puberty. *NEJM* 2003;**349**:1614−27.

50. de Roux N, Genin E, Carel JC, Matsuda F, Chaussain JL, Milgrom E. Hypogonadotropic hypogonadism due to loss of function of the KiSS1-derived peptide receptor GPR54. *Proc Natl Acad Sci USA* 2003;**100**:10972−6.

51. Topaloglu AK, Reimann F, Guclu M, Yalin AS, Kotan LD, Porte KM, et al. TAC3 and TACR3 mutations in familial hypogonadotropic hypogonadism reveal a key role for Neurokinin B in the central control of reproduction. *Nat Genet* 2009;**41**:354−8.

52. Oakley AE, Clifton DK, Steiner RA. Kisspeptin signaling in the brain. *Endocr Rev* 2009;**30**:713−43.

53. Garcia-Galiano D, Pinilla L, Tena-Sempere M. Sex steroids and the control of the Kiss1 system:

developmental roles and major regulatory actions. *J Neuroendocrinol* 2012;**24**:22–33.

54. Goodman RL, Lehman MN. Kisspeptin Neurons from Mice to Men: Similarities and Differences. *Endocrinology* 2012;**153**:5015–8.

55. Roseweir AK, Kauffman AS, Smith JT, Guerriero KA, Morgan K, Pielecka-Fortuna J, et al. Discovery of potent kisspeptin antagonists delineate physiological mechanisms of gonadotropin regulation. *J Neurosci* 2009;**29**:3920–9.

56. Foradori CD, Amstalden M, Goodman RL, Lehman MN. Colocalisation of dynorphin a and neurokinin B immunoreactivity in the arcuate nucleus and median eminence of the sheep. *J Neuroendocrinol* 2006;**18**:534–41.

57. Burke MC, Letts PA, Krajewski SJ, Rance NE. Coexpression of dynorphin and neurokinin B immunoreactivity in the rat hypothalamus: Morphologic evidence of interrelated function within the arcuate nucleus. *J Comp Neurol* 2006;**498**:712–26.

58. Lehman MN, Hileman SM, Goodman RL. Neuroanatomy of the kisspeptin signaling system in mammals: comparative and developmental aspects. *Adv Exp Med Biol* 2013;**784**:27–62.

59. Goodman RL, Lehman MN, Smith JT, Coolen LM, de Oliveira VR, Jafarzadehshirazi MR, et al. Kisspeptin neurons in the arcuate nucleus of the ewe express both dynorphin A and neurokinin B. *Endocrinology* 2007;**148**:5752–60.

60. Amstalden M, Coolen LM, Hemmerle AM, Billings HJ, Connors JM, Goodman RL. Neurokinin 3 receptor immunoreactivity in the septal region, preoptic area and hypothalamus of the female sheep: colocalisation in neurokinin B cells of the arcuate nucleus but not in gonadotrophin-releasing hormone neurones. *J Neuroendocrinol* 2010;**22**:1–12.

61. Wakabayashi Y, Yamamura T, Sakamoto K, Mori Y, Okamura H. Electrophysiological and morphological evidence for synchronized GnRH pulse generator activity among kisspeptin/neurokinin B/dynorphin A (KNDy) neurons in goats. *J Reprod Dev* 2013;**59**:40–8.

62. Krajewski SJ, Burke MC, Anderson MJ, McMullen NT, Rance NE. Forebrain projections of arcuate neurokinin B neurons demonstrated by anterograde tract-tracing and monosodium glutamate lesions in the rat. *Neuroscience* 2010;**166**:680–97.

63. Ramaswamy S, Seminara SB, Ali B, Ciofi P, Amin NA, Plant TM. Neurokinin B stimulates GnRH release in the male monkey (Macaca mulatta) and is colocalized with kisspeptin in the arcuate nucleus. *Endocrinology* 2010;**151**:4494–503.

64. Krajewski SJ, Anderson MJ, Iles-Shih L, Chen KJ, Urbanski HF, Rance NE. Morphologic evidence that neurokinin B modulates gonadotropin-releasing

hormone secretion via neurokinin 3 receptors in the rat median eminence. *J Comp Neurol* 2005;**489**:372–86.

65. Goodman RL, Coolen LM, Anderson GM, Hardy SL, Valen M, Connors J, et al. Evidence that dynorphin plays a major role in mediating progesterone negative feedback on gonadotropin-releasing hormone neurons in sheep. *Endocrinology* 2004;**145**:2959–67.

66. Cheng G, Coolen LM, Padmanabhan V, Goodman RL, Lehman MN. The kisspeptin/neurokinin B/dynorphin (KNDy) cell population of the arcuate nucleus: sex differences and effects of prenatal testosterone in sheep. *Endocrinology* 2010;**151**:301–11.

67. True C, Kirigiti M, Ciofi P, Grove KL, Smith MS. Characterisation of arcuate nucleus kisspeptin/neurokinin B neuronal projections and regulation during lactation in the rat. *J Neuroendocrinol* 2011;**23**:52–64.

68. Navarro VM, Gottsch ML, Chavkin C, Okamura H, Clifton DK, Steiner RA. Regulation of gonadotropin-releasing hormone secretion by kisspeptin/dynorphin/neurokinin B neurons in the arcuate nucleus of the mouse. *J Neurosci* 2009;**29**:11859–66.

69. Wakabayashi Y, Nakada T, Murata K, Ohkura S, Mogi K, Navarro VM, et al. Neurokinin B and dynorphin A in kisspeptin neurons of the arcuate nucleus participate in generation of periodic oscillation of neural activity driving pulsatile gonadotropin-releasing hormone secretion in the goat. *J Neurosci* 2010;**30**:3124–32.

70. Rance NE. Menopause and the human hypothalamus: evidence for the role of kisspeptin/neurokinin B neurons in the regulation of estrogen negative feedback. *Peptides* 2009;**30**:111–22.

71. Gianetti E, Tusset C, Noel SD, Au MG, Dwyer AA, Hughes VA, et al. TAC3/TACR3 mutations reveal preferential activation of gonadotropin-releasing hormone release by neurokinin B in neonatal life followed by reversal in adulthood. *J Clin Endocrinol Metab* 2010;**95**:2857–67.

72. Ohkura S, Takase K, Matsuyama S, Mogi K, Ichimaru T, Wakabayashi Y, et al. Gonadotrophin-releasing hormone pulse generator activity in the hypothalamus of the goat. *J Neuroendocrinol* 2009;**21**:813–21.

73. Rance NE, Krajewski SJ, Smith MA, Cholanian M, Dacks PA. Neurokinin B and the hypothalamic regulation of reproduction. *Brain Res* 2010;**1364**:116–28.

74. Maeda K, Ohkura S, Uenoyama Y, et al. Neurobiological mechanisms underlying GnRH pulse generation by the hypothalamus. *Brain Res* 2010;**1364**:103–15.

75. Irwig MS, Fraley GS, Smith JT, et al. Kisspeptin activation of Gonadotrophin-releasing hormone pulse generator activity in the hypothalamus of the goat. *J Neuroendocrinol* 2009;**21**:813–21.

76. Han SK, Gottsch ML, Lee KJ, et al. Activation of Gonadotrophin-releasing hormone pulse generator activity in the hypothalamus of the goat. *J Neuroendocrinol* 2009;**21**:813−21.

77. Smith JT, Li Q, Pereira A, Clarke IJ. Kisspeptin neurons in the ovine arcuate nucleus and preoptic area are involved in the preovulatory luteinizing hormone surge. *Endocrinology* 2009;**150**:5530−8.

78. Smith JT, Li Q, Yap KS, et al. Kisspeptin is essential for the full preovulatory LH surge and stimulates Gonadotrophin-releasing hormone pulse generator activity in the hypothalamus of the goat. *J Neuroendocrinol* 2009;**21**:813−21.

79. Sannella MI, Petersen SL. Dual label *in situ* hybridization studies provide evidence that luteinizing hormone-releasing hormone neurons do not synthesize messenger ribonucleic acid for mu, kappa, or delta opiate receptors. *Endocrinology* 1997;**138**:1667−72.

80. Keen KL, Wegner FH, Bloom SR, Ghatei MA, Terasawa E. An increase in kisspeptin-54 release occurs with the pubertal increase in luteinizing hormone-releasing hormone-1 release in the stalk-median eminence of female rhesus monkeys *in vivo*. *Endocrinology* 2008;**149**:4151−7.

81. Guerriero KA, Keen KL, Terasawa E. Developmental increase in kisspeptin-54 release in vivo is independent of the pubertal increase in estradiol in female rhesus monkeys (Macaca mulatta). *Endocrinology* 2012;**153**:1887−97.

82. Kinsey-Jones JS, Li XF, Luckman SM, O'Byrne KT. Effects of kisspeptin-10 on the electrophysiological manifestation of gonadotropin-releasing hormone pulse generator activity in the female rat. *Endocrinology* 2008;**149**:1004−8.

83. Goodman RL, Parfitt DB, Evans NP, Dahl GE, Karsch FJ. Endogenous opioid peptides control the amplitude and shape of gonadotropin-releasing hormone pulses in the ewe. *Endocrinology* 1995;**136**:2412−20.

84. Sakamoto K, Murata K, Wakabayashi Y, et al. Gonadotrophin-releasing hormone pulse generator activity in the hypothalamus of the goat. *J Neuroendocrinol* 2009;**21**:813−21.

85. Goodman RL, Hileman SM, Nestor CC, et al. Kisspeptin, neurokinin B, and dynorphin act in the arcuate nucleus to control activity of the GnRH pulse generator in ewes. *Endocrinology* 2013;**154**:4259−69.

86. Sandoval-Guzman T, Rance NE. Central injection of senktide, an NK3 receptor agonist, or neuropeptide Y inhibits LH secretion and induces different patterns of Fos expression in the rat hypothalamus. *Brain Res* 2004;**1026**:307−12.

87. Navarro VM, Castellano JM, McConkey SM, Pineda R, Ruiz-Pino R, Pinilla L, et al. Interactions between kisspeptin and neurokinin B in the control of GnRH secretion in the female rat. *Am J Physiol Endocrinol Metab* 2011;**300**:E202−10.

88. Grachev P, Li XF, Lin YS, Hu MH, Elsamani L, Paterson SJ, et al. GPR54-dependent stimulation of luteinizing hormone secretion by neurokinin B in prepubertal rats. *PloS One* 2012;**7**:e44344.

89. Kinsey-Jones JS, Grachev P, Li XF, Lin YS, Milligan SR, Lightman SL, et al. The inhibitory effects of neurokinin B on GnRH pulse generator frequency in the female rat. *Endocrinology* 2012;**153**:307−15.

90. Garcia-Galiano D, van Ingen Schenau D, Leon S, Krajnc-Franken MA, Manfredi-Lozano M, et al. Kisspeptin signaling is indispensable for neurokinin B, but not glutamate, stimulation of gonadotropin secretion in mice. *Endocrinology* 2012;**153**:316−28.

91. Navarro VM, Gottsch ML, Wu M, García-Galiano D, Hobbs SJ, Bosch MA, et al. Regulation of NKB pathways and their roles in the control of Kiss1 neurons in the arcuate nucleus of the male mouse. *Endocrinology* 2011;**152**:4265−75.

92. Ruka KA, Burger LL, Moenter SM. Regulation of Arcuate Neurons Coexpressing Kisspeptin, Neurokinin B, and Dynorphin by Modulators of Neurokinin 3 and kappa-Opioid Receptors in Adult Male Mice. *Endocrinology* 2013;**154**:2761−71.

93. de Croft S, Boehm U, Herbison AE. Neurokinin B activates arcuate kisspeptin neurons through multiple tachykinin receptors in the male mouse. *Endocrinology* 2013;**154**:2750−60.

94. Grachev P, Li XF, Kinsey-Jones JS, et al. Suppression of the Gonadotrophin-releasing hormone pulse generator activity in the hypothalamus of the goat. *J Neuroendocrinol* 2009;**21**:813−21.

95. Noritake K, Matsuoka T, Ohsawa T, et al. Involvement of Gonadotrophin-releasing hormone pulse generator activity in the hypothalamus of the goat. *J Neuroendocrinol* 2009;**21**:813−21.

96. Yang JJ, Caligioni CS, Chan YM, Seminara SB. Uncovering novel reproductive defects in neurokinin B receptor null mice: closing the gap between mice and men. *Endocrinology* 2012;**153**:1498−508.

97. Colledge WH. Transgenic mouse models to study Gpr54/kisspeptin physiology. *Peptides* 2009;**30**:34−41.

98. Pielecka-Fortuna J, Chu Z, Moenter SM. Kisspeptin acts directly and indirectly to increase gonadotropin-releasing hormone neuron activity and its effects are modulated by estradiol. *Endocrinology* 2008;**149**:1979−86.

99. Li XF, Kinsey-Jones JS, Cheng Y, Knox AMI, Lin Y, Petrou NA, et al. Kisspeptin signalling in the hypothalamic arcuate nucleus regulates GnRH pulse generator frequency in the rat. *PloS One* 2009;**4**:e8334.

100. Chan YM, Butler JP, Pinnell NE, et al. Kisspeptin resets the hypothalamic GnRH clock in men. *J Clin Endocrinol Metab* 2011;**96**:E908–15.

101. Ramaswamy S, Guerriero KA, Gibbs RB, Plant TM. Structural interactions between kisspeptin and GnRH neurons in the mediobasal hypothalamus of the male rhesus monkey (*Macaca mulatta*) as revealed by double immunofluorescence and confocal microscopy. *Endocrinology* 2008;**149**:4387–95.

102. Plant TM, Ramaswamy S, Dipietro MJ. Repetitive activation of hypothalamic G protein-coupled receptor 54 with intravenous pulses of kisspeptin in the juvenile monkey (*Macaca mulatta*) elicits a sustained train of gonadotropin-releasing hormone discharges. *Endocrinology* 2006;**147**:1007–13.

103. Ramaswamy S, Seminara SB, Pohl CR, DiPietro MJ, Crowley Jr. WF, Plant TM. Effect of continuous intravenous administration of human metastin 45-54 on the neuroendocrine activity of the hypothalamic-pituitary-testicular axis in the adult male rhesus monkey (*Macaca mulatta*). *Endocrinology* 2007;**148**:3364–70.

104. Seminara SB, Dipietro MJ, Ramaswamy S, Crowley Jr. WF, Plant TM. Continuous human metastin 45-54 infusion desensitizes G protein-coupled receptor 54-induced gonadotropin-releasing hormone release monitored indirectly in the juvenile male Rhesus monkey (*Macaca mulatta*): a finding with therapeutic implications. *Endocrinology* 2006;**147**:2122–6.

105. Roa J, Vigo E, Garcia-Galiano D, Castellano JM, Navarro VM, Pineda R, et al. Desensitization of gonadotropin responses to kisspeptin in the female rat: analyses of LH and FSH secretion at different developmental and metabolic states. *Am J Physiol Endocrinol Metab* 2008;**294**:E1088–96.

106. Caraty A, Smith JT, Lomet D, Ben Said S, Morrissey A, Cognie J, et al. Kisspeptin synchronizes preovulatory surges in cyclical ewes and causes ovulation in seasonally acyclic ewes. *Endocrinology* 2007;**148**:5258–67.

107. Jayasena CN, Nijher GM, Abbara A, et al. Twice-weekly administration of kisspeptin-54 for 8 weeks stimulates release of reproductive hormones in women with hypothalamic amenorrhea. J. *Clin Pharm Ther* 2010;**88**:840–7.

108. Jayasena CN, Nijher GM, Chaudhri OB, Murphy KG, Ranger A, Lim A, et al. Subcutaneous injection of kisspeptin-54 acutely stimulates gonadotropin secretion in women with hypothalamic amenorrhea, but chronic administration causes tachyphylaxis. *J Clin Endocrinol Metab* 2009;**94**:4315–23.

109. George JT, Veldhuis JD, Roseweir AK, Newton CL, Faccenda E, Millar RP, et al. Kisspeptin-10 is a potent stimulator of LH and increases pulse frequency in men. *J Clin Endocrinol Metab* 2011;**96**:E1228–36.

110. Young J, George JT, Tello JA, Fraunco B, Bouligand J, Guiochon-Mantel A, et al. Kisspeptin Restores Pulsatile LH Secretion in Patients with Neurokinin B Signaling Deficiencies: Physiological, Pathophysiological and Therapeutic Implications. *Neuroendocrinology* 2013;**97**:193–202.

111. Merkley CM, Porter KL, Coolen LM, Hileman SM, Billings HJ, Drews S, et al. KNDy (kisspeptin/neurokinin B/dynorphin) neurons are activated during both pulsatile and surge secretion of LH in the ewe. *Endocrinology* 2012;**153**:5406–15.

112. Matsuyama S, Ohkura S, Mogi K, Wakabayash Y, Mori Y, Tsukamura H, et al. Morphological evidence for direct interaction between kisspeptin and gonadotropin-releasing hormone neurons at the median eminence of the male goat: an immunoelectron microscopic study. *Neuroendocrinology* 2011;**94**:323–32.

113. Boukhliq R, Goodman RL, Berriman SJ, Adrian B, Lehman MN. A subset of gonadotropin-releasing hormone neurons in the ovine medial basal hypothalamus is activated during increased pulsatile luteinizing hormone secretion. *Endocrinology* 1999;**140**:5929–36.

114. Merkley CM, Coolen LM, Goodman RL, Lehman MN. Direct projections of arcuate KNDy (kisspeptin/neurokinin B/dynorphin) neurons to GnRH neruons in the sheep. *Program, Annual Meeting Society for Neuroscience* 2011: Abstr. 712.05.

115. Kallo I, Vida B, Deli L, Molnar CS, Hrabovszky E, Caraty A, et al. Co-localisation of kisspeptin with galanin or neurokinin B in afferents to mouse GnRH neurones. *J Neuroendocrinol* 2012;**24**:464–76.

116. Uenoyama Y, Inoue N, Pheng V, et al. Ultrastructural evidence of kisspeptin-gonadotrophin-releasing hormone (GnRH) interaction in the median eminence of female rats: implication of axo-axonal regulation of GnRH release. *J Neuroendocrinol* 2011;**23**:863–70.

117. Yeo SH, Herbison AE. Projections of arcuate nucleus and rostral periventricular kisspeptin neurons in the adult female mouse brain. *Endocrinology* 2011;**152**:2387–99.

118. d'Anglemont de Tassigny X, Fagg LA, Carlton MB, Colledge WH. Kisspeptin can stimulate gonadotropin-releasing hormone (GnRH) release by a direct action at GnRH nerve terminals. *Endocrinology* 2008;**149**:3926–32.

119. Hrabovszky E, Sipos MT, Molnar CS, Ciofi P, Borsay BÁ, Gergely P, et al. Low degree of overlap between kisspeptin, neurokinin B, and dynorphin immunoreactivities in the infundibular nucleus of

young male human subjects challenges the KNDy neuron concept. *Endocrinology* 2012;**153**:4978–89.

120. Rometo AM, Rance NE. Changes in prodynorphin gene expression and neuronal morphology in the hypothalamus of postmenopausal women. *J Neuroendocrinol* 2008;**20**:1376–81.

121. Rance NE, Young III WS. Hypertrophy and increased gene expression of neurons containing neurokinin-B and substance-P messenger ribonucleic acids in the hypothalami of postmenopausal women. *Endocrinology* 1991;**128**:2239–47.

122. Rometo AM, Krajewski SJ, Voytko ML, Rance NE. Hypertrophy and increased kisspeptin gene expression in the hypothalamic infundibular nucleus of postmenopausal women and ovariectomized monkeys. *J Clin Endocrinol Metab* 2007;**92**:2744–50.

123. Hrabovszky E, Ciofi P, Vida B, Horvath MC, Keller E, Caraty A, et al. The kisspeptin system of the human hypothalamus: sexual dimorphism and relationship with gonadotropin-releasing hormone and neurokinin B neurons. *Eur J Neurosci* 2010;**31**:1984–98.

124. Dacks PA, Krajewski SJ, Rance NE. Activation of neurokinin 3 receptors in the median preoptic nucleus decreases core temperature in the rat. *Endocrinology* 2011;**152**:4894–905.

125. Casper RF, Yen SS, Wilkes MM. Menopausal flushes: a neuroendocrine link with pulsatile luteinizing hormone secretion. *Science* 1979;**205**:823–5.

Proteomics in Reproduction: the Dialogue Between the Blastocyst and the Endometrium

Francisco Dominguez and Carlos Simon

INCLIVA, Instituto Universitario IVI (IUIVI), Valencia University, Valencia, Spain

INTRODUCTION

Implantation of the developing blastocyst is an absolute requirement for reproduction. From the viewpoint of the embryo, its goal is to invade the maternal tissue and to gain access to the nutrients that are essential for its survival and development. Implantation is a complex process in which a semi-allogeneic embryo must be accepted by the maternal endometrium. For this to occur, bi-directional communication between the blastocyst and the endometrium is required. This dialog enables synchronous development of the viable embryo and the development of endometrial receptivity, followed by embryo apposition, adhesion, and invasion into the stroma.[1]

Endometrial receptivity is a self-limited period in which the endometrium acquires a functional and transient ovarian steroid-dependent status; this allows a blastocyst to be received and further supports implantation by mediating the response to implantation by immune cells, and the production of cytokines, growth factors, chemokines, and adhesion molecules.[2–4] This specific period, known as "the window implantation," opens 4–5 days after endogenous or exogenous progesterone administration and closes 9–10 days afterwards.[5,6]

Implantation itself is governed by a collection of endocrine and autocrine signals of both embryonic and maternal origin, and also by the corresponding embryo–endometrial dialog. Understanding the activity and function of the molecules involved in this dialog will enable us to use them as predictors of either endometrial receptivity or embryo quality.

Recently, major advances in understanding the genomics of the endometrium[7] have been achieved with the use of the microarray and bioinformatics technologies now available, which have provided us with a vast amount of information regarding gene expression.

Cellular Endocrinology in Health and Disease.
DOI: http://dx.doi.org/10.1016/B978-0-12-408134-5.00010-X

However, gene expression is only one aspect of the complex regulatory network that allows cells to respond to intracellular and extracellular signals. Unlike the genome, the proteome is dynamic, complex and variable, and depends on the developmental stage of tissues containing the cells, reflecting the impact of both internal and external environmental stimuli. Proteomics is often considered the next step in the study of biological systems, but it is also more complicated than genomics, mostly because the proteome differs from cell to cell, while an organism's genome is much more constant. Lack of sensitivity still remains a major hurdle for the global introduction of proteomics into the field of human reproduction. However, new developments in mass spectrometry using protein profiling and peptide sequencing are being implemented to help to elucidate the underlying biological processes involved (Figure 10.1). In this chapter, we will review the state of the art technologies now available in proteomics, focusing on the two players involved in human implantation: the embryo and the endometrium.

PROTEOMICS/SECRETOMICS OF THE HUMAN EMBRYO

A crucial aspect in implantation is the concept of "embryo viability," that is, the embryo must acquire the ability to recognize, adhere, and invade endometrial tissue. For this reason the selection of appropriate embryos for transfer into the uterus is a critical issue in the field of reproductive medicine. Morphological evaluation remains the "gold standard" of embryo assessment during in vitro fertilization (IVF) cycles, but its limited predictive power and inherent inter- and intra-observer variability limits its value.[8] Consequently, there is a need to objectively identify embryos with the highest implantation potential based on specific genomic, proteomic, and/or metabolomic profiles.

Katz-Jaffe et al. developed a method to analyze the proteome of individual human blastocysts and were able to identify differentially expressed proteins prior to implantation using time-of-flight mass spectrometry.[9] Differential protein expression profiles were observed between early and expanded blastocysts, and also between developing blastocysts as opposed to degenerate embryos. Several upregulated and downregulated proteins were detected in degenerating embryos; in particular Tcf-4 (a transcription factor which mediates Wnt signaling) and an apoptotic protease-activating factor were highlighted in this study. Degenerating embryos displayed significant upregulation of several potential biomarkers that may be involved in apoptotic and growth-inhibiting pathways. Taken together, these data provide the first tentative proteomic profile links with embryo morphology.

Given the technical and ethical difficulties implicit in handling human embryos, there is intense research aimed at the application of non-invasive proteomics analysis. Specifically, many groups are studying the molecules produced by the embryo (most of which are secreted into the surrounding conditioned media) in order to identify novel biomarkers of embryo development and viability. However, perhaps surprisingly, still very little is known about peptide and/or protein production and consumption by human embryos.

Several groups have focused on identifying the specific molecules secreted by the embryo into conditioned media in vitro that are considered critical for embryo viability, including interleukin (IL)-1 and IL-6,[10] IL-1α,[11] and soluble human leukocyte antigen G (HLA-G).[12,13] These latter studies revealed that there were higher pregnancy rates when soluble HLA-G was detected in the conditioned media of day-3 embryos. However, the presence of the molecule was not diagnostic of pregnancy success

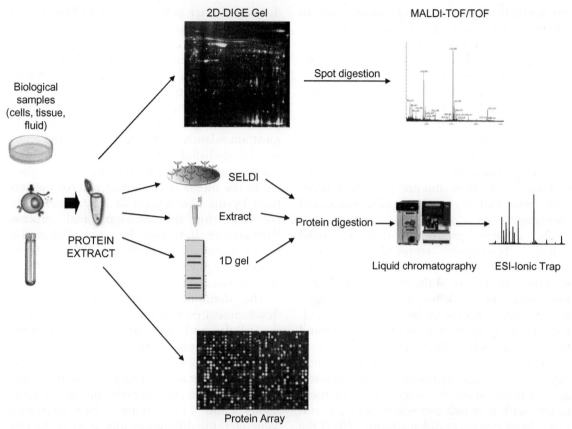

FIGURE 10.1 **Strategies of proteomic analysis.** Proteins are extracted from biological samples, separated and analyzed by differential techniques. In gel-based methods (top), different protein samples are labeled with different fluorescent dyes, and are then mixed together. Next, proteins are separated into two-dimensional difference gel electrophoresis (2D-DIGE) according to their isoelectric point and molecular weight. Gels are scanned by laser scanners and those spots corresponding to proteins with a differential expression are identified. Finally, these proteins are identified by mass spectrometry (MALDI-TOF/TOF). In the chromatographic separation methods (center), the extract proteins, protein fraction (SELDI) or one-dimensional gel bands (SDS-PAGE), are digested enzymatically, while the peptidic mix is separated by liquid chromatography (HPLC). Usually, peptides are analyzed and identified by mass spectrometry of electronebulization (ESI) and coupled with an ionic trap. Other methods are based on protein arrays (bottom). These arrays are membranes that contain a certain number of pre-absorbed antibodies that correspond to different proteins.

or failure, as pregnancies were also obtained from embryos whose conditioned media was HLA-G negative. This highlights the high variability in the developmental potential of mammalian embryos, even within the same cohort; hence, it will be necessary to evaluate several parameters in order to give a more definitive indication of embryonic developmental competence and viability.

The proteomics platform has been successfully used to analyze the secretome of mammalian embryos throughout preimplantation development, and a database of secretome profiles representing preimplantation development has been created.[14] This work revealed that human embryos produce a different protein profile roughly every 24 hours in their development ($p < 0.05$). Several proteins are differentially expressed, while others remain constant throughout the different embryonic stages. Correlation of day-5 secretome data with data from ongoing blastocyst development revealed an 8.5 kDa protein biomarker that was significantly upregulated ($p < 0.05$). The best candidate for this biomarker was ubiquitin, which has been implicated in the implantation process in some mammalian species. Current research is focusing on the identification of other proteins, and also on the correlation of these unique protein profiles with both viability and ongoing successful pregnancy.

The authors' group reported a partial human blastocyst secretome (proteins secreted/consumed) and related it to the implantation success of the embryos studied.[15] The aim of this work was to identify changes in the protein profile of the conditioned culture media from human blastocysts cultured for 24 hours, which later either implanted or did not implant, using protein-array technology (Figure 10.2). Furthermore, a statistical approach was used to compare each of these media with a matched-medium without cultured blastocysts (control medium). When the protein profile of the blastocyst culture medium was compared with the controls, soluble tumor necrosis factor receptor 1 (sTNFR1) and IL-10 was significantly increased, whereas macrophage-stimulating protein α (MSP)-α, stem cell factor (SCF), C-X-C motif chemokine 13 (CXCL13), TNF-related apoptosis-inducing ligand decoy receptor 3 (TRAILR3), and macrophage inflammatory protein (MIP)-1β were significantly decreased. Specifically, CXCL13 and granulocyte-macrophage colony-stimulating factor (GM-CSF) also decreased significantly in the implanted blastocyst media compared with the media from their non-implanted counterparts which had a similar morphology (Figure 10.2).

The authors also investigated the secretome profile of implanted blastocysts which developed after performing an embryo biopsy for preimplantation genetic diagnosis; these embryos were either grown in sequential media or co-cultured with endometrial epithelial cells.[16] When the protein-array technology was applied to the sequential and co-culture conditioned spent media, a different protein pattern emerged. Interestingly, IL-6 was the most abundantly secreted protein in the endometrial epithelial cell (EEC) co-cultures, which led the authors to conclude that the IL-6 present in the media is consumed and/or metabolized by the blastocyst, and may therefore be necessary for the developmental process. Furthermore, IL-6 could be a potential predictor of blastocyst implantation, which can be used alongside morphological criteria for embryo selection.

The definitive identification of the key development proteins will provide insight into the cellular and biochemical processes occurring during human embryonic development. In addition, these data could contribute to the development of a non-invasive viability assays, which could be used in both clinical IVF treatments and in animal biotechnologies. Moreover, the differences identified in the protein profile of the culture media in the presence of implanted versus non-implanted blastocysts may be useful as a marker profile to indicate embryo viability and, therefore, also serve as a useful tool to complement morphological criteria in the selection of the most appropriate blastocysts for transfer.

PROTEOMICS OF THE HUMAN ENDOMETRIUM

The dynamics of the endometrial transition from the non-receptive stage to the receptive stage at the proteomic level deserves further attention to identify potential molecules for closer analysis, which should help to improve our understanding of endometrial receptivity.

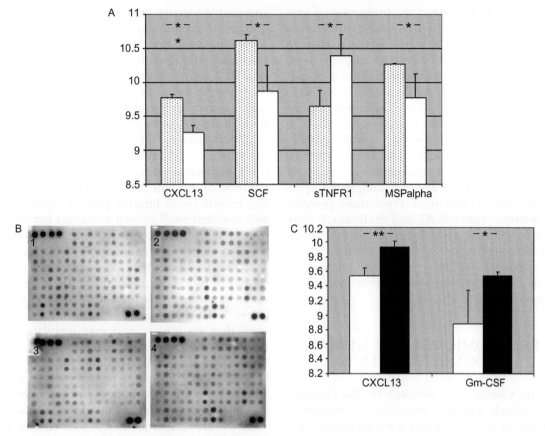

FIGURE 10.2 **Secretome and implantome of the human blastocyst.** (A) Densitometric value of all significant proteins in control medium (dotted bar) compared to conditioned media with a single blastocyst (white bar). Data are expressed as the mean of the normalized densitometry values (log2) ± s.d. * Statistical differences in the relative expression mean ($p < 0.05$); ** ($p < 0.01$) between control and blastocyst. (B) Membrane arrays VII (1 & 3) and VI (2 & 4) incubated with conditioned media from implanted (3 & 4) versus non-implanted blastocyst (1 & 2). Proteins included in the array are set in duplicate. Positive controls are included in quadruplicate. Also negative controls are included. Each spot pair corresponds to one protein studied. Protein spot intensity correlates with higher protein abundance in the medium analyzed. (C) Densitometric value of proteins in conditioned media from implanted blastocysts (white bar) compared to non-implanted blastocysts (black bar). Data are expressed as the mean of the normalized densitometry values (log2) ± s.d. Means and standard deviation were obtained from three experiments. * represents statistical differences in the relative expression mean ($p < 0.05$); ** ($p < 0.01$) between viable and non-viable.

DeSouza *et al.* employed a quantitative approach to assess the proliferative and secretory endometrial proteomic repertoire using isotope-coded affinity tags (ICAT), affinity purification, and liquid chromatography coupled online to mass spectrometry (LC-MS).[17] Only five proteins showed consistent differential expression, of which the glutamate NMDA receptor subunit zeta 1 precursor and FRAT1, a substrate-specific regulator of glycogen synthase kinase-3 activity, were the most interesting. The utility of these proteins as indicators of endometrial receptivity is subject to ongoing further research.

The group compared the proteomes of pre-receptive versus receptive endometrial biopsies obtained from the same fertile women in the same menstrual cycle. The biopsies were analyzed using two-dimensional fluorescence difference gel electrophoresis (2D-DIGE) and matrix-assisted laser desorption/ionization time-of-flight mass spectrometry (MALDI-TOF-MS). Seventy-eight differentially expressed proteins were found in the receptive versus pre-receptive endometrium, with 44 and 34 upregulated and downregulated spots respectively, (Table 10.1). From these proteins of interest, Annexin A2 and Stathmin 1 were two of the most consistently and differentially expressed proteins, which may prove to be important in predicting receptivity status and could, therefore, also be possible therapeutic targets.[18]

PROTEOMICS OF THE HUMAN ENDOMETRIAL FLUID

The viscous fluid secreted by the endometrial glands provides nutrients for blastocyst formation and constitutes a microenvironment where the embryo—endometrium dialog occurs prior to implantation. It is also an important compartment for the assessment of endometrial maturation.[19–25] Furthermore, uterine secretions are less complex in terms of their protein repertoire, and may represent a pool of biomarkers that can be interrogated in order to identify those which are essential for functional endometrial operation.

Endometrial secretions have been shown to contain: (i) proteins originating from the transudation of serum; (ii) leakage products from apoptotic epithelial cells; and (iii) proteins secreted from the glandular epithelium (Figure 10.3). This secretion undergoes significant changes in protein content during the transition from the proliferative phase to the secretory phase.[24] Endometrial secretion composition varies during the menstrual cycle as a result of the changes in the ovarian steroid serum concentration.[25] Estradiol (E_2) regulates transudation by modifying blood vessel dilatation and permeability, and progesterone (P) controls the secretory activity of the endometrial glands. Furthermore, endometrial secretions contain cytokines such as leukemia inhibitory factor (LIF),[26] glycodelin (PP14),[27] macrophage colony-stimulating factor (M-CSF), epidermal growth factor (EGF), vascular endothelial growth factor (VEGF),[28] insulin-like growth factor binding protein 1 (IGFBP-1), and interleukins,[29] as well as steroid hormones (estrogen, progesterone, prolactin, human chorionic gonadotropin and their precursors).[30,31]

In the past, the patterns of protein content in uterine secretions throughout the menstrual cycle have been analyzed by electrophoresis. These analyses revealed three different patterns that are typical, and may correlate with the equivalent phases in the menstrual cycle: the intermediate phase, proliferative phase, and secretory phase. The results showed characteristic "families" of protein bands, corresponding to 63 individual proteins, some which were identifiable by their molecular weight.[25]

In another paper, endometrial fluid was obtained transcervically by aspiration immediately prior to embryo transfer and the protein profile in each sample was determined, concluding that endometrial secretion aspiration prior to embryo transfer does not reduce implantation rates.[32] Although uterine fluid aspiration is a safe method, sometimes not enough material is obtained for analysis or it may become diluted as a result of uterine washing, making the results difficult to interpret. These studies also demonstrate that endometrial secretions can be obtained for analysis immediately prior to embryo transfer in IVF cycles without disrupting implantation.[33,34]

More recently, Van der Gaast et al. investigated the effect of ovarian stimulation in IVF

TABLE 10.1 Proteins Identified by MALDI-TOF/TOF Showing Significant Changes Between Human Prereceptive (LH + 2) and Receptive (LH + 7) Endometrium

Number	p-Value[a]	Fold Change[b]	Accession Code[c]	Protein Description	Mascot Score[d]	Mascot expect[e]	MW/pI Theor.[f]	Matched Peptides[g]	Sequence Cov. (%)[h]	Protein Function[i]
1	9.80E-04	−2.5	IBA2_HUMAN	Ionized calcium binding adapter molecule 2 isoform 1	115	2.00E-05	17.1/6.6	2	18	Protein binding
2	1.10E-03	−2.6	STMN1_HUMAN	Stathmin (phosphoprotein p19) (pp19) (oncoprotein 18)	249	7.90E-19	17.3/5.8	7	36	Cytoskeleton, intracellular signaling cascade
3	8.72E-04	−1.6	STMN1_HUMAN	Stathmin (phosphoprotein p19) (pp19) (oncoprotein 18)	118	1.10E-05	17.3/5.8	3	18	Cytoskeleton, intracellular signaling cascade
4	8.72E-04	−1.4	SYDC_HUMAN	Aspartyl-tRNA synthetase	424	2.50E-36	57.5/6.1	19	41	Translation process
5	1.01E-03	1.5	PARK7_HUMAN	DJ-1 protein	136	5.00E-09	20.1/6.3	2	10	Ras protein signal transduction
6	7.33E-04	1.5	PCOC1_HUMAN	Procollagen C-endopeptidase enhancer 1	213	3.10E-15	48.8/7.4	12	30	Cytoskeleton
7	9.80E-04	1.5	ANXA2_HUMAN	Annexin A2	117	1.20E-05	38.8/7.6	3	9	Skeletal development
8	6.53E-04	1.6	Q59GX5_HUMAN	L-plastin	137	1.20E-07	56.2/5.2	1	3	Cytoskeleton
9	7.33E-04	1.7	CATB_HUMAN	Liver cathepsin B, chain B	174	2.50E-11	23.0/5.2	3	18	Proteolysis, regulation of apoptosis
10	9.80E-04	1.7	TAGL2_HUMAN	Transgelin 2	155	2.00E-09	21.2/7.6	8	44	Protein binding, development
11	8.54E-04	1.7	APOH_HUMAN	Apolipoprotein H	103	3.10E-04	36.7/8.3	2	7	Cholesterol metabolism
12	6.53E-04	1.8	AL1A3_HUMAN	Aldehyde dehydrogenase 1A3	82	1.20E-02	56.9/7.0	1	2	Metabolic process

(Continued)

TABLE 10.1 Proteins Identified by MALDI-TOF/TOF Showing Significant Changes Between Human Prereceptive (LH + 2) and Receptive (LH + 7) Endometrium (Continued)

Number	p-Value[a]	Fold Change[b]	Accession Code[c]	Protein Description	Mascot Score[d]	Mascot expect[e]	MW/pI Theor.[f]	Matched Peptides[g]	Sequence Cov. (%)[h]	Protein Function[i]
13	7.33E-04	2.0	TPIS_HUMAN	Chain A, triosephosphate isomerase	238	9.90E-18	26.8/6.5	12	42	Metabolism of small molecules
14	9.65E-04	2.1	EI2BL_HUMAN	Translation initiation factor eIF-2B	121	5.00E-06	39.5/5.9	2	8	Translation initiation
15	1.06E-03	2.5	ALBU_HUMAN	Albumin, isoform 2, fragment	198	9.90E-14	23.2/8.2	4	18	Transport
16	8.72E-04	2.8	ANXA2_HUMAN	*Annexin A2*	388	9.90E-33	**38.8/7.6**	15	43	Skeletal development
17	9.80E-04	3.4	AOFA_HUMAN	Monoamine oxidase A	97	4.30E-05	60.2/7.9	1	2	Electron transport
18	8.72E-04	3.7	APOL2_HUMAN	Apolipoprotein L2	165	2.00E-10	37.1/6.3	4	14	Cholesterol metabolism
19	1.74E-02	−2.4	PGRC1_HUMAN	Membrane-associated progesterone receptor component 1	148	4.30E-10	21.8/4.6	2	7	Signaling
20	2.53E-02	−2.2	STMN1_HUMAN	Stathmin (phosphoprotein p19) (pp19) (oncoprotein 18)	219	3.40E-17	17.3/5.8	4	18	Cytoskeleton, intracellular signaling cascade
21	2.55E-03	−1.8	F13A_HUMAN	Coagulation factor XIII A	374	1.10E-32	83.7/5.8	17	27	Coagulation
22	1.90E-02	−1.7	SERPH_HUMAN	Serpin H1 precursor	156	6.80E-11	46.5/8.8	8	32	Response to unfolded protein
23	3.34E-02	−1.5	VIME_HUMAN	Vimentin	766	6.80E-72	53.7/5.1	23	52	Cytoskeleton
24	1.05E-02	−1.4	HNRL1_HUMAN	Heterogeneous nuclear ribonucleoprotein U-like 1	160	2.70E-11	96.3/6.5	11	19	RNA processing, response to virus
25	1.51E-02	−1.3	PPIB_HUMAN	Peptidyl-prolyl cis-trans isomerase B	475	8.50E-43	22.8/9.3	17	56	Protein synthesis

p-value[a]	Ratio[b]	Accession[c]	Protein	Score[d]	MW/pI[f]	Peptides[g]	Coverage[h]	Function[i]
26	1.23E-02	−1.3	RSSA_HUMAN	40S ribosomal protein SA (p40)	280	33.0/4.8	9	35	Cell adhesion
27	6.87E-04	1.5	TPM3_HUMAN	Tropomyosin-alpha 3	132	32.9/4.7	8	33	Cytoskeleton
28	2.10E-03	1.5	RUVB2_HUMAN	RuvB-like 2	253	51.3/5.5	9	20	Regulation of transcription, cell proliferation
29	2.26E-02	1.6	CO6A1_HUMAN	Collagen alpha-1(VI)	255	109.6/5.3	15	15	Skeletal development
29	2.26E-02	1.6	HYOU1_HUMAN	Hypoxia upregulated protein 1	188	111.5/5.2	10	11	Response to stress
30	2.20E-02	1.6	RUXG_HUMAN	Small nuclear ribonucleoprotein G	130	8.5/9.0	2	17	Spliceosome assembly
31	1.52E-03	1.9	ANXA4_HUMAN	Annexin A4	168	36.1/5.8	10	35	Signal transduction
32	1.10E-02	2.1	K1C19_HUMAN	Cytokeratin-19	292	44.1/5	17	43	Response to estrogen stimulus
33	5.10E-03	2.1	ANXA2_HUMAN	Annexin A2	250	38.8/7.6	14	38	Skeletal development
34	2.26E-02	2.2	SSRD_HUMAN	Translocon-associated protein subunit delta	250	19.2/5.8	7	41	Transport
35	2.10E-02	4.8	S10AA_HUMAN	Protein S100-A10	130	11.3/6.8	3	31	Signal transduction

[a]Student t-test p-value and
[b]average volume ratio (LH + 7/LH + 2) as calculated by the DeCyder BVA analysis.
[c]Protein accession code from SwissProt database.
[d]Mascot score.
[e]Expected value.
[f]Theoretical molecular weight (kDa) and pI
[g]Number of matched peptides and
[h]protein sequence coverage for the most probable candidate as provided by Mascot.
[i]Biological function retrieved from SwissProt.

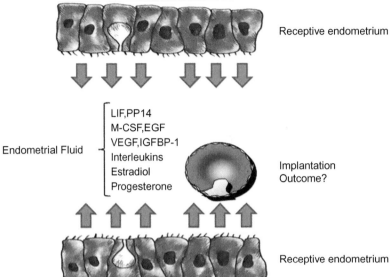

FIGURE 10.3 **Secretome of the human endometrial fluid during the window of receptivity.** Molecules secreted by the epithelial endometrium into the uterine cavity during the receptive stage.

Receptive endometrium

Endometrial Fluid
- LIF,PP14
- M-CSF,EGF
- VEGF,IGFBP-1
- Interleukins
- Estradiol
- Progesterone

Implantation Outcome?

Receptive endometrium

on endometrial secretion and receptivity markers in the mid-luteal phase.[35] The endometrial fluids obtained in this period in the stimulated cycle were compared with those from the spontaneous cycle, and the protein composition was analyzed by sodium dodecyl sulfate—polyacrylamide gel electrophoresis (SDS-PAGE), stained with Coomassie brilliant blue. The protein pattern was obtained by measuring the relative density of each band with a scanning laser densitometer and the GelScan XL software package. Although, in this pilot study, ovarian stimulation did not alter the mid-luteal phase endometrial maturation markers investigated.

Classically, two-dimensional electrophoresis (2D-PAGE), based on a combination of isoelectric focusing and sodium dodecyl sulfate—polyacrylamide gel electrophoresis, was the only method available to analyze the protein complement in a sample at a high resolution, although the introduction of protein chips and mass spectrometry has now facilitated protein identification. These approaches circumvent one of the major challenges in

endometrial research: investigating endometrial performance during the window of implantation without disrupting the endometrial function and the subsequent process of implantation. However, the application of proteomics to the study of protein patterns in endometrial fluid is still technically hindered because the majority of proteins identified correspond to serum proteins. This masks the identification of potentially more interesting lower-concentration proteins, such as biomarkers for endometrial receptivity, embryo development, diseases, and/or proteins which could be intercepted for disease treatment.

PROTEOMICS OF ENDOMETRIOSIS

Endometriosis is defined by the presence of endometrial tissue outside the uterus. This definition is based on Sampson's concept that the disease is caused by peritoneal regurgitation and implantation of viable endometrial cells in menstrual debris.[36] Consequently the diagnosis of endometriosis is based on histological identification of ectopic endometrial glands and stroma. Deep endometriosis, which is found along the outside of the Müllerian tract, is predominantly characterized by fibromuscular hyperplasia and the formation of an adenomyotic nodules and microendometriomas.[37,38] Peritoneal and ovarian endometriosis is characterized by chronic bleeding which results in the formation of hemorrhagic blisters, fibrosis, adhesions, and ovarian endometriomas. Endometriosis is further characterized by altered immune cell responses, inflammation, neoangiogenesis, and ovarian and uterine dysfunction.

Although laparoscopy is now the gold standard diagnostic tool for endometriosis it still has many limitations, hence novel invasive and non-invasive techniques are also used both for screening and for clinical purposes. Accurate invasive and non-invasive diagnostic techniques are urgently needed if the clinical management of women with endometriosis is to be more effective, especially given that the younger the patient is at the onset of symptoms, the longer it takes before the diagnosis of endometriosis is usually made.

In the past a number of different animal models, including baboons, rhesus monkeys, rats, and mice have also been used for endometriosis research. The difficulty with animal models is that many of the species studied have no menstrual cycle and only primates develop endometriosis spontaneously. Different standard methodologies, including suppression subtractive hybridization, differential display, and reverse transcription polymerase chain reaction, have proved to be powerful in detecting and characterizing differentially expressed genes, and emerging "—omics" technologies are now allowing us to analyze the expression of large numbers of genes and proteins simultaneously. With this approach we can contrast the expression of all the genes and/or proteins of a cell/tissue/organism in two conditions, for example in this case the gene expression profiles of endometriotic endometrium and eutopic endometrium. Proteome analysis is now widely accepted as a complementary technology to genetic profiling, and together these two areas will lead to a better understanding of diseases and the development of new treatments in clinical medicine.

2D-PAGE has been used since the 1990s to investigate the molecules involved in the pathogenesis of endometriosis. More recently, Tabibzadeh et al. compared the peritoneal fluid (PF) from women with and without endometriosis using this method. However, the gels exhibited a limited number of protein spots, and the majority of the abnormally expressed proteins in endometriosis were not identifiable by immunoblotting or mass spectrometry.[39]

Marked differences in both the amount and types of PF proteins present were noted between women with mild endometriosis, severe endometriosis, infertility but no endometriosis, and controls ($n = 6$ in each group) using 2D-PAGE. The proteins observed in women with infertility but without endometriosis did not differ from those of healthy controls, and mild endometriosis was associated with only a mild reduction in the number of proteins in the 35—40 kDa and 5.7—6.0 pH range compared with controls. However, in women with severe endometriosis a greater decrease in the same protein spots was observed, while there was also a 2—4-fold increase in the quantity of many other proteins when compared to controls. In terms of the published literature, these protein differences have not yet been characterized, hence these findings have not progressed to the development of a diagnostic test for endometriosis. However, this is certainly an area of considerable potential that merits validation and exploration with further studies.

Fowler et al. investigated the effects of endometriosis on the proteome of human eutopic endometrium by using 2D-PAGE and mass spectrometry;[40] several dysregulated proteins were identified, including molecular chaperones, proteins involved in the cellular redox state, molecules involved in protein and DNA formation and/or breakdown, and secreted proteins. In a similar study in 2006, designed to search for endometriosis-specific proteins,[41] Zhang et al. observed abnormal expression of proteins involved in the cell cycle, signal transduction, and immunological function using 2D-PAGE, Western blotting, and mass spectrometry. First, they aimed to find a difference in the way serum and eutopic endometrial proteins were expressed in women with and without endometriosis. They were also interested in searching for endometrial proteins that were specifically recognized by sera from patients with endometriosis in the hope that these potential markers might lead to novel diagnostic, therapeutic, and prognostic methods for the management of endometriosis and may also improve our understanding of the pathogenic mechanism of endometriosis. The 2D-PAGE profiles from the six endometriosis sera samples were very similar, and together all the endometriosis sera were used to construct an average 2D-PAGE electrophoresis map, which included 237 protein spots. Similarly, an average electrophoresis map of six normal sera was also established, which had 216 protein spots. They compared the protein patterns of the average gels of sera samples from women with and without endometriosis and detected protein spots with at least a three-fold discrepancy. After the comparative proteomic study, they found 13 protein spots from serum and 11 from endometrium (in both cases correlated with 11 known proteins), which were differently expressed between women with and without endometriosis. Some of these matched proteins were cytoskeletal, while others were known regulators of the cell cycle, signal transduction, and immunological function. Proteins from the G antigen family included (but were not limited to) B1 protein, actin-related protein 6, actin-like 7-anhydrase I, dentin matrix acidic phosphoprotein I, CD166 antigen, and cyclin A1.[24] Recently, protein chip technology has also been used to study proteins expressed by endometrium, endometrial tissues, and normal peritoneum obtained from women with and without endometriosis.[42]

PROTEOMICS IN OTHER ENDOMETRIAL PATHOLOGIES

Uterine Fibroids

Uterine leiomyomas are benign neoplasms arising from the myometrial compartment of the uterus and are the most common gynecological neoplasm in reproductive-age women. This benign tumor represents significant reproductive health problems, such as abnormal uterine bleeding, pelvic pain, constipation and reproductive dysfunction. Ahn et al. identified proteins that were differentially expressed in uterine leiomyoma-affected myometrium compared to normal myometrium using 2D-PAGE analysis.[43] These changes might hint at the cellular processes that occur in the biology of uterine leiomyoma, helping to define the complexity of the expression profiles, thus bringing real relevance to the description of disease-specific pathogeneses.

Several proteins of unknown function, such as GAGEC1 (a member of the GAGE family) and LAGY, were upregulated. Conversely, several functional proteins that were downregulated in leiomyoma, such as structural and extracellular matrix adhesion molecules, indicate that repression of cellular processes may have a significant impact on uterine leiomyoma pathogenesis. Downregulated cell adhesion activity indicates a decrease in adhesive properties in the leiomyoma disease state. For instance, downregulated eotaxin (CCL11), a potent eosinophil chemotaxis inducer which leads to eosinophil migration, is not able to take part in fine-tuning the cellular responses that occur at sites of allergic inflammation in leiomyoma-affected myometria, where both monocyte chemotacticprotein-1 (MCP1) and eotaxin are normally produced.

Of the completely downregulated genes, ANXA1 may cause specific inhibition of trans-endothelial neutrophil migration and desensitization of neutrophils towards chemoattractant challenges, resulting in the loss of anti-inflammatory effects. Another downregulated protein, XLKD1, appears to be involved in facilitating cell migration during wound healing, inflammation, and embryonic morphogenesis, which also leads to the loss of anti-inflammatory effects.

CONCLUSIONS

Although many efforts in the last years have been focused in transcriptomics, especially with the introduction of the array technology, proteomics shows a closer reflection of cellular function; thus, it focuses in the last step of protein production. Therefore, proteomics could represent an excellent tool not only to help to elucidate the physiopathology of certain diseases, but also to identify proteins that are potential biomarkers of various disease states, with the aim of improving their diagnosis and eventual treatment. Proteomics, together with transcriptomics and metabolomics, are complementary approaches that provide diverse but comparable perspectives that will improve our understanding of the complexity of the process of human implantation, identifying key biomarkers. The next step will be to integrate this information into a system biology approach to develop models for functions of interest, such as embryo viability, endometrial receptivity and the embryo—endometrial dialog.

References

1. Dominguez F, Pellicer A, Simon C. Paracrine dialogue in implantation. *Mol Cell Endocrinol* 2002;**186**:175—81.
2. Kämmerer U, von Wolff M, Markert UR. Inmunology of human endometrium. *Inmunobiology* 2004; **209**:569—74.
3. Giudice LC. *Implantation and endometrial function. Molecular biology in reproductive medicine.* New York: Parthenon Publishing Group; 1999.
4. Dimitriadis E, White CA, Jones R, Salamonsen LA. Cytokines, chemokines and growth factors in endometrium related to implantation. *Hum Reprod Update* 2005;**11**:613—30.
5. Finn CL, Martin L. The control of implantation. *J Reprod Fertil* 1974;**39**:195—206.
6. Martín J, Dominguez F, Avila S, Castrillo JL, Remohí J, Pellicer A, et al. Human endometrial receptivity: gene regulation. *J Reprod Immunol.* 2002;**55**:131—9.
7. Horcajadas JA, Pellicer A, Simón C. Wide genomic analysis of human endometrial receptivity: new times, new opportunities. *Hum Reprod Update* 2007;**13**:77—86.
8. Guerif F, Le Gouge A, Giraudeau B, Poindron J, Bidault R, Gasnier O, et al. Limited value of morphological assessment at days 1 and 2 to predict blastocyst development potential: a prospective study based on 4042 embryos. *Hum Reprod* 2007;**22**:1973—81.
9. Katz-Jaffe MG, Gardner DK, Schoolcraft WB. Proteomic analysis of individual human embryos to identify novel biomarkers of development and viability. *Fertil Steril* 2006;**85**:101—7.
10. Baranao RI, Piazza A, Rumi LS, Polak de Fried E. Determination of IL-1 and IL-6 levels in human embryo culture-conditioned media. *Am J Reprod Immunol* 1997;**37**:191—4.
11. Sheth KV, Roca GL, al-Sedairy ST, Parhar RS, Hamilton CJ, al-Abdul Jabbar F. Prediction of successful embryo implantation by measuring interleukin-1-alpha and immunosuppressive factor(s) in preimplantation embryo culture fluid. *Fertil Steril* 1991;**55**:952—7.
12. Desai N, Filipovits J, Goldfarb J. Secretion of soluble HLA-G by day 3 human embryos associated with higher pregnancy and implantation rates: assay of culture media using a new ELISA kit. *Reprod Biomed Online* 2006;**13**:272—7.
13. Fuzzi B, Rizzo R, Criscuoli L, Noci I, Melchiorri L, Scarselli B, et al. HLA-G expression in early embryos is a fundamental prerequisite for the obtainment of pregnancy. *Eur J Immunol* 2002;**32**:311—5.
14. Katz-Jaffe MG, Schoolcraft WB, Gardner DK. Analysis of protein expression (secretome) by human and mouse preimplantation embryos. *Fertil Steril* 2006; **86**:678—85.

15. Domínguez F, Gadea B, Esteban FJ, Horcajadas JA, Pellicer A, Simón C. Comparative protein-profile analysis of implanted versus non-implanted human blastocyst. *Human Reprod* 2008;**23**:1993−2000.

16. Dominguez F, Gadea B, Mercader A, Esteban F, Pellicer A, Simon C. Embryological outcome and secretome profile of implanted blastocyst obtained. *Fertil Steril* 2010;**93**(774−782):e1.

17. DeSouza L, Diehl G, Yang E, Guo J, Rodrigues MJ, Romaschin AD, et al. Proteomic analysis of the proliferative and secretory phases of the human endometrium: protein identification and differential protein expression. *Proteomics* 2005;**5**:270−81.

18. Dominguez F, Garrido-Gomez T, Lopez JA, Camafeita E, Quiñonero A, Pellicer A, et al. Proteomic analysis of the human receptive versus non-receptive endometrium using differential in-gel electrophoresis and MALDI-MS unveils stathmin 1 and annexin A2 as differentially regulated. *Hum Reprod* 2009;**24**(10):2607−17.

19. Beier-Hellwig K, Sterzik K, Bonn B, Beier HM. Contribution to the physiology and pathology of endometrial receptivity: the determination of proteins patterns in human uterine secretions. *Hum Reprod* 1989;**4**:115−20.

20. Giudice LC. Potencial biochemical markers of uterine receptivity. *Hum Reprod* 1999;**14**:3−16.

21. Lindhard A, Bentin-Ley U, Ravn V. Biochemical evaluation of endometrial function at the time of implantation. *Fertil Steril* 2002;**78**:221−33.

22. Herrler A, von Rango U, Bier HM. Embryo-maternal signalling: how the embryo starts talking to its mother to accomplish implantation. *Reprod BioMed Online* 2003;**6**:244−56.

23. Beier HM. Oviducal and uterine fluids. *J Reprod Fertil* 1974;**37**:221−37.

24. Maathuis JB, Aitken RJ. Protein patterns of human uterine flushings collected at various stages of the menstrual cycle. *J Reprod Fertil* 1978;**53**:343−8.

25. Beier HM, Beier-Hellwig K. Molecular and cellular aspects of endometrial receptivity. *Hum Reprod Update* 1998;**4**:448−58.

26. Laird SM, Tuckerman EM, Dalton CF. The production of leukaemia inhibitory factor by human endometrium: presence in uterine flushings and production by cells in culture. *Hum Reprod* 1997;**12**:569−74.

27. Li TC, Dalton C, Hunjan KS, Warren MA, Bolton AE. The correlation of placental protein 14 concentrations in uterine flushings and endometrial morphology in the peri-implantation period. *Hum Reprod* 1993;**8**:1923−7.

28. Classen-Linke I, Alfer J, Krusche CA, Chwalisz K, Rath W, Beier HM. Progestins, progesterone receptor modulators, and progesterone antagonist change VEGF release of endometrial cells in culture. *Steroids* 2000;**65**:763−71.

29. Simon C, Mercader A, Frances A, et al. Hormonal regulation of serum and endometrial IL-1α, IL-1β and IL-1ra: IL-1 endometrial microenvironment of the human embryo at the apposition phase under physiological and supraphysiological steroid level conditions. *J Reprod Immunol* 1996;**31**:165−84.

30. Stone BA, Petrucco OM, Seamark RF, Godfrey BM. Concentrations of steroid hormones, and of prolactin, in washings of the human uterus during the menstrual cycle. *J Reprod Fertil* 1986;**78**:21−5.

31. Licht P, Losch A, Dittrich R, et al. Novel insights into human endometrial paracrinology and embryo-maternal communication by intrauterine microdialysis. *Hum Reprod Update* 1998;**4**:532−8.

32. Van der Gaast MH, Beier-Hellwig K, Fauser BC, Beier HM, Macklon NS. Endometrial secretion aspiration prior to embryo transfer does not reduce implantation rates. *Reprod BioMed Online* 2003;**7**:105−9.

33. Li TC, MacKenna A, Roberts R. The techniques and complications of out-patient uterine washing in the assessment of endometrial function. *Hum Reprod* 1993;**8**:343−6.

34. Olivennes F, Ledee-Bataille N, Samama M, Kadoch J, Taupin JL, Dubanchet S, et al. Assessment of leukemia inhibitary factor levels by uterine flushing at the time of egg retrieval does not adversely affect pregnancy rates with in vitro fertilization. *Fertil Steril* 2003;**79**:900−4.

35. Van der Gaast MH, Classen-Linke I, Krusche CA, Beier-Hellwig K, Fauser BCJM, Beier HM, et al. Impact of ovarian stimulation on mid-luteal endometrial tissue and secretion markers of receptivity. *RBM Online* 2008;**4**:553−63.

36. Sampson JA. Peritoneal endometriosis due to the menstrual dissemination of endometrial tissues into the peritoneal cavity. *Am J Obstet Gynecol* 1927;**14**:422−69.

37. Brosens IA. Classification of endometriosis revisited. *Lancet* 1993;**341**:630.

38. Nisolle M, Donnez J. Peritoneal endometriosis, ovarian endometriosis, and adenomyotic nodules of the rectovaginal septum are three different entities. *Fertil Steril* 1997;**68**:585−96.

39. Tabibzadeh S, Becker JL, Parsons AK. Endometriosis is associated with alterations in the relative abundance of proteins and IL-10 in the peritoneal fluid. *Front Biosci* 2003;**8**:70−8.

40. Fowler PA, Tattum J, Bhattacharya S, Klonisch T, Hombach-Klonisch S, Gazvan R, et al. An investigation of the effects of endometriosis on the proteome of human eutopic endometrium: a heterogeneous tissue with a complex disease. *Proteomics* 2007;**7**:130−42.

41. Zhang H, Niu Y, Feng J, Guo H, Ye X, Cui H. Use of proteomic analysis of endometriosis to identify different protein expression in patients with endometriosis versus normal controls. *Fertil Steril* 2006;**86**:274–82.

42. Kyama CM, T'Jampens D, Mihalyi A, et al. ProteinChip technology is a useful method in the pathogenesis and diagnosis of endometriosis: a preliminary study. *Fertil Steril* 2006;**86**:203–9.

43. Ahn WS, Kim KW, Bae SM, et al. Targeted cellular process profiling approach for uterine leiomyoma using cDNA microarray, proteomics and gene ontologyanalysis. *Int J ExpPathol* 2003;**84**:267–79.

11

Transcriptome Analysis of Adrenocortical Cells in Health and Disease

Tomohiro Ishii

Department of Pediatrics, School of Medicine, Keio University, Tokyo, Japan

INTRODUCTION

The human and mouse genome projects are historical landmarks in which the entire genomic sequence of these species was published.[1–3] The projects identified 10,000 to 15,000 genes, far less than expected. Therefore, it was inferred that quantitative or qualitative variations in transcription, including the production, stability, degradation, and alternative splicing of RNA also contribute to phenotypic variation. Recently developed high-throughput techniques have helped identify nearly 40,000 predicted mRNAs in a mouse genome.[3] A study that examined 55 different tissues reported that approximately 2000 transcripts were detected in all tissues, suggesting 5% of all mRNAs are ubiquitously expressed.[4] In contrast, more than 4000 transcripts were expressed in only one tissue type, indicating 10% of mRNAs fulfill a specific function in a single tissue. The number of transcripts varies in each tissue. For example, approximately 10,000 of the 40,000 transcripts were expressed in the mouse eye,

but less than 6000 were expressed in the pancreas. More than 9000 transcripts were found in the adrenal glands, suggesting that over 7000 transcripts play a specific role in the adrenal cortex and medulla. In order to understand the significance of a set of transcripts, specific technology is required, especially that which relates to bioinformatics. This chapter provides an overview of transcriptome analysis, discusses recent progress, and introduces representative studies involving transcriptome analysis of adrenocortical cells.

STRATEGIES FOR DETERMINING A TRANSCRIPTOME

The transcriptome is the complete set of transcripts in a specific type of cells or tissue. In general, there are two strategies for determining a transcriptome (Table 11.1). cDNA microarray analysis is based on hybridization between fluorescent-labeled target cDNAs and probes representing known transcripts that are synthesized and spotted on the solid

Cellular Endocrinology in Health and Disease.
DOI: http://dx.doi.org/10.1016/B978-0-12-408134-5.00011-1

TABLE 11.1   Comparison of Two Strategies for
Determining Transcriptome

	cDNA Microarray	RNA-Seq
Technology	Hybridization	Next-generation sequencing
Quantification	Relative (ratio)	Absolute (molecule)
Unknown transcripts	Uncovered	Covered
Bioinformatics	Developed	In progress
Cost	Low	High

Abbreviation: RNA-Seq, RNA sequencing.

surface of a chip.[5] Hybridization chips are custom-made or commercially available for different species and platforms. Comparing cDNA microarray data from different platforms is sometimes difficult. MicroArray Quality Control (MAQC) recommends standards and quality control metrics.[6,7] National Center for Biotechnology Information's Gene Expression Omnibus (www.ncbi.nlm.nih.gov/geo) provides a public database of microarray data and allows only quality-controlled data to be uploaded. Microarray uses relative quantification in which the intensity of a feature is compared to that of the same feature under a different condition, and the identity of the feature is known by its position. In contrast, RNA sequence (RNA-Seq) has been recently developed based on a next-generation sequencing technique.[8,9] RNA-Seq is not limited to defined transcripts that correspond to an existing genomic sequence. RNA-Seq can identify rare or tissue-specific splicing variants that have never been reported. Although RNA-Seq is a very promising method with the potential to overtake cDNA microarrays, the technology is still under active development. Understanding the strategy used to determine a transcriptome is essential for determining functional elements of cells or tissues and for interpreting physiology and pathophysiology in health and disease, respectively.

DATA MINING

Generally, the goal of transcriptome analysis is to identify genes differentially expressed among different conditions, leading to a new understanding of genes or pathways associated with the conditions. Transcriptome analysis requires an appropriate statistical method with a multiple comparison test to interpret global changes in the expression of thousands of genes.[10] Analysis of variance (ANOVA) is one of the traditional methods used to analyze the significance of differential expression by comparing the variance across replicate groups to the variance within replicate groups. Additionally, the application of a multiple comparison test, such as the Bonferroni correction or false discovery rate (FDR), is recommended.[11] The Bonferroni correction is too strict for maintaining a low false negative rate. FDR represents a reasonable approach and controls the yield of false positives and false negatives with a variable cut-off between 0.05 and 0.25.[12] In addition to statistical significance, fold change is also important for assessing the degree of differential expression as a measure of biological significance. A volcano plot is a popular graphical display of p values from a statistical test against fold changes on a logarithmic two-dimensional plot (Figure 11.1). It provides an easy method for identifying differentially expressed genes (DEGs) that are both statistically and biologically significant in the condition. Pathway or ontology analysis enables researchers to interpret DEGs in the framework of biological processes and systems, rather than in a traditional gene-centric manner. Comparing the list of gene profiles of interest against previously assembled lists of genes grouped by function or cellular localization can provide useful insights. Several pathway databases are

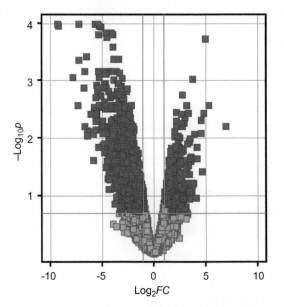

FIGURE 11.1 **Volcano plot.** Red squares indicate genes differentially expressed in adrenocortical cells of *Star*[−/−] mice; grey squares indicate genes not differentially expressed; green lines show cut-off lines of fold change (FC) = 2.0 (*x*-axis) and *p* value modified by false discovery rate = 0.2 (*y*-axis).

available, including the Gene Ontology (GO) Consortium (www.geneontology.org)[13] and the Kyoto Encyclopedia of Genes and Genomes (KEGG) (www.genome.jp/kegg).[14] Dedicated web-based applications for pathway analysis are available, including Ingenuity Pathways Analysis (IPA) and NextBio.[15,16] The results of transcriptome analysis must be carefully interpreted, and conclusions must not be drawn without using an appropriate data mining technique.

TRANSCRIPTOME ANALYSIS OF ADRENAL GLANDS AND ADRENOCORTICAL CELL LINES UNDER ADRENOCORTICOTROPIC HORMONE STIMULATION

Trophic hormone regulation is essential for adrenocortical cells to differentiate and produce sufficient amounts of steroid hormones.[17] Mineralcorticoid production by glomerulosa cells is controlled by renin and angiotensin, as well as by potassium level. Glucocorticoid synthesis by fasciculata cells is regulated by adrenocorticotropic hormone (ACTH). Lee and Widmaier reported the transcriptome of the adrenal glands in neonatal rats with or without chronic ACTH stimulation *in vivo*, using a rat cDNA microarray with 15,923 probes.[18] ACTH induced significant changes in the expression of 214 genes with an FDR of 0.25 or less and a fold change of 1.5 or greater, including downregulation of genes associated with *de novo* cholesterol biosynthesis and cholesterol trafficking and upregulation of genes associated with intracellular metabolism and glucocorticoid inactivation (Table 11.2). These data demonstrate that the developmental effects of ACTH alter the expression of a broad range of genes involved not only in steroid hormone synthesis but also in cellular functions related to the growth and differentiation of the adrenal glands. Additionally, the negative effects of ACTH on the genes required for cholesterol synthesis and the production of active glucocorticoids suggest a key adaptive feature of the developing adrenal glands, which minimizes excessive exposure of other tissues, such as brain, to glucocorticoids.[19] Schimmer *et al*. conducted transcriptome analysis of adrenocorticotropin action in the Y1 mouse adrenal tumor cell line *in vitro*, using a mouse cDNA microarray of 5655 probes.[20] ACTH affected the accumulation of 1289 transcripts (588 upregulated and 687 downregulated) with a FDR less than 0.05 (Table 11.3), a much larger number of genes than previously reported. The genes upregulated by ACTH were in the category of cholesterol biosynthesis, cholesterol mobilization, or corticosteroid biosynthesis and metabolism. As expected, ACTH regulated the expression of genes encoding transcription factors involved in the expression of steroidogenic enzymes and of signaling molecules involved in hormonal

TABLE 11.2   Genes Regulated by ACTH in Adrenal Glands of Neonatal Rats by cDNA Microarray

Functional Category	Gene	Description	FC
Signal transduction	Spp1	Secreted phosphoprotein 1	4.6
	Pla2g1b	Phospholipase A2, group IB	3.7
	Rdc1	Chemokine orphan receptor 1	3.2
	Gria3	Glutamate receptor, ionotropic. AMPA3 (alpha 3)	2.9
	Per2	Period homolog 2	2.6
	Mtap2	Microtubule-associated protein 2	2.5
	Pdk2	Pyruvate dehydrogenase kinase 2	2.5
	Pak1	p21 (CDKM1A)-activated kinase 1	2.4
	Cxcl10	Chemokine (C-X-C motif) ligand 10	2.3
	Cx3cl1	Chemokine (C-X3-C motif) ligand 1	2.2
	Il18	Interleukin-18	2.0
	Ghr	Growth hormone receptor	2.0
	Gprk5	G protein-coupled receptor kinase 5	0.5
	Emb	Embigin	0.48
	Gfra1	Glial cell line derived neurotrophic factor family receptor 1	0.45
	Inha	Inhibin alpha	0.45
	Pdyn	Prodynorphin	0.43
	Cd36	cd36 antigen	0.4
	Sfrp4	Secreted frizzled-related protein 4	0.37
	Tm4sf3	Transmembrane 4 superfamily member 3	0.16
Transcription factor activity	Dbp	D site albumin promoter binding protein	3.8
Cell growth/maintenance	Spp1	Secreted phosphoprotein 1	4.6
	Pla2g1b	Phospholipase A2, group IB	3.7
	Gria3	Glutamate receptor, ionotropic, AMPA3 (alpha 3)	2.9
	Ret	Ret proto-oncogene	2.8
	Apoc1	Apolipoprotein C-l	2.6
	Hpx	Hemopexin	2.5
	Mtap2	Microtubule-associated protein 2	2.5
	Cxcl10	Chemokine (C-X-C motif) ligand 10	2.3
	Loc65042	Tricarboxylate carrier-like protein	2.2
	Il18	Interleukin-18	2.0
	Ghr	Growth hormone receptor	2.0

(Continued)

TABLE 11.2   (Continued)

Functional Category	Gene	Description	FC
	Vamp1	Vesicle-associated membrane protein 1	0.5
	Gfra1	Glial cell line derived neurotrophic factor family receptor 1	0.45
	Inha	Inhibin alpha	0.45
	Cd36	cd36 antigen	0.4
	Slc1a3	Solute carrier family 1, member 3	0.34
Cell–cell signaling	Spp1	Secreted phosphoprotein 1	4.6
	Gria3	Glutamate receptor, ionotropic. AMPA3 (alpha 3)	2.9
	Inha	Inhibin alpha	0.45
	Pdyn	Prodynorphin	0.43
	Slc 1a3	Solute carrier famil 1, member 3	0.34
Development/morphogenesis	Srd5a1	Steroid 5 alpha-reductase 1	6.4
	Spp1	Secreted phosphoprotein 1	4.6
	Mtap2	Microtubule-associated protein 2	2.5
	Cxcl10	Chemokine (C-X-C motif) ligand 10	2.3
	Il18	Interleukin-18	2.0
	Ghr	Growth hormone receptor	2.0
	Ania4	Activity and neurotransmitter-induced early gene protein 4	2.0
	Gfra1	Glial cell line derived neurotrophic factor family receptor 1	0.45
	Inha	Inhibin alpha	0.45
	Sfrp4	Secreted frizzled-related protein 4	0.37
Structural/cell adhesion/extracellular matrix	Spp1	Secreted phosphoprotein 1	4.6
	C1s	Complement component 1, s subcomponent	3.0
	Lox	Lysyl oxidase	2.8
	Cx3cl1	Chemokine (C-X3-C motif) ligand 1	2.2
	Il18	Interleukin-18	2.0
	Vamp1	Vesicle-associated membrane protein 1	0.5
	Cd36	cd36 antigen	0.4
	Bcan	Brevican	0.23
	Mmp9	Matrix metalloproteinase 9	0.13
Lipid and cholesterol	Pla2g1b	Phospholipase A2, group IB	3.7
Metabolism/biosynthesis	Gria3	Glutamate receptor, ionolropic. AMPA3 (alpha 3)	2.9
	Mch	Medium-chain S-acyl fatty acid synthetase thio ester hydrolase	2.6

(Continued)

TABLE 11.2   (Continued)

Functional Category	Gene	Description	FC
	Dig1	Dithiolethione-inducible gene-1	2.3
	Sc4mol	Sterol-C4-methyl oxidase-like	0.5
	Hmgcs1	3-Hydroxy-3-methylglutaryl-coenzyme A synthase 1	0.48
	Cd36	cd36 antigen	0.4
	Idi1	Isopentenyl-diphosphate delta isomerase	0.4
	Cyp11b	Cytochrome P450, subfamily XIB gene cluster	0.37
	Pla2g4a	Phospholipase A2, group IVA	0.37
	Ptgs1	Prostaglandin-endoperoxide synthase 1	0.36
	Sqle	Squalene epoxidase	0.36
Immune response	*Spp1*	Secreted phosphoprotein 1	4.6
	Cxcl10	Chemokine (C-X-C motif) ligand 10	2.3
	Cx3cl10	Chemokine (C-X3-C motif) ligand 1	2.2
	RT1Aw2	RT1 class lb gene(Aw2)	2.2
	Il18	Interleukin-18	2.0
	Inha	Inhibin alpha	0.45

Abbreviation: FC, fold change.
*Modified data published by Lee JJ, Widmaie EP. J Steroid Biochem Mol Biol 2005; **96**: 31−44.*

regulation of steroidogenesis. The genes down-regulated in response to ACTH contained significantly enriched clusters with functions in DNA replication, mitotic cell cycle, cell division, nuclear transport, and RNA processing, consistent with the growth-inhibiting effect of ACTH in Y1 cells.[21] Schimmer *et al.* also determined the relative contributions of the cyclic adenosine monophosphate (cAMP) signaling pathway and the protein kinase C-dependent pathway to the ACTH effects as approximately 56% and 6%, respectively. Approximately 38% of the ACTH-affected transcripts could not be assigned to these signaling pathways and thus represent candidates for regulation via other mechanisms. The combined data from the two studies by Lee and Widmaier and Schimmer *et al.*, despite a certain level of inconsistency between these studies, provide useful reference sets for more detailed studies of the factors that contribute to regulation of steroidogenesis and growth and development in the adrenal cortex upon ACTH stimulation.

TRANSCRIPTOME ANALYSIS OF DIFFERENT ZONES

Adrenocortical zonation remains an unresolved aspect for researchers investigating the mechanism underlying and the control element of zonation.[22] Transcriptome analysis of different zones might help elucidate the molecular signature of functional zonation. The zona fasciculata and zona glomerulosa are present in rodents, whereas the zona reticularis is observed only in primates. Nishimoto *et al.* assessed the transcriptome of DEGs in the zona glomerulosa and zona

TABLE 11.3   Genes Regulated by ACTH in Y1 Adrenocortical Cell Line by cDNA Microarray

Gene	Description	FC	p	PKA	PKC
UPREGULATION					
Adamts4	A disintegrin-like and metalloprotease (reprolysin type) with thrombospondin type 1 motif, 4	8.68	0.0368	+	+
Orc4l	Origin recognition complex, subunit 4-like (S. cerevisiae)	5.55	0.0008	+	−
Procr	Protein C receptor, endothelial	4.96	0.0008	+	−
Spp1	Secreted phosphoprotein 1	3.98	0.0064	+	−
Scarb1	Scavenger receptor class B, member 1	3.85	0.0008	+	−
Cdk8	Cyclin-dependent kinase 8	3.68	0.0031	+	+
Ier3	Immediate early response 3	3.61	0.002	−	+
Star	Steroidogenic acute regulatory protein	2.92	0.0008	+	−
Elac2	ElaC homolog 2 (E. coli)	2.91	0.0008	−	+
Alas1	Aminolevulinic acid synthase 1	2.62	0.0008	+	−
Ucp2	Uncoupling protein 2, mitochondrial	2.42	0.0015	+	−
Cnnm2	Cyclin M2	2.40	0.0111	−	+
Ostf1	Osteoclast stimulating factor 1	2.32	0.0008	+	+
Prkar2b	Protein kinase, cAMP dependent regulatory, type II beta	2.26	0.0008	+	+
Eno1	Enolase 1, alpha non-neuron	2.22	0.0368	−	−
Hs1bp1	HS1 binding protein	2.17	0.0008	+	+
Gla	Galactosidase, alpha	2.17	0.0008	+	−
Sdc3	Syndecan 3	2.11	0.0008	+	+
Net1	Neuroepithelial cell transforming gene 1	2.08	0.0008	+	−
Mthfd2	Methylenetetrahydrofolate dehydrogenase (NAD+ dependent), methenyltetrahydrofolate cyclohydrolase	2.07	0.0008	+	−
Psap	Prosaposin	2.07	0.0147	+	−
Esd	Esterase D/formylglutathione hydrolase	2.06	0.0008	+	+
Dhcr7	7-Dehydrocholesterol reductase	2.06	0.002	+	−
Alas2	Aminolevulinic acid synthase 2, erythroid	2.03	0.0082	−	−
Odc1	Ornithine decarboxylase, structural 1	2.02	0.0008	−	+
Bzrp	Benzodiazepine receptor, peripheral	2.01	0.0008	+	−
Gnaq	Phosphatidylethanolamine binding protein	2.01	0.0287	+	−
Myo10	Myosin X	2.00	0.004	+	+
Psph	Phosphoserine phosphatase	2.00	0.0287	+	−

(Continued)

TABLE 11.3   (Continued)

Gene	Description	FC	p	PKA	PKC
DOWNREGULATION					
Hmgb2	High mobility group box 2	0.45	0.0008	+	−
Ptn	Pleiotrophin	0.45	0.0008	+	−
Aldh1a1	Aldehyde dehydrogenase family 1, subfamily A1	0.46	0.0008	+	+
Pbx3	Pre B-cell leukemia transcription factor 3	0.47	0.0008	+	+
Slc12a2	Solute carrier family 12, member 2	0.50	0.0008	−	−

Abbreviations: FC, fold change; PKA, protein kinase A; PKC, protein kinase C.
*Modified data published by Schimmer BP, Cordova M, Cheng H, et al. Endocrinology 2006; **147**: 2357−67.*

fasciculata of 9-week-old male rats, using laser capture microdissection (LCM) and a rat cDNA microarray with 22,523 probes.[23] They found 235 and 231 transcripts upregulated more than two-fold in the zona glomerulosa and zona fasciculata, respectively, with an FDR less than 0.05 (Table 11.4). *Frzb*, which inhibits Wnt signaling, was differentially expressed in the zona fasciculata. In contrast, *Wnt4* showed a trend of upregulation in zona glomerulosa cells. Lack of functional WNT4/Wnt4 results in adrenal dysgenesis and compromised aldosterone synthesis in human patients with the female sex reversal and dysgenesis of kidneys, adrenals, and lungs (SERKAL) syndrome and knockout mice lacking Wnt4, respectively.[24,25] It has been inferred that Frzb and Wnt4 inversely modulate adrenal zonation during organogenesis and aldosterone synthesis after birth. Information on zone-specific changes in transcripts will help clarify the development and maintenance of adrenal zonation.

TRANSCRIPTOME ANALYSIS OF ADRENOCORTICAL CELLS IN A MOUSE MODEL OF HUMAN LIPOID ADRENAL HYPERPLASIA

Undoubtedly, mouse models of human disorders would help identify the pathophysiology of disorders and even help determine possible approaches for potential new treatments. Human adrenal glands are not easily accessible and are rarely removed except in cancer, thus making a knockout animal with adrenal insufficiency particularly valuable. One such model, knockout mice lacking steroidogenic acute regulatory protein (StAR), has been described here in detail.

Congenital lipoid adrenal hyperplasia (lipoid CAH) is an autosomal recessive disease due to *STAR* gene mutations. StAR regulates the rate-limiting step of steroidogenesis to facilitate the translocation of newly synthesized cholesterol from the outer to inner mitochondrial membrane.[17,26−28] Patients with lipoid CAH and knockout mice lacking StAR (*Star*[−/−] mice) exhibit significant defects in steroid hormone biosynthesis and diffuse accumulation of cholesterol esters in the cytoplasm of adrenocortical or gonadal steroidogenic cells.[29−33] Bose *et al.*[29] proposed the "two-hit" model, in which the first hit is loss of StAR-dependent steroidogenesis due to StAR deficiency and the second hit is loss of StAR-independent steroidogenesis caused by accumulated cholesterol esters, thus explaining the varying phenotypes among steroidogenic tissues. However, the molecular mechanisms underlying the cholesterol translocation by StAR and the abrogation of StAR-independent steroidogenesis by cholesterol deposits require clarification. Thus, the transcriptome of steroidogenic cells in the adrenal glands of *Star*[−/−] mice

TABLE 11.4 Genes Significantly Upregulated in Rat Zona Glomerulosa or Zona Fasciculata by cDNA Microarray

Gene	Description	FC	*p*
ZONA GLOMERULOSA			
Cyp11b2	*Rattus norvegicus* cytochrome P450, family 11, subfamily B, polypeptide 2 (Cyp11b2), nuclear gene encoding mitochondrial protein, mRNA	214.2	0.00078
Rgs4	*Rattus norvegicus* regulator of G-protein signaling 4 (Rgs4), mRNA	68.4	0.00052
Smoc2_predicted	PREDICTED: *Rattus norvegicus* SPARC related modular calcium binding 2 (predicted), transcript variant 2 (Smoc2_predicted), mRNA	49.3	0.00777
Mia1	PREDICTED: *Rattus norvegicus* melanoma inhibitory activity 1 (Mia1), mRNA	43.1	0.00031
Dlk1	*Rattus norvegicus* delta-like 1 homolog (*Drosophila*) (Dlk1), mRNA	38.3	0.00531
Sstr2	*Rattus norvegicus* somatostatin receptor 2 (Sstr2), mRNA	37.2	0.00852
Cadps	*Rattus norvegicus* Ca2+-dependent secretion activator (Cadps), mRNA	33.7	0.00483
LOC311772	PREDICTED: *Rattus norvegicus* similar to nidogen 2 (LOC311772), mRNA	33.6	0.00083
Igsf1	*Rattus norvegicus* immunoglobulin superfamily, member 1 (Igsf1), mRNA	32.9	0.00078
LOC362564	PREDICTED: *Rattus norvegicus* hypothetical LOC362564 (LOC362564), mRNA	31.1	0.00296
Gpc3	*Rattus norvegicus* glypican 3 (Gpc3), mRNA	30.6	0.01434
Cpxm2_predicted	PREDICTED: *Rattus norvegicus* carboxypeptidase X 2 (M14 family) (predicted) (Cpxm2_predicted), mRNA	29.1	0.01163
Atp10a	PREDICTED: *Rattus norvegicus* ATPase, class V, type 10A, transcript variant 1 (Atp10a), mRNA	28.8	0.00118
Postn_predicted	PREDICTED: *Rattus norvegicus* periostin, osteoblast specific factor (predicted) (Postn_predicted), mRNA	24.1	0.02619
Boc_predicted	PREDICTED: *Rattus norvegicus* biregional cell adhesion molecule-related/downregulated by oncogenes (Cdon) binding protein (predicted) (Boc_predicted), mRNA	23.3	0.00604
RGD1566317_predicted	PREDICTED: *Rattus norvegicus* similar to Tescalcin (predicted) (RGD1566317_predicted), mRNA	22.9	0.00149
Ndn	*Rattus norvegicus* necdin (Ndn), mRNA	19.9	0.00318
Dpt_predicted	PREDICTED: *Rattus norvegicus* dermatopontin (predicted) (Dpt_predicted), mRNA	18.5	0.01768
Kcnn2	*Rattus norvegicus* potassium intermediate/small conductance calcium-activated channel, subfamily N, member 2 (Kcnn2), mRNA	17.5	0.00582
Rbp1	*Rattus norvegicus* retinol binding protein 1, cellular (Rbp1), mRNA	17.4	0.00375
Wfdc1	*Rattus norvegicus* WAP four-disulfide core domain 1 (Wfdc1), mRNA	15.7	0.01163
Acy3	*Rattus norvegicus* aspartoacylase (aminoacylase) 3 (Acy3), mRNA	15.6	0.01682
RGD1307506_predicted	PREDICTED: *Rattus norvegicus* similar to RIKEN cDNA 2310016C16 (predicted) (RGD1307506_predicted), mRNA	15.1	0.00299
Ptgis	*Rattus norvegicus* prostaglandin I2 (prostacyclin) synthase (Ptgis), mRNA	14.9	0.01147

(Continued)

TABLE 11.4   (Continued)

Gene	Description	FC	p
Reck_predicted	PREDICTED: *Rattus norvegicus* reversion-inducing-cysteine-rich protein with kazal motifs (predicted) (Reck_predicted), mRNA	14.5	0.01353
RGD1564008_predicted	PREDICTED: *Rattus norvegicus* similar to dapper 1 (predicted) (RGD1564008_predicted), mRNA	13.2	0.00483
Nr0b1	*Rattus norvegicus* nuclear receptor subfamily 0, group B, member 1 (Nr0b1), mRNA	12.7	0.01434
Fmod	*Rattus norvegicus* fibromodulin (Fmod), mRNA	12.6	0.02731
Pde2a	*Rattus norvegicus* phosphodiesterase 2A, cGMP-stimulated (Pde2a), mRNA	12.6	0.00184
Dab2	*Rattus norvegicus* disabled homolog 2, mitogen-responsive phosphoprotein (Drosophila) (Dab2), mRNA	12.4	0.00420

ZONA FASCICULATA

Gene	Description	FC	p
LOC363306	PREDICTED: *Rattus norvegicus* similar to RIKEN cDNA 4930555G01 (LOC363306), mRNA	19.30	0.01525
Ddah1	*Rattus norvegicus* dimethylarginine dimethylaminohydrolase 1 (Ddah1), mRNA	16.21	0.00369
LOC498373	PREDICTED: *Rattus norvegicus* similar to RIKEN cDNA 1700001E04 (LOC498373), mRNA	15.55	0.01342
Cidea_predicted	PREDICTED: *Rattus norvegicus* cell death-inducing DNA fragmentation factor, alpha subunit-like effector A (predicted) (Cidea_predicted), mRNA	15.54	0.00176
Hpx	*Rattus norvegicus* hemopexin (Hpx), mRNA	15.43	0.01921
Hamp	*Rattus norvegicus* hepcidin antimicrobial peptide (Hamp), mRNA	15.32	0.03461
RGD1562717_predicted	PREDICTED: *Rattus norvegicus* similar to ABI gene family, member 3 (NESH) binding protein (predicted) (RGD1562717_predicted), mRNA	14.90	0.00572
Cyp4f4	*Rattus norvegicus* cytochrome P450, family 4, subfamily f, polypeptide 4 (Cyp4f4), mRNA	14.60	0.03031
Serpina11	*Rattus norvegicus* serine (or cysteine) peptidase inhibitor, clade A (alpha-1 antiproteinase, antitrypsin), member 11 (Serpina11), mRNA	13.03	0.00670
LOC501497	PREDICTED: *Rattus norvegicus* LOC501497 (LOC501497), mRNA	11.73	0.00630
Fabp6	*Rattus norvegicus* fatty acid binding protein 6, ileal (gastrotropin) (Fabp6), mRNA	11.35	0.02633
Nkx6-2_predicted	PREDICTED: *Rattus norvegicus* NK6 transcription factor related, locus 2 (Drosophila) (predicted) (Nkx6-2_predicted), mRNA	10.04	0.00385
RGD1560609_predicted	PREDICTED: *Rattus norvegicus* similar to Vanin-3 (predicted) (RGD1560609_predicted), mRNA	9.79	0.03668
Frzb	PREDICTED: *Rattus norvegicus* frizzled-related protein (Frzb), mRNA	9.45	0.00565
LOC363060	*Rattus norvegicus* similar to RIKEN cDNA 1600029D21 (LOC363060), mRNA	9.31	0.03191
Ephb6	PREDICTED: *Rattus norvegicus* Eph receptor B6 (Ephb6), mRNA	9.07	0.00770
Fbxo17	*Rattus norvegicus* F-box only protein 17 (Fbxo17), mRNA	9.03	0.00483

(Continued)

TABLE 11.4   (Continued)

Gene	Description	FC	p
Rab33a_predicted	PREDICTED: *Rattus norvegicus* RAB33A, member of RAS oncogene family (predicted) (Rab33a_predicted), mRNA	8.99	0.01338
Hhex	*Rattus norvegicus* hematopoietically expressed homeobox (Hhex), mRNA	8.96	0.02417
Vnn1	*Rattus norvegicus* vanin 1 (Vnn1), mRNA	8.80	0.00777
Hsd11b2	*Rattus norvegicus* hydroxysteroid 11-beta dehydrogenase 2 (Hsd11b2), mRNA	8.26	0.00589
Mt1a	*Rattus norvegicus* metallothionein 1a (Mt1a), mRNA	8.17	0.00876
Macrod1	*Rattus norvegicus* MACRO domain containing 1 (Macrod1), mRNA	7.99	0.00247
Edg4_predicted	PREDICTED: *Rattus norvegicus* endothelial differentiation, lysophosphatidic acid G protein-coupled receptor 4 (predicted) (Edg4_predicted), mRNA	7.88	0.00483
Dhcr7	*Rattus norvegicus* 7-dehydrocholesterol reductase (Dhcr7), mRNA	7.15	0.00486
LOC681153	PREDICTED: *Rattus norvegicus* hypothetical protein LOC681153 (LOC681153), mRNA	6.96	0.03839
Arih1	*Rattus norvegicus* ariadne ubiquitin-conjugating enzyme E2 binding protein homolog 1 (*Drosophila*) (Arih1), mRNA	6.85	0.01501
F11r	*Rattus norvegicus* F11 receptor (F11r), mRNA	6.76	0.01682
Comtd1_predicted	PREDICTED: *Rattus norvegicus* catechol-O-methyltransferase domain containing 1 (predicted) (Comtd1_predicted), mRNA	6.70	0.02182
Amid_predicted	PREDICTED: *Rattus norvegicus* apoptosis-inducing factor (AIF)-like mitochondrion-associated inducer of death (predicted) (Amid_predicted), mRNA	6.49	0.04764

Abbreviation: FC, fold change.
Modified data published by Nishimoto K, Rigsby CS, Wang T, et al. Endocrinology 2012; 153: 1755–63.

could help identify related biological pathways that may help elucidate the pathophysiology of StAR deficiency. The following subsection is a review of findings published previously.[34]

Targeted Expression of eGFP in Adrenocortical Cells

In order to collect pure adrenocortical cells, Ishii *et al.* generated transgenic mice carrying a bacterial artificial chromosome (BAC) transgene targeting eGFP to steroidogenic cells under the control of endogenous regulatory sequences from the mouse *Star* gene (StAR/eGFP mice).[34] For transgene construction, eGFP cDNA was inserted into the coding region of *Star* in a BAC clone containing the mouse *Star* gene with 47 kb of the 5′-flanking region and 62 kb of the 3′-flanking regions (StAR/eGFP BAC) (Figure 11.2A). Anti-eGFP antibody immunoreactivity and intrinsic eGFP fluorescence were observed in the adrenocortical cells, the interstitial cells of testes, and the theca cells, luteal cells, and stromal cells of ovaries from mice carrying the StAR/eGFP BAC transgene (Figure 11.2B), consistent with the endogenous StAR expression. The adrenal glands were dissected from wild type (*Star$^{+/+}$*) embryos or *Star$^{-/-}$* embryos with the StAR/eGFP BAC transgene at embryonic day (E) 17.5 or E18.5. Single cell

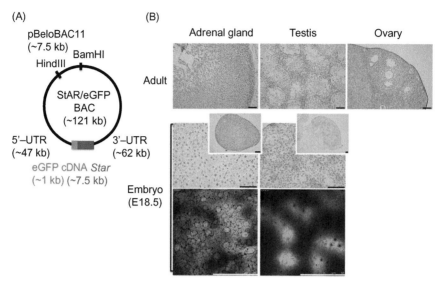

FIGURE 11.2   (A) Structure of the StAR/eGFP bacterial artificial chromosome (BAC) transgene. (B) Expression of the StAR/eGFP BAC transgene in adrenal glands and gonads from adults (top) or embryos at E18.5 (middle and bottom). Pictures in the top and middle show immunohistochemical staining for GFP expression in bright field lighting; those in the bottom demonstrate GFP fluorescence in dark field lighting. Each smaller picture in the middle shows a lower magnification view of the larger picture. Scale bars are 100 μm.

suspensions were prepared by enzymatic dissociation. eGFP-positive cells were selectively isolated by fluorescence-activated cell (FAC) sorting. Cell suspensions from the adrenal glands of $Star^{+/+}$ or $Star^{-/-}$ mice carrying no StAR/eGFP transgene were used as negative controls in FAC sorting to exclude the possibility of macrophage contamination due to autofluorescence.[35] StAR/eGFP mice provide a useful strategy for selective purification of StAR-expressing steroidogenic cells from the adrenal glands.

Differentially Regulated Genes in Mouse cDNA Microarrays

Ishii *et al.* determined the transcriptome of the adrenocortical cells of $Star^{+/+}$ or $Star^{-/-}$ mice using a Whole Mouse Genome Microarray G4122F, Cy3-labeled cRNAs (One-Color Quick Amp Labeling kit), and a Gene Array Scanner (all from Agilent Technologies, Inc., Palo Alto, CA).[34] The data were analyzed using the Gene Spring GX 11.5.1 software (Agilent Technologies, Inc.). The difference in gene expression was assessed by averaging the normalized values and performing a pairwise analysis between $Star^{+/+}$ mice and $Star^{-/-}$ mice. DEGs were defined as having a gene expression ratio of ≥2-fold and a p value evaluated by FDR <0.2.[12] The transcriptome analysis identified 1973 DEGs (1206 upregulated and 767 downregulated) between $Star^{+/+}$ and $Star^{-/-}$ mice (Figure 11.1).

The expression levels of the selected genes were validated by quantitative polymerase chain reaction (qPCR) using additional samples of FAC-sorted cells from the adrenal glands at E17.5 or E18.5. The relative gene expression in $Star^{+/+}$ and $Star^{-/-}$ mice was calculated using the ddCt method and examined using Student's t test, with a p value <0.05 considered significant. The transcriptome

TABLE 11.5 Genes Differentially Expressed in Knockout Mice Lacking StAR Compared with those in Wild-Type Mice by Quantitative Real-Time PCR

Functional Category	Gene	Description	FC (Mean ± SE)
Steroid hormone biosynthesis	Nr5a1	Nuclear receptor subfamily 5, group A, member 1	0.68 ± 0.52
	Mc2r	Melanocortin 2 receptor	0.29 ± 0.18
	Mrap	Melanocortin 2 receptor accessory protein	0.26 ± 0.51
	Star	Steroidogenic acute regulatory protein	0.02 ± 0.01*
	Cyp11a1	Cytochrome P450, family 11, subfamily a, polypeptide 1	0.63 ± 0.48
	Fdx1	Ferredoxin 1	0.63 ± 0.69
	Fdxr	Ferredoxin reductase	0.69 ± 0.56
	Hsd3b1	Hydroxy-delta-5-steroid dehydrogenase, 3 beta- and steroid delta-isomerase 1	0.51 ± 0.53
	Agtr1a	Angiotensin II receptor, type 1a	0.76 ± 0.58
	Ldlr	Low-density lipoprotein receptor	1.58 ± 1.19
	Scarb1	Scavenger receptor class B, member 1	1.39 ± 2.91
	Hmgcr	3-Hydroxy-3-methylglutaryl-Coenzyme A reductase	0.77 ± 0.31
	Lipe	Lipase, hormone sensitive	0.93 ± 0.45
Cholesterol biosynthesis and influx	Npc2	Niemann Pick type C2	3.46 ± 4.7
	Stard3	START domain containing 3	4.69 ± 3.15*
	Bzrpl1	Benzodiazapine receptor, peripheral-like 1	0.82 ± 0.94
	Srebf1	Sterol regulatory element binding factor 1	1.9 ± 0.93
	Acat2	Acetyl-Coenzyme A acetyltransferase 2	1.19 ± 1.08
	Nr1h2	Nuclear receptor subfamily 1, group H, member 2	3.04 ± 1.21*
	Nr1h3	Nuclear receptor subfamily 1, group H, member 3	7.76 ± 1.38*
	Abca1	ATP-binding cassette, sub-family A (ABC1), member 1	2.92 ± 1.59*
Cholesterol metabolism and efflux	Abcg1	ATP-binding cassette, sub-family G (WHITE), member 1	5.69 ± 3.1*
	Abcb1a	ATP-binding cassette, sub-family B (MDR/TAP), member 1A	0.37 ± 0.93
	Dhcr24	24-Dehydrocholesterol reductase	0.54 ± 0.62
	ch25h	Cholesterol 25-hydroxylase	0.57 ± 1.69
	Wt1	Wilms tumor homolog	1.83 ± 1.26
	Nr0b1	Nuclear receptor subfamily 0, group B, member 1	0.09 ± 0.10*
Adrenocortical development	Shh	Sonic hedgehog	0.02 ± 0.02*
	Wnt4	Wingless-related MMTV integration site 4	0.08 ± 0.06*

(Continued)

TABLE 11.5　(Continued)

Functional Category	Gene	Description	FC (Mean ± SE)
	Cd86	CD86 antigen	6.44 ± 3.39*
	Cd36	CD36 antigen	3.29 ± 2.03*
	Cd5l	CD5 antigen-like	130.95 ± 273.32*
	Spp1	Secreted phosphoprotein 1	259.15 ± 133.85*
	Ccl5	Chemokine (C-C motif) ligand 5	113.14 ± 144.27*
	Cxcl9	Chemokine (C-X-C motif) ligand 9	0.28 ± 1.23
	Clec7a	C-type lectin domain family 7, member a	100.69 ± 65.1*
Inflammatory response	H2-Aa	Histocompatibility 2, class II antigen A, alpha	36.41 ± 147.76
	H2-Ab1	Histocompatibility 2, class II antigen A, beta 1	24.29 ± 70.12
	H2-Ea	Histocompatibility 2, class II antigen E alpha	37.09 ± 475.29
	Tlr1	Toll-like receptor 1	4.12 ± 3.97
	Tlr2	Toll-like receptor 2	2.82 ± 2.31
	Tlr4	Toll-like receptor 4	2.08 ± 1.24
	Tlr6	Toll-like receptor 6	8.2 ± 8.33
	Tlr7	Toll-like receptor 7	4.59 ± 3.41

Asterisks (*) indicate statistically significant change ($p < 0.05$).
Modified data published by Ishii T, Mitsui T, Suzuki S, et al. Endocrinology 2012; 153: 2714–23.

analysis and subsequent qPCR validation did not show significant differences between $Star^{+/+}$ and $Star^{-/-}$ mice in the expression of genes related to steroid hormone biosynthesis, other than *Star*, or cholesterol biosynthesis and influx. However, significant upregulation of the genes involved in cholesterol efflux, such as those encoding liver X receptors (LXRs) and adenosine triphosphate (ATP)-binding cassettes, was observed; significant downregulation of genes involved in adrenocortical development, such as dosage-sensitive sex reversal-adrenal hypoplasia critical region on X chromosome 1 (*Dax-1*; officially *Nr0b1*), wingless-related MMTV integration site 4 (*Wnt4*) and sonic hedgehog (*Shh*), was also observed (Table 11.5). These data reflect the pathophysiological features of adrenocortical cells upon StAR deficiency.

Pathway Analysis

Ishii *et al.* performed pathway analysis through IPA and NextBio.[34] IPA returned a list of canonical pathways most relevant to the transcriptome data set by a right- or two-tailed Fisher's exact probability test.[16] IPA identified significant associations between the DEGs and the canonical pathway of inflammatory or immune response, including dendritic cell maturation, antigen presentation pathway, allograft rejection

TABLE 11.6 Canonical Pathways Significantly Altered in Knockout Mice Lacking StAR Compared with those in Wild-Type Mice by Ingenuity Pathway Analysis

Canonical Pathway	p	Ratio
Dendritic cell maturation	2.13796E-10	2.27E-01
Antigen presentation pathway	1.41254E-09	3.72E-01
Allograft rejection signaling	1.62181E-08	2.00E-01
Role of NFAT in regulation of the immune response	1.90546E-08	2.21E-01
Altered T cell and B cell signaling in rheumatoid arthritis	2.51189E-08	2.86E-01
Graft-versus-host disease signaling	7.07946E-08	3.54E-01
PKCθ signaling in T lymphocytes	8.91251E-08	2.41E-01
Communication between innate and adaptive immune cells	0.0000001	2.23E-01
Systemic lupus erythematosus signaling	2.0893E-07	1.79E-01
CD28 signaling in T helper cells	2.51189E-07	2.48E-01
OX40 signaling pathway	6.76083E-07	2.04E-01
Autoimmune thyroid disease signaling	7.94328E-07	2.54E-01
TREM1 signaling	8.31764E-07	3.16E-01
Natural killer cell signaling	9.12011E-07	2.55E-01
B cell receptor signaling	1.62181E-06	2.33E-01
IL-4 signaling	1.99526E-06	3.00E-01
FcγRIIB signaling in b lymphocytes	2.51189E-06	3.02E-01
iCOS-iCOSL signaling in T helper cells	3.46737E-06	2.35E-01
Atherosclerosis signaling	4.36516E-06	2.40E-01
Glycosaminoglycan degradation	5.7544E-06	4.44E-01
Cytotoxic T lymphocyte-mediated apoptosis of target cells	1.02329E-05	1.88E-01
B cell development	1.54882E-05	2.97E-01
Fcγ receptor-mediated phagocytosis in macrophages and monocytes	1.77828E-05	2.53E-01
T helper cell differentiation	2.39883E-05	2.68E-01
Role of pattern recognition receptors in recognition of bacteria and viruses	3.16228E-05	2.56E-01
PI3K signaling in B lymphocytes	3.46737E-05	2.13E-01
Crosstalk between dendritic cells and natural killer cells	5.49541E-05	2.17E-01
Type I diabetes mellitus signaling	5.7544E-05	2.14E-01
LXR/RXR activation	6.76083E-05	2.41E-01
Leukocyte extravasation signaling	7.24436E-05	1.91E-01

(Continued)

TABLE 11.6   (Continued)

Canonical Pathway	p	Ratio
IL-8 signaling	8.31764E-05	1.94E-01
Production of nitric oxide and reactive oxygen species in macrophages	0.000147911	1.98E-01
IL-10 signaling	0.000165959	2.50E-01
Role of BRCA1 in DNA damage response	0.000165959	2.71E-01
CTLA4 signaling in cytotoxic T lymphocytes	0.000190546	2.16E-01
IL-12 signaling and production in macrophages	0.000190546	1.98E-01
T cell receptor signaling	0.000436516	2.14E-01
fMLP signaling in neutrophils	0.000489779	1.95E-01
Sphingolipid metabolism	0.000537032	2.14E-01
NF-κB activation by viruses	0.000630957	2.28E-01
Renin-angiotensin signaling	0.000977237	2.02E-01
NF-κB signaling	0.00128825	1.71E-01
Role of JAK1 and JAK3 in γc cytokine signaling	0.001513561	2.27E-01
MSP-RON signaling pathway	0.001584893	2.50E-01
Phospholipase C signaling	0.001659587	1.53E-01
Primary immunodeficiency signaling	0.001778279	1.75E-01
Erythropoietin signaling	0.001778279	2.16E-01
Virus entry via endocytic pathways	0.001862087	2.04E-01
Role of macrophages, fibroblasts and endothelial cells in rheumatoid arthritis	0.002511886	1.41E-01
Phospholipid degradation	0.002754229	2.10E-01

p values are calculated by Fischer's exact probability test.
Ratios indicate the number of differentially expressed genes by total number of genes that map to the pathway.

signaling, communication between innate and adaptive immune cells, and interleukin (IL)-4 signaling (Table 11.6). NextBio identified a list of biogroups or cell types most significantly correlated with the transcriptome data set by Fisher's exact probability test.[15] The NextBio analysis identified immune response as the most significant biogroup. The NextBio analysis identified macrophages from the peripheral tissues of C57BL/6 mice, microglial cells from C57BL/6 mice, and follicular dendritic cells from the lymph nodes of BALB/cAnNCrlCrlj mice as the most significant cell types positively correlated with the DEGs (Table 11.7). Both pathway analyses consistently indicated that the inflammatory or immune response was significantly altered in the adrenocortical cells of $Star^{-/-}$ mice.

TABLE 11.7   Cell types Significantly Altered in Knockout Mice Lacking StAR Compared with those in Wild-Type Mice by NextBio

Cell Type	Body System	Score	Correlation	p
Macrophage of peripheral tissue of C57BL/6 strain	Immune	263.88	+	2.50E-115
Microglial cell of C57BL/6 strain	Immune	255.83	+	7.8E-112
Dendritic cell follicular of lymph node of BALB/cAnNCrlCrlj strain	Immune	238.97	+	1.60E-104
Macrophage of bone marrow of C57BL/6 strain	Immune	217.19	+	4.7E-95
Dendritic cell (CD11c+ CD11b_hi CD103-) of lung of BALB/cByJ strain	Immune	214.96	+	4.40E-94
Myeloid-derived suppressor cell (CD11b+) of spleen of BALB/c strain	Immune	208.02	+	4.6E-91
Macrophage of peritoneum of C57BL/6J strain	Immune	187.31	+	4.50E-82
Osteoclast of C57BL/6 strain	Musculoskeletal	180.83	+	2.9E-79
Dendritic cell of bone marrow of 129SvJae x C57BL/6 strain	Immune	173.11	+	6.60E-76
Macrophage of peritoneum of C57BL/6 strain	Immune	170.00	+	1.5E-74
Dendritic cell of bone marrow of C57BL/6 strain	Immune	160.33	+	2.30E-70
Monocyte (Ly6c_hi) of peripheral blood of C57BL/6 strain	Immune	156.61	+	9.60E-69
Macrophage (CD11c_int CD11b + B220- CD4- CD8- CD169_hi F4/80+) of lymph node medulla of C57BL/6 strain	Immune	154.21	+	1.1E-67
Monocyte (Ly6c_lo) of peripheral blood of C57BL/6 strain	Immune	150.70	+	3.60E-66
Hematopoietic stem cell (CD45+ Sca1+) of muscle of C57BL/6 strain	Immune	146.86	+	1.7E-64
Macrophage (CD11b+ CD11c-) of lamina propria of small intestine of C57BL/6 strain	Immune	146.40	+	2.60E-64
Dendritic cell (CD11c+) of bone marrow of C57BL/6J strain	Immune	140.43	+	1E-61
Dendritic cell (myeloid CD8a-) of C57BL/6 strain	Immune	133.03	+	1.70E-58
Macrophage of mammary gland of FVB/N strain	Exocrine	130.98	+	1.3E-57

(Continued)

TABLE 11.7   (Continued)

Cell Type	Body System	Score	Correlation	p
Dendritic cell of bone marrow of BALB/c strain	Immune	127.28	+	5.30E-56
Macrophage (CD11b+ CD11c-) cell of spleen of C57BL/6 strain	Immune	126.92	+	7.6E-56
Neutrophil (lineage- c-kit- Gr1+ Mac1+) of bone marrow of C57BL/6 strain	Immune	125.64	+	2.70E-55
Monocyte (Mac1 +) of peripheral blood of C57BL/6 strain	Immune	123.10	+	3.5E-54
Dendritic cell of lymph node of BALB/c strain	Immune	120.18	+	6.40E-53
Dendritic cell (CD11c_hi CD11b+ CD19- CD3- NK1.1- 120G8_lo) of spleen of C57BL/6 strain	Immune	114.76	+	1.5E-50
Dendritic cell (CD11c + CD11b_lo CD103 +) of lung of BALB/cByJ strain	Immune	112.02	+	2.30E-49
Dendritic cell (CD8- 33D1 + CD205- CD11c +) of spleen of C57BL/6 strain	Immune	110.27	+	1.3E-48
Dendritic cell of spleen of BALB/c strain	Immune	110.03	+	1.60E-48
Spermatagonial cell (Thy1-) of testis of C57BL/6 strain	Urogenital	108.87	−	5.2E-48
Dendritic cell (plasmacytoid CD11c_int CD19- CD3- NK1.1- 120G8_hi) of spleen of C57BL/6 strain	Immune	105.74	+	1.20E-46
B lymphocyte (B220 +) of spleen of C57BL/6 strain	Immune	103.87	+	7.7E-46
Synovial fibroblast of knee joint of BALB/c strain	Musculoskeletal	103.11	+	1.70E-45
Dendritic cell (CD8 + 33D1- CD205 + CD11c +) of spleen of C57BL/6 strain	Immune	101.76	+	6.4E-45
Macrophage of bone marrow of BALB/c strain	Immune	101.51	+	8.20E-45
Myeloid-derived suppressor cell (CD11b +) of bone marrow of BALB/c strain	Immune	100.95	+	1.4E-44
Stem cell of spermatagonia of C57BL/6J strain	Urogenital	98.10	−	2.50E-43
Side population cell of tibialis anterior muscle of C57BL/6 strain	Musculoskeletal	93.86	+	1.7E-41
Cardiac progenitor cell of ROSA26 x Isl1 strain	Cardiovascular	91.09	−	2.80E-40

(*Continued*)

TABLE 11.7   (Continued)

Cell Type	Body System	Score	Correlation	p
Type II cell of lung of ICR10 strain	Respiratory	87.77	+	7.7E-39
Dendritic cell (IFN-producing killer) of lymph node of BALB/c strain	Immune	87.54	+	9.60E-39
Dendritic cell (CD11c+) of lamina propria of C57BL/6 strain	Immune	87.03	+	1.6E-38
B lymphocyte (B220+ CD19+ IgM+) of spleen of C57BL/6 strain	Immune	85.93	+	4.80E-38
Dendritic cell of Peyer's patch of BALB/c strain	Immune	84.02	+	3.2E-37
Mesenchymal cell of proximal forelimb of of CD-1 strain	Integumentary	83.27	−	6.90E-37
Granulocyte (lineage- Gr-1+ clone7.4+) of bone marrow of C57BL/6 strain	Immune	82.65	+	1.3E-36
T lymphocyte (CD4+) of undetermined strain	Immune	82.12	+	2.20E-36
Dendritic cell (CD8 −) of spleen of C57BL/6 strain	Immune	81.97	+	2.5E-36
Granulosa cell of ovary of C57BL/6J x 129S5/SvEvBrd strain	Urogenital	81.23	−	5.30E-36
Mesenchymal cell of distal hindlimb of CD-1 strain	Integumentary	81.03	−	6.4E-36
Mesothelial cell (Mesothelin +) of liver of C57BL/6 strain	Digestive	80.84	+	7.80E-36

The correlation score represents the magnitude of correlation between gene expression profiles, computed with the Running Fisher algorithm, a non-parametric rank-based statistical approach.
Symbols of + and − indicate positive and negative correlation, respectively.
p values are calculated by Fischer's exact probability test.

Pathophysiological Hallmarks Identified by the Transcriptome Analysis of the Adrenocortical Cells of $Star^{-/-}$ Mice

In the study by Ishii $et\ al.$, genes involved in the inflammatory or immune response, the LXR signaling, and the adrenocortical zonation show a significant positive correlation with the DEGs in $Star^{-/-}$ mice-derived adrenocortical cells.[34] The results provide useful information to elucidate the pathophysiology of adrenocortical cells in patients with lipoid CAH.

Inflammatory or Immune Response

The inflammatory or immune response may play a physiological role in adrenocortical cells, especially under stress conditions. Woods and Judd reported that IL-4 signaling inhibited the expression of ACTH-induced steroidogenic

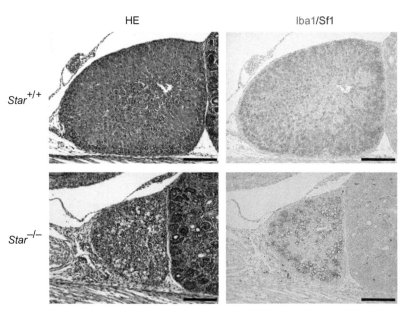

FIGURE 11.3   Immunohistochemistry for Iba1 (a macrophage marker) and *in situ* hybridization for Nr5a1 (a steroidogenic cell marker) in the adrenal glands at E18.5. Red and blue staining represents Iba1 immunoreactivity and Nr5a1 transcripts, respectively. Scale bars are 200 μm. HE, hematoxylin and eosin staining.

enzymes in bovine zona reticularis cells.[36] IL-4 signaling is one of the canonical pathways in which genes are significantly upregulated in *Star*[−/−] mice (Table 11.6). The increased inflammatory or immune response, including IL-4 signaling, could be the second "hit" of the two-hit model for lipoid CAH, attenuating the signal transduction of melanocortin type 2 receptor under high levels of circulating ACTH. The actual role of the inflammatory or immune response in the adrenocortical cells of *Star*[−/−] mice remains unclear.

Based on the results of transcriptome analysis, the expression pattern of a specific macrophage marker, Iba1, was examined in the adrenal glands by immunohistochemical analysis.[37] Fixed whole embryos at E17.5−18.5 were double-stained by immunohistochemistry and *in situ* hybridization for markers of macrophages (Iba1; officially Aif1, allograft inflammatory factor 1) and steroidogenic cells (Sf1; officially Nr5a1), respectively. Iba1 immunoreactivity was

detected in only several *Sf1*-negative resident macrophages within the adrenal cortex of *Star*[+/+] mice, but in a larger number of *Sf1*-negative resident macrophages within the adrenal cortex of *Star*[−/−] mice (Figure 11.3).

These findings suggest a link between the inflammatory or immune system and the adrenal glands. Adrenocortical cells produce various cytokines during acute stress.[38] Additionally, intra-adrenal macrophages influence the function of the adrenal gland in an autocrine or paracrine manner.[38] As shown in Figure 11.3, many macrophages infiltrated into the adrenal glands of *Star*[−/−] mice.[34] *Star*[−/−] mice-derived adrenocortical cells exhibited significantly higher mRNA levels of chemokine ligand 5 (Ccl5) and CD36 antigen (Cd36) than cells from *Star*[+/+] mice.[34] Ccl5, an adipocytokine, recruits macrophages into adipose tissues.[39] CD36, a receptor for apoptosis inhibitor of macrophage (AIM), induces lipolysis and chemokine production in adipocytes.[40] These data suggest a reciprocal

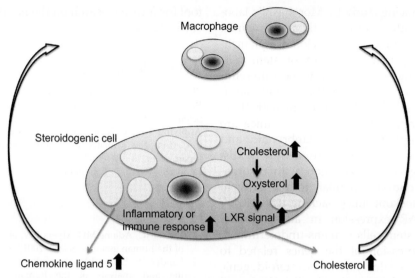

FIGURE 11.4 **Schematic model of "adrenal remodeling" in *Star*$^{-/-}$ mice.** Upward arrow indicates an increased number of molecules or transcripts involved in the index pathways. Blue ovals indicate nuclei; yellow ovals indicate lipid droplets of cholesterol ester in the cytoplasm of adrenocortical cells or macrophages. LXR, liver X receptor.

interaction between adrenocortical cells and infiltrated macrophages upon disruption of StAR. The adrenal glands of *Star*$^{-/-}$ mice may share some features of homeotic inflammation with white adipose tissue in patients with obesity and with vascular walls in those with atherosclerosis,[41−43] which can be called "adrenal remodeling" (Figure 11.4).

LXR Signaling

Significant upregulation of LXRs α and β (Nr1h3 and Nr1h2, respectively) has been observed in *Star*$^{-/-}$ mice-derived adrenocortical cells whose cytoplasms are filled with cholesterol esters.[34] Cummins *et al.* reported that knockout mice lacking LXRs α and β (*Lxrαβ*$^{-/-}$ mice) accumulated cholesterol esters.[44] Hence, it has been inferred that LXRs α and β play an important role in protecting adrenal glands from the accumulation of free cholesterol. Accumulated oxysterol intermediates could function as ligands for LXRs in the adrenocortical cells of *Star*$^{-/-}$ mice and induce cholesterol efflux by increasing the expression of

ATP-binding cassettes (Abca1 and Abcg1). Furthermore, enhanced LXR signaling could inhibit the expression of multiple steroidogenic genes, as reported in the H295R human adrenocortical cell line.[45] Cummins *et al.* showed increased adrenal steroidogenesis even with a decline of StAR expression in *Lxrαβ*$^{-/-}$ mice,[44] indicating a possible link between the repression of LXR signaling and the increase in StAR-independent steroidogenesis. These findings give additional credence to the role of LXRs in maintaining cholesterol homeostasis in adrenocortical cells and imply that LXR upregulation is relevant to the lost residual capacity for StAR-independent steroidogenesis as the second hit in the two-hit model for lipoid CAH.

Adrenocortical Zonation

Genes related to the transition from fetal to adult adrenocortical cells are significantly downregulated in *Star*$^{-/-}$ mice-derived adrenocortical cells.[34] The adult cortex ultimately develops from precursor cells in the fetal cortex, as shown by a lineage tracing study by Morohashi's laboratory.[46] The adult cortex is then maintained through proliferation and the clonal replenishment of subcapsular progenitor or stem cells that probably move centripetally and differentiate into steroidogenic cells in the zona glomerulosa or fasciculata.[22,47−49] Consistent with these ideas, the adrenal glands of $Star^{-/-}$ mice are severely hypoplastic during embryonic and early neonatal stages, and are definitely disturbed during cortical zonation with disorganized mitochondria,[31] similar to the adrenal glands of knockout mice lacking Cyp11a1.[50] Although StAR expression in adrenocortical progenitor or stem cells remains undetermined, decreased expression of the genes related to adrenocortical cell development in steroidogenic cells could directly or indirectly disturb normal proliferation and differentiation in the adrenal cortex of $Star^{-/-}$ mice.

PERSPECTIVE

Transcriptome analysis allows for simultaneous investigation of an entire set of transcripts and permits parallel comparisons between different cells, such as healthy and diseased cells. As shown in this chapter, transcriptome analysis has helped identify significant alteration of gene expression associated with the inflammatory or immune response, the cholesterol efflux, and the adrenocortical zonation in the adrenocortical cells of $Star^{-/-}$ mice. However, mRNA level may not always correlate with the biological function or phenotype of the cells. Biological function or phenotype is also affected by the corresponding protein level and modification. Bearing these limitations in mind, transcriptome analysis is a useful strategy for identifying hitherto undescribed functions or pathways to identify new pathophysiology or therapeutic approaches. If appropriately applied, global profiling of transcripts will provide an essential tool for future research on the nature of adrenocortical cells.

Acknowledgements

This work was supported by a Grant-in-Aid for Scientific Research (C) from the Japan Society for the Promotion of Science, a Health Science Research Grant for Research on Applying Health Technology (Jitsuyoka (Nanbyo)-Ippan-014) from the Ministry of Health, Labor and Welfare, Japan, and a Research Grant from the Yamaguchi Endocrine Research Foundation.

References

1. Venter JC, Adams MD, Myers EW, et al. The sequence of the human genome. *Science* 2001;**291**:1304−51.
2. Lander ES, Linton LM, Birren B, et al. Initial sequencing and analysis of the human genome. *Nature* 2001;**409**:860−921.
3. Mouse Genome Sequencing Consortium, Waterston RH, Lindblad-Toh K, et al. Initial sequencing and comparative analysis of the mouse genome. *Nature* 2002;**420**:520−62.
4. Zhang W, Morris QD, Chang R, et al. The functional landscape of mouse gene expression. *J Biol* 2004;**3**:21.
5. Shalon D, Smith SJ, Brown PO. A DNA microarray system for analyzing complex DNA samples using two-color fluorescent probe hybridization. *Genome Res* 1996;**6**:639−45.
6. MAQC Consortium, Shi L, Reid LH, et al. The MicroArray Quality Control (MAQC) project shows inter- and intraplatform reproducibility of gene expression measurements. *Nat Biotechnol* 2006;**24**:1151−61.
7. Shi L, Campbell G, Jones WD, et al. The MicroArray Quality Control (MAQC)-II study of common practices for the development and validation of microarray-based predictive models. *Nat Biotechnol* 2010;**28**:827−38.
8. Wang Z, Gerstein M, Snyder M. RNA-Seq: a revolutionary tool for transcriptomics. *Nat Rev Genet* 2009;**10**:57−63.
9. Costa V, Aprile M, Esposito R, Ciccodicola A. RNA-Seq and human complex diseases: recent accomplishments and future perspectives. *Eur J Hum Genet* 2013;**21**:134−42.
10. Grewal A, Lambert P, Stockton J. Analysis of expression data: an overview. In: Baxevanis AD, editor. *Current protocols in bioinformatics*. New York: John Wiley; 2007.
11. Benjamini Y, Hochberg Y. Controlling the false discovery rate: a practical and powerful approach to multiple testing. *J R Stat Soc Series B Stat Methodol* 1995;**45**:486−90.

12. Reiner A, Yekutieli D, Benjamini Y. Identifying differentially expressed genes using false discovery rate controlling procedures. *Bioinformatics* 2003;**19**: 368—75.

13. Gene Ontology Consortium. The Gene Ontology project in 2008. *Nucleic Acids Res* 2008;**36**:D440—4.

14. Kanehisa M, Goto S, Kawashima S, Okuno Y, Hattori M. The KEGG resource for deciphering the genome. *Nucleic Acids Res* 2004;**32**:D277—80.

15. Kupershmidt I, Su QJ, Grewal A, et al. Ontology-based meta-analysis of global collections of high-throughput public data. *PLoS One* 2010;**5**:e13066.

16. Salomonis N, Hanspers K, Zambon AC, et al. GenMAPP 2: new features and resources for pathway analysis. *BMC Bioinformatics* 2007;**8**:217.

17. Miller WL, Auchus RJ. The molecular biology, biochemistry, and physiology of human steroidogenesis and Its disorders. *Endocr Rev* 2011;**32**:81—151.

18. Lee JJ, Widmaier EP. Gene array analysis of the effects of chronic adrenocorticotropic hormone in vivo on immature rat adrenal glands. *J Steroid Biochem Mol Biol* 2005;**96**:31—44.

19. Meyer JS. Biochemical effects of corticosteroids on neural tissues. *Physiol Rev* 1985;**65**:946—1020.

20. Schimmer BP, Cordova M, Cheng H, et al. Global profiles of gene expression induced by adrenocorticotropin in Y1 mouse adrenal cells. *Endocrinology* 2006;**147**:2357—67.

21. Lotfi CFP, Todorovic Z, Armelin HA, Schimmer BP. Unmasking a growth-promoting effect of the adrenocorticotropic hormone in Y1 mouse adrenocortical tumor cells. *J Biol Chem* 1997;**272**:29886—91.

22. Kim AC, Barlaskar FM, Heaton JH, et al. In search of adrenocortical stem and progenitor cells. *Endocr Rev* 2009;**30**:241—63.

23. Nishimoto K, Rigsby CS, Wang T, et al. Transcriptome analysis reveals differentially expressed transcripts in rat adrenal zona glomerulosa and zona fasciculata. *Endocrinology* 2012;**153**:1755—63.

24. Heikkilä M, Peltoketo H, Leppäluoto J, Ilves M, Vuolteenaho O, Vainio S. Wnt-4 deficiency alters mouse adrenal cortex function, reducing aldosterone production. *Endocrinology* 2002;**143**:4358—65.

25. Mandel H, Shemer R, Borochowitz ZU, et al. SERKAL syndrome: an autosomal-recessive disorder caused by a loss-of-function mutation in WNT4. *Am J Hum Genet* 2008;**82**:39—47.

26. Artemenko IP, Zhao D, Hales DB, Hales KH, Jefcoate CR. Mitochondrial processing of newly synthesized steroidogenic acute regulatory protein (StAR), but not total StAR, mediates cholesterol transfer to cytochrome P450 side chain cleavage enzyme in adrenal cells. *J Biol Chem* 2001;**276**:46583—96.

27. Jefcoate C. High-flux mitochondrial cholesterol trafficking, a specialized function of the adrenal cortex. *J Clin Invest* 2002;**110**:881—90.

28. Clark BJ, Wells J, King SR, Stocco DM. The purification, cloning, and expression of a novel luteinizing hormone-induced mitochondrial protein in MA-10 mouse Leydig tumor cells. Characterization of the steroidogenic acute regulatory protein (StAR). *J Biol Chem* 1994;**269**:28314—22.

29. Bose HS, Sugawara T, Strauss JF, Miller WL. The pathophysiology and genetics of congenital lipoid adrenal hyperplasia. *N Engl J Med* 1996;**335**:1870—8.

30. Hasegawa T, Zhao L, Caron KM, et al. Developmental roles of the steroidogenic acute regulatory protein (StAR) as revealed by StAR knockout mice. *Mol Endocrinol* 2000;**14**:1462—71.

31. Ishii T, Hasegawa T, Pai C-I, et al. The roles of circulating high-density lipoproteins and trophic hormones in the phenotype of knockout mice lacking the steroidogenic acute regulatory protein. *Mol Endocrinol* 2002;**16**:2297—309.

32. Sasaki G, Ishii T, Jeyasuria P, et al. Complex role of the mitochondrial targeting signal in the function of steroidogenic acute regulatory protein revealed by bacterial artificial chromosome transgenesis *in vivo*. *Mol Endocrinol* 2008;**22**:951—64.

33. Caron KM, Soo SC, Wetsel WC, Stocco DM, Clark BJ, Parker KL. Targeted disruption of the mouse gene encoding steroidogenic acute regulatory protein provides insights into congenital lipoid adrenal hyperplasia. *Proc Natl Acad Sci USA* 1997;**94**:11540—5.

34. Ishii T, Mitsui T, Suzuki S, Matsuzaki Y, Hasegawa T. A genome-wide expression profile of adrenocortical cells in knockout mice lacking steroidogenic acute regulatory protein. *Endocrinology* 2012;**153**:2714—23.

35. ten Hagen TL, van Vianen W, Bakker-Woudenberg IA. Isolation and characterization of murine Kupffer cells and splenic macrophages. *J Immunol Methods* 1996;**193**:81—91.

36. Woods AM, Judd AM. Interleukin-4 increases cortisol release and decreases adrenal androgen release from bovine adrenal cells. *Domest Anim Endocrinol* 2008;**34**:372—82.

37. Kanazawa H. Macrophage/microglia-specific protein Iba1 enhances membrane ruffling and Rac activation via phospholipase C-γ-dependent pathway. *J Biol Chem* 2002;**277**:6—32.

38. Bornstein SR, Rutkowski H, Vrezas I. Cytokines and steroidogenesis. *Mol Cell Endocrinol* 2004;**215**:135—41.

39. Keophiphath M, Rouault C, Divoux A, Clement K, Lacasa D. CCL5 promotes macrophage recruitment and survival in human adipose tissue. *Arterioscler Thromb Vasc Biol* 2009;**30**:39—45.

40. Kurokawa J, Nagano H, Ohara O, et al. Apoptosis inhibitor of macrophage (AIM) is required for obesity-associated recruitment of inflammatory macrophages into adipose tissue. *Proc Natl Acad Sci USA* 2011;**108**:12072–7.

41. Gregor MF, Hotamisligil GS. Inflammatory mechanisms in obesity. *Annu Rev Immunol* 2011;**29**:415–45.

42. Rocha VZ, Libby P. Obesity, inflammation, and atherosclerosis. *Nat Rev Cardiol* 2009;**6**:399–409.

43. Suganami T, Ogawa Y. Adipose tissue macrophages: their role in adipose tissue remodeling. *J Leukoc Biol* 2010;**88**:33–9.

44. Cummins C, Mangelsdorf D. Liver X receptors and cholesterol homoeostasis: spotlight on the adrenal gland. *Biochem Soc Trans* 2006;**34**:1110–3.

45. Nilsson M, Stulnig TM, Lin C-Y, et al. Liver X receptors regulate adrenal steroidogenesis and hypothalamic-pituitary-adrenal feedback. *Mol Endocrinol* 2007;**21**:126–37.

46. Zubair M, Parker KL, Morohashi K-I. Developmental links between the fetal and adult zones of the adrenal cortex revealed by lineage tracing. *Mol Cell Biol* 2008;**28**:7030–40.

47. King P, Paul A, Laufer E. Shh signaling regulates adrenocortical development and identifies progenitors of steroidogenic lineages. *Proc Natl Acad Sci USA* 2009;**106**:21185–90.

48. Huang C-CJ, Miyagawa S, Matsumaru D, Parker KL, Yao HH-C. Progenitor cell expansion and organ size of mouse adrenal is regulated by sonic hedgehog. *Endocrinology* 2010;**151**:1119–28.

49. Ching S, Vilain E. Targeted disruption of Sonic Hedgehog in the mouse adrenal leads to adrenocortical hypoplasia. *Genesis* 2009;**47**:628–37.

50. Hu M-C, Hsu N-C, El Hadj NB, et al. Steroid deficiency syndromes in mice with targeted disruption of Cyp11a1. *Mol Endocrinol* 2002;**16**:1943–50.

Bone as an Endocrine Organ

Gerard Karsenty

Columbia University Medical Center, New York, NY, USA

INTRODUCTION

Bone is a mineralized tissue that has two characteristics, both unique and revealing. The first one is that it is the only tissue in the body that contains a cell type, the osteoclast, whose function is to destroy (resorb) the host tissue. In that sense, osteoclasts are totally different from lymphocytes and macrophages that exist to remove foreign bodies, not to constantly attack our own organs. This destruction of bone that amounts to what can be viewed as a physiological auto-immune reaction does not occur at random, but is part of two, closely related, physiological functions: bone modeling during childhood and remodeling during adulthood.[1-3] In these two functions resorption of bone by osteoclasts is followed by *de novo* bone formation by osteoblasts. This allows longitudinal growth during childhood and renewal of bone and maintenance of bone mass during adulthood.[2-4]

Throughout evolution and to this day bone modeling has fulfilled a function that is a survival one. This is because without bone modeling, the function that holds most of the keys to an in-depth understanding of bone physiology,

there cannot be longitudinal growth, and in absence of growth there is no possibility of walking. It is unlikely that vertebrates would have survived very long without the ability to walk, i.e., to move. This remains true nowadays as well. As for bone remodeling, without it there would have been no way to repair micro- and macrodamages, i.e., fractures, until the 20th century, when orthopedic surgery became readily accessible. That these two functions are both conserved and necessary to survive shed a different light on bone biology, and explain why bone physiology is so intimately connected to the functions of other physiological processes. The osteoclast, like any other cell type, requires a constant flow of energy to perform its daily task of destroying bone. The same is true for osteoblasts, which are responsible for bone formation afterwards. Thus, bone modeling and remodeling require a constant supply of energy to bone cells.[1] The second revealing feature of bone is that it is one of the tissues that covers the largest surface in the body of vertebrates. This is important if one considers that bone modeling and remodeling do not occur in just one, but in multiple, locations at a time everyday from

Cellular Endocrinology in Health and Disease.
DOI: http://dx.doi.org/10.1016/B978-0-12-408134-5.00012-3

birth to death. Since the energetic toll of these two functions presented above is proportional to the surface covered by bones, this leads to the assumption that it must be quite significant. Unfortunately, this is a difficult parameter to measure *in vivo*.

This analysis of bone biology is fully supported by clinical observations. Specifically, anorexia nervosa in children leads to a complete arrest of skeletal growth. This means that in absence of energy intake, since food is a source of energy, there is no possibility of bone modeling. Likewise, adult anorectic patients often develop an osteoporosis. At the other end of the spectrum, adult obese patients often display a higher bone mass that protects them from osteoporosis.[5–9] No matter how one tries to rationalize these observations, taken at face value they say something simple and important: there is a relationship that deserves to be studied between bone mass accrual and food intake. Hence, this conceptual view of bone (re)modeling is supported by these clinical observations, and suggests the following working hypothesis: there may be a coordinated regulation, possibly of endocrine nature, of bone mass accrual and energy metabolism, so that bone modeling and remodeling occur only when energy is available.[1,10]

Besides these biological and clinical observations, one of the most established features of bone pathology is that osteoporosis, a low bone mass disease, appears after menopause.[11–13] In other words, there is a regulation of bone mass by sex steroid hormones. Thus, one could ask: is there an endocrine regulation of sex steroid hormone synthesis by bone-derived hormones? This now broadens the working hypothesis that now becomes that *endocrine regulation of bone mass, energy metabolism, and reproduction maybe coordinated*. This is the hypothesis that serves as the basis of our current work.

Because it is justified by the appearance of bone modeling and remodeling during evolution,

this hypothesis infers that these hormones, if they exist, would preferably appear during evolution with bone. It also implies that the appearance of bone during evolution has changed substantially the physiology of vertebrates throughout their bodies. As a result, if this hypothesis is true, the study of whole-organism vertebrate physiology is best conducted in vertebrates. In the initial phase of testing this hypothesis, the approach of the author's laboratory has been to place emphasis on identifying hormones regulating energy metabolism and showing that indeed they regulate bone mass. This aspect of the work led to the discovery that leptin is a powerful regulator of bone mass and, more importantly, reveals the existence of a central control of bone mass,[1] a topic now studied in many laboratories around the world. More recently, it led to the demonstration that adiponectin antagonizes this aspect of leptin biology.[13] However, if the overarching hypothesis was to be correct, bone must not only be a recipient of hormonal inputs it has to also be an endocrine organ affecting energy metabolism and reproduction. This chapter will summarize the current state of knowledge about this aspect of the general hypothesis.

THE REGULATION OF ENERGY METABOLISM BY BONE

As a laboratory dedicated to the study of bone and osteoblast biology the author's team were interested in elucidating the function of genes that are expressed *exclusively* in osteoblasts, whether they encode secreted or intracellular signaling molecules. *Esp*, the gene that eventually revealed the endocrine nature of bone, was such a gene.

Esp (*embryonic stem cell phosphatase*) encodes a large protein called osteo-testicular tyrosine phosphatase (OST-PTP) containing a long extracellular domain, a transmembrane one and an intracellular tyrosine phosphatase moiety in search of substrates.[14–16] This gene is expressed in only two cell types, the osteoblast and the Sertoli cell of the testis, for the purposes of a laboratory devoted to the study of bone biology. Its pattern of expression by itself justified the study of the function of this gene *in vivo*. This was done through two complementary strategies. The laboratory of Austin Smith knocked in a *Lac Z* allele in the *Esp* locus while we removed, in an osteoblast-specific manner, the phosphatase domain of OST-PTP.[14,17] That the phenotypes developed by each mouse mutant strain were identical made the important point that *Esp* fulfills the functions described in this chapter through its expression in osteoblasts.

The first phenotype that was observed in both $Esp^{-/-}$ and $Esp_{osb}^{-/-}$ mice is that, although they were born at the expected Mendelian ratio, a significant proportion of them died in the first 2 weeks of life. So much so that at weaning, and unlike at birth, we never obtained 25% of homozygous mutant mice when heterozygous mutant mice were intercrossed, while there was no obvious developmental defect of any kind that could explain these postnatal deaths,[15] an extensive biochemical analysis showed that $Esp^{-/-}$ and $Esp_{osb}^{-/-}$ mice were hypoglycemic and hyper-insulinemic.[15,18] A more in-depth analysis showed that glucose stimulated insulin secretion and was increased, compared to what was seen in wild-type (WT) littermates. The same was true for insulin sensitivity that was increased in mice lacking *Esp* in osteoblasts only. Conversely, mice overexpressing *Esp* only in osteoblasts were glucose intolerant because of a decrease in glucose stimulated insulin secretion and sensitivity. In summary, the analysis of *Esp* function showed unambiguously that the osteoblast influences insulin secretion by β-cells of the pancreas and insulin sensitivity in liver, muscle and white adipose tissue.[15] As important as these observations

were, *Esp* could not be responsible alone for these functions of the osteoblasts since it encodes for an intracellular enzyme. Thus these results inferred that osteoblasts must secrete one or several proteins signaling in pancreas and other organs. This was confirmed by a simple and yet powerful experiment. In a co-culture assay in which cells were separated by a filter, osteoblasts, but not the most closely related cell type to the osteoblasts (i.e., fibroblasts), enhanced insulin secretion by islets or β-cells.[15] This experiment established formally that the osteoblast must be an endocrine cell favoring insulin secretion.

The search for what is for now the only known hormone made by osteoblasts regulating glucose metabolism was facilitated by using our understanding of what a polypeptide hormone should be and what we knew about osteoblast biology. The main feature of polypeptide hormones is that they often are cell-specific molecules. When it comes to osteoblasts that narrowed the search rather drastically; there are only two known secreted proteins that are made only by osteoblasts, FGF23 and osteocalcin. What made osteocalcin a credible candidate to be a hormone regulating energy metabolism is that we had noticed at the time we generated them that $Osteocalcin^{-/-}$ mice had an obvious increase in abdominal fat mass. Given the osteoblast-specific nature of osteocalcin and the fact that it is secreted, this observation simply meant that bone affects energy metabolism possibly and at least in part through osteocalcin.[15]

Osteocalcin is a small (45 amino acid-long) protein that is carboxylated on 3 and 4 glutamic acid residues in mice and humans, respectively, and is extremely abundant in the bone extracellular matrix (ECM).[19] Carboxylation of glutamic acid residues transforming is a post-translational modification. conferring to proteins a high affinity for mineral ions. This feature of osteocalcin, and the fact that it is so abundant in an ECM that is mineralized, had long been interpreted as implying that this protein is involved in bone ECM mineralization.[20] Yet, loss-of-function mutation in *Osteocalcin* has unambiguously established that this is not the case.[20]

Besides being present in the bone ECM, osteocalcin can also be found in the general circulation. So it was not unconceivable that osteocalcin might be an osteoblast-derived hormone, regulating glucose metabolism and possibly other aspects of energy metabolism. This hypothesis was verified by showing that, unlike their WT counterparts, $Osteocalcin^{-/-}$ osteoblasts were unable to induce insulin secretion by β-cells of the pancreas. Consistent with the finding, $Osteocalcin^{-/-}$ mice had a metabolic phenotype that is essentially the mirror image of the one observed in $Esp^{-/-}$ mice. Specifically they are hyperglycemic, hypoinsulinemic and insulin resistant in liver, muscle and white adipose tissue. These features are readily seen in $Osteocalcin^{-/-}$ mice that are fed a normal diet. That the glucose intolerance phenotype of $Osteocalcin^{-/-}$ mice was corrected by removing one allele of Esp from these mice established that Esp acts upstream of *Osteocalcin* or, more precisely, that the $Esp^{-/-}$ mice are a model of a gain-of-function for osteocalcin. Remarkably, $Esp^{-/-}$ mice or WT mice receiving exogenous osteocalcin do not develop an obesity or a glucose intolerance phenotype when fed a high fat diet.[20] These results raise the prospect that osteocalcin may be useful as a treatment for type 2 diabetes, a hypothesis that is being tested currently. In the last part of this initial investigation, what was established through multiple *ex vivo* and *in vivo* assays is that the form of osteocalcin that is responsible of its metabolic function is not the carboxylated form but the undercarboxylated one, the least abundant form of circulating osteocalcin, that lack a single gla residue.

In summary, this part of the work demonstrated that bone is an endocrine organ regulating energy metabolism, a function that is itself critical for bone (re)modeling to happen. Furthermore, that *Esp* and *Osteocalcin* expression are regulated by ATF4, a transcription factor highly abundant in osteoblasts, established the importance of the osteoblast and more generally of the skeleton as a determinant of whole-body glucose metabolism. Since the initial description of the osteocalcin metabolic functions was reported in the mouse, numerous studies have indicated that in humans circulating levels of undercarboxylated osteocalcin is a marker of glucose tolerance[21–30] just as it is in the mouse.

COORDINATED REGULATION OF BONE MASS AND ENERGY METABOLISM: THE VIEW POINT OF THE PANCREAS

That there were not one but two genes expressed in osteoblasts regulating glucose metabolism was unexpected, but that one of them encodes an intracellular phosphatase while the other one encodes a hormone whose processing does not require phosphorylation was plain puzzling. In addition, the regulation of insulin secretion by osteocalcin raised another question: does a feedback or feedforward loop of insulin signaling in osteoblasts regulate the expression, secretion or activation of osteocalcin? This is an important question to address for another reason. Mice lacking the insulin receptor in myoblasts or white adipocytes do not develop any glucose intolerance phenotype, when fed a normal diet. This observation could be viewed as a suggestion that insulin may signal in additional organs besides these two to regulate whole-organism glucose homeostasis.

The insulin receptor, which is a tyrosine kinase receptor, does not escape this rule and its activity is negatively regulated, in many insulin target cells, by a tyrosine phosphatase,

PTP-1B; this observation suggested that, if expressed in osteoblasts, the insulin receptor could be a substrate of OST-PTP,[31,32] the tyrosine phosphatase encoded by *ESP*. An implication of this hypothesis is that insulin signaling in osteoblasts may be necessary for glucose homeostasis in animals fed a normal diet. As hypothesized, the insulin receptor is robustly expressed in osteoblasts and is a substrate of ESP. Moreover, selective inactivation of the insulin receptor in osteoblasts results in a glucose intolerance and a decrease in insulin secretion[33] in animals fed a normal diet. Various biochemical and genetic evidence showed that insulin signaling in osteoblasts favors osteocalcin activation by decreasing its carboxylation by hijacking the cellular events that are integral to bone (re)modeling. Osteoblasts are multifunctional cells that are not only responsible for bone formation but also determine osteoclast differentiation through two genes. Those two genes are *RankL*, a positive regulator of osteoclast differentiation, and *Osteoprotegrin* (*Opg*), a soluble receptor sequestering RankL and thus a negative regulator of this process.[2,4] The analysis of mice lacking the insulin receptor in osteoblasts only showed that insulin signaling in this cell type favors bone resorption by inhibiting the expression of *Opg*. Moreover, the expression of two genes expressed in osteoclasts and regulated by RankL signaling, *Cathepsin* K and *Tcirg1*, and contributing to the acidification of the extracellular space around the osteoclasts[34,35] was increased by insulin signaling in osteoblasts. Thus insulin signaling in osteoblasts favors acidification of the bone ECM, a necessary component of bone resorption.

In a superb example of functional dependence between organs, this bone (re)modeling-related function of insulin signaling in osteoblasts has, in fact, major implications for glucose homeostasis. The reason for that is that the only known mechanism to decarboxylate a protein outside a cell is to lower the pH. Hence, by favoring bone resorption, and even

occurring at pH 4.5, insulin signaling in osteoblasts favors decarboxylation, i.e., activation of osteocalcin. This ultimately favors insulin secretion. Thus, in a feedforward loop insulin signals in osteoblasts to enhance bone resorption, which activates osteocalcin and upregulates *Insulin* expression and secretion.

Because of its impact on whole-body glucose homeostasis, the demonstration of this function of insulin signaling in osteoblasts immediately raised another question, which is: does endocrine function of bone also exist in humans? Since bone is the latest organ to appear during evolution, it was intuitively unlikely that functions, or regulation of functions of bone, would differ between mice and humans, especially when they are so important. As a matter of fact, there is no example of that. This is not the case for leptin or serotonin and it is not the case either for the insulin/phosphatase/osteocalcin pathway. An analysis, of osteopetrotic patients showed that a decrease in osteoclast function results in a decrease in the active form of osteocalcin and hypoinsulinemia.[33] The only change between mouse and humans is that *Esp*, which is a pseudogene in humans, is replaced in human osteoblasts by *PTP1B*, a tyrosine phosphatase well known for dephosporylating the insulin receptor.[31,32] The relevance of osteocalcin to human biology was revisited and, more directly demonstrated, once its reproductive function was established (see below).

OSTEOCALCIN REGULATION OF TESTOSTERONE PRODUCTION BY THE LEYDIG CELLS OF THE TESTIS

As mentioned above, the well-known regulation of bone remodeling by gonads[36,37] had suggested that bone may, in turn, through its endocrine function affect reproductive functions. The suspicion was that if this was the case it would preferably be in female animals, since osteoporosis is a disease that affects women more often and more severely than men. Verifying this hypothesis is, in fact, needed to verify the concept that bone energy metabolism and reproduction are coordinately regulated.

To test the validity of this hypothesis, a cell biology approach was again first used. Surprisingly, this approach failed to detect any regulation by osteoblasts of the synthesis of sex steroids hormones in ovaries. In contrast, however, it showed that osteoblasts secrete factor(s) that could markedly increase testosterone production by testis explants and primary Leydig cells.[38] The specificity of this function was verified in three ways: first, supernatant of osteoblast culture could not enhance sex steroid production in ovary explants; second, they could not induce estradiol production by Leydig cells; and, third, no other mesenchymal cell type shared this ability with osteoblasts. This novel role of osteoblasts was verified recently *in vivo*.[39]

Taken together, the facts that osteocalcin is a bone-derived hormone and that *Osteocalcin*$^{-/-}$ mice bred poorly suggested that the afore-described ability of supernatants of osteoblast culture could bind to a novel function of osteocalcin. Testing this hypothesis relied again on the use of a gain-of-function model for osteocalcin (*Esp*$^{-/-}$ mice) and a loss-of-function one (*Osteocalcin*$^{-/-}$ mice).[38] *Osteocalcin*-deficient mice showed a decrease in testes, epididymedes and seminal vesicles weights, whereas the weight of these organs was increased in *Esp*-deficient mice. The spermogram of male *osteocalcin*-deficient mice showed a 50% decrease in sperm count; in contrast, the one of the male *Esp*-deficient mice showed a 30% increase in this parameter and Leydig cells maturation appears to be halted in absence of *Osteocalcin*.[38] These features suggested that osteocalcin might favor testosterone synthesis.

Again, this was first verified by the simple but powerful co-culture assay, and then *in vivo*.[38]

Circulating testosterone levels are low in *Osteocalcin*$^{-/-}$ and high in *Esp*$^{-/-}$ mice. Osteocalcin promotes the expression of all the genes needed for testosterone biosynthesis, but it does not affect expression of the genes encoding the aromatase enzymes and circulating estrogens, which are within the normal range in *Esp*$^{-/-}$ and *Osteocalcin*$^{-/-}$ mice. These observations were consistent with the fact that supernatants of osteoblast culture do not affect estradiol production.[38] To formally establish that osteocalcin regulates testosterone production as a bone-derived hormone and not as a testis-secreted growth factor, mice lacking *Osteocalcin* only in osteoblasts were generated. Male *Osteocalcin*$_{osb}$$^{-/-}$ mice had the same testosterone production defect as the classical *Osteocalcin*$^{-/-}$ mice, while deletion of *Osteocalcin* in Leydig cells did not affect male fertility.[38] Taken together, these experiments established that osteocalcin is a bone-derived hormone favoring fertility in male mice by promoting Leydig cell maturation and testosterone production in the mouse (Figures 12.1 and 12.2). In other words, it verified that for at least one gender there is an endocrine regulator of reproduction by the skeleton. It also illustrates the existence of major differences in the regulation of fertility between male and female mice.

With hindsight, these observations were both surprising and expected. They were surprising because bone is not classically seen as an endocrine organ, much less one regulating reproduction; they were surprising also because of the absence of regulation of fertility in females. On the other hand, they were expected because the feedback rule that applies to most endocrine regulations described to date, given the fact that sex steroid hormones regulate bone mass in both genders, suggesting that such feedback may exist. In broader terms, the existence of this function

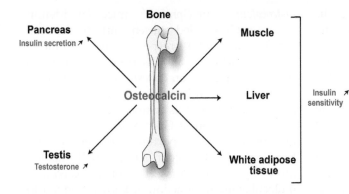

FIGURE 12.1 **Endocrine regulation of energy metabolism by bone.** The endocrine regulation of energy metabolism by the bone is mediated by an osteoblast-specific secreted molecule, osteocalcin, that when undercarboxylated acts as a hormone favoring β-cell proliferation and insulin secretion in pancreas. The mechanism by which osteocalcin could be activated is regulated in osteoblasts by insulin signaling favoring osteocalcin bioavailability by promoting its undercarboxylation. In contrast, the sympathetic tone regulated centrally by leptin, decreases osteocalcin bioactivation.

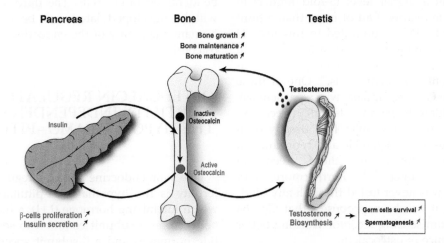

FIGURE 12.2 **Osteocalcin-stimulated testosterone biosynthesis is positively regulated by insulin signaling in osteoblasts.** Insulin signaling in osteoblasts stimulates the bioactivation of osteocalcin. Under feedback loop control, undercarboxylated active osteocalcin then stimulates insulin secretion by the β-cells of the pancreatic islets, promotes insulin sensitivity in peripheral organs and favors testosterone biosynthesis in Leydig cells of the testis. Testosterone, in turn, favors bone growth, maintenance, and maturation.

increases the importance of osteocalcin as a hormone, and of bone as an endocrine organ.

OSTEOCALCIN REPRODUCTIVE FUNCTION IS MEDIATED BY A G-COUPLED RECEPTOR, Gprc6a

In the molecular era, the identification of a novel hormone immediately begs the question of its mechanism of action. A prerequisite to

answering this question is to characterize a receptor to which this hormone would bind specifically on its target cells. In the case of osteocalcin, this was achieved through a two-step strategy, taking advantage of the fact that osteocalcin regulates sex steroid biosynthesis in testes but not in ovaries.[38]

In the first step, what was asked was: What is the signal transduction pathway affected by osteocalcin in two target cells, the β-cell of the pancreas and the Leydig cell of the testis? This approach identified the production of cyclic adenosine monophosphate (cAMP) as the only intracellular signaling event triggered reproducibly by osteocalcin in these two cell types. We interpreted this result as suggesting that an osteocalcin receptor is probably a G protein-coupled receptor (GPCR) linked to adenylate cyclase. Therefore, in the second step of this experimental strategy, and taking full advantage of the dichotomy of function of osteocalcin between males and females, we asked whether there were orphan GPCRs expressed at a higher level (5-fold higher) in testes than in ovaries. Out of more than a hundred orphan GPCRs submitted to this test, 22 of them were more expressed in testes than in ovaries and only four were expressed predominantly or only in Leydig cells.[40] One of these four orphan GPCRs, Gprc6a, was a particularly good candidate to be an osteocalcin receptor, since its inactivation in mice had already been shown to result in metabolic and reproduction phenotypes similar to those seen in *Osteocalcin*$^{-/-}$ mice.[40–42] Furthermore, and although it was never tested through any binding assays, it has been proposed that Gprc6a was a calcium-sensing receptor working better in the presence of osteocalcin.

Although the afore-mentioned result could not be reproduced, several criteria formally identified Gprc6a as an osteocalcin receptor present in Leydig cells. First, there is direct binding of osteocalcin to WT but not to *Gprc6a*-deficient Leydig cells; second, osteocalcin increases cAMP production in WT but not in *Gprc6a*-deficient Leydig cells; third, and more to the point, a Leydig cell-specific deletion of *Gprc6a* revealed a reproduction phenotype caused by low testosterone production similar, if not identical, to the one seen in the case of osteocalcin inactivation; fourth, in an even more convincing experiment, compound heterozygous mice lacking one copy of *Osteocalcin* and one copy of *Gprc6a* had a reproduction phenotype identical in all aspects to the one seen in

Osteocalcin$^{-/-}$ or *Gprc6a*$^{-/-}$ mice. The identification of Gprc6a led subsequently to the realization that cAMP response element-binding protein (CREB) is a transcriptional effector of osteocalcin regulation of testosterone biosynthesis by favoring in Leydig cells the expression of key enzymes of this biosynthetic pathway.[38] The identification of Gprc6a will, in the future, give us a greater opportunity to identify novel functions of osteocalcin. Also, it will enable a more sophisticated dissection of the osteocalcin molecular mode of action in known and yet-to-be-identified target cells. The third aspect, that will be developed later, will be to verify in humans the reality of the endocrine function of osteocalcin.

OSTEOCALCIN REGULATES MALE FERTILITY INDEPENDENTLY OF THE HYPOTHALAMO–PITUITARY AXIS

The main endocrine pathway regulating male fertility is the hypothalamo–pituitary axis, in which luteinizing hormone (LH), a heterodimer between an α-subunit common to several peptide hormones and a β-subunit specific to LH, favors testosterone biosynthesis. Although less severe, the reproductive phenotype of *Osteocalcin*$^{-/-}$ and *Gprc6a*$^{-/-}$ male mice bears resemblance to the one seen in *Lhb*$^{-/-}$ (LH-deficient) male mice, as they are both characterized by a defect in testosterone synthesis and testosterone-dependent events.[43,44] Yet, a remarkable feature of the reproduction phenotype observed in *Osteocalcin*$^{-/-}$ or *Gprc6a*$^{-/-}$ mice is that it develops in the face of an increase in circulating levels of LH. This situation raised the following question: Does osteocalcin act downstream of LH or does the realization that osteocalcin regulates male fertility reveal the existence of two different pathways, both necessary for male fertility, one pituitary-dependent and one bone-dependent? To address this question, rescue analysis was undertaken of the $Lhb^{-/-}$ or $Osteocalcin^{-/-}$ mice with either osteocalcin or human chorionic gonadotropin (hCG), respectively.

First, it was shown that circulating levels of the active form (undercarboxylated) of osteocalcin were not lower in $Lhb^{-/-}$ than in WT male mice, and daily injections of osteocalcin for 1 month in 6 week-old $Lhb^{-/-}$ male mice did not normalize circulating testosterone levels.[39] Second, histological analysis of testes of 10 week-old $Lhb^{-/-}$ male mice injected with osteocalcin failed to show any improvement in spermatogenesis, testis size and weight or a reversal of their Leydig cell hypoplasia.[39] Taken together, these experiments indicate that the regulation of testosterone synthesis by osteocalcin does not depend on a measurable influence of Gprc6a on Lh expression.

Conceivably, however, LH could be required for osteocalcin stimulation of testosterone biosynthesis by Leydig cells. Yet, two experimental sets of evidence suggested that it is not the case. First, the positive effect of osteocalcin on testosterone synthesis in Leydig cells was recorded when cells were maintained in serum-free medium, i.e., in total absence of LH.[39] Second, in cell culture LH does not regulate expression of $Osteocalcin$ or the gene modifying it in osteoblasts. In summary, these results support the notion that osteocalcin regulates male fertility independently of the hypothalamo–pituitary axis; they also failed to provide any evidence that LH regulates $Osteocalcin$ expression.[39]

INSULIN SIGNALING IN OSTEOBLASTS AS A DETERMINANT OF OSTEOCALCIN REPRODUCTIVE FUNCTION IN THE MOUSE

Dissociating pituitary-dependent from bone-dependent regulation of male fertility suggests the existence of a second axis regulating this function, and raised the question of the identity of upstream regulators of osteocalcin reproductive function. That the ability of osteocalcin to favor glucose homeostasis is determined by osteoclastic bone resorption raised the question of whether male fertility is another physiological function to be added to the credit of bone resorption. Testing this hypothesis using a loss-of-function and gain-of-function mouse models of bone resorption, it was shown that bone resorption is a physiological determinant of osteocalcin's regulation of testosterone production and male reproductive function through its ability to activate osteocalcin.[39] In addition, it was recently shown in humans that osteocalcin and the bone turnover is associated with testosterone circulating levels in general population and in patients with bone disorders. The data presented so far support a model in which osteoclast-mediated bone resorption regulates male fertility in mice through the decarboxylation and the activation of osteocalcin. This model combined with previously published observations, also implies that bone resorption is required, not only for male fertility, but also for the control of energy metabolism.

The cardinal role of bone resorption in the regulation of testosterone production provided an opportunity to look for additional upstream regulators of osteocalcin reproductive function. As described previously, insulin signaling in osteoblasts enhances osteocalcin activity, which in turn favors insulin secretion. Consequently, insulin signaling in osteoblasts might influence testosterone biosynthesis in an osteocalcin-dependent manner.[39]

This possibility was addressed by analyzing male mice lacking the gene encoding for insulin receptor selectively in osteoblasts ($InsR_{osb}^{-/-}$). These animals, that have less active osteocalcin, demonstrated a decrease in testes size and weight, in epididymides and seminal vesicle weights, in sperm count and circulating testosterone levels. Lastly, this observation was firmly confirmed generating and testing compound mutant mice lacking one allele of *InsR* in osteoblasts and one allele of either *Osteocalcin* or *Gprc6a*. Here again, with regard to testis, epididymides, and seminal vesicle weights, sperm count and circulating levels of testosterone, these compound mutant mice demonstrated abnormalities that were similar to those seen in $InsR_{osb}^{-/-}$, *Osteocalcin*$^{-/-}$, or even more relevant, in *Gprc6a*$^{-/-}$ mice.[38] Taken together, these observations suggested strongly the existence of a pancreas–bone–testis axis in the control of male reproductive functions that acts in parallel to the hypothalamus–pituitary–testis axis.[39]

CONSERVATION OF THE REPRODUCTIVE FUNCTION OF OSTEOCALCIN IN HUMANS

A second legitimate question that has plagued osteocalcin research since this molecule was recognized to be a hormone in rodents has been to determine whether it also has an endocrine function in humans. The function of osteocalcin as a regulator of testosterone production has been recently extended to humans. It was recently shown in humans that osteocalcin and bone turnover is associated with testosterone circulating levels in the general population and in patients with bone disorders.[45] Moreover, Dr. Khosla's group also showed that there is a significant association between serum osteocalcin and testosterone levels during mid-puberty in males.[46] They postulate that this axis may be most relevant during rapid skeletal growth in adolescent human males to help maximize bone size. However, while there is a growing body of evidence that osteocalcin serum levels are a reliable indicator of the degree of insulin secretion, insulin sensitivity, and circulating serum testosterone levels in humans,[24,47] there was,

until recently, no genetic evidence establishing that osteocalcin fulfills its endocrine functions in humans.

As mentioned earlier in this chapter, the identification of an osteocalcin receptor and the realization that osteocalcin influences male fertility, provided an opportunity to tackle this issue. In fact, the fertility phenotype of the *Osteocalcin*$^{-/-}$ mice, mainly characterized by subfertility, mediocre spermogram, low circulating testosterone levels, and high circulating LH levels,[38] is the exact phenocopy of a rare but well-defined syndrome in humans called peripheral testicular insufficiency.[48,49] Thus, a systematic genomic analysis of *Osteocalcin* and *GPRC6A* loci of a cohort of 59 patients presenting this syndrome was initiated, with the goal to identify loss-of-function mutation in *Osteocalcin* or its receptor that would explain their clinical presentation.[39]

As a result of this genomic analysis, two patients in this cohort harbored a point mutation $T > A$ transversion in exon 4 of GPRC6A.[39] This missense mutation was shown to result in an amino acid substitution (F464Y) in a highly conserved region of one of the transmembrane domain of GPRC6A and prevent its localization to the cell membrane and, therefore, resulted in a loss of function of this receptor.[39] Furthermore, three different cell-based assays indicated that this mutation also acts in a dominant negative manner in cells.[68] Lastly, it was also noted that both patients harboring this substitution-mutation in *GPRC6A* originated from the same region of the globe, shared a history of glucose intolerance, and displayed similar defects in reproductive hormones. Indeed, these patients presented a metabolic syndrome characterized by an increase in body mass index, as well as a glucose intolerance determined by hyperinsulinemia after fasting, a glucose tolerance test, and an insulin tolerance test.[39] Many of these features are seen in mice lacking *Osteocalcin* or *Gprc6a* in all cells.[38] It is, thus, interesting that the F464Y variant was not observed in 1000 controls. Careful phenotypic analysis of individuals carrying the F464Y allele may clarify the spectrum of associated metabolic, cardiovascular and reproductive defects. Taken together, these results indicate the importance of osteocalcin signaling in humans and suggest that GPRC6A may be a new susceptibility locus for primary testicular failure in humans, a disease the cause of which is often unidentified. Results of this initial foray in the genetic analysis of osteocalcin functions in humans should be viewed as a stepping-stone to perform a more systematic analysis in a larger patient population with primary testicular failure as well as in patients with glucose intolerance or metabolic syndromes.

The functions of the skeleton are not all known yet, but the ones we know indicate that skeleton physiology affects many more organs and functions than skeleton itself, including glucose homeostasis, energy expenditure, and fertility. These novel functions of bone underscore the notion that skeleton is an important member of an endocrine network, affecting multiple functions in the body. As such they raise the prospect that all endocrine functions of the skeleton have not been described yet.

References

1. Karsenty G. Convergence between bone and energy homeostases: leptin regulation of bone mass. *Cell Metab* 2006;**4**:341–8.
2. Rodan GA, Martin TJ. Therapeutic approaches to bone diseases. *Science* 2000;**289**:1508–14.
3. Teitelbaum SL. Osteoclasts, integrins, and osteoporosis. *J Bone Miner Metab* 2000;**18**:344–9.
4. Harada S, Rodan GA. Control of osteoblast function and regulation of bone mass. *Nature* 2003;**423**:349–55.
5. Reid IR. Relationships among body mass, its components, and bone. *Bone* 2002;**31**:547–55.
6. Rigotti NA, Nussbaum SR, Herzog DB, Neer RM. Osteoporosis in women with anorexia nervosa. *N Engl J Med* 1984;**311**:1601–6.
7. Wolfert A, Mehler PS. Osteoporosis: prevention and treatment in anorexia nervosa. *Eat Weight Disord* 2002;**7**:72–81.
8. Zhao LJ, Liu YJ, Liu PY, Hamilton J, Recker RR, Deng HW. Relationship of obesity with osteoporosis. *J Clin Endocrinol Metab* 2007;**92**:1640–6.
9. Zipfel S, Seibel MJ, Lowe B, Beumont PJ, Kasperk C, Herzog W. Osteoporosis in eating disorders: a follow-up study of patients with anorexia and bulimia nervosa. *J Clin Endocrinol Metab* 2001;**86**:5227–33.
10. Ducy P, Amling M, Takeda S, Priemel M, Schilling AF, et al. Leptin inhibits bone formation through a hypothalamic relay: a central control of bone mass. *Cell* 2000;**100**:197–207.
11. Khosla S. Update on estrogens and the skeleton. *J Clin Endocrinol Metab* 2010;**95**:3569–77.
12. Riggs BL, O'Fallon WM, Muhs J, O'Connor MK, Kumar R, Melton III LJ. Long-term effects of calcium supplementation on serum parathyroid hormone level, bone turnover, and bone loss in elderly women. *J Bone Miner Res* 1998;**13**:168–74.
13. Kajimura D, Lee HW, Riley Kyle J, Arteaga-Solis E, Ferron M, Zhou B, et al. Adiponectin regulates bone mass via opposite central and peripheral mechanisms through FoxO1. *Cell Metabolism* 2013;**17**(6):901–15.
14. Lee K, Nichols J, Smith A. Identification of a developmentally regulated protein tyrosine phosphatase in embryonic stem cells that is a marker of pluripotential epiblast and early mesoderm. *Mech Dev* 1996;**59**:153–64.
15. Lee NK, Sowa H, Hinoi E, Ferron M, Ahn JD, et al. Endocrine regulation of energy metabolism by the skeleton. *Cell* 2007;**130**:456–69.
16. Mauro LJ, Olmsted EA, Skrobacz BM, Mourey RJ, Davis AR, Dixon JE. Identification of a hormonally regulated protein tyrosine phosphatase associated with bone and testicular differentiation. *J Biol Chem* 1994;**269**:30659–67.
17. Dacquin R, Mee PJ, Kawaguchi J, Olmsted-Davis EA, Gallagher JA, et al. Knock-in of nuclear localised β-galactosidase reveals that the tyrosine phosphatase Ptprv is specifically expressed in cells of the bone collar. *Dev Dyn* 2004;**229**:826–34.
18. Ferron M, Hinoi E, Karsenty G, Ducy P. Osteocalcin differentially regulates β cell and adipocyte gene expression and affects the development of metabolic diseases in wild-type mice. *Proc Natl Acad Sci USA* 2008;**105**:5266–70.
19. Hauschka PV, Lian JB, Cole DE, Gundberg CM. Osteocalcin and matrix Gla protein: vitamin K-dependent proteins in bone. *Physiol Rev* 1989;**69**:990–1047.
20. Ducy P, Desbois C, Boyce B, Pinero G, Story B, et al. Increased bone formation in osteocalcin-deficient mice. *Nature* 1996;**382**:448–52.
21. Aonuma H, Miyakoshi N, Hongo M, Kasukawa Y, Shimada Y. Low serum levels of undercarboxylated

osteocalcin in postmenopausal osteoporotic women receiving an inhibitor of bone resorption. *Tohoku J Exp Med* 2009;**218**:201−5.

22. Fernandez-Real JM, Izquierdo M, Ortega F, Gorostiaga E, Gomez-Ambrosi J, et al. The relationship of serum osteocalcin concentration to insulin secretion, sensitivity, and disposal with hypocaloric diet and resistance training. *J Clin Endocrinol Metab* 2009;**94**:237−45.

23. Hwang YC, Jeong IK, Ahn KJ, Chung HY. The uncarboxylated form of osteocalcin is associated with improved glucose tolerance and enhanced β-cell function in middle-aged male subjects. *Diabetes Metab Res Rev* 2009;**25**:768−72.

24. Im JA, Yu BP, Jeon JY, Kim SH. Relationship between osteocalcin and glucose metabolism in postmenopausal women. *Clin Chim Acta* 2008;**396**:66−9.

25. Kanazawa I, Yamaguchi T, Yamamoto M, Yamauchi M, Kurioka S, et al. Serum osteocalcin level is associated with glucose metabolism and atherosclerosis parameters in type 2 diabetes mellitus. *J Clin Endocrinol Metab* 2009;**94**:45−9.

26. Kindblom JM, Ohlsson C, Ljunggren O, Karlsson MK, Tivesten A, et al. Plasma osteocalcin is inversely related to fat mass and plasma glucose in elderly Swedish men. *J Bone Miner Res* 2009;**24**:785−91.

27. Levinger I, Zebaze R, Jerums G, Hare DL, Selig S, Seeman E. The effect of acute exercise on undercarboxylated osteocalcin in obese men. *Osteoporos Int* 2011;**2**(5):1621−66.

28. Pittas AG, Harris SS, Eliades M, Stark P, Dawson-Hughes B. Association between serum osteocalcin and markers of metabolic phenotype. *J Clin Endocrinol Metab* 2009;**94**:827−32.

29. Winhofer Y, Handisurya A, Tura A, Bittighofer C, Klein K, et al. Osteocalcin is related to enhanced insulin secretion in gestational diabetes mellitus. *Diabetes Care* 2010;**33**:139−43.

30. Yeap BB, Chubb SA, Flicker L, McCaul KA, Ebeling PR, et al. Reduced serum total osteocalcin is associated with metabolic syndrome in older men via waist circumference, hyperglycemia, and triglyceride levels. *Eur J Endocrinol* 2010;**163**:265−72.

31. Delibegovic M, Bence KK, Mody N, Hong EG, Ko HJ, et al. Improved glucose homeostasis in mice with muscle-specific deletion of protein-tyrosine phosphatase 1B. *Mol Cell Biol* 2007;**27**:7727−34.

32. Delibegovic M, Zimmer D, Kauffman C, Rak K, Hong EG, et al. Liver-specific deletion of protein-tyrosine phosphatase 1B (PTP1B) improves metabolic syndrome and attenuates diet-induced endoplasmic reticulum stress. *Diabetes* 2009;**58**:590−9.

33. Ferron M, Wei J, Yoshizawa T, Del Fattore A, DePinho RA, et al. Insulin signaling in osteoblasts integrates bone remodeling and energy metabolism. *Cell* 2010;**142**:296−308.

34. Saftig P, Hunziker E, Wehmeyer O, Jones S, Boyde A, et al. Impaired osteoclastic bone resorption leads to osteopetrosis in cathepsin-K-deficient mice. *Proc Natl Acad Sci USA* 1998;**95**:13453−8.

35. Scimeca JC, Franchi A, Trojani C, Parrinello H, Grosgeorge J, et al. The gene encoding the mouse homologue of the human osteoclast-specific 116-kDa V-ATPase subunit bears a deletion in osteosclerotic (*oc/oc*) mutants. *Bone* 2000;**26**:207−13.

36. Khosla S, Riggs BL. Pathophysiology of age-related bone loss and osteoporosis. *Endocrinol Metab Clin North Am* 2005;**34**:1015−30.

37. Khosla S, Melton LJ, Atkinson EJ, O'Fallon WM. Relationship of serum sex steroid levels to longitudinal changes in bone density in young versus elderly men. *J Clin Endocrinol Metab* 2001;**86**:3555−61.

38. Oury F, Sumara G, Sumara O, Ferron M, Chang H, Smith CE, et al. Endocrine regulation of male fertility by the skeleton. *Cell* 2011;**144**:796−809.

39. Oury F, et al. Osteocalcin regulates murine and human fertility through a pancreas-bone-testis axis. *J Clin Invest* 2013;**123**:1−13.

40. Pi M, et al. GPRC6A null mice exhibit osteopenia, feminization and metabolic syndrome. *PloS One* 2008;**3**: e3858.

41. Pi M, Quarles LD. Multiligand specificity and wide tissue expression of GPRC6A reveals new endocrine networks. *Endocrinology* 2012;**153**:2062−9.

42. Pi M, Wu Y, Quarles LD. GPRC6A mediates responses to osteocalcin in beta-cells in vitro and pancreas *in vivo*. *J Bone Miner Res* 2011;**26**:1680−3.

43. Kumar TR. Functional analysis of LHbeta knockout mice. *Mol Cell Endocrinol* 2007;**269**:81−4.

44. Burns KH, Matzuk MM. Minireview: genetic models for the study of gonadotropin actions. *Endocrinology* 2002;**143**:2823−35.

45. Saleem U, Mosley Jr. TH, Kullo IJ. Serum osteocalcin is associated with measures of insulin resistance, adipokine levels, and the presence of metabolic syndrome. *Arterioscler Thromb Vasc Biol* 2010;**30**:1474−8.

46. Kirmani S, et al. Relationship of testosterone and osteocalcin levels during growth. *J Bone Miner Res* 2011;**26**: 2212−6.

47. Fernandez-Real JM, et al. The relationship of serum osteocalcin concentration to insulin secretion, sensitivity, and disposal with hypocaloric diet and resistance training. *J Clin Endocrinol Metab* 2009;**94**:237−45.

48. Glass AR, Vigersky RA. Testicular reserve of testosterone precursors in primary testicular failure. *Fertil Steril* 1982;**38**:92–6.

49. Winters SJ, Troen P. A reexamination of pulsatile luteinizing hormone secretion in primary testicular failure. *J Clin Endocrinol Metab* 1983;**57**:432–5.

50. Kong YY, Boyle WJ, Penninger JM. Osteoprotegerin ligand: a common link between osteoclastogenesis, lymph node formation and lymphocyte development. *Immunol Cell Biol* 1999;**77**:188–93.

51. Teitelbaum SL, Ross FP. Genetic regulation of osteoclast development and function. *Nat Rev Genet* 2003;**4**:638–49.

13

Regulation of Steroidogenesis

Andrew A. Bremer* and Walter L. Miller[†]

*Vanderbilt University, Nashville, TN, USA, [†]University of California, San Francisco, CA, USA

INTRODUCTION

Steroidogenesis is regulated at multiple levels, principally by transcription of genes encoding steroidogenic enzymes and co-factors, and by their post-translational modification. An understanding of steroidogenesis and its regulation first requires an understanding of the biochemistry and genetics of these enzymes and co-factors. The patterns of gland and cell type-specific steroidogenesis reflect variations in these regulatory mechanisms. Understanding the roles of steroidogenic factors has been facilitated by identifying their genetic lesions, which cause rare disorders of steroidogenesis. Understanding steroidogenesis and its regulation are important for understanding disorders of sexual differentiation, reproduction, fertility, hypertension, obesity, and physiologic homeostasis.

CLASSES OF STEROIDOGENIC ENZYMES

Most steroidogenic enzymes are either cytochrome P450s or hydroxysteroid dehydrogenases (HSDs). These enzymes are functionally unidirectional, so the accumulation of product does not drive flux back to the precursor. All P450-mediated hydroxylations and carbon—carbon bond cleavage reactions are mechanistically irreversible. Alternatively, HSD reactions are mechanistically reversible and can run in either direction under certain conditions *in vitro*, but each HSD enzyme drives steroid flux predominantly in either the oxidative or reductive mode *in vivo*.[1] At least two HSD enzymes drive the flux of a hydroxysteroid and its cognate ketosteroid in opposite directions, favoring either ketosteroid reduction or hydroxysteroid oxidation.

Cytochrome P450 Enzymes

Most steroidogenic enzymes are members of the cytochrome P450 group of ezymes,[2] a generic term for a group of oxidases, all of which have about 500 amino acids and contain a single heme group. The enzymes are termed P450 (pigment 450) because they absorb light at 450 nm in their reduced states. The genes for these enzymes are formally termed *CYP* genes, and a systematic nomenclature has

Cellular Endocrinology in Health and Disease.
DOI: http://dx.doi.org/10.1016/B978-0-12-408134-5.00013-5

been described (http://drnelson.uthsc.edu/cytochromeP450.html); the encoded proteins may be given the same name without italics. Cytochrome P450 enzymes bind their substrates and achieve catalysis in an active site associated with the heme group. The human genome has 57 *CYP* genes; 7 encode "Type 1" P450s targeted to mitochondria; the other 50 encode "Type 2" P450s targeted to the endoplasmic reticulum. Although the liver metabolizes countless endogenous and exogenous toxins, drugs, xenobiotics, and environmental pollutants, these reactions are catalyzed by only about eight P450s, and most P450s participate in biosynthetic processes.[3] Each P450 enzyme can metabolize multiple substrates, catalyzing a broad array of oxidations.

Six P450 enzymes are involved in steroidogenesis (Figure 13.1). Mitochondrial P450scc (CYP11A1) is the cholesterol side-chain cleavage enzyme catalyzing the series of reactions

formerly termed "20,22 desmolase." Two distinct isozymes of P450c11, P450c11β (11β-hydroxylase; CYP11B1) and P450c11AS (aldosterone synthase; CYP11B2), also found in mitochondria, catalyze 11β-hydroxylase, 18-hydroxylase, and 18-methyl oxidase activities. P450c17 (CYP17A1), found in the endoplasmic reticulum, catalyzes both 17α-hydroxylase and 17,20-lyase activities, and P450c21 (CYP21A2) catalyzes the 21-hydroxylation of both glucocorticoids and mineralocorticoids. In the gonads and elsewhere, P450aro (CYP19A1) in the endoplasmic reticulum catalyzes aromatization of androgens to estrogens.

Hydroxysteroid Dehydrogenase Enzymes

The HSD enzymes have molecular masses of ~35–45 kDa, do not have heme groups, and require nicotine adenine dinucleotide or its phosphate (NADH/NAD$^+$ or NADPH/NADP$^+$)

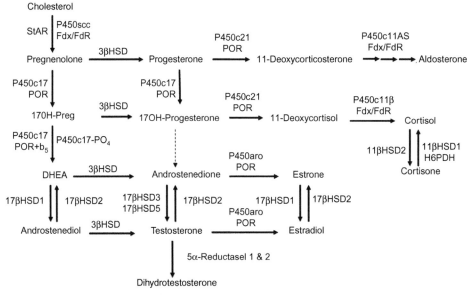

FIGURE 13.1 **The major human steroidogenic pathways.** Key enzymes and cofactors are shown near arrows indicating chemical reactions. The steroids in the first column are Δ^5-steroids, which constitute the preferred pathway to C$_{19}$ steroids in humans. The dashed arrow indicates poor flux from 17α-hydroxyprogesterone to androstenedione via P450c17, and the three small arrows below P450c11AS emphasizes the three discrete steps with intermediates corticosterone and 18-hydroxycorticosterone. Note: not all the intermediate steroids, pathways, and enzymes are shown.

as co-factors to either reduce or oxidize a steroid by two electrons via a hydride transfer mechanism. Unlike most steroidogenic reactions catalyzed by P450 enzymes, which are due to the action of a single form of P450, each HSD reaction can be catalyzed by at least two, often very different, isozymes. Members of this family include the 3α- and 3βHSDs, the two 11βHSDs, and a series of 17βHSDs; the 5α-reductases are unrelated to this family.

The HSD enzymes are categorized into two groups: the short-chain dehydrogenase reductase (SDR) family, characterized by a "Rossman fold," and the aldo-keto reductase (AKR) family, characterized by a triosephosphate isomerase (TIM) barrel motif.[1] The SDR enzymes include 11βHSDs 1 and 2, and 17βHSDs 1, 2, 3, and 4; the AKR enzymes include 17βHSD5, which is important in extra-glandular activation of androgenic precursors, and the 3αHSDs that participate in the so-called "backdoor pathway" of fetal androgen synthesis (see below). Based on their activities, it is physiologically more useful to classify these enzymes as dehydrogenases or reductases: the dehydrogenases use NAD^+ as their cofactor to oxidize hydroxysteroids to ketosteroids; the reductases mainly use NADPH to reduce ketosteroids to hydroxysteroids. Although these enzymes are typically bidirectional *in vitro*, they physiologically tend to function in only one direction *in vivo*, with the direction determined by the available intracellular cofactor(s).[1]

STEROID HORMONE BIOSYNTHESIS

Cholesterol Uptake, Storage, Transport, and Delivery to the Mitochondria

Steroidogenic cells can synthesize cholesterol *de novo* from acetate, but most of their cholesterol supply comes from dietary-derived low-density lipoproteins (LDL). The intracellular cholesterol pool is then regulated by the sterol

response element binding proteins (SREBPs), a group of transcription factors involved in the biosynthesis of cholesterol and fatty acids.[4] Adequate concentrations of LDL suppress 3-hydroxy-3-methylglutaryl co-enzyme A (HMG CoA) reductase, the rate-limiting enzyme in cholesterol synthesis. However, tropic hormones that act via cyclic adenosine monophosphate (cAMP) (adrenocorticotropic hormone (ACTH) in the adrenal, luteinizing hormone (LH) in the gonad) stimulate the synthesis of HMG CoA reductase and LDL receptors and uptake of LDL cholesterol (LDL-C). LDL-C esters are taken up by receptor-mediated endocytosis, and are then stored directly or converted to free cholesterol for use in steroid hormone biosynthesis. Free cholesterol can be esterified by acyl-CoA:cholesterol transferase (ACAT) and stored in lipid droplets. Cholesterol is accessed by the NPC proteins (named for their causative role in Niemann–Pick type C disease) and by hormone-sensitive lipase (HSL) in late endosomes. ACTH stimulates HSL and inhibits ACAT, thus increasing the availability of free cholesterol for steroid hormone synthesis. The unesterified "free" cholesterol may travel from endosomes to the mitochondria by either vesicular transport, involving membrane fusion, or by non-vesicular transport, bound to cytosolic START (StAR-related lipid transfer) proteins that are structurally related to the steroidogenic acute regulatory protein, StAR. However, StAR itself plays a minor role in cholesterol transport to the mitochondria, but instead functions to move cholesterol from the outer mitochondrial membrane (OMM) to the inner mitochondrial membrane (IMM).[5]

The Steroidogenic Acute Regulatory Protein, StAR

Whereas cells that produce polypeptide hormones can store hormone in secretory vesicles for immediate release in response to altered membrane potentials, steroidogenic cells store very little steroid, so that an acute demand for more steroid requires the rapid synthesis of steroid. ACTH, a proteolytic product of pituitary pro-opiomelanocortin (POMC), and possibly other POMC-peptides are *trophic* factors that stimulate adrenal cellular growth via fibroblast growth factor (FGF), epidermal growth factor (EGF), and insulin-like growth factor 2 (IGF2). ACTH is also a *tropic* factor that stimulates the chronic regulation of steroidogenesis by stimulating the transcription of genes for steroidogenic enzymes, principally *CYP11A1* encoding P450scc. The rapid induction of steroidogenesis, such as in the classic "fight-or-flight" response or in response to intravenously administered ACTH, is mediated by regulating the movement of cholesterol from the OMM to the IMM.[6] This mitochondrial inflow of cholesterol is regulated by the steroidogenic acute regulatory protein, StAR, which thus regulates steroidogenesis by regulating the access of cholesterol substrate to the P450scc enzyme residing on the IMM.[7]

Historically, when steroidogenic cells or intact rats were treated with inhibitors of protein synthesis such as cycloheximide, the acute steroidogenic response was eliminated, suggesting that a short-lived protein acts at the level of the mitochondrion as the specific trigger to the acute steroidogenic response. This factor was identified as short-lived 30- and 37-kDa phosphoproteins that were rapidly synthesized when steroidogenic cells were stimulated with tropic hormones, then cloned from mouse Leydig MA-10 cells and named the steroidogenic acute regulatory protein, StAR.[6,7] The central role of StAR in steroidogenesis was proven by finding that StAR mutations caused congenital lipoid adrenal hyperplasia.[8,9] Thus, StAR is the acute trigger that is required for the rapid flux of cholesterol from the outer to the inner mitochondrial membrane that is needed for the acute response of

aldosterone to angiotensin II, of cortisol to ACTH, and of sex steroids to an LH pulse.

P450scc

The first and rate-limiting step in steroidogenesis is the conversion of cholesterol to pregnenolone by cytochrome P450scc (encoded by *CYP11A1*). The expression of this enzyme renders a cell "steroidogenic" and determines its steroidogenic capacity, thus acting as the long-term chronic regulator of steroidogenesis.[2,5] P450scc catalyzes three distinct chemical reactions, 20α-hydroxylation, 22-hydroxylation, and scission of the cholesterol side-chain to yield pregnenolone and isocaproic acid. No other enzyme has been found that can produce pregnenolone from any source. The dissociation constants for the 20-OH-cholesterol and 20,22-(OH)-cholesterol intermediates are high, so that these steroids tend to remain bound to the active site, awaiting the next reaction. The transcription of the single *CYP11A1* gene on chromosome 15 is regulated differently in different cell types. In the adrenal zona fasciculata and gonad, ACTH and LH act via cAMP utilizing specific *cis*-active promotor regions; in the adrenal zona glomerulosa, angiotensin II acts via the calcium/calmodulin pathway utilizing wholly different *cis*-elements of the same promoter, but in both cases steroidogenic factor-1 (SF1) is essential. By contrast, placental expression of *CYP11A1*, which is essential for placental synthesis of the progesterone that suppresses uterine contractility and hence maintains pregnancy, is wholly different. Placental *CYP11A1* expression is independent of SF1, and utilizes TReP-132 and CP2 members of the *grainyhead* family of transcription factors that act at wholly distinct *cis*-elements of *CYP11A1*.[2] Because placental progesterone is needed to maintain pregnancy, it is counterintuitive that there should be mutations in human CYP11A1; nevertheless, at the time of writing, 19 such patients have been reported with mutations ranging from complete loss-of-function, to retention of substantial activity.[10]

Transport of Electrons to P450scc: Ferredoxin Reductase and Ferredoxin

P450scc functions as the terminal oxidase in a mitochondrial electron transport system.[11] Electrons from NADPH are accepted by a flavoprotein termed ferredoxin reductase (also known as adrenodoxin reductase) that is loosely associated with the IMM. Ferredoxin reductase subsequently transfers the electrons to an iron/sulfur protein termed ferrodoxin (also known as adrenodoxin), which is found in the mitochondrial matrix or loosely adherent to the IMM. Ferredoxin then transfers the electrons to P450scc (Figure 13.2). The same ferredoxin and ferredoxin reductase also transfer electrons to

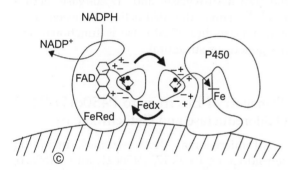

FIGURE 13.2 Electron transport to mitochondrial forms of cytochrome P450. The flavin group (FAD) of ferredoxin reductase (FeRed), which is bound to the inner mitochondrial membrane (IMM), accepts two electrons from NADPH, converting it to NADP$^+$. These electrons pass to the iron—sulfur (Fe$_2$S$_2$, diamond with dots) cluster of ferredoxin (Fedx), which is found either in the mitochondrial matrix, as shown, or loosely associated with the IMM. Fedx then donates the electrons to the heme of the P450 (square with Fe). Negatively charged residues in Fedx (−) guide docking and electron transfer with positively charged residues (+) in both FeRed and the P450. For P450scc, three pairs of electrons must be transported to the P450 to convert cholesterol to pregnenolone.

the P450c11β and P450c11AS isozymes of 11β-hydroxylase and to other mitochondrial P450s such as the vitamin D 1α- and 24-hydroxylases. Each of the three reactions catalyzed by P450scc requires a pair of electrons transferred by this system; consequently, the rate of conversion of cholesterol to pregnenolone is determined by the diffusion of ferredoxin through the mitochondrial matrix. Mutations have not been described in ferredoxin or ferredoxin reductase.

3β-Hydroxysteroid Dehydrogenase/ $\Delta^5 \rightarrow \Delta^4$ Isomerase (3βHSD)

Once pregnenolone is produced from cholesterol, it may undergo 17α-hydroxylation by P450c17 to yield 17α-hydroxypregnenolone (17OH-Preg), or it may be converted to progesterone, the first biologically important steroid in the pathway. A single 42-kDa 3βHSD enzyme catalyzes both the conversion of the hydroxyl group to a keto group on carbon 3 and the isomerization of the double bond from the B ring (Δ^5 steroids) to the A ring (Δ^4 steroids).[2] This single enzyme, 3βHSD, converts pregnenolone to progesterone, 17OH-Preg to 17α-hydroxyprogesterone (17OHP), dehydroepiandrosterone (DHEA) to androstenedione, and androstenediol to testosterone, all with similar efficiency (K$_m$ and V$_{max}$). As is typical of HSD enzymes, there are two isozymes of 3βHSD, encoded by separate genes (*HSD3B1* and *HSD3B2*).[12] These isozymes share 93.5% amino acid sequence identity and are biochemically and enzymatically very similar. The enzyme catalyzing 3βHSD activity in the adrenals and gonads is the type-2 enzyme, while the type-1 enzyme catalyzes 3βHSD activity in placenta, breast, liver and other "extraglandular" tissues. Structural data show that 3βHSD can be found in both the endoplasmic reticulum and in mitochondria. It is not clear whether the subcellular distribution of 3βHSD differs in various types of steroidogenic cells, but, this could be a novel mechanism in regulating the direction of steroidogenesis.[2]

P450c17: 17α-Hydroxylase/17,20-lyase

Microsomal P450c17 catalyzes both 17α-hydroxylase and 17,20-lyase activities in the adrenals and gonads, and hence is the principal qualitative regulator of steroidogenesis, determining the class of steroid that will be produced. In the absence of P450c17, the adrenal zona glomerulosa, ovarian granulosa cells, and placenta produce 21-carbon (C_{21}) 17-deoxy steroids such as progesterone and aldosterone; in the presence of 17-hydroxylase activity, the adrenal zona fasciculata produces the C_{21} 17-hydroxy steroid, cortisol; and when both 17-hydroxylase and 17,20-lyase activities are present, the adrenal zona reticularis, ovarian theca cells, and testicular Leydig cells produce C_{19} androgens.[2] Rodents fail to express P450c17 in their adrenals, and consequently must utilize corticosterone as their glucocorticoid. The 17α-hydroxylase reaction occurs more readily than the 17,20-lyase reaction. Pregnenolone and progesterone may undergo 17α-hydroxylation to 17OH-Preg and 17OHP, respectively. 17OH-Preg may also undergo scission of the C17,20 carbon bond to yield DHEA; but, since human P450c17 catalyzes the conversion of 17OHP to androstenedione with only 3% of the rate for conversion of 17OH-Preg to DHEA, this reaction makes a negligible contribution to human androgen synthesis.[13] This strong "preference" for 17OH-Preg as the substrate for 17,20-lyase activity is typical of primates but not other mammals, and is consistent with the large amounts of DHEA/S produced by primate adrenals. P450c17 is encoded by the CYP17A1 gene on chromosome 10q24.3, which is structurally related to the CYP21A2 gene for P450c21 (21-hydroxylase). Adrenal transcription of CYP17A1 is principally regulated by

the transcription factors NF-1C, Sp1, Sp3 and GATA4/6.

Like all Type 2 P450 enzymes, P450c17 is bound to the smooth endoplasmic reticulum where it mediates catalysis with electrons from NADPH donated by P450 oxidoreductase (POR); the efficiency of electron transport from NADPH is the principal factor regulating the 17,20-lyase reaction (Figure 13.1).[14] Serum cortisol concentrations, an index of adrenal 17α-hydroxylase activity are fairly consistent throughout life, whereas DHEA and DHEAS concentrations, which reflect adrenal 17,20-lyase activity, are low in early childhood but rise abruptly during adrenarche, which is contemporaneous with, but independent of, puberty. While this suggested that distinct enzymes performed the hydroxylase and lyase activities, the cloning of P450c17 permitted the rigorous demonstration that it catalyzed both the 17α-hydroxylase and 17,20-lyase activities.[14,15] Thus, the distinction between 17α-hydroxylase and 17,20-lyase is functional and not genetic or structural.

Transport of Electrons to P450c17: P450 Oxidoreductase and Cytochrome b_5

All microsomal P450 enzymes, including steroidogenic P450c17, P450c21, and P450aro, receive electrons from POR, a membrane-bound flavoprotein that is distinct from mitochondrial ferrodoxin reductase.[11] POR is a butterfly-shaped protein that has a flavin adenine dinucleotide (FAD) moiety in one "wing" and an flavin mononucleotide (FMN) moiety in the other "wing." The FAD moiety receives two electrons from NADPH, transfers them to the FMN, which then transfers them one at a time to the P450 enzyme. For P450c17, electron transfer for the 17,20-lyase reaction is promoted by the action of the hemoprotein cytochrome b_5 as an allosteric factor rather than as an alternate electron donor.[13] The 17,20-lyase activity of P450c17 also requires the phosphorylation of serine residues on P450c17 by a cAMP-dependent protein kinase (Figure 13.3).[14] Thus the availability of electrons determines whether P450c17 performs only 17α-hydroxylation, or also performs 17,20 bond scission; increasing the ratio of POR or cytochrome b_5 to P450c17 *in vitro* or *in vivo* increases the ratio of 17,20-lyase activity to 17α-hydroxylase activity. Thus, the regulation of 17,20-lyase activity, and consequently of DHEA production, depends on factors that facilitate the flow of electrons to P450c17: high concentrations of POR, the presence of cytochrome b_5, and serine phosphorylation of P450c17.[14,15]

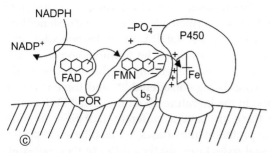

FIGURE 13.3 **Electron transport to microsomal forms of cytochrome P450.** NADPH interacts with P450 oxidoreductase (POR), bound to the endoplasmic reticulum, and gives up a pair of electrons (e⁻), which are received by the FAD moiety. Electron receipt induces a conformational change, permitting the isoalloxazine rings of the FAD and FMN moieties to come close together, so that the electrons pass from the FAD to the FMN. Following another conformational change that returns the protein to its original orientation, the FMN domain of POR interacts with the redox-partner binding site of the P450. Electrons from the FMN domain of POR then reach the heme group to mediate catalysis. The interaction of POR and the P450 is coordinated by negatively charged acidic residues on the surface of the FMN domain of POR, and positively charged basic residues in the concave redox-partner binding site of the P450, similar to the interaction of Fedx with mitochondrial P450s. The active site containing the steroid lies on the side of heme ring (Fe) opposite from the redox-partner binding site.

P450c21: Steroid 21-Hydroxylase

Microsomal P450c21 catalyzes the 21-hydroxylation of progesterone to deoxycorticosterone (DOC) and 17OHP to 11-deoxycortisol in the biosynthesis of mineralocorticoids and glucocorticoids, respectively (Figure 13.1).[2] The nature of the 21 hydroxylation reaction has been of great clinical interest because 21-hydroxylase deficiency causes about 95% of all cases of congenital adrenal hyperplasia (CAH), which has an incidence of about 1:15,000 births. The clinical symptoms associated with this common genetic disease are complex and potentially devastating.[16] Decreased aldosterone and cortisol synthesis can lead to sodium loss, potassium retention, and hypotension, which can cause cardiovascular collapse and death in infancy if not treated appropriately. Decreased synthesis of cortisol *in utero* leads to overproduction of ACTH and consequent overstimulation of fetal adrenal steroid synthesis; as the 21-hydroxylase step is impaired, 17OHP accumulates because P450c17 converts only miniscule amounts of 17OHP to androstenedione. However, 17OH-Preg also accumulates and is efficiently converted to DHEA, and subsequently to androstenedione and testosterone, resulting in severe prenatal virilization of female fetuses.[16]

Clinical variations in this disease, especially the identification of patients without apparent defects in mineralocorticoid activity, initially suggested that there were two separate 21 hydroxylases that were differentially expressed in the adrenal zona glomerulosa and fasciculata. However, gene cloning has shown that there is a single adrenal 21-hydroxylase encoded by a single functional gene (*CYP21A2*) on chromosome 6p21.[2,17,18] As this gene lies in the middle of the major histocompatibility locus, disorders of adrenal 21-hydroxylation are closely linked to specific human leukocyte antigens (HLA) types.

P450c21 is found in smooth endoplasmic reticulum, where it employs the same POR used by P450c17 to receive electrons from NADPH.

21-hydroxylase activity has also been described in a broad range of extra-adrenal tissues, but extra-adrenal 21-hydroxylation is not catalyzed by P450c21. Hepatic 21-hydroxylation is mediated by several enzymes, notably CYP2C19 and CYP3A4, which are principally involved in drug metabolism.[19] These enzymes can 21-hydroxylate progesterone, but not 17OHP, and hence may contribute to the synthesis of mineralocorticoids but not glucocorticoids and possibly account for the diminished mineralocorticoid requirements in adult patients. Thus, some 21-hydroxylase activity may be regulated by the pattern of drug and xenobiotic exposure.

Isozymes of P450c11: P450c11β and P450c11AS

The final steps in the synthesis of glucocorticoids and mineralocorticoids are catalyzed by two closely related mitochondrial enzymes, P450c11β and P450c11AS.[2,20] These two isozymes have 93% amino acid sequence identity and are encoded by tandemly duplicated genes (CYP11B1 and CYP11B2, respectively) on chromosome 8q21-22. Like P450scc, these two enzymes are found on the IMM, and use ferredoxin and ferredoxin reductase to receive electrons from NADPH to mediate catalysis. By far the more abundant of the two isozymes is P450c11β, which is the classical 11β-hydroxylase that converts 11-deoxycortisol to cortisol and 11-deoxycorticosterone (DOC) to corticosterone. The less abundant isozyme, P450c11AS, is found only in the zona glomerulosa, where it has 11β-hydroxylase, 18-hydroxylase and 18-methyl oxidase (aldosterone synthase) activities; thus P450c11AS is able to catalyze all the reactions needed to convert DOC to aldosterone. Both enzymes can convert DOC to corticosterone and corticosterone to 18OH-corticosterone, but only P450c11AS can synthesize aldosterone from 18OH-corticosterone. Patients with disorders in P450c11β have classical 11β-hydroxylase deficiency but can still produce aldosterone, while patients with disorders in P450c11AS have rare forms of aldosterone deficiency (so-called corticosterone methyl oxidase deficiency) while retaining the ability to produce cortisol.[2,20] Transcription of the CYP11B1 gene encoding P450c11β is induced by ACTH via cAMP and is suppressed by glucocorticoids, whereas transcription of the CYP11B2 gene encoding P450c11AS is induced by potassium and angiotensin II via the protein kinase C pathway.[21]

Isozymes of 17β-Hydroxysteroid Dehydrogenase (17βHSD)

The 17β-hydroxysteroid dehydrogenases (17βHSD), sometimes also termed 17-ketosteroid reductases, principally convert androstenedione to testosterone, DHEA to androstenediol, and estrone to estradiol. The terminologies for these enzymes vary depending on the direction of the reaction being considered,[1,22] and there is confusion in the literature about the 17βHSDs because: i) there are several different 17βHSDs; ii) some are preferential oxidases, while others are preferential reductases; iii) they differ in their substrate preference and sites of expression; iv) there is inconsistent nomenclature (especially with rodent enzymes); and v) some proteins termed 17βHSD actually have very little 17βHSD activity, and are principally involved in other reactions.

17βHSD1, encoded by HSD17B1 is a cytosolic reductive SDR enzyme first isolated and cloned from the placenta, where it produces estriol; it is expressed in ovarian granulosa cells where it produces estradiol.[1,22] 17βHSD1 uses NADPH as its co-factor to catalyze reductase activity. It acts as a dimer and only accepts steroid substrates with an aromatic A ring; thus, its activity is confined to activating estrogens. To date, no genetic deficiency syndrome for 17βHSD1 has been described.

17βHSD2, encoded by *HSD17B2*, is a microsomal oxidase that uses NAD^+ to inactivate both estradiol to estrone and testosterone to androstenedione. 17βHSD2 is found in the placenta, liver, small intestine, prostate, secretory endometrium, and ovary. In contrast to 17βHSD1, which is found in placental syncytiotrophoblast cells, 17βHSD2 is expressed in endothelial cells of placental intravillous vessels, consistent with its apparent role in defending the fetal circulation from transplacental passage of maternal estradiol or testosterone.[1,22] No genetic deficiency syndrome for 17βHSD2 has been reported.

17βHSD3, encoded by *HSD17B3*, is a microsomal enzyme that is expressed in the testis, and is the principal androgenic form of 17βHSD. 17βHSD3 is the enzyme that is disordered in the classic syndrome of male pseudohermaphroditism that is often termed 17-ketosteroid reductase deficiency.[1,22]

17βHSD4 was initially identified as an NAD^+-dependent oxidase with activities similar to 17βHSD2, and thus given 17βHSD status. However, this protein is located in peroxisomes and is primarily an enoyl-CoA hydratase and 3-hydroxyacyl-CoA dehydrogenase.[2] Deficiency of 17βHSD4 causes a form of Zellweger syndrome, in which bile acid biosynthesis is affected but steroidogenesis is not.

17βHSD5, originally cloned as a 3αHSD, is an AKR enzyme (in contrast to 17βHSD types 1–4, which are SDR enzymes) termed AKR1C3. 17βHSD5/AKR1C3 catalyzes the reduction of androstenedione to testosterone.[2,23] Interestingly, 17βHSD5/AKR1C3 is more highly expressed in the human fetal adrenal gland during the time of sexual differentiation than 17βHSD3; thus, it may participate in adrenal testosterone production, particularly in virilizing CAHs.[24] The postnatal adrenal zona reticularis also expresses 17βHSD5/AKR1C3 at low levels, accounting for the small amount of testosterone produced by the adrenal gland.[25]

P450aro: Aromatase

Estrogens are produced by the aromatization of androgens, including those produced by the adrenal gland, by a complex series of reactions catalyzed by a single microsomal aromatase, P450aro.[26,27] This microsomal P450 is encoded by the *CYP19A1* gene on chromosome 15q21.1. This gene uses several different promoter sequences, transcriptional start sites, and alternatively chosen first exons to encode aromatase mRNA in different tissues under different hormonal regulation. Aromatase expression in "extraglandular" tissues, especially adipose tissue, can convert adrenal androgens to estrogens, and aromatase expression in the epiphyses of growing bone converts testosterone to estradiol. Males with aromatase deficiency have tall stature, delayed epiphyseal maturation, and osteopenia, which are rapidly reversed by estrogen replacement, which shows that estrogen, not testosterone, is primarily responsible for epiphyseal maturation. Although estrogens were once considered necessary for embryonic and fetal development, fetuses with genetic lesions in *CYP19A1*, who cannot produce estrogens, and fetuses with genetic lesions in *CYP11A1*, who cannot produce any steroids, have normal fetal development and normal parturition showing that feto−placental estrogen is not essential.[2,27]

Isozymes of 5α-Reductase

Testosterone is converted to the more potent androgen dihydrotestosterone (DHT) by two forms of 5α-reductase. The type-1 enzyme, encoded by the *SRD5A1* gene on chromosome 5p15, is found in the scalp and other peripheral tissues; the type-2 enzyme, encoded by the structurally related *SRD5A2* gene on chromosome 2p23, is the predominant form found in male reproductive tissues.[28] Mutations in the *SRD5A2* gene cause 5α-reductase deficiency, a disorder of male sexual differentiation that has broad phenotypic variation, depending on the causative mutation(s); mutations in *SRD5A1* have not been described. The 5α-reductase genes exhibit distinct patterns of developmental regulation. The type-1 gene is not widely expressed in the fetus, but is expressed in the fetal testis, where it participates in the alternative "backdoor" pathway of androgen synthesis;[29] it is then briefly expressed in newborn skin, and then remains unexpressed again until after puberty. The type-2 gene is expressed in fetal genital skin, where it participates in normal male sexual differentiation, and is also expressed in adult prostate and prostatic adenocarcinoma cells. Thus, the type-1 enzyme may be responsible for the pubertal virilization seen in patients with classic 5α-reductase deficiency, and the type-2 enzyme may be involved in male pattern baldness.[30]

Isozymes of 11β-Hydroxysteroid Dehydrogenase (11βHSD)

Although certain steroids are typically categorized as glucocorticoids or mineralocorticoids, the "mineralocorticoid" (glucocorticoid type 2) receptor has equal affinity for both aldosterone and cortisol. However, because mineralocorticoid-responsive tissues (such as the kidney) convert cortisol to cortisone, which does not bind to receptors, cortisol does not act as a mineralocorticoid *in vivo*, even though its concentrations can exceed those of aldosterone by 100- to 1000-fold. The interconversion of cortisol and cortisone is mediated by two isozymes of 11β-hydroxysteroid dehydrogenase (11βHSD). Both isozymes can catalyze both oxidase and reductase activities, depending on the co-factor available ($NADP^+$ or NADPH, respectively)[31]; the ratio of $NADP^+$ to NADPH is regulated by hexose-6-phosphate dehydrogenase (H6PDH).[32]

The type-1 enzyme (11βHSD1), encoded by the *HSD11B1* gene, is expressed mainly in glucocorticoid-responsive tissues such as the liver, testis, lung, and proximal convoluted tubule. 11βHSD1 is located on the luminal side of the endoplasmic reticulum, and is not in contact with the cytoplasm; in this unusual cellular location, 11βHSD1 receives NADPH provided by H6PDH and links the enzyme to the pentose monophosphate shunt, providing a direct paracrine connection between local glucocorticoid production and energy storage as fat.[33−35] 11βHSD1 can catalyze both the oxidation of cortisol to cortisone using $NADP^+$ as its co-factor (K_m 1.6 μM), or the reduction of cortisone to cortisol using NADPH as its co-factor (K_m 0.14 μM); the direction of the reaction catalyzed depends on which co-factor is available, but 11βHSD1 can only function with high (micromolar) concentrations of steroid. By contrast, 11βHSD2, encoded by the *HSD11B2* gene, only catalyzes the oxidation of cortisol to cortisone using NADH, and can function with low (nanomolar) concentrations of steroid (K_m 10−100 nM). 11βHSD2 is expressed in mineralocorticoid-responsive tissues and thus serves to "defend" the mineralocorticoid receptor by inactivating cortisol to cortisone, so that only "true" mineralocorticoids, such as aldosterone or DOC, can exert a mineralocorticoid effect. Thus, 11βHSD2 prevents cortisol from overwhelming renal mineralocorticoid receptors and inactivates cortisol in the placenta and other fetal tissues. Importantly, the placenta also has abundant $NADP^+$ favoring the oxidative action of 11βHSD1, so that in the placenta both 11βHSD1 and 11βHSD2 protect the fetus from high maternal concentrations of cortisol (but not from maternally administered betamethasone or dexamethasone, which are often used to enhance fetal lung development because they are ineffective substrates for 11βHSD2).

Isozymes of 3α-Hydroxysteroid Dehydrogenase (3αHSD)

The four major human 3α-hydroxysteroid dehydrogenases (3αHSDs) are AKR enzymes of the AKR1C family; 3αHSD types 1, 2, 3, and 4 are also termed AKR1C4, 1C3, 1C2, and 1C1, respectively.[22] These enzymes are structurally very similar, are encoded by a gene cluster on chromosome 10p14−15, and catalyze a wide array of steroidal conversions and other reactions.[2] The 3αHSDs are essential constituents of the so-called "backdoor pathway" of steroidogenesis.[36] This remarkable pathway (discussed below), first discovered in studies of steroidogenesis in the marsupial fetal testis,[36,37] plays a central role in human male sexual differentiation.[29] In this pathway, 17OHP is converted to DHT without going through DHEA, androstenedione, or testosterone, and hence provides a mechanism by which 17OHP can contribute to the virilization of female fetuses with 21-hydroxylase deficiency.[16,38] The "backdoor pathway" is characterized by both reductive and oxidative 3αHSD activities; the reductive activity can be catalyzed by either AKR1C2 (3αHSD3) or AKR1C4 (3αHSD1),[29] but the nature of the oxidative enzyme(s) is uncertain.[2] AKR1C3 (3αHSD2), which converts androstenedione to testosterone in the adrenals, is also known as 17βHSD5.

Steroid Sulfotransferase and Sulfatase

Steroid sulfates may be synthesized directly from cholesterol sulfate or formed via the sulfation of existing steroids by cytosolic sulfotransferase (SULT) enzymes.[39,40] At least 44 distinct isoforms of these enzymes have been identified belonging to five families of *SULT* genes. Many of these genes yield alternately spliced products accounting for the large number of enzymes,

and the obligatory sulfate donor for all the SULT enzymes is 3'-phosphoadenosine-5'-phosphosulfate (PAPS). The SULT enzymes that sulfonate steroids include SULT1E (which sulfates estrogens), SULT2A1 (which sulfates non-aromatic steroids), and SULT2B1 (which sulfates sterols). SULT2A1 is the principal SULT enzyme expressed in the adrenal gland, where it sulfates the 3β-hydroxyl group of Δ^5 steroids (pregnenolone, 17OH-pregnenolone, DHEA, and androsterone) but not of cholesterol. SULT2B1a will also sulfonate pregnenolone but not cholesterol; alternatively, cholesterol is the principal substrate for SULT2B1b in the skin, liver, and elsewhere. Whether steroid sulfates are simply inactivated forms of steroid or whether they serve specific hormonal roles is unclear.

Knockout of the mouse *SULT1E1* gene causes elevated estrogen levels, increased expression of tissue factor in the placenta, and increased platelet activation, leading to placental thrombi and fetal loss. Mutations ablating human SULT enzymes have not been described, but some single nucleotide polymorphisms that alter the amino-acid sequences and catalytic activity affect drug activity. African-Americans have a high rate of *SULT2A1* polymorphisms, which influences plasma ratios of DHEA:DHEAS and may correlate with risk of prostatic and other cancers.

Steroid sulfates may also be hydrolyzed to the native steroid by steroid sulfatase; deletions in the steroid sulfatase gene on chromosome Xp22.3 cause X-linked ichthyosis, due to the accumulation of steroid sulfates in the stratum corneum of the skin. The fact that males have a single copy of this gene also probably accounts for the higher DHEAS levels in males than females of the same age. In the fetal adrenals and placenta, diminished or absent sulfatase deficiency reduces the pool of free DHEA available for placental conversion to estrogen, resulting in low estriol concentrations in the maternal blood and urine.

TISSUE-SPECIFIC PATHWAYS OF STEROIDOGENESIS

Adrenal Pathways

Diagrams of steroidogenic pathways, such as the one shown in Figure 13.1 typically combine the pathways from multiple cell types to provide an overview of all steroidogenic processes. However, such diagrams are misleading, as the pathways differ in each cell type. The three major pathways of steroidogenesis in the human adrenal glands are shown in Figure 13.4. The adrenal zona glomerulosa (ZG) is characterized by three distinct features: i) it expresses angiotensin II receptors; ii) it expresses P450c11AS; and iii) it fails to express P450c17. As a result, the ZG produces aldosterone under regulation by the renin/angiotensin system. By contrast, the adrenal zona fasciculata (ZF) does not express angiotensin II receptors or P450c11AS, but instead expresses MC2R (the ACTH receptor) and P450c11β, which cannot convert 18-hydroxycorticosterone to aldosterone and has minimal capacity to convert corticosterone to 18-hydroxycorticosterone. Both the ZG and ZF express P450c21, but the ZF also expresses P450c17, permitting cortisol synthesis. The ZF, however, expresses little (if any) cytochrome b_5;[41] as a result, P450c17 in the ZF catalyzes 17α-hydroxylation but very little 17,20-lyase activity. The ZF thus produces cortisol and corticosterone under the influence of ACTH, but very little DHEA. Patients with severe mutations in P450c17 cannot synthesize cortisol but instead increase corticosterone production (as do rodent adrenals, which lack P450c17), which explains why they are not glucocorticoid-deficient despite their lack of cortisol production (Figure 13.1). The adrenal zona reticularis (ZR) also expresses MC2R but very little P450c21 or

P450c11β, so that the ZR produces very little cortisol. By contrast, the ZR expresses large amounts of P450c17 and cytochrome b_5, maximizing 17,20-lyase activity, so that DHEA is produced, much of which is sulfated by SULT2A1 to DHEAS. Furthermore, the ZR expresses relatively little 3βHSD2, and the K_m of 3βHSD2 is ~5 μM for pregnenolone and 17OH-Preg, whereas the K_m for both the 17α-hydroxylase and 17,20-lyase activities of P450c17 is ~1 μM, favoring the production of DHEA.[2] As DHEA accumulates, small amounts are converted to androstenedione; very small amounts of this androstenedione are then converted to testosterone by AKR1C3/17βHSD5. The pattern of steroid products secreted by each adrenal zone is therefore determined by the enzymes produced in that zone and may be logically deduced from an understanding of their specific enzymatic properties.[2]

Gonadal Pathways

The synthesis of testosterone in the testis follows a pathway that is similar to C_{19}-steroid production in the adrenal ZR, with the notable exceptions that the stimulus for steroidogenesis is transduced by the LH receptor rather than MC2R and that Leydig cells express abundant 3βHSD2 and 17βHSD3 but no SULT2A1. Thus, the DHEA produced in the testis is not sulfated but is rather readily converted to androstenedione and then testosterone (Figure 13.5A). As in the adrenal, the principal pathway to C_{19}-steroids is via Δ^5 steroids to DHEA; the Δ^4 pathway from 17OHP to androstenedione makes only a minimal contribution.

Steroidogenesis in the ovary is partitioned between the granulosa and theca cells, which surround the oocyte and form a follicle. The patterns of steroidogenesis also vary during the menstrual cycle: Estradiol is the principal product in the follicular phase, whereas progesterone is produced in the luteal phase (Figure 13.5B).

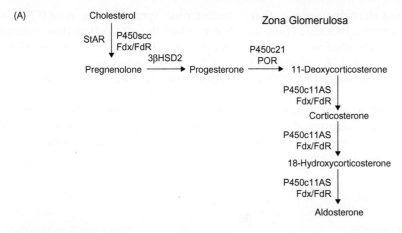

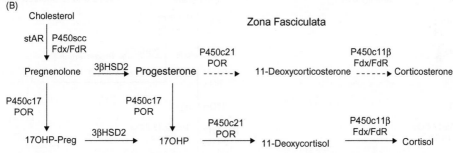

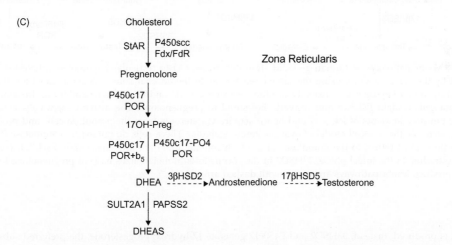

FIGURE 13.4 **Major steroidogenic pathways in the human adrenal cortex.** The conversion of cholesterol to pregnenolone by P450scc is common to all 3 zones of the adrenal cortex. (A) In the zona glomerulosa (ZG), 3βHSD2 converts pregnenolone to progesterone. P450c17 is absent, but P450c21 produces 11-deoxycorticosterone, which is a substrate for P450c11AS. P450c11AS catalyzes 11-hydroxylation and two 18-oxygenations, which completes aldosterone synthesis. (B) The zona fasciculata (ZF) expresses P450c17, so pregnenolone is hydroxylated to 17α-hydroxypregnenolone, (or progesterone to 17-hydroxyprogesterone); but, the ZF contains little cytochrome b_5, minimizing the 17,20-lyase activity of P450c17,

The key point in ovarian steroidogenesis is that granulosa cells do not express P450c17. Thus, in general, steroidogenesis is initiated in granulosa cells under the influence of LH, which, via cAMP, stimulates the expression of P450scc. Pregnenolone and progesterone from granulosa cells then diffuse into the adjacent theca cells, where they are converted to androstenedione by P450c17 and 3βHSD2. Small amounts of this androstenedione are secreted or converted to testosterone (probably by AKR1C3/17βHSD5), but most of the androstenedione returns to the granulosa cells where it is converted to estrone and then to estradiol by P450aro and 17βHSD1, under the influence of follicle-stimulating hormone (FSH). Thus, as with the three zones of the adrenal, the patterns of gonadal steroidogenesis in the ovary are dictated by the cell-specific expression of specific steroidogenic enzymes.[2]

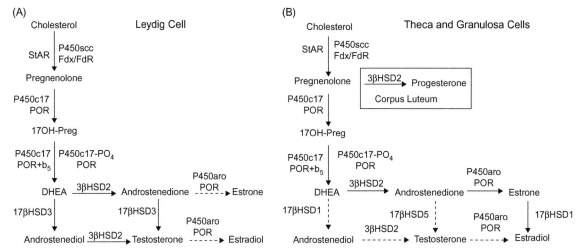

FIGURE 13.5 **Major pathways of human gonadal steroidogenesis.** (A) In testicular Leydig cells, cholesterol is converted to DHEA by the same enzymes using the same co-factors as in the adrenal zona reticularis. Leydig cells contain abundant 17βHSD3, so that Leydig cells efficiently produce testosterone, via androstenedione and/or androstenediol. (B) Ovarian granulosa cells contain P450scc and convert cholesterol to pregnenolone. The ovarian theca cells express low levels of P450scc but high amounts of P450c17 and hence acquire C_{21} steroids from the granulosa cells and produce C_{19} precursors of sex steroids (the two-cell model of ovarian steroidogenesis). Theca cells do not express aromatase (P450aro); thus, androstenedione must return to the granulosa cells, which contain abundant aromatase and 17βHSD1, completing the synthesis of estradiol. In the luteal phase, 3βHSD2 in the corpus luteum metabolizes nascent pregnenolone to progesterone, the final product. Minor pathways are shown with dashed arrows.

and little DHEA is produced. Instead, 3βHSD2 and P450c17 generate 17-hydroxyprogesterone, the preferred substrate for P450c21, yielding 11-deoxycortisol. P450c11β, which is unique to the ZF, completes the synthesis of cortisol. Corticosterone is normally a minor product (dashed arrows) derived from a parallel pathway without the action of P450c17. (C) The zona reticularis (ZR) has large amounts of P450c17 and cytochrome b_5 but little 3βHSD2, so that pregnenolone is sequentially oxidized to 17-hydroxypregnenolone and then DHEA. SULT2A1, using PAPS synthesized by PAPSS2 (see text), sulfates DHEA, and DHEAS is exported to the circulation. Testosterone synthesis is a very minor pathway (dashed arrows).

The "Backdoor Pathway" to DHT

The alternative or "backdoor" pathway leads from 17OHP to DHT without going through androstenedione or testosterone as intermediate steroids. This pathway is initiated by the 5α-reduction of either progesterone or 17OHP by type-1 5α-reductase. The resulting 5α-reduced C_{21} steroids, dihydroprogesterone (5α-pregnane-3,20-dione) and 5α-pregnane-17α-ol-3,20-dione, are then readily catalyzed by reductive 3αHSDs to yield allopregnanolone (5α-pregnan-3α-ol-20-one; Allo) and 17α-hydroxylated Allo (5α-pregnane-3α,17α-diol-20-one; 17OH-Allo). Dihydroprogesterone and Allo are excellent substrates for the 17α-hydroxylase activity of P450c17, and 17OH-Allo is the most efficient substrate known for the 17,20-lyase activity of human P450c17. Furthermore, unlike the conversion of 17OH-Preg to DHEA, the cleavage of 17OH-Allo to androsterone is minimally dependent on cytochrome b_5. The resulting androsterone may then be 3α-oxidized to DHT by retinol dehydrogenase (RoDH), the microsomal 3αHSD, 3 (α→β)-hydroxysteroid epimerase (also known as 17βHSD6) (Figure 13.6). The presence of 5α-reductases in steroidogenic cells does not preclude the production of C_{19} steroids, but rather paradoxically enhances the production of DHT by directing steroid flux to 5α-reduced precursors of DHT. The "backdoor pathway" thus enables production of C_{19} steroids from 17OHP, despite the poor 17,20-lyase activity of human P450c17 for 17OHP, by using 17OH-Allo as the substrate for the 17,20-lyase reaction. The "backdoor pathway" is also relevant to normal and abnormal human steroidogenesis. In normal male sexual development, both the conventional pathway of androgen production (via DHEA, androstenedione, and testosterone to DHT) and the fetal testicular "backdoor pathway" are required for development of normal male external genitalia.[29] Similarly, when 17OHP accumulates in 21-hydroxylase deficiency and POR deficiency, the "backdoor pathway" is responsible for a portion of the overproduction of androgens.[38,42]

CHRONIC REGULATION OF STEROIDOGENESIS

Whereas the acute regulation of steroidogenesis is determined by the action of StAR, P450scc is the enzymatic rate-limiting step in steroidogenesis. Thus, the chronic regulation of steroidogenesis is quantitatively (i.e., how much steroid is produced) determined by P450scc gene expression and qualitatively (i.e., which steroids are produced) determined by the expression of downstream enzymes, especially P450c17. Patients with inactivating mutations in the ACTH receptor (MC2R)[43] or the LH/hCG receptor[44] make negligible steroids from the adrenals or gonads. Conversely, activating mutations of the $G\alpha_s$ protein, which couples G protein-coupled receptor (GPCR) binding to cAMP generation, and activating mutations of the LH receptor cause hypersecretion of steroids,[45] and cAMP-responsive elements have been identified in the genes for most of the human steroidogenic P450 enzymes. However, the regulation of cAMP generation alone does not explain the diversity of steroid production observed in the various zones of the adrenal cortex and the gonads of both sexes. Other transcription factors, including AP2, SP1, SP3, NF1C, NR4A1, NR4A2, GATA4, and GATA6, also participate in regulating the basal- and cAMP-stimulated transcription of each gene. SF1 coordinates the expression of steroidogenic enzymes in the adrenals and gonads. By contrast, steroidogenesis in the brain and placenta is independent of SF1. Disruption of the mouse SF1 gene disrupts steroid biosynthesis and blocks the development of the adrenals, gonads, and

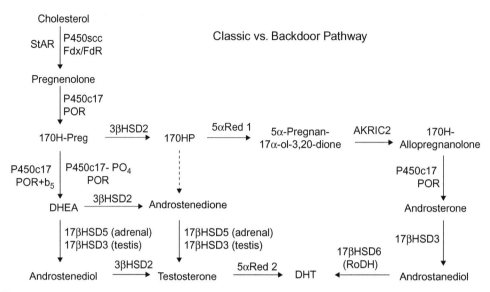

FIGURE 13.6 **Synthesis of dihydrotestosterone via the classic and alternative ("backdoor") pathways.** The classic pathway of steroidogenesis leading to dihydrotestosterone (DHT) is shown on the left; the alternative ("backdoor") pathway is shown on the right. The factors in the classic pathway are CYP11A1 (cholesterol side-chain cleavage enzyme, P450scc), StAR (steroidogenic acute regulatory protein), CYP17A1 (17α-hydroxylase/17,20-lyase, P450c17), HSD3B2 (3β-hydroxysteroid dehydrogenase, type 2), HSD17B3 (17β-HSD3 (17β-hydroxysteroid dehydrogenase, type 3)), and 5α-reductase, type 2 (5α-reductase 2, encoded by SRD5A2). The alternative ("backdoor") pathway is characterized by the presence of additional enzymes: 5α-reductase, type 1 (5α-reductase 1, encoded by SRD5A1), AKR1C2 3 (3α-reductase, type 3) and possibly AKR1C4 (3α-reductase, type 1), and RoDH (3-hydroxyepimerase, encoded by HSD17B6). Most steroids are identified by their common names; 17-hydroxy-dihydroprogesterone (17OH-DHP) is 5α-pregnane-17α-ol-3,20-dione; 17-hydroxy-allopregnanolone (17OH-allo) is 5α-pregnan-3α,17α-diol-20-one; 5α-dihydroprogesterone (5α-DHP) is 5α-pregnane-3,20-dione, and allopregnanolone is 3α-hydroxy-dihydroprogesterone (3α-OH-DHP) or 5α-pregnane-3α-ol-20-one.

ventromedial hypothalamus. The action of SF1 is further modified by other transcription factors (e.g., WT1 and DAX1) and by phosphorylation or sumoylation. Thus, the development of steroidogenic organs is intimately related to the capacity to produce steroids, and multiple factors acting on the genes for steroidogenic enzymes yield both common features and diversity among the steroidogenic tissues.[2]

ACUTE REGULATION OF STEROIDOGENESIS

Unlike cells that produce polypeptide hormones, which store large amounts of hormone in secretory vesicles ready for rapid release, steroidogenic cells store very little steroid. Thus, a rapid steroidogenic response (e.g., secretion of aldosterone and cortisol in response to stress or the "pulsing" of sex steroids in response to an LH surge) requires rapid synthesis of new steroid. ACTH promotes steroidogenic cell growth and maintains the steroidogenic machinery at three distinct levels (LH probably acts similarly on gonadal steroidogenic cells, but has not been studied as thoroughly). First, acting over weeks or months, ACTH promotes adrenal growth, primarily by ACTH-stimulated synthesis of IGF-2, basic fibroblast growth factor, and EGF. These growth factors subsequently stimulate adrenal cellular hypertrophy and hyperplasia and thus regulate the amount of steroidogenic tissue. Second, acting over days, ACTH (through the cAMP pathway) and angiotensin II (through the calcium/calmodulin pathway) promote the transcription of genes encoding various steroidogenic enzymes and electron-donating co-factor proteins, thus determining the amount of steroidogenic machinery in the cell. Third, ACTH rapidly stimulates StAR gene transcription and the post-translational phosphorylation of Ser195 in existing StAR to increase the flow of cholesterol from the OMM to the IMM where it becomes substrate for P450scc.[5,7]

Some adrenal steroidogenesis occurs independent of StAR. When non-steroidogenic cells are transfected with StAR and the P450scc system, they convert cholesterol to pregnenolone at ~14% of the StAR-induced rate. Furthermore, the placenta utilizes mitochondrial P450scc to initiate steroidogenesis but does not express StAR. The specific mechanism(s) of StAR-independent steroidogenesis is unclear; it may occur without a triggering protein, or some other protein may exert StAR-like activity to promote cholesterol flux, but without StAR's rapid kinetics; for example a protein called N-218MLN64 is found in mitochondria and has StAR-like activity.[5] The exact mechanism of StAR's action is also unclear, but it is well-established that StAR acts on the OMM, does not need to enter the mitochondria to be active, and undergoes conformational changes on the OMM that are required for its activity.[5-7] Mechanistically, StAR functions as a component of a molecular machine termed a "transduceosome" on the OMM that consists of StAR, TSPO (the translocator protein formerly known as the peripheral benzodiazepine receptor), TSPO-associated protein 7 (PAP7; ACBD3 for acyl-CoA-binding-domain 3), the voltage-dependent anion channel (VDAC-1), and protein kinase A regulatory subunit 1α (PKAR1A).[46] The mechanism by which these proteins interact and move cholesterol from the outer mitochondrial membrane to P450scc, and the means by which cholesterol is loaded into the OMM, remains unclear.

POST-TRANSLATIONAL DIFFERENTIAL REGULATION OF STEROIDOGENIC ENZYMES

Contemporary work is focused on identifying post-translational mechanisms regulating steroidogenesis. The phosphorylation of StAR, phosphorylation and sumoylation of SF1, and glycosylation of P450aro are well-described examples. The increase in 17,20-lyase activity of P450c17, but not its 17-hydroxylase activity by phosphorylation of P450c17 is an especially interesting example.

Adjusted for body surface area, the human secretion of cortisol remains fairly constant from infancy to old age. However, the secretion of C_{19} adrenal steroids is low in childhood begins to rise just before puberty, reaches maximum levels in young adulthood (well after the completion of puberty), and then falls slowly to childhood levels in the elderly.[47] The peripuberal rise in DHEA, DHEAS, and androstenedione[48,49] is referred to as "adrenarche." Although these C_{19} steroids are commonly referred to as "adrenal androgens," they are androgen precursors rather than true androgens because they do not activate the androgen receptor. Primates are unique among mammals in having high concentrations of adrenal C_{19} steroids; the pattern of human age-related rise and fall in C_{19} steroids is also unique.[49] Adrenarche is of interest for many reasons: i) evolutionarily, because it appears to be a recent innovation; ii) physiologically, because the function, if any, of adrenal C_{19} steroids is controversial; iii) endocrinologically, because its regulatory mechanisms are unknown; and iv) biochemically, because it represents an unusual example of post-translational differential regulation of P450c17's two enzymatic activities.

The traditional questions concerning adrenarche have concerned its regulation, intracellular mechanisms, and role in human physiology. Investigators have long searched for a hypothetical, specific Adrenal Androgen Stimulating Hormone that would regulate C_{19} steroid synthesis by the adrenal ZR, analogous to angiotensin II acting on the ZG or ACTH acting on the ZF; however, no such factor has been identified. Adrenarche that begins earlier and that results in higher concentrations of C_{19} steroids also frequently precedes the development of polycystic ovary syndrome (PCOS), which is characterized by adrenal and ovarian hyperandrogenism, insulin resistance, and obesity, although many affected women have only some of these features;[50] many now regard premature exaggerated adrenarche as an early form of PCOS.[51–53] Because of the many connections between adrenal C_{19} steroid production and metabolic regulation, factors not specific to the adrenal glands, including nutrition, insulin, insulin-like factors, leptin, and FGF have also been considered as potential triggers of adrenarche.[54]

Because Ser/Thr phosphorylation of P450c17 augments its 17,20-lyase activity, and because Ser phosphorylation of the beta-chain of the insulin receptor causes insulin resistance, it has been proposed that the hyperandrogenism and insulin resistance in PCOS may be connected by a signal transduction pathway that ultimately increases the Ser/Thr phosphorylation of both P450c17 and either the insulin receptor or its substrate.[55,56] Rho-associated, coiled coil-containing protein kinase 1 (ROCK1) can promote the 17,20-lyase activity of P450c17, but how ROCK1 promotes this activity is unclear. Recent work shows that P450c17 can be phosphorylated by p38α in a fashion that selectively augments its 17,20-lyase activity, thus providing a proof-of-principle for the regulation of 17,20-lyase activity by P450c17 phosphorylation.[57]

p38α is a mitogen-activated kinase (MAPK14) that is the terminal component of a typical three-component MAP kinase cascade; the phosphorylation of P450c17 by p38α can be reversed by protein phosphatase 2A, which in turn can be regulated by phosphoprotein SET, permitting the modeling of the relevant cellular pathways (Figure 13.7). Hormones (e.g., insulin, IGFs, leptin), environmental agents, and dietary factors could potentially activate a pathway that may involve ROCK1, which might act as an upstream scaffolding protein in a MAPK pathway.[57] The identification of this pathway is incomplete, but promises to help define the mechanisms regulating adrenal C_{19} steroid synthesis, and may also reveal links to PCOS.

CONCLUSION

Steroidogenesis involves the conversion of cholesterol to glucocorticoids, mineralocorticoids, and sex steroids, and is regulated at multiple levels, principally by transcription of genes encoding steroidogenic enzymes and co-factors, and by their post-translational modification, in a tissue-specific fashion. Most steroidogenic enzymes are either HSD or cytochrome P450 enzymes, the activities of which are modulated by post-translational modifications and co-factors, especially electron-donating redox partners. The first, rate-limiting step in steroidogenesis is catalyzed by P450scc in all steroidogenic tissues, determining steroidogenic capacity; the qualitative regulation of steroidogenesis, determining the class of steroid produced, is determined by the expression of downstream enzymes, principally P450c17. Steroidogenesis regulates development and physiology, and its understanding is necessary to fully understand disorders of sexual differentiation, reproduction, fertility, hypertension, obesity, and physiologic homeostasis. Moreover, an understanding of steroidogenesis is essential for rational steroid therapies.

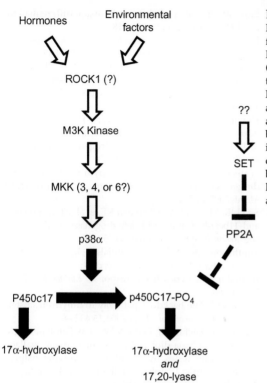

FIGURE 13.7 **A hypothetical model of signaling leading to P450c17 phosphorylation.** We propose that hormones and other factors activate an intracellular pathway that may include ROCK1, eventually leading to a MAP kinase kinase kinases (M3K); the number of steps involved and potential sites of cross-talk remain undetermined. An activated M3K would activate a MAP kinase kinase, such as MKK3, 4, or 6, which would then activate p38α, permitting phosphorylation of P450c17 and the acquisition of 17,20-lyase activity. P450c17 is dephosphorylated by protein phosphatase 2A (PP2A), which in turn can be inhibited by phosphoprotein SET. Potential cross-talk with other second messenger signaling pathways is not shown. Steps that have been established experimentally are shown with closed arrows; hypothetical steps are shown with open arrows; inhibitory steps are shown with dashed lines.

References

1. Agarwal AK, Auchus RJ. Minireview: cellular redox state regulates hydroxysteroid dehydrogenase activity and intracellular hormone potency. *Endocrinology* 2005;**146**:2531−8.
2. Miller WL, Auchus RJ. The molecular biology, biochemistry, and physiology of human steroidogenesis and its disorders. *Endocr Rev* 2011;**32**:81−151.
3. Nebert DW, Wikvall K, Miller WL. Human cytochromes P450 in health and disease. *Philos Trans R Soc Lond B Biol Sci* 2013;**368**:20120431.
4. Horton JD, Goldstein JL, Brown MS. SREBPs: activators of the complete program of cholesterol and fatty acid synthesis in the liver. *J Clin Invest* 2002;**109**:1125−31.
5. Miller WL, Bose HS. Early steps in steroidogenesis: intracellular cholesterol trafficking. *J Lipid Res* 2011;**52**:2111−35.
6. Stocco DM, Clark BJ. Regulation of the acute production of steroids in steroidogenic cells. *Endocr Rev* 1996;**17**:221−44.
7. Miller WL. StAR search—what we know about how the steroidogenic acute regulatory protein mediates mitochondrial cholesterol import. *Mol Endocrinol* 2007;**21**:589−601.
8. Lin D, Sugawara T, Strauss III JF, et al. Role of steroidogenic acute regulatory protein in adrenal and gonadal steroidogenesis. *Science* 1995;**267**:1828−31.
9. Bose HS, Sugawara T, Strauss III JF, et al. The pathophysiology and genetic of congenital lipoid adrenal hyperplasia. International congenital lipoid adrenal hyperplasia consortium. *N Engl J Med* 1996;**335**:1870−8.
10. Tee MK, Abramsohn M, Loewenthal N, et al. Varied clinical presentations of seven patients with mutations in CYP11A1 encoding the cholesterol side-chain cleavage enzyme, P450scc. *J Clin Endocrinol Metab* 2013;**98**:713−20.
11. Miller WL. Minireview: regulation of steroidogenesis by electron transfer. *Endocrinology* 2005;**146**:2544−50.
12. Simard J, Ricketts ML, Gingras S, et al. Molecular biology of the 3beta-hydroxysteroid dehydrogenase/delta5-delta4 isomerase gene family. *Endocr Rev* 2005;**26**:525−82.
13. Auchus RJ, Lee TC, Miller WL. Cytochrome b5 augments the 17,20-lyase activity of human P450c17 without direct electron transfer. *J Biol Chem* 1998;**273**:3158−65.

14. Miller WL. The syndrome of 17,20 lyase deficiency. *J Clin Endocrinol Metab* 2012;**97**:59—67.

15. Miller WL, Auchus RJ, Geller DH. The regulation of 17,20 lyase activity. *Steroids* 1997;**62**:133—42.

16. Speiser PW, Azziz R, Baskin LS, et al. Congenital adrenal hyperplasia due to steroid 21-hydroxylase deficiency: an Endocrine Society clinical practice guideline. *J Clin Endocrinol Metab* 2010;**95**:4133—60.

17. Morel Y, Miller WL. Clinical and molecular genetics of congenital adrenal hyperplasia due to 21-hydroxylase deficiency. *Adv Hum Genet* 1991;**20**:1—68.

18. Speiser PW, White PC. Congenital adrenal hyperplasia. *New England Journal of Medicine* 2003;**349**:776—88.

19. Gomes LG, Huang N, Agrawal V, et al. Extraadrenal 21-hydroxylation by CYP2C19 and CYP3A4: effect on 21-hydroxylase deficiency. *J Clin Endocrinol Metab* 2009;**94**:89—95.

20. White PC, Curnow KM, Pascoe L. Disorders of steroid 11 beta-hydroxylase isozymes. *Endocr Rev* 1994;**15**:421—38.

21. Clyne CD, Zhang Y, Slutsker L, et al. Angiotensin II and potassium regulate human CYP11B2 transcription through common cis-elements. *Mol Endocrinol* 1997;**11**:638—49.

22. Penning TM. Molecular endocrinology of hydroxysteroid dehydrogenases. *Endocr Rev* 1997;**18**:281—305.

23. Penning TM, Burczynski ME, Jez JM, et al. Human 3alpha-hydroxysteroid dehydrogenase isoforms (AKR1C1-AKR1C4) of the aldo-keto reductase superfamily: functional plasticity and tissue distribution reveals roles in the inactivation and formation of male and female sex hormones. *Biochem J* 2000;**351**:67—77.

24. Goto M, Piper Hanley K, Marcos J, et al. In humans, early cortisol biosynthesis provides a mechanism to safeguard female sexual development. *J Clin Invest* 2006;**116**:953—60.

25. Nakamura Y, Hornsby PJ, Casson P, et al. Type 5 17beta-hydroxysteroid dehydrogenase (AKR1C3) contributes to testosterone production in the adrenal reticularis. *J Clin Endocrinol Metab* 2009;**94**:2192—8.

26. Simpson ER, Mahendroo MS, Means GD, et al. Aromatase cytochrome P450, the enzyme responsible for estrogen biosynthesis. *Endocr Rev* 1994;**15**:342—55.

27. Grumbach MM, Auchus RJ. Estrogen: consequences and implications of human mutations in synthesis and action. *J Clin Endocrinol Metab* 1999;**84**:4677—94.

28. Thigpen AE, Silver RI, Guileyardo JM, et al. Tissue distribution and ontogeny of steroid 5 alpha-reductase isozyme expression. *J Clin Invest* 1993;**92**:903—10.

29. Fluck CE, Meyer-Boni M, Pandey AV, et al. Why boys will be boys: two pathways of fetal testicular androgen biosynthesis are needed for male sexual differentiation. *Am J Hum Genet* 2011;**89**:201—18.

30. Russell DW, Wilson JD. Steroid 5 alpha-reductase: two genes/two enzymes. *Annu Rev Biochem* 1994;**63**:25—61.

31. White PC, Mune T, Agarwal AK. 11 beta-Hydroxysteroid dehydrogenase and the syndrome of apparent mineralocorticoid excess. *Endocr Rev* 1997;**18**:135—56.

32. Hewitt KN, Walker EA, Stewart PM. Minireview: hexose-6-phosphate dehydrogenase and redox control of 11{beta}-hydroxysteroid dehydrogenase type 1 activity. *Endocrinology* 2005;**146**:2539—43.

33. Walker EA, Stewart PM. 11beta-hydroxysteroid dehydrogenase: unexpected connections. *Trends Endocrinol Metab* 2003;**14**:334—9.

34. Seckl JR, Morton NM, Chapman KE, et al. Glucocorticoids and 11beta-hydroxysteroid dehydrogenase in adipose tissue. *Recent Prog Horm Res* 2004;**59**:359—93.

35. Tomlinson JW, Walker EA, Bujalska IJ, et al. 11beta-hydroxysteroid dehydrogenase type 1: a tissue-specific regulator of glucocorticoid response. *Endocr Rev* 2004;**25**:831—66.

36. Auchus RJ. The backdoor pathway to dihydrotestosterone. *Trends Endocrinol Metab* 2004;**15**:432—8.

37. Wilson JD, Auchus RJ, Leihy MW, et al. 5alpha-androstane-3alpha,17beta-diol is formed in tammar wallaby pouch young testes by a pathway involving 5alpha-pregnane-3alpha,17alpha-diol-20-one as a key intermediate. *Endocrinology* 2003;**144**:575—80.

38. Kamrath C, Hochberg Z, Hartmann MF, et al. Increased activation of the alternative "backdoor" pathway in patients with 21-hydroxylase deficiency: evidence from urinary steroid hormone analysis. *J Clin Endocrinol Metab* 2012;**97**:E367—75.

39. Falany CN. Enzymology of human cytosolic sulfotransferases. *FASEB J* 1997;**11**:206—16.

40. Strott CA. Sulfonation and molecular action. *Endocr Rev* 2002;**23**:703—32.

41. Suzuki T, Sasano H, Takeyama J, et al. Developmental changes in steroidogenic enzymes in human postnatal adrenal cortex: immunohistochemical studies. *Clin Endocrinol (Oxf)* 2000;**53**:739—47.

42. Homma K, Hasegawa T, Nagai T, et al. Urine steroid hormone profile analysis in cytochrome P450 oxidoreductase deficiency: implication for the backdoor pathway to dihydrotestosterone. *J Clin Endocrinol Metab* 2006;**91**:2643—9.

43. Tsigos C, Arai K, Hung W, et al. Hereditary isolated glucocorticoid deficiency is associated with abnormalities of the adrenocorticotropin receptor gene. *J Clin Invest* 1993;**92**:2458—61.

44. Martens JW, Verhoef-Post M, Abelin N, et al. A homozygous mutation in the luteinizing hormone receptor causes partial Leydig cell hypoplasia: correlation between receptor activity and phenotype. *Mol Endocrinol* 1998;**12**:775—84.

45. Shenker A. G protein-coupled receptor structure and function: the impact of disease-causing mutations. *Baillieres Clin Endocrinol Metab* 1995;**9**:427—51.

46. Papadopoulos V, Miller WL. Role of mitochondria in steroidogenesis. *Best Pract Res Clin Endocrinol Metab* 2012;**26**:771—90.

47. Orentreich N, Brind JL, Rizer RL, et al. Age changes and sex differences in serum dehydroepiandrosterone sulfate concentrations throughout adulthood. *J Clin Endocrinol Metab* 1984;**59**:551—5.

48. Miller WL. Androgen synthesis in adrenarche. *Rev Endocr Metab Disord* 2009;**10**:3—17.

49. Auchus RJ, Rainey WE. Adrenarche - physiology, biochemistry and human disease. *Clin Endocrinol (Oxf)* 2004;**60**:288—96.

50. Ehrmann D. Polycystic ovary syndrome. *N Engl J Med* 2005;**352**:1223—36.

51. Oppenheimer E, Linder B, DiMartino-Nardi J. Decreased insulin senstivity in prepubertal girls with premature pubarche and acanthosis nigricans. *J Clin Endocrinol Metab* 1995;**80**:614—8.

52. Ibanez L, Dimartino-Nardi J, Potau N, et al. Premature adrenarche—normal variant or forerunner of adult disease? *Endocr Rev* 2000;**21**:671—96.

53. Idkowiak J, Lavery GG, Dhir V, et al. Premature adrenarche: novel lessons from early onset androgen excess. *Eur J Endocrinol* 2011;**165**:189—207.

54. Auchus RJ. The physiology and biochemistry of adrenarche. *Endocr Dev* 2011;**20**:20—7.

55. Zhang LH, Rodriguez H, Ohno S, et al. Serine phosphorylation of human P450c17 increases 17,20-lyase activity: implications for adrenarche and the polycystic ovary syndrome. *Proc Natl Acad Sci USA* 1995;**92**:10619—23.

56. Dunaif A, Xia J, Book CB, et al. Excessive insulin receptor serine phosphorylation in cultured fibroblasts and in skeletal muscle. A potential mechanism for insulin resistance in the polycystic ovary syndrome. *J Clin Invest* 1995;**96**:801—10.

57. Tee MK, Miller WL. Phosphorylation of human P450c17 by p38alpha selectively increases 17,20 lyase activity and androgen biosynthesis. *J Biol Chem* 2013;**288**:23903—13.

Adipose Tissue as an Endocrine Organ

Nicolas Musi* and Rodolfo Guardado-Mendoza[†]

*Barshop Institute for Longevity and Aging Studies and Geriatric Research, San Antonio, TX, USA,
[†]University of Guanajuato, León, Mexico

ADIPOSE TISSUE BIOLOGY AND FUNCTION

For many years, it was considered that the only function of the adipose tissue was to store energy in the form of fat. However, evidence accumulated over the last two decades has demonstrated that adipose tissue plays other important roles.[1] Adipose tissue works as an endocrine organ capable of synthesizing and secreting a large number of substances that regulate energy balance and metabolic homeostasis. In addition to fat cells, adipose tissue also contains a stroma—vascular fraction that includes blood cells, endothelial cells, pericytes and adipose precursor cells.[2,3] In order for an adipocyte to become a mature cell capable of carrying out its metabolic functions it has to undergo adipogenesis, a highly regulated process involving the coordinated activation of numerous transcription factors (Table 14.1).[4]

The morphological and functional changes that occur during adipogenesis correspond to a shift in transcription factor expression and activity leading the formation of a mature cell phenotype from an early multipotent state.[1]

When increased storage requirements are needed, immature cells differentiate into mature adipocytes, facilitating the hyperplasic expansion of adipose tissue. Also, mature adipocytes can expand in size to store more lipids and even become hypertrophic under conditions of overnutrition. The adipose tissue is highly adaptable and is able to modify adipocyte number and morphology in response to alterations in energy balance through changes in free fatty acid (FFA) uptake, esterification, and lipolysis.[5,6]

In humans, adipose tissue is broadly classified into 2 types, white and brown; some propose that there is a third type, named beige adipose tissue (Figure 14.1).[3] White adipose tissue (WAT) is much more abundant than brown adipose tissue (BAT). WAT has extensive distribution in the body, including most of the subcutaneous region, abdominal cavity, mediastinum, and areas between muscle groups. Due to its ability to accumulate and provide energy when needed, WAT is the most important buffering system for lipid energy balance in the body.[7] While BAT can also store lipids, it plays an important role for

TABLE 14.1   Transcriptional Regulators of Adipogenesis

Negative Regulators	Positive Regulators
Zfp521	EBF1
KLF 2 and 7	Zfp423
PREF-1	AP-1
Wnt 10b	KLF 4 and 6
Wnt 5a	C/EBPs β and δ
	KLFs 5 and 15
	PPARγ
	C/EBPα
	STAT 5A
	SREBP-1

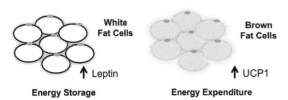

FIGURE 14.1   **Types of adipose tissue.** Adipose tissue can be functionally classified as white (WAT) and brown (BAT). WAT is largely responsible for the synthesis and storage of triglycerides. WAT also secretes numerous adipokines, such as leptin and adiponectin. BAT is responsible for heat production and energy expenditure due to its high content of uncoupling protein 1(UCP1).

body temperature maintenance. Therefore, BAT is more abundant in infants who are more susceptible to hypothermia, and its mass diminishes as age advances.[8] A key function of BAT is heat production secondary to its high content of uncoupling protein 1 (UCP1).[9] UCPs are mitochondrial proteins that dissipate the proton gradient before it is used to provide energy for oxidative phosphorylation, generating heat. BAT derives its color from the extensive vascularization and the presence of many densely packed mitochondria. The physiologic

relevance of BAT is underscored by experiments in which transplantation of a small amount of BAT from a lean mouse to a high-fat fed mouse causes significant weight loss and improves glucose tolerance.[10] It has been assumed for several years that adult humans have a negligible amount of BAT. However, with the advent of new imaging technologies (positron emission tomography (PET) scanning) that can identify BAT in humans *in vivo*, it is becoming apparent that adult humans have substantial depots of BAT, particularly in the anterior neck and thoracic areas.[8] Although the total amount of BAT is much smaller than WAT, its very high metabolic rate and capacity to consume enormous amounts of energy suggests that BAT has important physiologic implications. The origin of beige adipose tissue is less clear compared with WAT and BAT. Initially it was thought that beige adipocytes arise from unique precursor cells,[3] although recent evidence suggests that these cells also can arise from white adipocytes through transdifferentiation.[2] The function of beige adipose tissue is not well known; it is possible that it has some of the properties of brown adipose tissue, including the ability to dissipate energy.[3]

In addition to the functional classification (WAT vs. BAT) discussed above, adipose tissue is classified according to its physical location in the body into subcutaneous and visceral. Visceral fat, also known as intra-abdominal fat, surrounds the viscera in the abdominal cavity. This physical classification also has important functional implications. For example, visceral fat produces a large amount of pro-inflammatory proteins such as interleukin (IL)-6, tumor necrosis factor (TNF) α and plasminogen activator inhibitor 1 (PAI-1), and harbors a larger amount of inflammatory cells (i.e., monocytes/macrophages) than subcutaneous fat. In contrast, subcutaneous fat produces more adiponectin and leptin, adipokines that play important roles in the regulation of glucose and lipid metabolism, energy balance, and appetite.

The discovery of leptin in 1994 was the initial demonstration that the adipose tissue functions as an endocrine organ that synthesizes and secretes proteins that affect other tissues/organs. Many of the proteins secreted by adipose tissue are synthesized by the adipocyte *per se*, which composes approximately 80% of adipose tissue. Yet, other cell types present within adipose tissue, including pericytes, endothelial cells, monocytes, macrophages, preadipocytes, and stem cells, can also produce and secrete proteins. The humoral products of adipose tissue are involved in various processes such as inflammation, lipid metabolism, energy balance, vascular tone, glucose homeostasis, insulin sensitivity, and atherosclerosis. The expression of different receptors and the production of a variety of substances allow the adipose tissue to cross-talk with other tissues and regulate systemic energy balance and metabolism.[11]

ADIPOKINE SECRETION AND INTERACTION WITH OTHER TISSUES

The list of newly discovered adipocyte-derived factors is rapidly growing.[12,13] Below, some of the better characterized proteins secreted by fat are discussed (Table 14.2).

Leptin

This adipokine is a small peptide (16 kDa) that belongs to the IL-6 family of cytokines and is encoded by the *ob* gene. Plasma leptin concentration directly correlates with the amount of adipose tissue present in the body. Leptin is an anorexigenic peptide that interacts with its receptor in the central nervous system to regulate food intake and energy expenditure.[14] The leptin receptor also is expressed in hematopoietic and immune cells where it has immunomodulating properties. Plasma leptin is increased by glucocorticoids, acute infection and proinflammatory cytokines, and it is reduced by cold exposure, adrenergic stimulation, growth hormone, thyroid hormone, melatonin, smoking and thiazolidinediones.

Adiponectin

This adipokine circulates in three isoforms: a trimer of low molecular weight, a hexamer of medium molecular weight, and a multimeric of high molecular weight. Two adiponectin receptors have been identified, AdipoR1 and AdipoR2; skeletal muscle contain both type of receptors whereas liver primarily expresses AdipoR2. Adiponectin improves insulin sensitivity and stimulates fatty acid oxidation in skeletal muscle, liver, and adipose tissue. The cellular effects of adiponectin are mediated, in part, through the activation of adenosine monophosphate (AMP)-activated protein kinase (AMPK). In addition to its effect on the liver and muscle, adiponectin regulates energy expenditure through activation of AMPK in the hypothalamus, where AdipoR1 and AdipoR2 co-localize with the leptin receptor.[15,16]

TNFα

TNFα is a 26 kDa transmembrane protein that undergoes cleavage by a metalloproteinase to be released into the circulation as a 17-kDa soluble protein. Although adipocytes produce some TNFα, it is thought that macrophages from the stromal vascular fraction are the primary source of adipose-derived TNFα. TNFα impairs insulin signaling in liver, adipose tissue and skeletal muscle[17] by activation of serine kinases such as the c-Jun-N-terminal kinase (JNK) and inhibitor of NF-kB kinase (IKK), which serine phosphorylate insulin receptor substrate 1 (IRS1).[18] In hepatocytes

TABLE 14.2   Proteins Secreted by Adipose Tissue

Molecule	Cellular Communication Mechanism	Target Tissue/Organ	Biological Effect
Leptin	Endocrine	Hypothalamus/central nervous system	Food intake and energy expenditure regulation
Adiponectin	Endocrine	Hypothalamus	Energy expenditure
	Paracrine	Skeletal muscle	Glucose uptake/insulin action
		Adipose tissue	Fatty acid oxidation
Tumor necrosis factor α	Endocrine	Liver	Reduces insulin signaling and fatty acid oxidation
	Paracrine	Skeletal muscle	
		Adipose tissue	
Interleukin-6	Endocrine	Liver	Reduces insulin signaling
	Paracrine	Skeletal muscle	Increases fatty acid oxidation and glucose uptake
		Islet of Langerhans	A-cell proliferation and survival
Plasminogen activator inhibitor 1 (PAI-1)	Autocrine	Vascular/endothelial cells	Fibrinolysis
	Paracrine		
	Endocrine		
Resistin	Endocrine	Liver	Decreases insulin signaling
	Paracrine	Adipose tissue	
		Skeletal muscle	
Visfatin	Endocrine	Liver	Increases production of TNFα and IL-6
	Paracrine	Adipose tissue	Insulinomimetic?
		Skeletal muscle	
Angiotensin	Endocrine	Vascular cells	Blood pressure regulation
	Paracrine	Adipocytes	Adipocytogenesis

and muscle, TNFα also can inhibit AMPK, consequently reducing fatty acid oxidation.[19,20]

Interleukin-6

In humans, approximately 30% of circulating IL-6 originates from adipose tissue. Whether IL-6 is beneficial or detrimental to glucose metabolism is controversial. Data from some studies indicate that IL-6 impairs insulin action, whereas other studies have shown that IL-6 stimulates fatty acid oxidation and glucose uptake in skeletal muscle.[21,22] There is also evidence indicating that IL-6 increases pancreatic α cell proliferation and survival with a consequent increase in glucagon secretion.[23]

Plasminogen Activator Inhibitor 1 (PAI-1)

Endothelial and vascular smooth muscle cells are important sources of PAI-1, but other cells such as adipocytes, macrophages, monocytes, fibroblasts, mesangial cells, hepatocytes, and platelets also secrete PAI-1. The plasma level of PAI-1 directly correlates with adipose tissue mass. PAI-1 synthesis is upregulated by insulin, glucocorticoids, angiotensin II, fatty acids, TNFα, and transforming growth factor-β, while catecholamines reduce its production.[24] PAI-1 functions as a serine protease inhibitor that lowers tissue plasminogen activator (tPA). tPA activates plasminogen to promote fibrinolysis. Consequently, increases in PAI-1 level enhance clotting and contributes to the remodeling of vascular architecture and atherosclerosis.

Resistin

Resistin shares structural similarities with adiponectin. However, in contrast to adiponectin, resistin is associated with insulin resistance. Resistin inhibits insulin signaling and promotes insulin resistance by increasing hepatic gluconeogenesis. Pre-adipocytes express more resistin compared with mature adipocytes, suggesting a role in adipogenesis.[25]

Visfatin

Visfatin is mainly produced by visceral tissue adipocytes, but macrophages and subcutaneous fat also secrete this adipokine to a lesser degree. Visfatin has a mimetic insulin effect and it is capable of lowering glucose levels.[26] Visfatin also works as a pro-inflammatory cytokine that activates leukocytes and stimulates production of TNFα and IL-6.[27]

Angiotensin

Adipose tissue expresses all of the components of the renin–angiotensin system. Adipose tissue angiotensinogen mRNA and protein levels are regulated by nutrition, with decreased levels at fasting and increased levels after feeding. Angiotensin II stimulates prostacyclin synthesis, adipocyte differentiation, and lipogenesis. It is possible that adipocyte-derived components of this system play a role in the cardiovascular alterations of obesity and type 2 diabetes.

ADIPOSE TISSUE DYSFUNCTION DURING OBESITY

An important biological property of adipose tissue is its capacity to rapidly respond to fluctuations in nutrient and energy supply through adipocyte hypertrophy and hyperplasia. In most obese subjects, the expansion/remodeling of the adipose tissue is thought to be pathological because it leads to a metabolic phenotype that promotes metabolic alterations and cardiovascular disease. However, not all the adipose tissue expansion is necessarily pathological. For example, there is an obese phenotype that is "metabolically healthy;" these are obese individuals who have normal or near-normal insulin sensitivity and lipid metabolism. The molecular basis that underlies the difference between a pathological versus a non-pathological form of obesity is not known.

An interesting observation regarding adipose tissue expansion is the role of macrophage infiltration. Macrophage infiltration into adipose tissue may occur through different mechanisms:[28] 1) macrophage infiltration in order to phagocytose dead adipocytes;

2) chemokine-induced macrophage mobilization from bone marrow; 3) local hypoxia, which induces release of chemoattractant cyokines;[29] and 4) FFA activation of toll-like receptor 4 (TLR4) that causes a local inflammatory state and macrophage infiltration/activation.[30,31]

In obesity, multiple inflammatory inputs contribute to metabolic dysfunction, including increases in circulating cytokines, decreases in protective factors (adiponectin), and cross-talk between inflammatory and metabolic cells (Figure 14.2). Upon stimulation, macrophages assume a classical pro-inflammatory activation state (M1) that generates a Th1 response. On the other hand, Th2 cytokines such as IL-4 and IL-3 generate an alternative macrophage activation state (M2) that attenuates the classical NF-kB-dependent pathway. Adipose tissue macrophages assume different states along the M1/M2 spectrum depending on fat depot location and nutritional status; adiposity results in a shift toward a pro-inflammatory signaling pathway where classically activated M1 cells predominate.[32-34] This specific adipose tissue network and cross-talk indicates that maintaining metabolic homeostasis requires a balanced immune response and integrated signals of multiple cell types.

Adipose tissue insulin resistance and dysfunctional lipid storage play a key role in the progression toward metabolic dysregulation in obesity. Visceral adipose tissue has a higher capacity to secrete pro-inflammatory cytokines and to become a hypertrophic cell prone to macrophage infiltration. With the prevalent inflammatory status, this causes endoplasmic reticulum (ER) stress, adipose tissue hypoxia and adipocyte death.[29,35] Similar to the effects in adipose tissue, liver is also affected by this inflammatory response. The steatotic liver is characterized by an elevation of many of the signaling pathways involved in both inflammation and metabolism (JNK, TLR4, ER

stress). In the skeletal muscle of obese patients, there is activation of inflammatory signals through TLR4 and downstream pathways.[36] During adipose tissue expansion, also there is activation of hypothalamic signaling pathways that induce food intake and nutrient storage.[37,38] This effect may be mediated by saturated fatty acids, which activate neuronal JNK and NF-kB signaling pathways with direct effects on leptin and insulin signaling.[39] In addition to the inflammatory state that occurs in insulin target tissues (muscle, liver, fat), during obesity also there is an increase in cytokine production and macrophage accumulation in the islets of Langerhans, which promote β cell dysfunction.[40,41]

ADIPOSE TISSUE AS A THERAPEUTIC TARGET AND FUTURE PERSPECTIVES

Adipose tissue plays a key role in maintaining energy balance and metabolic homeostasis through its ability to store energy and to secrete a large number of substances (FFA, adipokines, etc.). Accordingly, adipose tissue is an attractive target for pharmacological treatment of obesity, diabetes, and other metabolic diseases. Examples of strategies that could be employed include: 1) regulation of transcriptional factors that control adipogenesis to promote the development of "healthy" adipocytes; 2) modification of the adipokine secretion profile to favor beneficial factors (adiponectin) and decrease release of potentially deleterious ones (TNFα, IL-1, resistin); 3) inhibition of intracellular pathways activated by adipokines (TNFα receptor, IKKβ/NF-kB, JNK); 4) increasing WAT lipolysis; 5) WAT transdifferentiation to BAT, which promotes energy expenditure; and 6) activation of BAT.

Targeting the adipose tissue to treat metabolic diseases is a reality; for example, the

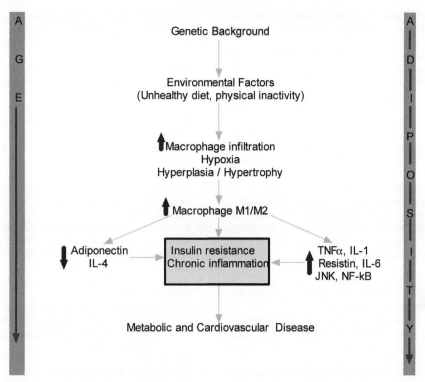

FIGURE 14.2 Role of adipose tissue expansion on metabolic and cardiovascular disease. The interplay between genetic and environmental factors (sedentary lifestyle, high caloric intake) leads to obesity, characterized by adipocyte hyperplasia and hypertrophy. Obesity also is characterized by the development of a pro-inflammatory state in adipose tissue, leading to macrophage infiltration, hypoxia, and macrophage polarization (M1/M2). During obesity, there is also an increase in the production and secretion of pro-inflammatory cytokines that impair glucose homeostasis and promote cardiovascular disease, including TNFα, IL-1, IL-6, and resistin, whereas cytokines that improve glucose metabolism and promote cardiovascular health such as adiponectin are decreased. Aging also induces inflammation and promotes adiposity, further worsening cardiometabolic diseases.

antidiabetic agents thiazolidenediones (TZDs) promote differentiation of preadipocytes into white adipocytes and decrease the expression of TNFα, IL-1, and resistin while increasing adiponectin. Currently, major efforts are underway to develop novel strategies to prevent and treat metabolic diseases by targeting the adipose tissue and its products.[42,43] Consequently, it is likely that in the near future new agents will be available for the management of obesity, type 2 diabetes, and associated cardiovascular diseases.

References

1. Stephens JM. The fat controller: adipocyte development. *PLoS Biol* 2012;**10**(11):e1001436.
2. Cinti S. The adipose organ at a glance. *Dis Model Mech* 2012;**5**(5):588–94.
3. Wu J, Bostrom P, Sparks LM, et al. Beige adipocytes are a distinct type of thermogenic fat cell in mouse and human. *Cell* 2012;**150**(2):366–76.
4. Kang S, Akerblad P, Kiviranta R, et al. Regulation of early adipose commitment by Zfp521. *PLoS Biol* 2012;**10**(11):e1001433.
5. Coelho M, Oliveira T, Fernandes R. Biochemistry of adipose tissue: an endocrine organ. *Arch Med Sci* 2013;**9**(2):191–200.

6. Gray SL, Vidal-Puig AJ. Adipose tissue expandability in the maintenance of metabolic homeostasis. *Nutr Rev* 2007;**65**(6 Pt 2):S7−12.

7. Kiess W, Petzold S, Topfer M, et al. Adipocytes and adipose tissue. *Best Pract Res Clin Endocrinol Metab* 2008;**22**(1):135−53.

8. Cypess AM, Lehman S, Williams G, et al. Identification and importance of brown adipose tissue in adult humans. *N Engl J Med* 2009;**360**(15):1509−17.

9. Yoneshiro T, Aita S, Matsushita M, et al. Brown adipose tissue, whole-body energy expenditure, and thermogenesis in healthy adult men. *Obesity (Silver Spring)* 2011;**19**(1):13−6.

10. Stanford KI, Middelbeek RJ, Townsend KL, et al. Brown adipose tissue regulates glucose homeostasis and insulin sensitivity. *J Clin Invest* 2013;**123**(1):215−23.

11. Sethi JK, Vidal-Puig AJ. Thematic review series: adipocyte biology. Adipose tissue function and plasticity orchestrate nutritional adaptation. *J Lipid Res* 2007;**48**(6):1253−62.

12. Halberg N, Wernstedt-Asterholm I, Scherer PE. The adipocyte as an endocrine cell. *Endocrinol Metab Clin North Am* 2008;**37**(3):753−68, x−xi.

13. Lafontan M. Historical perspectives in fat cell biology: the fat cell as a model for the investigation of hormonal and metabolic pathways. *Am J Physiol Cell Physiol* 2012;**302**(2):C327−59.

14. Galic S, Oakhill JS, Steinberg GR. Adipose tissue as an endocrine organ. *Mol Cell Endocrinol* 2010;**316**(2):129−39.

15. Kadowaki T, Yamauchi T. Adiponectin and adiponectin receptors. *Endocr Rev* 2005;**26**(3):439−51.

16. Kubota N, Yano W, Kubota T, et al. Adiponectin stimulates AMP-activated protein kinase in the hypothalamus and increases food intake. *Cell Metab* 2007;**6**(1):55−68.

17. Monroy A, Kamath S, Chavez AO, et al. Impaired regulation of the TNF-alpha converting enzyme/tissue inhibitor of metalloproteinase 3 proteolytic system in skeletal muscle of obese type 2 diabetic patients: a new mechanism of insulin resistance in humans. *Diabetologia* 2009;**52**(10):2169−81.

18. Hotamisligil GS, Peraldi P, Budavari A, Ellis R, White MF, Spiegelman BM. IRS-1-mediated inhibition of insulin receptor tyrosine kinase activity in TNF-alpha- and obesity-induced insulin resistance. *Science* 1996;**271**(5249):665−8.

19. Nachiappan V, Curtiss D, Corkey BE, Kilpatrick L. Cytokines inhibit fatty acid oxidation in isolated rat hepatocytes: synergy among TNF, IL-6, and IL-1. *Shock* 1994;**1**(2):123−9.

20. Steinberg GR, Michell BJ, van Denderen BJ, et al. Tumor necrosis factor alpha-induced skeletal muscle insulin resistance involves suppression of AMP-kinase signaling. *Cell Metab* 2006;**4**(6):465−74.

21. Carey AL, Steinberg GR, Macaulay SL, et al. Interleukin-6 increases insulin-stimulated glucose disposal in humans and glucose uptake and fatty acid oxidation in vitro via AMP-activated protein kinase. *Diabetes* 2006;**55**(10):2688−97.

22. Steinberg GR, Watt MJ, Ernst M, Birnbaum MJ, Kemp BE, Jorgensen SB. Ciliary neurotrophic factor stimulates muscle glucose uptake by a PI3-kinase-dependent pathway that is impaired with obesity. *Diabetes* 2009;**58**(4):829−39.

23. Ellingsgaard H, Ehses JA, Hammar EB, et al. Interleukin-6 regulates pancreatic alpha-cell mass expansion. *Proc Natl Acad Sci USA* 2008;**105**(35):13163−8.

24. Correia ML, Haynes WG. A role for plasminogen activator inhibitor-1 in obesity: from pie to PAI?. *Arterioscler Thromb Vasc Biol* 2006;**26**(10):2183−5.

25. Steppan CM, Bailey ST, Bhat S, et al. The hormone resistin links obesity to diabetes. *Nature* 2001;**409**(6818):307−12.

26. Wang P, Xu TY, Guan YF, Su DF, Fan GR, Miao CY. Perivascular adipose tissue-derived visfatin is a vascular smooth muscle cell growth factor: role of nicotinamide mononucleotide. *Cardiovasc Res* 2009;**81**(2):370−80.

27. Moschen AR, Kaser A, Enrich B, et al. Visfatin, an adipocytokine with proinflammatory and immunomodulating properties. *J Immunol* 2007;**178**(3):1748−58.

28. Sun K, Kusminski CM, Scherer PE. Adipose tissue remodeling and obesity. *J Clin Invest* 2011;**121**(6):2094−101.

29. Hosogai N, Fukuhara A, Oshima K, et al. Adipose tissue hypoxia in obesity and its impact on adipocytokine dysregulation. *Diabetes* 2007;**56**(4):901−11.

30. Reyna SM, Ghosh S, Tantiwong P, et al. Elevated toll-like receptor 4 expression and signaling in muscle from insulin-resistant subjects. *Diabetes* 2008;**57**(10):2595−602.

31. Shi H, Kokoeva MV, Inouye K, Tzameli I, Yin H, Flier JS. TLR4 links innate immunity and fatty acid-induced insulin resistance. *J Clin Invest* 2006;**116**(11):3015−25.

32. Lumeng CN, Bodzin JL, Saltiel AR. Obesity induces a phenotypic switch in adipose tissue macrophage polarization. *J Clin Invest* 2007;**117**(1):175−84.

33. Shaul ME, Bennett G, Strissel KJ, Greenberg AS, Obin MS. Dynamic, M2-like remodeling phenotypes of CD11c + adipose tissue macrophages during high-fat diet—induced obesity in mice. *Diabetes* 2010;**59**(5):1171−81.

34. Zeyda M, Gollinger K, Kriehuber E, Kiefer FW, Neuhofer A, Stulnig TM. Newly identified adipose

tissue macrophage populations in obesity with distinct chemokine and chemokine receptor expression. *Int J Obes (Lond)* 2010;**34**(12):1684−94.

35. Ozcan U, Cao Q, Yilmaz E, et al. Endoplasmic reticulum stress links obesity, insulin action, and type 2 diabetes. *Science* 2004;**306**(5695):457−61.

36. Frisard MI, McMillan RP, Marchand J, et al. Toll-like receptor 4 modulates skeletal muscle substrate metabolism. *Am J Physiol Endocrinol Metab* 2010;**298**(5):E988−98.

37. German JP, Thaler JP, Wisse BE, et al. Leptin activates a novel CNS mechanism for insulin-independent normalization of severe diabetic hyperglycemia. *Endocrinology* 2011;**152**(2):394−404.

38. Thaler JP, Schwartz MW. Minireview: inflammation and obesity pathogenesis: the hypothalamus heats up. *Endocrinology* 2010;**151**(9):4109−15.

39. Zhang X, Zhang G, Zhang H, Karin M, Bai H, Cai D. Hypothalamic IKKbeta/NF-kappaB and ER stress link overnutrition to energy imbalance and obesity. *Cell* 2008;**135**(1):61−73.

40. Ehses JA, Perren A, Eppler E, et al. Increased number of islet-associated macrophages in type 2 diabetes. *Diabetes* 2007;**56**(9):2356−70.

41. Benoit SC, Kemp CJ, Elias CF, et al. Palmitic acid mediates hypothalamic insulin resistance by altering PKC-theta subcellular localization in rodents. *J Clin Invest* 2009;**119**(9):2577−89.

42. Haas B, Schlinkert P, Mayer P, Eckstein N. Targeting adipose tissue. *Diabetol Metab Syndr* 2012;**4**(1):43.

43. Lumeng CN, Saltiel AR. Inflammatory links between obesity and metabolic disease. *J Clin Invest* 2011;**121**(6):2111−7.

Insulin-Secreting Cell Lines: Potential for Research and Diabetes Therapy

Shanta J. Persaud, Astrid C. Hauge-Evans and Peter M. Jones

King's College London, London, UK

INTRODUCTION

Islets of Langerhans are three-dimensional clusters of approximately 1000 cells that constitute the endocrine portion of the pancreas, and each islet is around 50–500 μm in diameter. The most abundant islet cell type in all species is the insulin-secreting β-cell, although there is some variation in the proportion of β-cells between species, with estimates that mouse islets comprise 80–90% β-cells, while in human islets the β-cells contribute 60–70% to the islet mass.[1] There are many similarities in the functional characteristics of rodent and human islets, and since islets isolated from cadaver organ donors are not widely available for research many studies over the past 50 years have made use of mouse and rats islets. Islets from these rodents are similar in size to human islets. However, a mouse pancreas is considerably less than 0.1% of the volume of a human pancreas, so while 250,000–500,000 islets can be obtained from a human pancreas the yield from a mouse pancreas is only 200–250 islets. In addition, islet isolation from all species by collagenase digestion of the exocrine pancreas and purification by handpicking or density gradients is time consuming. Furthermore, while some primary cells, such as those derived from smooth muscle, proliferate in culture to produce additional cells for experimental use, islet cells do not readily proliferate. Therefore, considerable effort has been expended since the 1970s to generate insulin-secreting cells that proliferate in culture and show functional characteristics of primary β-cells. These immortalized insulin-secreting cell lines, which can be maintained in continuous cultures, have been developed by a number of methodologies and they are the subject of this chapter.

KEY FEATURES OF PRIMARY β-CELLS

Islet β-cells are designed to respond to elevations in circulating glucose with metabolic coupling that culminates in the regulated release of insulin from secretory vesicles. Insulin-secreting

Cellular Endocrinology in Health and Disease.
DOI: http://dx.doi.org/10.1016/B978-0-12-408134-5.00015-9

239

cell lines must recapitulate, as closely as possible, the key features of primary β-cells that allow them to synthesize and store insulin, recognize glucose and metabolize it, and secrete insulin when required. The essential features of β-cells are summarized below, and Table 15.1 defines the characteristics of the main cell lines in terms of glucose transport and phosphorylation, and also with respect to their insulin content and secretory responsiveness to glucose.

Glucose Transport and Metabolism

Circulating glucose enters β-cells via facilitated, insulin-independent transporters. In rodent islets it has been established that the high-K_m ($\sim$15 mM) GLUT2 transporter is responsible for glucose entry into β-cells, but it has been proposed that the low-K_m ($\sim$3 mM) transporters, GLUT1 and/or 3, are sufficient for glucose uptake by human β-cells.[2] However, a more recent study has demonstrated that loss of function mutations in the gene encoding GLUT2 in humans can cause insulin-dependent neonatal diabetes,[3] suggesting that GLUT2 may play an important role in glucose transport into β-cells in humans as well as in rodents.

Glucose uptake is not rate limiting for β-cell responses to glucose since it greatly exceeds glucose usage and there is rapid equilibration of extracellular and intracellular glucose concentrations. However, it has long been known that glucose phosphorylation to glucose-6-phosphate by hexokinase IV (glucokinase) is a key regulatory step in β-cell stimulus–response coupling, and it is acknowledged that this high-K_m ($\sim$10 mM), high specificity hexokinase is an important β-cell glucose sensor.

Insulin Synthesis, Processing, Storage and Secretion

Islet β-cells are the only cells in the body that synthesize insulin, which is why β-cell

destruction in type 1 diabetes has such devastating metabolic consequences. The insulin precursor peptide, proinsulin, is folded in the endoplasmic reticulum such that the A- and B-chains of insulin are aligned and connected by disulphide bonds between cysteine residues. Conversion of proinsulin to insulin takes place in β-cell secretory vesicles by removal of the C (connecting) peptide by the endoproteases PC2 and PC3. Islet β-cells essentially function as insulin synthesising factories, and electron micrographs demonstrate that β-cells contain numerous secretory vesicles that are characterized by electron-dense cores of insoluble insulin complexed with zinc in hexameric crystals (Figure 15.1, left panel).

Insulin synthesis is regulated by glucose, and the concentration–response curve is left-shifted compared to that for stimulation of insulin secretion by glucose, perhaps providing a mechanism of maintaining adequate supplies of stored insulin. Exocytotic release of fully processed insulin from the secretory vesicles occurs following granule fusion with the β-cell plasma membrane by a process that is dependent on elevations in intracellular calcium subsequent to glucose metabolism. This regulated release of insulin allows fluctuations in blood glucose levels to be rapidly restored following insulin-dependent uptake and storage of glucose.

Proliferation

The current consensus suggests that primary β-cells within islets have a very low proliferative capacity. A limited increase in the rate of proliferation *in vitro* of rodent β-cells can be induced by maintenance in culture with growth factors, but human β-cells are largely non-proliferative under similar conditions.[4] High-throughput screening of large chemical libraries has identified

TABLE 15.1   Insulin-Secreting Cell Line Characteristics

Cell Type	Source	Doubling Time	Glucose Transporter	Hexokinase	Insulin[a] (pg/cell)	Glucose-Stimulated Insulin Secretion	Other Islet Hormones
Rodent islets	Rodent pancreas	Non-proliferative	GLUT2	GK >> HK	~20	5−20 mM	Glucagon, SST, PP, IAPP
RIN	Transplantable insulinoma	~50−80 h	GLUT2	HK >> GK	0.2	No	Glucagon, SST
INS-1	Transplantable insulinoma	~100 h	GLUT2	GK >> HK	~8	2.8−11.2 mM	
INS-1E	INS-1 cell dilution cloning	NA	GLUT2	GK	NA	5−20 mM	
INS-1 832/13	INS-1 cells with human insulin	NA	NA	NA	~1.5	4−20 mM	
HIT-T15	SV40-tranfected hamster islet cells	~30 h	GLUT1	GK	~0.3	1−10 mM	Glucagon
BRIN-BD11	RIN/rat islet cell electrofusion	~20 h	GLUT2	GK > HK	~0.08	4.2−16.7 mM	
βTC1	Transgenic insulinoma	~60 h	GLUT1	HK > GK	~0.2	0.5−1.25 mM	Glucagon
βTC6-F7	Transgenic insulinoma	NA	GLUT2	GK >> HK	~3	5−30 mM	
MIN6	Transgenic insulinoma	~60 h	GLUT2	GK >> HK	~6	5−25 mM	
NIT-1	Transgenic insulinoma	~50 h	NA	NA	~1.5	11−16.5 mM	Glucagon
βHC9	Transgenic hyperplastic islets	70−100 h	GLUT2 (also some GLUT1)	GK >> HK	~5	7−15 mM	PP[βHC8]
Human islets	Human pancreas	Non-proliferative	GLUT1	GK >> HK	~10	5−20 mM	Glucagon, SST, PP, IAPP
CM	Human insulinoma	NA	GLUT1 and GLUT2	GK	~0.003	No	
HP-62	SV40-transfected human islet cells	20−30 h	NA	NA	~0.004	2−17.6 mM[b]	Glucagon, SST
βlox5-PDX-1[c]	SV40-, H-ras- and hTERT-transformed human β-cells	NA	NA	GK	~1	3−17 mM	IAPP
1.1B4	PANC-1/human islet cell electrofusion	~20 h	GLUT1	GK	~0.004	5.6−16.7 mM	IAPP
EndoC-βH1	Human fetal pancreas	~120 h	NA	GK	~0.5	2.8−20 mM	SST (<0.05% of cells)

Abbreviations: IAPP, islet amyloid polypeptide; NA, information not available; PP, pancreatic polypeptide; SST, somatostatin.
[a]*Insulin content usually decreases in cell lines with prolonged culture: information in the table has generally been obtained from papers where initial characterization of the line was carried out.*
[b]*Insulin expression is lost by passage 9.*
[c]*Cells maintained in clusters in the presence of exendin-4.*

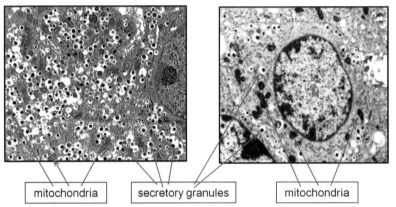

FIGURE 15.1 **Electron micrographs of insulin-secreting cells.** Left: numerous electron dense insulin secretory granules are visible within rodent primary β-cells. Right: insulin secretory granules are also present in MIN6 cells, but they are considerably less abundant. N, nucleus.

small-molecule inducers of rodent β-cell replication but the effects are relatively modest, inducing at best a doubling in the low basal rates of proliferation.[5,6] Some of the effective small molecules were identified as Wnt agonists,[5] and manipulation of Wnt signaling pathways has recently been reported to enhance proliferation of human β-cells.[7] However, even with optimized treatment only 10−15% of the β-cell population expressed markers of proliferation after five days in culture, suggesting that the overall increase in β-cell numbers was modest.

It is obvious that β-cells within insulin-secreting lines must replicate readily in culture so that new cells are generated for experimental and, potentially, clinical use. There is therefore a balance with successful lines, where the doubling time is sufficient to produce new daughter cells without proliferation occurring so rapidly that the cells lose their differentiated functions. An ideal cell line should maintain a stable proliferative and functional phenotype in culture such that it retains secretory responsiveness through prolonged passaging, and some strategies, described below, have been developed to induce growth arrest of expanded cell populations to enhance differentiation.

RODENT AND HUMAN INSULIN-SECRETING CELL LINES

The earliest insulin-secreting cell lines were generated in rodents by transplantation of pancreatic tumors, and the most recent line has been developed by transplanting genetically modified human fetal pancreatic buds into immunocompromised mice. As more sophisticated approaches have been utilized in recent years the possibility of cells being developed as potentially transplantable material for the treatment of type 1 diabetes becomes ever closer.

Rodent Cell Lines

Transplantable Insulinomas

The first attempts at generating insulin-secreting cell lines were published nearly 50 years ago when islet cell tumors were transplanted into the golden hamster.[8,9] The resulting insulinomas were not particularly successful, having very low insulin content, but a decade later rat insulinoma cell lines were established from tumors developed in rats (RIN-r) or nude mice (RIN-m) that acted as recipients of an X-ray-induced rat islet cell tumor.[10,11] RIN cells proliferate readily in culture and are reported to secrete somato-statin in addition to insulin.[11] However, the usefulness of RIN cells as primary β-cell surrogates is severely compromised by their inability to secrete insulin in response to mM concentrations of glucose.[12,13] This is a consequence of the expression of a low-K_m (hexokinase) glucose-phosphorylating activity,[14] rather than the high-K_m (glucokinase) activity found in primary β-cells. RIN 1046-38 cells that underwent several rounds of transfection with the human preproinsulin gene, showed up to 60-fold higher insulin content than the parent cells,[15] and co-expression of glucokinase and GLUT2 genes resulted in at least 6-fold increases in insulin secretion at 5 mM glucose.[16] However, even though these cells were transfected to express glucokinase, the high native expression of hexokinase caused maximal insulin secretory responses at glucose concentrations as low as 250 μM.[16]

A more successful β-cell line, termed INS-1, has also been generated from a rat transplantable radiation-induced insulinoma.[17] As can be seen from Table 15.1, INS-1 cells have substantially higher insulin content than RIN cells, they express glucokinase and show glucose-stimulated insulin secretion, albeit with a magnitude much less than that obtained with primary rat islets.[17] However, INS-1 cells demonstrate reduced glucose-responsiveness over time, which may be a consequence of preferential expansion of poorly glucose-responsive cells.[18] Attempts at overcoming this reduced performance of later-passage INS-1 cells led to cloning of new INS-1 lines (INS-1D and INS-1E) that show improved insulin secretion and are glucose-responsive in the physiological range,[19] and the INS-1E clone is still widely used. Another INS-1 line showing more robust glucose-induced insulin secretory responses has been generated by stable introduction of the human preproinsulin gene, and these clonal INS-1 832/13 cells show

10-fold elevations in insulin release in response to glucose, which is reportedly sustained with continued culture.[18]

SV40-Transformed β-Cells

The simian virus 40 large T antigen (SV40LT) inactivates tumor suppressor proteins to induce tumors in hosts infected with SV40. This function of SV40LT has been used to induce unregulated DNA replication in β-cells through SV40 infection of hamster islet cells.[20] The proliferative cells that were generated by this strategy are known as HIT cells and they secrete insulin in response to glucose. However, they show a left-shifted response to glucose compared to primary β-cells, such that maximal responses are obtained at 10 mM glucose,[21] and this is most likely a consequence of expression of a low-K_m (~4 mM) glucose transporter that does not allow sufficiently rapid glucose uptake into β-cells.[22] HIT-T15 cells are the most commonly used HIT clone, and like other cell lines they show a reduction in insulin content and insulin secretion with continued propagation.[20,23] Nonetheless, they have been used widely in studies on the regulation of insulin secretion, oxidative stress and apoptosis, with over 100 studies published using this line in the past decade.

Electrofusion-Derived Insulin-Secreting Cells

The expression of hexokinase by RIN cells, as described above, has limited their usefulness for studies on glucose-induced insulin secretion, and this led to a strategy of electrofusing RIN insulinoma cells with rat islet cells to determine whether the hybrid fused cells showed characteristics closer to primary β-cells than RIN cells. The resulting BRIN-BD11 cells show improvements over the parental RIN cells, most importantly in terms of glucokinase expression and glucose-stimulated insulin secretion.[24,25] However, like native INS-1 cells,

BRIN-BD11 cells show a rather small increment in insulin output in response to glucose, and their insulin content is considerably lower than that of primary β-cells (Table 15.1).

Targeted Oncogenesis

Targeted oncogenesis makes use of the insulin gene promoter to direct expression of an oncogene, most commonly SV40LT, to β-cells so that tumors develop only in the pancreas, and β-cell oncogene expression may be conditionally regulated by the bacterial tetracycline operon (tet). This confers the advantage of growth arresting the cells by maintaining them in the presence[26] or absence[27] of tetracycline or its derivative, doxycycline. In this way proliferation of the insulin-secreting cells can be tightly regulated and switched off when required to generate cells that have a more differentiated phenotype.

The βTC insulin-secreting cell lines were established from insulin promoter/SV40LT-driven insulinomas developed in transgenic mice.[28] The βTC1 line expresses GLUT1 transporters and low-K_m hexokinases, and although these cells are glucose responsive insulin secretion is maximal at 1.25 mM, most likely a consequence of the preferential expression of hexokinase over glucokinase.[28,29] A glucose-responsive βTC line (βTC6-F7) that expresses GLUT2 and glucokinase and maintains normal glucose-induced insulin secretion with prolonged culture was derived by clonal selection of βTC6 cells.[30] Although these cells show promising characteristics, they have not been widely adopted for β-cell research.

The MIN6 insulin-secreting cell line was also generated by targeted β-cell oncogenesis using the insulin promoter to drive SV40LT expression.[31] MIN6 cells transport and phosphorylate glucose via GLUT2 and glucokinase, and have higher insulin content than most of the other cell lines (Table 15.1). However, the insulin content is still less than that of primary β-cells, and this is reflected by the presence of fewer insulin secretory granules (Figure 15.1). MIN6 cells show appropriate glucose-induced insulin secretion,[31,32] but, in common with other cell lines that demonstrate significantly increased insulin release in response to a glucose challenge at low passage, this property declines with continued monolayer culture.[33]

Spontaneous insulinoma development has also been induced in non-obese diabetic (NOD) mice by the insulin promoter/SV40LT transgene approach. The NIT-1 cells derived from these insulinomas show increased insulin secretion at 11 and 16.5 mM glucose when used at low passage, but they demonstrate decreased sensitivity to glucose with prolonged maintenance in culture.[34] Despite their poorly sustained glucose responsiveness, NIT-1 cells are still widely used, both for studies aimed at identifying regulation of β-cell function and for assessing cytotoxic T cell-induced β-cell death.

Targeted oncogenesis to induce β-cell hyperproliferation and retrieval of hyperplastic islets before solid tumors form in the pancreas has been used as a strategy to generate insulin-secreting cell lines with properties more closely akin to primary β-cells. The βHC9 line was established from hyperplastic SV40LT-expressing islets and βHC9 cells showed a 40-fold stimulation of insulin secretion at 25 mM glucose, most likely due to the preferential expression of glucokinase.[35] However, the capacity of βHC9 cells to respond to glucose is not maintained with continued culture and is absent by passage 40,[36] which limits their usefulness for functional studies.

Human Cell Lines

While most of the rodent-derived insulin-secreting cell lines described above have been widely used as surrogates for primary islet β-cells in numerous studies, there have been parallel, mostly less successful, attempts to generate immortalized, stable lines from human pancreatic tissue. The starting material has ranged from insulinomas and adult islets to fetal pancreatic buds, and the characteristics of the resulting cells are summarized below and in Table 15.1.

Human Insulinomas

The earliest attempts to immortalize human β-cells used a dissociated human insulinoma, but the resulting cells did not show glucose-dependent insulin secretion and insulin content declined with continued maintenance of the cells in culture.[37] Further attempts at generating β-cell lines from insulinomas have been marginally more successful, with reports of small insulin secretory responses to glucose,[38] maintained insulin-secreting capability with prolonged culture[39,40] and the CM cell line showed glucose-dependent preproinsulin gene expression.[41] However, the insulinoma-derived cell lines reported to date suffer from the major disadvantages of lack of glucose-regulated insulin secretion,[37,40,41] fibroblast overgrowth,[37,39] or proliferation failure,[38] and none is a suitable model for primary human β-cells.

Human Islet-Derived Cells

Although adult human β-cells normally show very limited division, there have been reports that proliferation may be induced by maintenance of human islet cells in the presence of some growth factors. For example, hepatocyte growth factor, an agonist at β-cell-expressed c-met mitogenic receptors, stimulated a 30,000-fold increase in islet cell number, but the proliferation was accompanied by a rapid loss of insulin expression.[42] Attempts have also been made to immortalize human islet cells using the SV40LT-transfection approach that had been used successfully to generate insulin-secreting HIT-T15 cells from hamster islets, as described above. The HP-62 cell line, derived from SV40-transformed human islet cells, replicated readily in culture and early passage cells showed glucose-induced insulin secretion over the physiological range.[43] However, although these cells continued to proliferate in culture for at least 2 years they no longer produced insulin after passage 8, so are not useful for β-cell research. Another strategy has been to infect purified adult human islet β-cells with retroviral vectors encoding SV40LT, the oncogene H-ras^{val12} and human telomerase reverse transcriptase (hTERT), a catalytic subunit of telomerase. The resulting cell line, βlox5, only expressed insulin following the forced expression of the transcription factor PDX-1 and maintenance as three-dimensional clusters in the presence of the GLP-1 analog exendin-4.[44] These PDX-1-expressing cells exhibited glucose-dependent insulin secretion *in vitro* and when transplanted into nude mice, but PDX-1 expression (and therefore insulin synthesis) was not maintained with continued culture.[44]

Electrofusion-Derived Insulin-Secreting-Cells

Three human insulin-secreting cell lines have been generated by electrofusion of human islet cells with a proliferative pancreatic epithelial cell line (PANC-1), all of which respond to elevated glucose with increased insulin secretion. The 1.4E7 line is of limited use as it showed only a 50% increase in insulin release, and the glucose concentration—response curve was left-shifted such that maximal stimulation was obtained at 5.6 mM glucose.[45] However, the 1.1B4 line had more promising characteristics, showing a maximal 2.3-fold increase in insulin at 11.1 mM glucose, and these cells expressed the GLUT1 glucose transporter and glucokinase.[45] Although the insulin content of 1.1B4 cells is maintained with prolonged culture, they contain at least 100-fold less insulin per cell than the recently developed EndoC-βH1 cells (see below).

Human Fetal Pancreas

Human fetal pancreas is a potentially suitable source of proliferative β-cells since the replicative capacity of β-cells is highest during fetal development. Epithelial cells obtained from a human fetal pancreas at 18 weeks have been transformed by expression of SV40LT and H-ras[val12], but although the resulting cell line initially expressed insulin, this was not maintained with passage.[46] Another strategy of inducing fetal human β-cell proliferation has been by targeted oncogenesis, using the insulin promoter to drive SV40LT expression in β-cells of fetal pancreatic buds, which were allowed to mature and form insulinomas under the kidney capsule of immunocompromised mice.[47] The retrieved insulinoma cells were transduced with hTERT to limit senescence, expanded further *in vivo*, and cell lines were then generated from proliferating cells in culture. One such line (EndoC-βH1) shows appropriate characteristics in terms of reasonably high insulin content, glucose-dependent insulin secretion over the physiological range, and a stable phenotype for at least 80 passages.[47] These cells offer much promise to the β-cell research community, but there have been no further publications on EndoC-βH1 cells since their development 3 years ago. This is most likely a consequence of their slower proliferative rate than other insulin-secreting cell lines, which limits the amount of material that is readily available for experimental use. Nonetheless, the investigators have made the line freely available to numerous laboratories worldwide, so it is likely that additional functional studies will be published soon.

STEM CELL-DERIVED INSULIN-SECRETING CELLS

Sources of Stem Cells

There has been a great deal of activity over the past decade in developing protocols that use progenitor or stem cell populations as the starting material for generating functional insulin-secreting cells for experimental or therapeutic purposes. The initial cell populations for most of these studies can be broadly subdivided into either *tissue stem cells*, defined as multipotent progenitor cells found in fetal and adult tissues; *embryonic stem cells*, defined as pluripotent undifferentiated cells generated from the inner cell mass of a developing blastocyst; or *induced pluripotent stem cells*, defined as pluripotent cells generated by reprogramming differentiated adult cells by forced expression of pluripotency genes.

Optimizing Differentiation Protocols

Tissue Stem Cells

Tissue stem cells are often considered as lineage-restricted progenitor cells that mature into the differentiated cells of the host tissue, but experimental studies have suggested that progenitor cells from a wide variety of tissues may have the potential to become insulin-expressing cells (reviewed in refs[48–50]).

Several studies have reported that bone marrow (BM) stem cells can be driven towards an insulin-expressing phenotype, either after *in vivo* administration,[51] by selective culture conditions *in vitro*,[52] or by the forced expression of β-cell transcription factors.[53] However, some reports of *in vitro* differentiation to a β-cell phenotype have proven difficult to reproduce,[54] and other studies suggest that BM stem cells reverse experimental diabetes *in vivo* by enhancing the regeneration and survival of endogenous β-cells rather than re-populating the islets with trans-differentiated β-cells.[55] The pancreas is another promising source of tissue stem cells. Although adult β-cells show limited proliferation *in vitro*,[4] β-cell mass increases during development, in pregnancy, and with obesity. Lineage tracing studies in mice suggest that these new insulin-secreting cells can arise from a number of sources, including self-renewal of existing β-cells,[56] transdifferentiation from glucagon-expressing α-cells[57] and differentiation from progenitor cells located in pancreatic ducts,[58] exocrine pancreas,[59] and islets.[60] It remains to be demonstrated whether it is technically feasible to isolate islet progenitor populations, expand them significantly *ex vivo* and differentiate them efficiently into functional insulin-secreting cells. Stem cells isolated from a range of other tissues have also been reported to differentiate into insulin-expressing cells,[61,62] but there is no evidence that these cells are capable of the *in vitro* expansion required to generate significant numbers of functional β-cells.

In summary, tissue stem cells from a wide range of sources have the potential to differentiate into β-cells, but reliable protocols are not yet available for generating *in vitro* large numbers of functional β-cells for experimental or therapeutic purposes.

Embryonic Stem Cells

Embryonic stem (ES) cells have two intrinsic qualities that make them ideal candidates as starting material from which to generate β-cells: when maintained in their undifferentiated, pluripotent state they have a limitless capacity to expand *in vitro*; and their developmental plasticity means that they can, in principle, generate any cell type, including β-cells.[63]

The first description of mouse ES (mES) cells differentiating spontaneously into insulin-expressing cells[64] was soon followed by a similar report using human ES (hES) cells.[65] These studies provided proof-of-concept, but spontaneous differentiation into insulin-expressing cells occurred at a very low frequency, with only ~1% of the initial cell population expressing insulin. Subsequent studies focused on finding more efficient ways of driving ES cell differentiation towards the β-cell phenotype, using a variety of directed differentiation

protocols. The most successful of these protocols are based on our detailed understanding of pancreas development (see refs[66,67]), and are designed to recapitulate *in vitro* the important developmental cues involved in the formation of the endocrine pancreas *in vivo*. Thus, sequential exposure of pluripotent ES cells to mitogens drives them towards fully differentiated pancreatic endocrine cells via definitive endoderm, posterior foregut cells, pancreatic endoderm, and endocrine progenitor cells. Measurement of important staging markers, usually transcription factors, enables an assessment of the effectiveness of the differentiation protocols in driving differentiation down the appropriate lineages.

Culture conditions that promoted differentiation of hES cells into definitive endoderm (e.g., refs[68,69]) offered a starting point from which to progress to islet endocrine precursors and hormone expressing cells.[70,71] However, cells generated by these *in vitro* protocols were functionally restricted, showing polyhormonal phenotypes and/or poor nutrient-induced insulin secretory responses. Several studies have demonstrated that transplantation of hES cell-derived progenitor cells into an *in vivo* environment in mice resulted in enhanced glucose-responsive insulin secretion[70] and a greatly increased efficiency of differentiation that produced >80% endocrine cells from the initial hES cell population.[72] The mechanistic basis of this *in vivo* differentiation remains unknown and inconsistencies have been reported in the *in vivo* functional maturation of partially differentiated hES cells, with a failure of hES cell-derived pancreatic progenitors to differentiate further on implantation in nude rats.[73] These observations suggest that an important maturation stage(s) is lacking from the current *in vitro* differentiation protocols. Identification of the factors involved in the *in vivo* maturation may inform the last stages of an entirely *in vitro* differentiation protocol for functional β-cells.

Much progress has been made over the past few years, but there remain important caveats to an over-optimistic interpretation of the use of directed differentiation protocols to generate functional insulin-secreting cells from ES cells. Thus, similar *in vitro* protocols are reported to have very different differentiation efficiencies, without obvious reasons. One likely cause is the differences in the differentiation potential between hES cell lines, emphasizing the importance of a systematic evaluation of available hES cell lines to identify the most suitable starting material.

Induced Pluripotent Stem Cells

Induced pluripotent stem (iPS) cells generated from fibroblasts or other somatic cell types offer an alternative, pluripotent starting material to ES cells for differentiation into functional insulin-secreting cells without the ethical dilemma of destroying a fertilized blastocyst, and with the potential therapeutic advantage of generating an autologous cell population. iPS cells show many phenotypic similarities to ES cells, and the application of directed differentiation protocols developed for hES cells can generate insulin-expressing cells from human iPS cells.[74,75] iPS cells offer the additional advantage of generating disease-specific β-cells for experimental studies by using somatic cells isolated from people with diabetes. Thus, iPS cells derived from skin biopsies of patients with type 1 diabetes[76] or maturity onset diabetes of the young (MODY) type 2[77] have been used to generate insulin-expressing cells via directed differentiation protocols.

In summary, current evidence suggests that, as for hES cells, the available *in vitro* differentiation protocols demonstrate proof-of-concept for the derivation of insulin-expressing cells from iPS cells, but improved differentiation protocols are required to generate large numbers of functionally competent β-cells.

USE OF INSULIN-SECRETING CELL LINES FOR RESEARCH

Advantages of Insulin-Secreting Cell Lines

Insulin-secreting cell lines offer several key advantages over the use of primary islets for research, and the main beneficial features are summarized in Table 15.2. It is clear that their ready availability, ease of use and (at least for the stable, differentiated lines) close similarities to primary β-cells has led to widespread adoption of a range of cell lines, and in the past 30 years over 3000 papers have been published using the cells reviewed here. The most

TABLE 15.2   Advantages of insulin-secreting cell lines for research

Attribute	Outcome
Recapitulate key features of primary β-cells	Screening β-cell stimulus−response coupling events Minimizing experimental animal use for islet isolation
Clonal β-cell lines	Identifying β-cell function in the absence of paracrine influences
Amenable to stable transfection	Generation of homogeneous modified β-cell populations for analysis Improving characteristics of cell lines to more closely recapitulate native β-cells
Large-scale availability of β-cells	Provision of enriched subcellular fractions for analysis

widely used are those that most closely recapitulate the key characteristics of primary islet β-cells, especially with respect to glucose-stimulated insulin secretion. Those derived from rodent insulinomas have been the most successful, although there has been recent promising progress in the generation of stable, glucose-responsive human cell lines.[45,47] However, as outlined above, stem cell-derived differentiation protocols are not yet sufficiently robust for reproducible differentiation towards a β-cell phenotype in sufficient numbers for experimental use.

Optimizing Function through Cell—Cell Contact

The anatomical organization of islets appears to be important for their correct functioning, and there is much evidence to suggest that cell—cell interactions within islets are crucial for normal function. Dissociation of rat islets into single cell suspensions resulted in a loss of regulated secretory function *in vitro*, while spontaneous re-aggregation of the cells in culture was associated with a return towards a more physiological pattern of basal and glucose-stimulated insulin release.[78–80] Studies using purified β-cell populations have shown that homologous β-cell contact plays a critical role in mediating the superior secretory response of aggregated islet cells. Thus, aggregated β-cells produced a greater insulin secretory response than their non-aggregated counterparts,[81] an effect that was dependent upon direct β-cell-to-β-cell contact.

These qualities are highly relevant when considering the use of cell lines for research, since these cells normally proliferate in culture as monolayers that spread out as a sheet of cells on the tissue culture plastic. Anatomically they are configured very differently to the three-dimensional, spherical structure of primary islets and therefore exhibit reduced

levels of contact between individual cells. The ability of primary β-cells to re-aggregate following dispersal does, however, also hold true for a number of β-cell lines given the appropriate conditions. The authors found that MIN6 β-cells can be induced to form islet-like structures (pseudoislets) when maintained in culture on gelatin-coated plastic for 6–8 days. MIN6 pseudoislets are very similar in size to primary islets (100–200 μm) and exhibit a high degree of cell—cell contact, similar to primary tissue.[33] Their formation is dependent on the expression of the adhesion molecule E-cadherin, and MIN6 pseudoislets exhibit improved Ca^{2+} signaling and a biphasic secretory response to glucose that is similar to that of primary mouse islets.[33,82] Secretory responses to a wide range of other insulin secretagogues, including metabolizable nutrients, neurotransmitters and sulphonylureas are also significantly enhanced from pseudoislets.[82–84] This effect is not a result of increased insulin content, or changes in PDX-1, glucokinase or GLUT2 expression,[33,83–85] although recent findings suggests that glucose oxidation and a range of genes involved in mitochondrial metabolism are upregulated in pseudoislets.[84] The increase in insulin release is rapidly reversible on dissociation of the islet-like structures into single cell suspensions,[82] suggesting that the improved secretory function is dependent on homotypic β-cell-to-β-cell interactions.

The ability to form islet-like structures is not restricted to MIN6 cells, and pseudoislet formation has also been reported using βTC6, INS-1 and RINm5F cells.[86–88] Some, but not all, rodent insulin-secreting cells improve functionally with this conformational change, suggesting that the positive effect of increased cell—cell contact still depends on the intrinsic properties of the individual cell lines. In addition, the functional improvement is not restricted to rodent cells since the 1.1B4 human cell line exhibited enhanced secretory characteristics and upregulated expression of adhesion molecules when cultured as pseudoislets, in a manner very similar to that of MIN6 cells.[89]

Following the authors' initial report on the functional benefits of pseudoislets, a number of studies have sought to optimize the model by modifying the culture conditions. Simple prevention of cell attachment to the tissue culture plastic, as provided by ultra-low attachment plates or bacterial dishes, induces pseudoislet formation,[83,90] although the presence of a scaffold containing extracellular matrix proteins (such as collagen or elastin derivatives) reportedly further improves cell viability and secretory function.[88,91] Similarly, stirred suspension or rotating microgravity bioreactors promote the viability of the pseudoislets and generate more uniformly sized structures.[92,93]

As a direct replacement for primary islets, pseudoislets have been used in a range of studies including assessments of insulin granule trafficking,[94] screening of herb extracts for insulinotropic effects[95] as well as investigations into the roles of G protein-coupled receptors[90,96] and intracellular effectors[97] in the regulation of insulin secretion. Due to their morphology, pseudoislets can, like primary islets, be successfully transplanted either under the kidney capsule or into the hepatic portal vein of mice.[91,93] As indicated in Table 15.2, cell lines have the advantage of being amenable to stable genetic modification and the MIN6 pseudoislet model therefore also enables the combination of *in vivo* transplantation studies with genetic over- or under-expression of particular candidates of interest,[98] an aspect that is technically difficult with primary islets unless transgenic animal models are employed.

MIN6 pseudoislets consist of transformed β-cells only and they are therefore a good model system in which to study the importance of direct β-cell-to-β-cell contact, since any input from other non-β-cells within the islet are excluded. The roles of E-cadherin, gap junctions and autocrine agents such as Ca^{2+}

and adenosine triphosphate (ATP) in intercellular communication and islet function have therefore been investigated in this model system.[33,96,99,100] Non-β-cells within primary islets do modulate insulin secretion, with glucagon-secreting α-cells stimulating, and somatostatin-producing δ-cells inhibiting insulin release. To further optimize the pseudoislet model for direct islet replacement studies and to investigate the importance of heterotypic cell interaction and islet architecture for islet function, studies have been performed where pseudoislets were generated from a combination of β- and non-β-cell lines to mimic the architectural arrangement of the primary rodent islet. Combining MIN6 β-cells with glucagon-secreting αTC1 cells had no effect on insulin secretion,[101] and heterotypic MIN6 pseudoislets that contained the somatostatin-producing TGP52 cells as well as αTC1.9 cells responded less well to glucose,[102] although it is unclear whether this was due to proportionally less MIN6 cells per pseudoislet or a potential inhibitory effect of somatostatin from the TGP52 cells. These non-β-cell lines contain significantly less hormone than primary islets (<0.1%, Hauge-Evans, unpublished data), which limits the likelihood of functionally relevant paracrine interactions and may explain the limited modification of glucose-induced insulin secretion from these heterotypic pseudoislets. The models do, however, provide valuable information regarding the intrinsic capacity of the cells to assemble into an anatomically correct islet-like structure, since both mixed MIN6/αTC1 and INS-1/αTC pseudoislets as well as MIN6/αTC1.9/TGP52 aggregates retained a segregated morphology with centrally located β-cells surrounded by peripheral non-β-cells similar to primary tissue.[101–103] The molecular ability to segregate differentially is thus retained in the cell lines and is thought to be linked to a variation in cell-specific expression of adhesion molecules, in particular but not exclusively, E-CAD and N-CAM.[103]

In summary, insulin-secreting cell lines can be induced to form spherical pseudoislets that are anatomically very similar to primary islets. They often exhibit enhanced secretory characteristics compared to monolayer cells as a consequence of the increased degree of cell—cell contact within these structures. Optimization and development of pseudoislets is ongoing, but they have already proven to be useful research models as replacements for primary tissue and in direct investigations into the importance of cell-to-cell interactions for islet development and function.[104]

Limitations of Insulin-Secreting Cell Lines

Although insulin-secreting cell lines may be used as β-cell substitutes for research, it should be borne in mind that they may differ from primary islet β-cells in terms of glucose transport and phosphorylation, insulin content, and glucose-stimulated insulin secretion (Table 15.1). In addition, while the intrinsic proliferation of cell lines is advantageous in terms of ready availability of cells, it leads to dedifferentiation that can contribute to the unstable phenotypes of most of the cell lines that have been characterized. There may also be differences in expression of receptor subtypes or protein isoforms, leading to altered signal transduction (e.g. ref[105]). It is, therefore, important that primary islets are used to confirm (or otherwise) observations that are made using insulin-secreting cell lines.

POTENTIAL USE OF INSULIN-SECRETING CELL LINES FOR DIABETES THERAPY

Progress to Date

Type 1 diabetes has been treated for over 90 years by delivery of insulin, but in the past decade islet transplantation has become more widely used to allow physiological patterns of circulating insulin. However, supply of human islets through organ donation is insufficient to provide islets for all type 1 diabetics who require β-cell replacements to adequately control plasma glucose levels. The recent generation of glucose-sensitive human insulin-secreting cell lines is a promising development, but the introduction of oncogenes that cause unregulated proliferation of these cells precludes their clinical use as tumor formation in recipients is inevitable. Nonetheless, there is interest in developing conditionally responsive human cell lines that can be induced to cease replication, for example in the absence of tetracycline, which has been shown to be feasible with mouse βTC lines.[27] Functional β-cells derived from stem cell populations may provide unlimited amounts of graft material for transplantation therapy of type 1 diabetes,[106] and progress is being made in this area.

Preclinical trials using β-cells derived from hES cells are currently in progress, with expectation of Phase I clinical trials to assess product safety scheduled to commence in 2014.[107] However, there remain some obstacles to the wider clinical application of β-cells derived from stem cells. Thus, iPS cells are a clinically attractive starting material because they offer the possibility of generating patient-specific autologous insulin-secreting cells for transplantation, but evidence is emerging that iPS cells are prone to genetic abnormalities such as copy number variation,[108] accumulation of somatic coding mutations,[109] and environment-dependent epigenetic modifications.[110] A systematic evaluation of the genetic and epigenetic stability of iPS cells derived from different somatic cells is required to identify the most suitable starting material from which to generate clinical-grade material.

The qualities that make stem cells attractive — their pluripotency and proliferative potential — raise the undesirable possibility of uncontrolled cellular proliferation and formation of teratomas after transplantation into graft recipients, as has been reported in mouse recipients of insulin-secreting cells derived from hES cells.[70] The scheduled Phase I clinical trials will use an artificial semi-permeable containment device implanted in the subcutaneous site to enable graft retrieval.[111] However, in current clinical human islet transplantation programs graft delivery is almost exclusively via infusion into the hepatic portal vein.[106] This ensures an excellent distribution of the islets throughout the liver, which is a major target organ for insulin, but also renders the transplanted material essentially irretrievable in the event of an adverse outcome. The distribution of $5-10 \times 10^8$ potentially teratogenic cells throughout the liver is a major safety issue, and the future clinical usefulness of substitute insulin-secreting cells derived from human stem cells or endocrine pancreas will depend not only on their functional competence but also on the development of foolproof methods of ensuring their safety after transplantation.

Future Perspectives

It is clear that the range of insulin-secreting cell lines that have been generated in the past thirty years have been beneficial to β-cell research by providing large populations of well-characterized cells with which to define stimulus−response coupling pathways. The identification that insulin secretory responses are often improved when these cells are formed into three-dimensional clusters provides key functional information about how best to configure cells that are being developed for transplantation therapy for type 1 diabetes. The future is promising, and there is a real possibility that well-defined human β-cells, either derived from stem cells or human endocrine pancreas, will be available to treat diabetes within the next decade.

References

1. Cabrera O, Berman DM, Kenyon NS, et al. The unique cytoarchitecture of human pancreatic islets has implications for islet cell function. *Proc Natl Acad Sci USA* 2006;10:2334−9.
2. De Vos A, Heimberg H, Quartier E, et al. Human and rat beta cells differ in glucose transporter but not in glucokinase gene expression. *J Clin Invest* 1995;96:2489−95.
3. Sansbury FH, Flanagan SE, Houghton JA, et al. SLC2A2 mutations can cause neonatal diabetes, suggesting GLUT2 may have a role in human insulin secretion. *Diabetologia* 2012;55:2381−5.
4. Parnaud G, Bosco D, Berney T, et al. Proliferation of sorted human and rat beta cells. *Diabetologia* 2008;51:91−100.
5. Wang W, Walker JR, Wang X, et al. Identification of small-molecule inducers of pancreatic β-cell expansion. *Proc Natl Acad Sci USA* 2009;106:1427−32.
6. Shen W, Tremblay MS, Deshmukh VA, Wang W, et al. Small-molecule inducer of β cell proliferation identified by high-throughput screening. *JACS* 2013;135:1669−72.
7. Aly H, Rohatgi N, Marshall C, et al. A novel strategy to increase the proliferative potential of adult human β-cells while maintaining their differentiated phenotype. *PLOS One* 2013;8:e66131.
8. Grillo TAI, Whitty AJ, Kirkman H, et al. Biological properties of a transplantable islet-cell tumor of the golden hamster. I. Histology and histochemistry. *Diabetes* 1967;16:409−14.
9. Sodoyez JC, Luyckx AS, Lefebvre PJ. Biological properties of a transplantable islet-cell tumor of the golden hamster. II. Insulin content of the tumor and some metabolic characteristics of the tumor-bearing animals. *Diabetes* 1967;16:415−7.
10. Chick WL, Shields W, Chute RN, et al. A transplantable insulinoma in the rat. *Proc Natl Acad Sci USA* 1977;74:628−33.
11. Gazdar AF, Chick WL, Oie HK, et al. Continuous, clonal, insulin- and somatostatin-secreting cell lines established from a transplantable rat islet cell tumor. *Proc Natl Acad Sci USA* 1980;77:3519−23.
12. Praz GA, Halban PA, Wollheim CB, et al. Regulation of immunoreactive-insulin release from a rat cell line (RINm5F). *Biochem J* 1983;210:345−52.
13. Clark SA, Burnham BL, Chick WL. Modulation of glucose-induced insulin secretion from a rat clonal beta-cell line. *Endocrinology* 1990;127:2779−88.
14. Halban PA, Praz GA, Wollheim CB. Abnormal glucose metabolism accompanies failure of glucose to stimulate insulin release from a rat pancreatic cell line (RINm5F). *Biochem J* 1983;212:439−43.

15. Clark SA, Quaade C, Constandy H, et al. Novel insulinoma cell lines produced by iterative engineering of GLUT2, glucokinase, and human insulin expression. *Diabetes* 1997;**46**:958–67.

16. Hohmeier HE, BeltrandelRio H, Clark SA, et al. Regulation of insulin secretion from novel engineered insulinoma cell lines. *Diabetes* 1997;**46**:968–77.

17. Asfari M, Janjic D, Meda P, et al. Establishment of 2-mercaptoethanol-dependent differentiated insulin-secreting cell lines. *Endocrinology* 1992;**130**:167–78.

18. Hohmeier HE, Mulder H, Chen G, et al. Isolation of INS-1-Derived cell lines with robust ATP-sensitive K^+ channel-dependent and-independent glucose-stimulated insulin secretion. *Diabetes* 2000;**49**:424–30.

19. Jancic D, Maechler P, Sekine N, et al. Free radical modulation of insulin release in INS-1 cells exposed to alloxan. *Biomed Pharmacol* 1999;**57**:639–48.

20. Santerre RF, Cook RA, Crisel RMD, et al. Insulin synthesis in a clonal cell line of simian virus 40-transformed hamster pancreatic beta cells. *Proc Natl Acad Sci USA* 1981;**78**:4339–43.

21. Ashcroft SJH, Hammonds P, Harrison DE. Insulin secretory responses of a clonal cell line of simian virus 40-transformed B cells. *Diabetologia* 1986;**29**:727–33.

22. Ashcroft SJH, Stubbs M. The glucose sensor in HIT cells is the glucose transporter. *FEBS Lett* 1987;**219**:311–5.

23. Welsh N, Bendtzen K, Welsh M. Expression of an insulin/interleukin-1 receptor antagonist hybrid gene in insulin-producing cell lines (HIT-T15 and NIT-1) confers resistance against interleukin-1-induced nitric oxide production. *J Clin Invest* 1995;**95**:1717–22.

24. McClenaghan NH, Barnett CR, Ah-Sing E, et al. Characterization of a novel glucose-responsive insulin-secreting cell line, BRIN-BD11, produced by electrofusion. *Diabetes* 1996;**45**:1132–40.

25. McClenaghan NH, Gray AM, Barnett CR, et al. Hexose recognition by insulin-secreting BRIN-BD11 cells. *Biochem Biophys Res Commun* 1996;**223**:724–8.

26. Efrat S, Fusco-DeMane D, Lemberg H, et al. Conditional transformation of a pancreatic β-cell line derived from transgenic mice expressing a tetracycline-regulated oncogene. *Proc Natl Acad Sci USA* 1995;**92**:3576–80.

27. Milo-Landesman D, Surana M, Berkovich I, et al. Correction of hyperglycemia in diabetic mice transplanted with reversibly immortalized pancreatic beta cells controlled by the tet-on regulatory system. *Cell Transplant* 2001;**10**:645–50.

28. Efrat S, Linde S, Kofod H, et al. Beta-cell lines derived from transgenic mice expressing a hybrid insulin gene-oncogene. *Proc Natl Acad Sci USA* 1988;**85**:9037–41.

29. Whitesell RR, Powers AC, Regen DM, et al. Transport and metabolism of glucose in an insulin-secreting cell line, βTC-1. *Biochemistry* 1991;**30**:11560–6.

30. Knaack D, Fiore DM, Surana M, et al. Clonal insulinoma cell line that stably maintains correct glucose responsiveness. *Diabetes* 1994;**43**:1413–7.

31. Miyazaki JI, Araki K, Yamato E, et al. Establishment of a pancreatic β cell line that retains glucose-inducible insulin secretion: special reference to expression of glucose transporter isoforms. *Endocrinology* 1990;**127**:126–32.

32. Ishihara H, Asano T, Tsukuda K, et al. Pancreatic beta cell line MIN6 exhibits charactetristics of glucose metabolism and glucose-stimulated insulin secretion similar to those of normal islets. *Diabetologia* 1993;**36**:1139–45.

33. Hauge-Evans AC, Squires PE, Persaud SJ, et al. Pancreatic β-cell-to-β-cell interactions are required for integrated responses to nutrient stimuli: enhanced Ca^{2+} and insulin secretory responses of MIN6 pseudoislets. *Diabetes* 1999;**48**:1402–8.

34. Hamaguchi K, Gaskins HR, Edward HL. NIT-1, a pancreatic β-cell line established from a transgenic NOD/Lt mouse. *Diabetes* 1991;**40**:842–9.

35. Radvanyi F, Christgau S, Baekkeskov S, et al. Pancreatic β cells cultured from individual preneoplastic foci in a multistage tumorigenesis pathway: a potentially general technique for isolating physiologically representative cell lines. *Mol Cell Biol* 1993;**13**:4223–32.

36. Noda M, Komatsu M, Sharp GWG. The βHC-9 pancreatic β-cell line preserves the characteristics of progenitor mouse islets. *Diabetes* 1996;**45**:1766–73.

37. Chick WL, Lauris V, Soeldner JS, et al. Monolayer culture of a human pancreatic beta-cell adenoma. *Metabolism* 1973;**22**:1217–24.

38. Thivolet CH, Demidem A, Haftek M, et al. Structure, function, and immunogenicity of human insulinoma cells. *Diabetes* 1988;**37**:1279–86.

39. Gartner W, Koc F, Nabokikh A, et al. Long-term in vitro growth of human insulin-secreting insulinoma cells. *Neuroendocrinology* 2006;**83**:123–30.

40. Labriola L, Peters MG, Krogh K, et al. Generation and characterization of human insulin-releasing cell lines. *BMC Cell Biology* 2009;**10**.10.1186/1471-2121-10-49

41. Baroni MG, Cavallo MG, Mark M, et al. Beta-cell gene expression and functional characterization of the human insulinoma cell line CM. *J Endocrinol* 1999;**161**:59–68.

42. Beattie GM, Itkin-Ansari P, Cirulli V, et al. Sustained proliferation of PDX-1$^+$ cells derived from human islets. *Diabetes* 1999;**48**:1013–9.

43. Soldevila G, Buscema M, Marini V, et al. Transfection with SV40 gene of human pancreatic endocrine cells. *J Autoimmun* 1991;**4**:381–96.

44. Dufayet D, Halvorsen T, Demeterco C, et al. β-cell differentiation from a human pancreatic cell line in vitro and in vivo. *Mol Endocrinol* 2001;**15**:476–83.

45. McCluskey JT, Hamid M, Guo-Parke H, et al. Development and functional characterization of insulin-releasing human pancreatic beta cell lines produced by electrofusion. *J Biol Chem* 2011;**286**:21982−92.

46. Wang S, Beattie GM, Mally MI, et al. Isolation and characterization of a cell line from the epithelial cells of the human fetal pancreas. *Cell Transplant* 1997;**6**:59−67.

47. Ravassard P, Hazhouz Y, Pechberty S, et al. A genetically engineered human pancreatic β cell line exhibiting glucose-inducible insulin secretion. *J Clin Invest* 2011;**121**:3589−97.

48. Efrat S, Russ HA. Making β-cells from adult tissues. *Trends Endocrinol Metab* 2012;**23**:278−85.

49. Chhabra P, Brayman KL. Stem cell therapy to cure type 1 diabetes: from hype to hope. *Stem Cells Transl Med* 2013;**2**:328−36.

50. Dominguez-Bendala J, Lanzoni G, Inverardi L, et al. Concise review: mesenchymal stem cells for diabetes. *Stem Cells Transl Med* 2012;**1**:58−63.

51. Ianus A, Holz GG, Theise ND, et al. In vivo derivation of glucose-competent pancreatic endocrine cells from bone marrow without evidence of cell fusion. *J Clin Invest* 2003;**111**:843−50.

52. Tang DQ, Cao LZ, Burkhardt BR, et al. In vivo and in vitro characterization of insulin-producing cells obtained from murine bone marrow. *Diabetes* 2004;**53**:1721−32.

53. Karnieli O, Izhar-Prato Y, Bulvik S, et al. Generation of insulin-producing cells from human bone marrow mesenchymal stem cells by genetic manipulation. *Stem Cells* 2007;**25**:2837−44.

54. Taneera J, Rosengren A, Renstrom E, et al. Failure of transplanted bone marrow cells to adopt a pancreatic beta-cell fate. *Diabetes* 2006;**55**:290−6.

55. Hasegawa Y, Ogihara T, Yamada T, et al. Bone marrow (BM) transplantation promotes beta-cell regeneration after acute injury through BM cell mobilization. *Endocrinology* 2007;**148**:2006−15.

56. Dor Y, Brown J, Martinez OI, et al. Adult pancreatic beta-cells are formed by self-duplication rather than stem-cell differentiation. *Nature* 2004;**429**:41−6.

57. Thorel F, Nepote V, Avril I, et al. Conversion of adult pancreatic alpha-cells to beta-cells after extreme beta-cell loss. *Nature* 2010;**464**:1149−54.

58. Bonner-Weir S, Taneja M, Weir GC, et al. In vitro cultivation of human islets from expanded ductal tissue. *Proc Natl Acad Sci U S A* 2000;**97**:7999−8004.

59. Zhao M, Amiel SA, Christie MR, et al. Evidence for the presence of stem cell-like progenitor cells in human adult pancreas. *J Endocrinol* 2007;**195**:407−14.

60. Zulewski H, Abraham EJ, Gerlach MJ, et al. Multipotential nestin-positive stem cells isolated from adult pancreatic islets differentiate ex vivo into pancreatic endocrine, exocrine, and hepatic phenotypes. *Diabetes* 2001;**50**:521−33.

61. Efrat S. Beta-cell replacement for insulin-dependent diabetes mellitus. *Adv Drug Deliv Rev* 2008;**60**:114−23.

62. Wagner RT, Lewis J, Cooney A, et al. Stem cell approaches for the treatment of type 1 diabetes mellitus. *Transl Res* 2010;**156**:169−79.

63. Naujok O, Burns C, Jones PM, et al. Insulin-producing surrogate β-cells from embryonic stem cells: are we there yet? *Mol Therapy* 2011;**19**:1759−68.

64. Soria B, Roche E, Berna G, et al. Insulin-secreting cells derived from embryonic stem cells normalize glycemia in streptozotocin-induced diabetic mice. *Diabetes* 2000;**49**:157−62.

65. Assady S, Maor G, Amit M, et al. Insulin production by human embryonic stem cells. *Diabetes* 2001;**50**:1691−7.

66. Jorgensen MC, Ahnfelt-Ronne J, Hald J, et al. An illustrated review of early pancreas development in the mouse. *Endocr Rev* 2007;**28**:685−705.

67. Stanger BZ, Hebrok M. Control of cell identity in pancreas development and regeneration. *Gastroenterology* 2013;**144**:1170−9.

68. D'Amour KA, Agulnick AD, Eliazer S, et al. Efficient differentiation of human embryonic stem cells to definitive endoderm. *Nat Biotechnol* 2005;**23**:1534−41.

69. Xu X, Kahan B, Forgianni A, et al. Endoderm and pancreatic islet lineage differentiation from human embryonic stem cells. *Cloning Stem Cells* 2006;**8**:96−107.

70. Kroon E, Martinson LA, Kadoya K, et al. Pancreatic endoderm derived from human embryonic stem cells generates glucose-responsive insulin-secreting cells in vivo. *Nat Biotechnol* 2008;**26**:443−52.

71. Xu X, Browning VL, Odorico JS. Activin, BMP and FGF pathways cooperate to promote endoderm and pancreatic lineage cell differentiation from human embryonic stem cells. *Mech Dev* 2011;**128**:412−27.

72. Bruin JE, Rezania A, Xu J, et al. Maturation and function of human embryonic stem cell-derived pancreatic progenitors in macroencapsulation devices following transplant into mice. *Diabetologia* 2013;**56**:1987−98.

73. Matveyenko AV, Georgia S, Bhushan A, et al. Inconsistent formation and nonfunction of insulin-positive cells from pancreatic endoderm derived from human embryonic stem cells in athymic nude rats. *Am J Physiol Endocrinol Metab* 2010;**299**:E713−20.

74. Tateishi K, He J, Taranova O, et al. Generation of insulin-secreting islet-like clusters from human skin fibroblasts. *J Biol Chem* 2008;**283**:31601−7.

75. Zhang D, Jiang W, Liu M, et al. Highly efficient differentiation of human ES cells and iPS cells into mature pancreatic insulin-producing cells. *Cell Res* 2009;**19**:429−38.

76. Maehr R, Chen S, Snitow M, et al. Generation of pluripotent stem cells from patients with type 1 diabetes. *Proc Natl Acad Sci U S A* 2009;**106**:15768–73.

77. Hua H, Shang L, Martinez H, et al. iPSC-derived β cells model diabetes due to glucokinase deficiency. *J Clin Invest* 2013;**123**:3146–53.

78. Halban PA, Wollheim CB, Blondel B, et al. The possible importance of contact between pancreatic islet cells for the control of insulin release. *Endocrinology* 1982;**111**:86–94.

79. Halban PA, Powers SL, George KL, et al. Spontaneous reassociation of dispersed adult rat pancreatic islet cells into aggregates with three-dimensional architecture typical of native islets. *Diabetes* 1987;**36**:783–90.

80. Hopcroft DW, Mason DR, Scott RS. Insulin secretion from perifused rat pancreatic pseudoislets. *In Vitro Cell Dev Biol* 1985;**2**:421–7.

81. Bosco D, Orci L, Meda P. Homologous but not heterologous contact increases the insulin secretion of individual pancreatic β-cells. *Exp Cell Res* 1989;**184**:72–80.

82. Luther MJ, Hauge-Evans A, Souza KL, et al. MIN6 beta-cell-beta-cell interactions influence insulin secretory responses to nutrients and non-nutrients. *Biochem Biophys Res Commun* 2006;**343**:99–104.

83. Kelly C, Guo H, McCluskey JT, et al. Comparison of insulin release from MIN6 pseudoislets and pancreatic islets of Langerhans reveals importance of homotypic cell interactions. *Pancreas* 2010;**39**:1016–23.

84. Chowdhury A, Dyachok O, Tengholm A, et al. Functional differences between aggregated and dispersed insulin-producing cells. *Diabetologia* 2013;**56**:1557–68.

85. Hauge-Evans AC, Squires PE, Belin VD, et al. Role of adenine nucleotides in insulin secretion from MIN6 pseudoislets. *Mol Cell Endocrinol* 2002;**191**:167–76.

86. Maillard E, Sencier MC, Langlois A, et al. Extracellular matrix proteins involved in pseudoislets formation. *Islets* 2009;**1**:232–41.

87. Gerbino A, Maiellaro I, Carmone C, et al. Glucose increases extracellular Ca^{2+} in rat insulinoma (INS-1E) pseudoislets as measured with Ca^{2+}-sensitive microelectrodes. *Cell Calcium* 2012;**5**:393–401.

88. Lee KM, Jung GS, Park JK, et al. Effects of Arg-Gly-Asp-modified elastin-like polypeptide on pseudoislet formation via up-regulation of cell adhesion molecules and extracellular matrix proteins. *Acta Biomater* 2013;**9**:5600–8.

89. Guo-Parke H, McCluskey JT, Kelly C, et al. Configuration of electrofusion-derived human insulin-secreting cell line as pseudoislets enhances functionality and therapeutic utility. *J Endocrinol* 2012;**214**:257–65.

90. Li C, Jones PM, Persaud SJ. Cannabinoid receptors are coupled to stimulation of insulin secretion from mouse MIN6 beta-cells. *Cell Physiol Biochem* 2010;**26**:187–96.

91. Yang KC, Wu CC, Yang SH, et al. Investigating the suspension culture on aggregation and function of mouse pancreatic beta-cells. *J Biomed Mater Res A* 2013;**101**:2273–82.

92. Lock LT, Laychock SG, Tzanakakis ES. Pseudoislets in stirred-suspension culture exhibit enhanced cell survival, propagation and insulin secretion. *J Biotechnol* 2011;**151**:278–86.

93. Tanaka H, Tanaka S, Sekine K, et al. The generation of pancreatic beta-cell spheroids in a simulated microgravity culture system. *Biomaterials* 2013;**34**:5785–91.

94. Hatlapatka K, Matz M, Schumacher K, et al. Bidirectional insulin granule turnover in the submembrane space during K^+ depolarization-induced secretion. *Traffic* 2011;**12**:1166–78.

95. Govindarajan R, Vijayakumar M, Rao C, et al. Antidiabetic activity of Croton klozchianus in rats and direct stimulation of insulin secretion in-vitro. *J Pharm Pharmacol* 2008;**60**:371–6.

96. Kitsou-Mylona I, Burns CJ, Squires PE, et al. A role for the extracellular calcium-sensing receptor in cell-cell communication in pancreatic islets of Langerhans. *Cell Physiol Biochem* 2008;**22**:557–66.

97. Milne HM, Burns CJ, Squires PE, et al. Uncoupling of nutrient metabolism from insulin secretion by overexpression of cytosolic phospholipase A_2. *Diabetes* 2005;**54**:116–24.

98. Fornoni A, Jeon J, Varona SJ, et al. Nephrin is expressed on the surface of insulin vesicles and facilitates glucose-stimulated insulin release. *Diabetes* 2010;**59**:190–9.

99. Squires PE, Hauge-Evans AC, Persaud SJ, et al. Synchronization of Ca^{2+} signals within insulin-secreting pseudoislets: effects of gap-junctional uncouplers. *Cell Calcium* 2000;**27**:287–96.

100. Carvell MJ, Marsh PJ, Persaud SJ, et al. E-cadherin interactions regulate beta-cell proliferation in islet-like structures. *Cell Physiol Biochem* 2007;**20**:617–26.

101. Brereton H, Carvell MJ, Persaud SJ, Jones PM. Islet alpha-cells do not influence insulin secretion from beta-cells through cell-cell contact. *Endocrine* 2007;**31**:61–5.

102. Kelly C, Parke HG, McCluskey JT, et al. The role of glucagon- and somatostatin-secreting cells in the regulation of insulin release and beta-cell function in heterotypic pseudoislets. *Diabetes Metab Res Rev* 2010;**26**:525–33.

103. Jia D, Dajusta D, Foty RA. Tissue surface tensions guide in vitro self-assembly of rodent pancreatic islet cells. *Dev Dyn* 2007;**236**:2039–49.

104. Persaud SJ, Arden C, Bergsten P, et al. Pseudoislets as primary islet replacements for research. *Islets* 2010;**2**:236–9.

105. Hauge-Evans AC, Richardson CC, Milne HM, et al. A role for kisspeptin in islet function. *Diabetologia* 2006;**49**:2131–5.

106. de Kort H, de Koning EJ, Rabelink TJ, et al. Islet transplantation in type 1 diabetes. *BMJ* 2011;**342**:426–32.

107. Viacyte 2013 <http://viacyte.com/clinical/clinicaltrials/>.

108. Hussein SM, Batada NN, Vuoristo S, et al. Copy number variation and selection during reprogramming to pluripotency. *Nature* 2011;**471**:58–62.

109. Gore A, Li Z, Fung HL, et al. Somatic coding mutations in human induced pluripotent stem cells. *Nature* 2011;**471**:63–7.

110. Lister R, Pelizzola M, Kida YS, et al. Hotspots of aberrant epigenomic reprogramming in human induced pluripotent stem cells. *Nature* 2011;**471**:68–73.

111. Viacyte 2013 <http://viacyte.com/products/vc-01-diabetes-therapy/>.

16

Architecture and Morphology of Human Pancreatic Islets

Alvin C. Powers

Vanderbilt University Medical Center, Nashville, TN, USA

INTRODUCTION

The pancreatic islet plays an essential role in regulating the metabolic environment of organisms by sensing nutrient and environmental stimuli and then responding with regulated secretion of a variety of bioactive compounds, most notably islet hormones. This chapter will focus on recent advances in our understanding of the anatomic structure and architecture of the human pancreatic islet in health and in disease with a focus on islet cell composition and arrangement, non-islet cells within the islet, and the overall supporting infrastructure. After a review of normal islet morphology and structure, the chapter will describe changes seen in human type 1 or type 2 diabetes. The physiology and molecular events within islet cells will be mentioned as related to islet architecture but will not be discussed in detail. Likewise, other critical structures in the pancreas, namely pancreatic exocrine cells and ductal structures that only rarely impact the function of the endocrine islet cells will not be discussed.

UNDERSTANDING OF HUMAN BIOLOGY HAS GREATLY INCREASED

Knowledge about the pancreatic islet, and especially its most numerous cell type, the beta cell, has increased dramatically over the past 25 years with an increased understanding of the critical genes, regulatory pathways, and molecular events. Much of this research has used rodent islets, transformed rodent islet cell lines, or mice with genetically modified islets. These advances have transformed how scientists think about the beta cell and its role in diabetes, and have discovered new causes of diabetes.[1-4] Furthermore, these discoveries have laid the foundation for efforts to create beta cells from embryonic stem cells and inducible pleuripotent stem cells with numerous scientists and a National Institutes of Health (NIH)-supported consortium, Beta Cell Biology Consortium (http://www.betacell.org/), working toward this goal.[1-4]

During the last decade, there has been a gradual realization that while human pancreatic

Cellular Endocrinology in Health and Disease.
DOI: http://dx.doi.org/10.1016/B978-0-12-408134-5.00016-0

islets have many similarities to islets from other species and transformed islet cell lines, there are clear differences that have not been fully appreciated. The prior focus on islets from animals was driven in large part by rationale experimental needs to control for genetic background, diet, etc., and the considerable difficulty in obtaining human pancreatic tissue and islets under conditions that would allow for molecular analysis.

Two forces have combined in the last decade to greatly increase the interest in and knowledge of human islet biology. First, scientists and physicians have sought ways to translate the new knowledge about the genetics of human physiology and rodent islet biology to human diabetes. Islet investigators now often strive to address human islet biology in their experimental design or in the interpretation of their results. Second, newly available resources now allow an increasing number of scientists to have access to human pancreatic tissue and human pancreatic islets for their research. For example, considerable activity in the human islet transplantation arena beginning with the investigators in Edmonton report in 2000 and continuing to the present has greatly increased the number of human islets isolated each year and the expertise of the islet isolation centers.[5–8] An unexpected benefit of this increased islet transplantation activity has been a dramatic increase in human islet biology research. For example, sometimes the yield of islets isolated for clinical transplantation or the match with the transplant recipient are not optimal, allowing the NIH/National Institute of Diabetes and Digestive and Kidney Diseases (NIDDK) and the Juvenile Diabetes Research Foundation (JDRF) to establish highly efficient programs to match these human islets with previously vetted investigators throughout the United States. This program, the Integrated Islet Distribution Program (IIDP; http://iidp. coh.org/) has created an automated matching system and an extremely efficient shipping mechanism to quickly ship viable human islets to multiple investigators.[9,10] More recently, the JDRF-supported nPOD program (http://www. jdrfnpod.org/) has made human pancreatic specimens (normal and diabetic) available to a growing number of investigators.

MORPHOLOGY AND CELLULAR COMPOSITION OF THE HUMAN ISLET

Distinctive clusters of cells within the pancreas were first noted by the German student Paul Langerhans in his doctoral thesis published in 1869.[11,12] Langerhans did not assign a function to this collection of cells since the role of the pancreas in diabetes had not yet been defined and insulin was discovered much later (1921). Nevertheless, Langerhans became associated with diabetes when others later termed these structures, "islands of Langerhans." These pancreatic islands or islets contain five endocrine cell types: alpha cells, which produce glucagon and other proglucagon-derived peptides; beta cells, which produce insulin and C-peptide from the parent proinsulin and islet-associated polypeptide (IAPP); delta cells, which produce somatostatin; PP cells, which produce pancreatic polypeptide; and epsilon cells which produce ghrelin.

The human pancreas has approximately 1 million islets that are dispersed throughout the organ and occupy 1–2% of the pancreatic volume. The range in islet number or beta cell mass (weight or volume of beta cells) cannot be determined prior to death using current methodology, but is likely considerable since cross-sectional studies of the human pancreas at autopsy have noted a 3–5-fold difference in beta cell mass.[13,14] Human islets range from 50–250 microns in diameter and may contain scores to more than a thousand endocrine cells. A thin, fibrous capsule, consisting of mostly collagen VI with smaller amounts of collagen I and IV, separates the endocrine portion from exocrine pancreas.[15] While a role for this capsule is undefined, it is the reason that islets can be isolated from the exocrine portion of the pancreas by collagenase digestion and mechanical shaking followed by differential centrifugation. Within the islet, collagen, lamins, and integrins are clearly important in providing structure to the islet cell clusters, but also provide instructive signals that influence endocrine cell differentiation, function, and possibly proliferation.[2,4,16−21] However, the precise role and identity of the intra-islet, extracellular matrix in the human islets *in vivo* remains to be defined.

The cellular composition and arrangement within the human islet has been defined in much greater detail in the past decade. It is clear that adult human islets being studied by many laboratories and adult rodents differ in the arrangement of islet cells.[22−25] Figure 16.1 shows alpha cells throughout human islets, whereas in mouse islets these are confined to the perimeter of the islet. There are caveats related to these differences that must be mentioned − differences in age between adult humans and rodents: some human islets may have a similar alpha/beta arrangement as rodent islets; older rodents or pregnant rodents may have similar alpha/beta cell arrangements to human islets; and islet size may influence cell arrangement.[9,10,26−28] Even when these are considered, it seems clear that the endocrine cell arrangement in rodent islets (core of beta cells with other cell types in the perimeter) is much more uniform and consistent than in human islets.

In addition to differences in morphology and cell composition between human and rodent islets, human adult islets differ in their basal glucose-stimulated insulin secretion rate and in the expression of key islet-enriched transcription factors like Pdx-1, MafA, and MafB.[24] Furthermore, human beta cells have a much lower basal proliferation rate and, in contrast with rodent islets, stimuli like obesity,

insulin resistance, and pregnancy either cause a modest increase or no increase in human beta cell proliferation.[13,14,29−31]

An additional caveat is that there is little information about islet cell arrangement in infants or children, so it is important to study such tissues to determine whether the adult human cell arrangement is present *in utero* or early neonatal life, or develops as humans age. Other changes in islet cells occur during this time (such as a decline in proliferation rate of beta cells within the first 2−3 years of life).[15,32,33]

The relative proportion of endocrine cell types in human islets varies among islets and in different regions of the pancreas. For example, most islet endocrine cells in the uncinate region are PP cells with few alpha or beta cells.[26] Hara and colleagues have developed a systemic approach to map both islet location and islet cell composition,[25,26] and have noted that prior studies in which only certain regions of the pancreas are sampled may be misleading. Using this approach, these investigators found that the tail of the pancreas has two-fold more islets than the head and body of the pancreas. Human islets have a range of beta (55−70%) and alpha cells (in certain islets the alpha cells may be 50% of the islet endocrine cells).[12,22,23,27] Whether the islets from different regions of the pancreas have distinctive physiologic properties is uncertain, but the available evidence suggests the islets have similar glucose-stimulated insulin secretion.[25]

VASCULARIZATION AND INNERVATION WITHIN THE ISLET

Pancreatic islets are highly vascularized and innervated, in keeping with their physiologic role to quickly sense changes in nutrients like glucose and then secrete the appropriate hormone. Vascularization and innervation are examples of characteristics of which much is known in rodent islets but considerably less

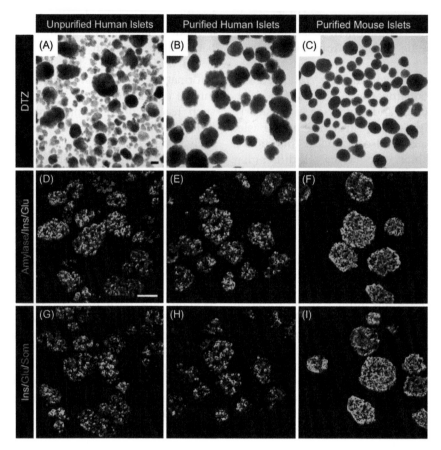

FIGURE 16.1   **Morphology of isolated human and mouse islets.** Human islets from the IIDP program (see text) were further purified by handpicking under microscopic guidance. (A) An example of a human islet preparation received from an islet isolation facility and stained by dithizone (DTZ) before, and (B) after, the handpicking procedure. This was a relatively impure preparation (stated purity 50%), but all islet preparations contain ductal and acinar fragments. Scale bar (A−C) 100 μm. To further evaluate the purity of handpicked islets (D−F), selected islet preparations were processed for cryosections and labeled for insulin (Ins, green), glucagon (Glu, green) and α-amylase (red). Adjacent sections (G−I) were labeled for insulin (green), glucagon (red) and somatostatin (Som, blue). Note the difference in islet cell distribution between human and mouse islets. Scale bar (D−I) 100 μm. Figure and figure legend are adapted from work by Dai et al.[24]

is known in human islets. For example, in several animal models, islet blood flow is several times greater than the surrounding exocrine tissue.[34,35] This high degree of vascularization in rodent islets is dependent on the islet cell-derived production of vascular endothelial growth factor-A (VEGF-A) beginning during islet development and continuing into adulthood.[36] Furthermore, the pattern of blood flow in mouse islets is that blood flows initially to the beta cell-rich core of the islet and then to cells on the islet periphery.[37] In this arrangement, compounds secreted by the beta cell (γ-aminobutyric acid (GABA) and zinc, in addition to insulin) may influence the downstream alpha and delta cells. However, little is known about whether similar directions in blood flow are present in human islets.

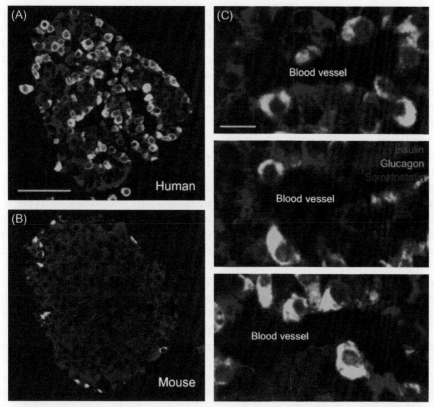

FIGURE 16.2 **Alpha and beta cells are in close proximity in human islets.** Staining in Panel A (human) and Panel B (mouse) shows insulin (red), glucagon (green), and somatostatin (blue). Note that the mouse beta cells are touching mostly other beta cells (because of the uniform central core of beta cells). In contrast, human beta cells are mostly touching alpha cells (green), or delta cells (blue). Scale bar = 50 μm. Panel C shows a confocal image of that human islet cells (stained as in Panel A) reside along blood vessels (dark spaces). Scale bar = 10 μm. Figure and figure legend are adapted from work by Caicedo et al.[38]

The islet vasculature is know to be lined by fenestrated endothelial cells and the basement membrane in human islets is much thicker than in rodents and has been termed a "double basement membrane."[21]

Neither the degree of vascularization nor the pattern of blood flow has been defined in human islets. Furthermore, the arrangement of alpha and beta cells within the human islet makes it difficult to envision how a product of one cell type delivered vascularly could influence another islet cell. In the rodent islet, where beta cells are "upstream" of alpha cells,

this seems quite plausible, but this seems unlikely in human islets. Conversely, evidence is beginning to emerge that human islet cells are in very close proximity to capillaries and endothelial cells and that alpha and beta cells are frequently in cell-to-cell contact (Figure 16.2). This suggests that paracrine or electrical activity may be the more common mechanism for communication among human islet cell types.[38]

Islet physiology and hormone secretion is regulated by both sympathetic and parasympathetic neuronal input in several species. This yin/yang of insulin and glucagon secretion extends to neuronal input, with numerous studies showing that sympathetic input inhibits insulin while stimulating glucagon secretion, and that parasympathetic inputs stimulates insulin but inhibits glucagon secretion.[39] For example, during hypoglycemia, the output of glucagon is thought in large part to be neurally mediated, with the early phase of feeding-induced insulin secretion modified when vagal nerve or parasympathetic input was interrupted. Immunohistochemical studies in animal models have shown the presence of both types of nerve fibers within the islet, leading to the assumption that neurotransmitters released from intra-islet nerve endings act directly on cell-surface receptors on the respective islet cell type. Caicedo and colleagues examined and compared innervation in human and mouse pancreatic sections. As predicted from prior studies, they found that mouse islets were highly innervated with sympathetic fibers near alpha cells and parasympathetic fibers near both alpha cells and beta cells. Surprisingly, they found in the human pancreatic sections that sympathetic fibers were alongside the smooth muscle cells rather than near islet cells.[38,40] The number of vascular-associated smooth muscle cells was much greater in human islets. Even more surprising was the paucity of intra-islet parasympathetic fibers in human islets as detected by vesicular acetylcholine transporter (vAChT) expression (Figure 16.3). Instead, they found this marker of parasympathetic expression was highly expressed in human islet endocrine cells. These dramatic differences with mouse islets led these investigators to propose a new paradigm for neuronal input into insulin and glucagon secretion in humans, including: 1) sympathetic input modulates islet endocrine cells not through direct signaling, but by influencing intra-islet smooth muscle cells that line intra-islet capillaries and this alters local islet blood flow, leading to changes in islet endocrine cell activity; and 2) parasympathetic influence on insulin secretion in humans may be relatively small or nonexistent. Importantly, these observations demonstrate the critical need for studies on human pancreas to understand human physiology and/or pathophysiology.

Islets in Type 1 Diabetes

Conventional wisdom holds that cell-mediated autoimmunity of type 1 diabetes is initiated by exposure to an unknown triggering event in genetically susceptible individuals and this leads to a gradual destruction of beta cells of islet until all beta cells have been destroyed. Two new findings about human islet morphology are beginning to challenge this conventional wisdom. First, our concept of insulitis is possibly incorrect.[41,42] "Insulitis," the inflammatory process that targets the beta cells, is often considered a definitive part of the pathogenesis. This is consistent with the belief that type 1 diabetes is a cell-mediated disease with the islet antibodies being secondary; so by definition insulitis would be expected. Insulitis was recognized early in the 20th century (before insulin was discovered) by several pathologists who studied the pancreas of children who died of diabetic ketoacidosis.[42] However, even in the 1920s, some pathologists noted that insulitis was not a universal finding at autopsy. Of course, this was long before the concept of type 1 and type 2 diabetes, but these were primarily children or adolescents who died of fulminant diabetes. Seminal works by Gepts and Foulis in the 1950s to 1970s, which shared some pathologic cases, led to gradual acceptance of the concept of insulitis.

As more pancreases from individuals who died near the time of onset of type 1 diabetes are studied, it is becoming increasing clear that the insulitis is neither as severe nor as uniform as previously thought. In fact, the degree of

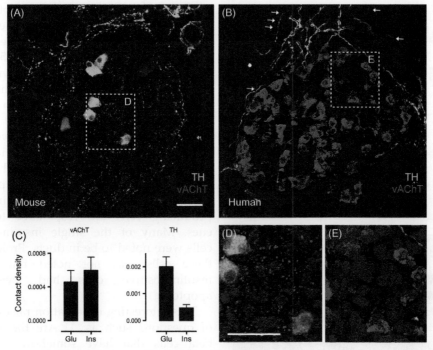

FIGURE 16.3 **Cholinergic innervation differs in mouse and human islets.** Confocal images in Panel A (mouse) and Panel B (human) shows immunostaining for the cholinergic marker, vesicular acetylcholine transporter (vAChT; red) and the catecholaminergic marker tyrosine hydroxylase (TH; green). Arrows in (B) point at vAChT-stained axons which are present in human exocrine tissue, but not in human islets. Panel C shows more vAChT-labeled axons contacting endocrine cells in mouse islets (alpha and beta) whereas catecholaminergic axons contact more alpha cells. Figure and figure legend are adapted from work by Caicedo et al.[40]

insulitis is sometimes so mild that there has been some controversy as to what actually constitutes the lesion and it may not affect all islets – only certain lobes of the pancreas. Under the auspices of nPOD, a consensus definition has been issued and states (Figure 16.4):[43]

Patients with insulitis are defined by the presence of a predominantly lymphocytic infiltration specifically targeting the islets of Langerhans. The infiltrating cells may be found in the islet periphery (peri-insulitis), often showing a characteristic tight focal aggregation at one pole of the islet that is in direct contact with the peripheral islet cells. The infiltrate may also be diffuse and present throughout the islet parenchyma (intra-insulitis). The lesion mainly affects islets containing insulin-positive cells and is always accompanied by the presence of (pseudo) atrophic islets devoid of beta cells. The fraction of infiltrated islets is generally low (<10% of islet profiles). The lesion should be established in a minimum of three islets, with a threshold level of ≥15 CD45+ cells/islet before the diagnosis can be made. The pathology report should include the total number of islets analyzed, the fraction of islets affected by insulitis, the fraction of (pseudo)atrophic islets, and a description of the spatial relationship of the infiltrate to the insulin-positive islet cells

A second observation about the islet cell morphology has come from examining the pancreas of individuals with long-standing type 1 diabetes. The Joslin Diabetes Center has been following a large cohort of patients with

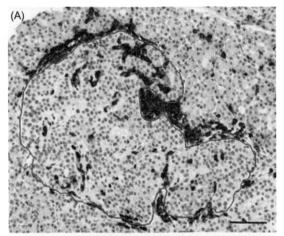

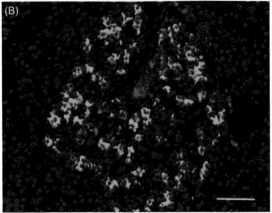

FIGURE 16.4 **Insulitis in human type 1 diabetes.** (A) Islet in the pancreatic body region of a 12-year-old boy with type 1 diabetes of 1 year's duration (case nPOD 6052), showing 249 CD45[+] nucleated cells (brown) surrounding and infiltrating the islet (islet periphery outlined in red on the basis of a consecutive slide stained for islet hormones; single color peroxidase staining). (B) Islet in the pancreatic tail region of a 66-year-old woman with type 1 diabetes of 20 years' duration (case BCBB 3450), showing 49 CD45[+] nucleated cells (red) surrounding and infiltrating the islet (insulin staining in green; two-color immunofluorescence staining). Scale bars, 50 μm. Figure and figure legend are from work by Campbell-Thompson et al.[43]

type 1 diabetes for more than 50 years ("Joslin Medalists"). In a collaborative partnership between some of these patients (N = 411) and Joslin physicians and scientists, the patients

agree to annual testing (presence of complications related to diabetes, C-peptide, etc.) and to donate his/her pancreas upon death. In this cohort with a mean duration of 56 years of type 1 diabetes, 67% of the individuals had low, but detectable C-peptide.[44] The pancreas from nine individuals have been examined (Figure 16.5) and in all nine, small clusters or single insulin-positive cells were seen. Furthermore, others of these individuals with the highest C-peptide levels had a further increase following mixed-meal provocative testing, indicating that the remaining insulin-producing cells are responsive to nutrient cues. Many of the single insulin-producing cells were noted to be in ducts. Proliferation of these cells was not noted but occasional insulin-positive cells had evidence of apoptosis.

This interesting observation raises a number of questions such as: 1) Are these "normal" beta cells that have somehow escaped the autoimmune process? 2) Do these cells reflect a regenerative process? 3) Did the autoimmune process cease in these individuals? Future studies seek to address these questions.

Islets in Type 2 Diabetes

Impaired insulin secretion is a hallmark of type 2 diabetes. Of the more than 70 genetic loci identified by genome-wide association studies (GWAS), most are likely to have an impact on beta cell development, function, or survival. Furthermore, almost all of the monogenic forms of diabetes likely result from impaired beta cell function, development, or differentiation. Numerous defects or deficits in beta function and mass have been postulated and the debate about reduced mass, reduced function, or both continues.[27,45] The pathologic changes in type 2 diabetes are likely heterogeneous, not only because of disease heterogeneity, but also because the onset of the disease is difficult to

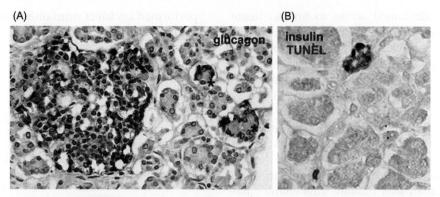

FIGURE 16.5 **Persistent insulin-positive cells in type 1 diabetes.** Histologic findings in pancreases from Joslin Medalists.[44] In seven of nine pancreases, there were mainly atrophic islets positive for glucagon but not insulin (Panel A). All nine pancreas had rare small clusters or scattered single insulin-positive cells (Panel B). In one antibody-positive Medalist, some insulin-positive cells were terminal deoxynucleotidyl transferase dUTP nick end labeling (TUNEL)-positive, indicating apoptosis (Panel B). Figure and figure legend are from work by Keenan *et al.*[44]

date and the progressive nature of the beta cell impairment makes it difficult to classify at what stage in the disease the pancreas is studied. Bluntly stated, the islet/beta cell pathology is likely quite different in an individual with relatively short disease duration who is being treated with oral agents and an individual with more than two decades of disease who requires insulin treatment. Unfortunately, many of the pathologic examinations of the pancreas tend to combine such disparate samples. Nevertheless, some consistent findings in the islets of individuals with type 2 diabetes include:[27,45]

- Modest reduction in beta cell mass
- Presence of islet amyloid, which is usually perivascular in location
- Little sign of proliferation or regeneration
- Some degree of apoptosis.

Because of the nature of studying a human tissue that cannot be assessed over time, all data is cross-sectional with no ability to quantify beta cell mass non-invasively or to biopsy the pancreas as type 2 diabetes progresses. Nevertheless, new experimental approaches may provide new insight into the pathogenesis of human type 2 diabetes. First, experimental approaches to study gene/protein expression in human tissue is evolving rapidly and being applied to the study of the pancreas. Second, more standardized cataloging and collection of the pancreas to allow for studies of formalin-fixed tissue are being planned. Third, islets from humans with type 2 diabetes are now becoming available through the IIDP program and some human islet isolation centers.

SUMMARY AND FUTURE DIRECTIONS FOR RESEARCH

Over the last decade, there has been a remarkable amount of new knowledge about human islet biology and this should lead to new ways to preserve or restore beta cell function. Much of this progress is due to increased efforts by scientists to translate findings in cells and animal models to human islets. Equally important for this progress has been the availability of human islets (to investigators who are not at islet isolation centers) and human pancreas sections. Continued progress will require continued access to human tissue and islets and the development of more robust, better annotated, human pancreatic and islet specimens that are collected in a way that allows for sophisticated research, but are also tied to clinical history of the patient in a way that preserves patient privacy but allows improved correlation of laboratory research and clinical history.

As is often the case, the more one knows about a subject, the more questions arise. In terms of human islet biology, this chapter has suggested several areas of fertile investigation, including:

- As the morphology and architecture of the normal adult human islet is being increasingly defined, how does this change from *in utero* to neonatal to adolescence to adulthood?
- Does the morphology and architecture of the human islet portend changes in function, survival, or proliferation?
- What are the morphology and architecture consequences of genes that are associated by GWAS or monogenic diabetes?
- What is the vascularization state of normal human islets? Is this altered in type 2 diabetes?
- How can infrastructure be developed that allows the studies of the pancreatic events in the pre-diabetic period of type 1 diabetes?
- What is the nature of the insulin-producing cells that are found in the pancreas of individuals with long-standing type 1 diabetes?
- What islet changes occur longitudinally in type 2 diabetes? How do the scientists study these processes? What infrastructure and tissue resources are needed?
- As normal human islets are sorted into individual cell populations for transcriptional epigenetic, and non-coding RNA profiling, can similar studies be

performed on islets from obese individuals or those with forms of diabetes?

Acknowledgements

The author's work was supported by a Merit Review Award from the VA Research Service (BX000666), by NIH grants (DK69603, DK68764, DK89572, DK66636, DK63439, DK72473, DK89538), by grants from the Juvenile Diabetes Research Foundation International, by the Vanderbilt Mouse Metabolic Phenotyping Center (NIH grant DK59637), by the Vanderbilt Diabetes Research and Training Center (DK20593) and by the Vanderbilt University Medical Center Cell Imaging Shared Resource (NIH grants CA68485, DK20593, DK58404, HD15052, DK59637, and EY08126). The author does not have relevant conflict of interest to disclose.

References

1. Greeley SAW, Tucker SE, Naylor RN, Bell GI, Philipson LH. Neonatal diabetes mellitus: a model for personalized medicine. *Trends Endocrinol Metab* 2010;**21**:464–72.
2. Ashcroft FM, Rorsman P. Diabetes mellitus and the β cell: the last ten years. *Cell* 2012;**148**:1160–71.
3. Pagliuca FW, Melton DA. How to make a functional β-cell. *Development* 2013;**140**:2472–83.
4. McCarthy MI, Hattersley AT. Learning from molecular genetics: novel insights arising from the definition of genes for monogenic and type 2 diabetes. *Diabetes* 2008;**57**:2889–98.
5. Shapiro AM, Lakey JR, Ryan EA, et al. Islet transplantation in seven patients with type 1 diabetes mellitus using a glucocorticoid-free immunosuppressive regimen. *N Engl J Med* 2000;**343**:230–8.
6. Robertson RP. Islet transplantation as a treatment for diabetes—a work in progress. *N Engl J Med* 2004;**350**:694–705.
7. Rickels MR, Liu C, Shlansky-Goldberg RD, et al. Improvement in β-cell secretory capacity after human islet transplantation according to the CIT07 protocol. *Diabetes* 2013;**62**:2890–7.
8. Ricordi C. Islet transplantation: a brave new world. *Diabetes* 2003;**52**:1595–603.
9. Kaddis JS, Olack BJ, Sowinski J, Cravens J, Contreras JL, Niland JC. Human pancreatic islets and diabetes research. *Jama* 2009;**301**:1580–7.
10. Kaddis JS, Hanson MS, Cravens J, et al. Standardized transportation of human islets: an islet cell resource

center study of more than 2,000 shipments. *Cell Transplant* 2013;**22**:1101−11.

11. Volk BW, Arquilla ER, editors. *The diabetic pancreas*. Boston, MA: Springer US; 1985. Available from: http://dx.doi.org/10.1007/978-1-4757-0348-1.

12. In't Veld P, Marichal M. Microscopic anatomy of the human islet of Langerhans. *Adv Exp Med Biol* 2010;**654**:1−19.

13. Rahier J, Guiot Y, Goebbels RM, Sempoux C, Henquin JC. Pancreatic beta-cell mass in European subjects with type 2 diabetes. *Diabetes Obes Metab* 2008;**10**(Suppl. 4):32−42.

14. Saisho Y, Butler AE, Meier JJ, et al. Pancreas volumes in humans from birth to age one hundred taking into account sex, obesity, and presence of type 2 diabetes. *Clin Anat* 2007;**20**:933−42.

15. Hughes SJ, Clark A, McShane P, Contractor HH, Gray DWR, Johnson PRV. Characterisation of collagen VI within the islet-exocrine interface of the human pancreas: implications for clinical islet isolation? *Transplantation* 2006;**81**:423−6.

16. Banerjee M, Virtanen I, Palgi J, Korsgren O, Otonkoski T. Proliferation and plasticity of human beta cells on physiologically occurring laminin isoforms. *Mol Cell Endocrinol* 2012;**355**:78−86.

17. Naylor R, Philipson LH. Who should have genetic testing for maturity-onset diabetes of the young? *Clin Endocrinol (Oxf)* 2011;**75**:422−6.

18. Kaido TJ, Yebra M, Kaneto H, Cirulli V, Hayek A, Montgomery AM. Impact of integrin-matrix interaction and signaling on insulin gene expression and the mesenchymal transition of human beta-cells. *J Cell Physiol* 2010;**224**:101−11.

19. Riopel M, Krishnamurthy M, Li J, Liu S, Leask A, Wang R. Conditional beta1-integrin-deficient mice display impaired pancreatic beta cell function. *J Pathol* 2011;**224**:45−55.

20. Daoud J, Petropavlovskaia M, Rosenberg L, Tabrizian M. The effect of extracellular matrix components on the preservation of human islet function in vitro. *Biomaterials* 2010;**31**:1676−82.

21. Virtanen I, Banerjee M, Palgi J, et al. Blood vessels of human islets of Langerhans are surrounded by a double basement membrane. *Diabetologia* 2008;**51**:1181−91.

22. Brissova M. Assessment of human pancreatic Islet architecture and composition by laser scanning confocal microscopy. *J Histochem Cytochem* 2005;**53**:1087−97.

23. Cabrera O, Berman DM, Kenyon NS, Ricordi C, Berggren P-O, Caicedo A. The unique cytoarchitecture of human pancreatic islets has implications for islet cell function. *Proc Natl Acad Sci USA* 2006;**103**:2334−9.

24. Dai C, Brissova M, Hang Y, et al. Islet-enriched gene expression and glucose-induced insulin secretion in human and mouse islets. *Diabetologia* 2012;**55**:707−18.

25. Wang X, Misawa R, Zielinski MC, et al. Regional differences in islet distribution in the human pancreas—preferential beta-cell loss in the head region in patients with type 2 diabetes. *PLoS ONE* 2013;**8**: e67454.

26. Wang X, Zielinski MC, Misawa R, et al. Quantitative analysis of pancreatic polypeptide cell distribution in the human pancreas. *PLoS ONE* 2013;**8**:e55501.

27. Bonner-Weir S, O'Brien TD. Islets in type 2 diabetes: in honor of Dr. Robert C. Turner. *Diabetes* 2008;**57**:2899−904.

28. Farhat B, Almelkar A, Ramachandran K, et al. Small human islets comprised of more beta-cells with higher insulin content than large islets. *Islets* 2013;**5**:87−94.

29. Butler AE, Cao-Minh L, Galasso R, et al. Adaptive changes in pancreatic beta cell fractional area and beta cell turnover in human pregnancy. *Diabetologia* 2010;**53**:2167−76.

30. Butler AE, Galasso R, Meier JJ, Basu R, Rizza RA, Butler PC. Modestly increased beta cell apoptosis but no increased beta cell replication in recent-onset type 1 diabetic patients who died of diabetic ketoacidosis. *Diabetologia* 2007;**50**:2323−31.

31. Matveyenko AV, Butler PC. Relationship between beta-cell mass and diabetes onset. *Diabetes, Obes Metab* 2008;**10**(**Suppl. 4**):23−31.

32. Gregg BE, Moore PC, Demozay D, et al. Formation of a human β-cell population within pancreatic islets is set early in life. *J Clin Endocrinol Metab* 2012;**97**:3197−206.

33. Meier JJ, Butler AE, Saisho Y, et al. Beta-cell replication is the primary mechanism subserving the postnatal expansion of beta-cell mass in humans. *Diabetes* 2008;**57**:1584−94.

34. Bonner-Weir S, Orci L. New perspectives on the microvasculature of the islets of Langerhans in the rat. *Diabetes* 1982;**31**:883−9.

35. Brunicardi FC, Stagner J, Bonner-Weir S, et al. Microcirculation of the islets of Langerhans. Long beach veterans administration regional medical education center symposium. *Diabetes* 1996;**45**:385−92.

36. Brissova M, Shostak A, Shiota M, et al. Pancreatic islet production of vascular endothelial growth factor-a is essential for islet vascularization, revascularization, and function. *Diabetes* 2006;**55**:2974−85.

37. Nyman LR, Wells KS, Head WS, et al. Real-time, multidimensional in vivo imaging used to investigate blood flow in mouse pancreatic islets. *J Clin Invest* 2008;**118**:3790−7.

38. Caicedo A. Paracrine and autocrine interactions in the human islet: more than meets the eye. *Semin Cell Dev Biol* 2013; **24**:11−21.

39. Taborsky GJJ. Islets have a lot of nerve! Or do they? *Cell Metab* 2011;**14**:5–6.
40. Rodriguez-Diaz R, Abdulreda MH, Formoso AL, et al. Innervation patterns of autonomic axons in the human endocrine pancreas. *Cell Metab* 2011;**14**:45–54.
41. Rowe PA, Campbell-Thompson ML, Schatz DA, Atkinson MA. The pancreas in human type 1 diabetes. *Semin Immunopathol* 2011;**33**:29–43.
42. In't Veld P. Insulitis in human type 1 diabetes: the quest for an elusive lesion. *Islets* 2011;**3**:131–8.
43. Campbell-Thompson ML, Atkinson MA, Butler AE, et al. The diagnosis of insulitis in human type 1 diabetes. *Diabetologia* 2013;1–3.
44. Keenan HA, Sun JK, Levine J, et al. Residual insulin production and pancreatic ß-cell turnover after 50 years of diabetes: Joslin medalist study. *Diabetes* 2010;**59**:2846–53.
45. Costes S, Langen R, Gurlo T, Matveyenko AV, Butler PC. beta-Cell failure in type 2 diabetes: a case of asking too much of too few? *Diabetes* 2013;**62**:327–35.

Computational Models to Decipher Cell-Signaling Pathways

Anne Poupon and Eric Reiter

INRA, Nouzilly, France, CNRS, Nouzilly, France, François Rabelais University, Tours, France

INTRODUCTION

An impressive amount of data and knowledge has been accumulated over the past decades on how hormonal signals activate their cognate receptors, for those present at the plasma membrane, how they trigger the activation of intracellular signaling cascades and how they eventually modify cell fate. As a consequence, the representation of hormone-triggered intracellular pathways has progressively shifted from a system where discrete linear signaling cascades connect a given plasma membrane receptor to transcription factors to a more dynamic and complex model, in which communication networks made of multi-protein ensembles process and integrate the signal fluxes into adapted biological outcomes. Moreover, with the advent of "-omics" technologies, inventories of transcripts, proteins, or metabolites regulated by hormones have been generated at an increasing rate.[1] The question of how these entities operate together to respond to hormones clearly represents a major challenge for the whole field,

because deciphering the molecular details of hormones' actions on their target cells could pave the road to the development of new generations of pathway-selective drugs.[2] However, cellular endocrinologists are facing an enormous task as they try to understand the behavior of these intertwined communication networks. It is becoming increasingly clear that mathematical and computational approaches are necessary to tackle such a high level of complexity.[3] In line with this idea, this chapter reviews the existing modeling strategies and how they can be combined with experimental biology in order to build predictive computational models of intracellular networks.

THE GROWING COMPLEXITY OF HORMONE-INDUCED SIGNALING MECHANISMS

Classically, upon ligand binding, hormone receptors are primed to transduce signals inside the cell. Two main classes of receptors

Cellular Endocrinology in Health and Disease.
DOI: http://dx.doi.org/10.1016/B978-0-12-408134-5.00017-2

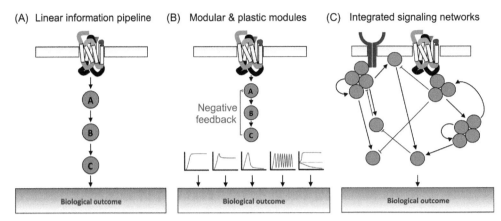

FIGURE 17.1 **Modular, plastic, and integrated nature of signaling networks.** (A) Signaling pathways have long been thought of as linear information pipelines connecting the plasma membrane to the nucleus. (B) More recently, signaling cascades have been shown to be highly modular and plastic. In this simple example, three proteins activate each other sequentially, with the third one acting as a negative regulator on the first one. Such a module, which is commonly encountered in signaling networks, can generate a wide variety of signals, depending on the kinetic parameters that are controlling each elementary reaction. (C) Signaling pathways are organized as integrated networks involving receptors from different families, acting together in signaling platforms. Feedback and cross-talk, either positive or negative, contribute to signal transmission and processing.

are relaying hormones' effects inside the cells: membrane receptors that are bound and activated by hydrophilic (peptidic/polypeptidic) hormones and hydrophobic hormones' receptors whose subcellular localization is wider (nucleus, cytosol, plasma membrane). All these receptors belong to different families (G protein-coupled receptors (GPCR), tyrosine kinase receptors (TRK), cytokine receptors, nuclear receptors) and specifically bind hormones characterized by vastly different chemical properties. It is increasingly recognized that a given plasma membrane receptor can be connected to multiple transduction mechanisms, thereby leading to the activation of intertwined signaling networks (Figure 17.1).[4] Moreover, different receptor subtypes can be associated into signaling platforms, increasing even more the possibilities for pathways' recruitment.

Adding to this combinatorial complexity is the fact that intracellular signals are spatially and temporally encoded. Modulation of signal duration increases the range of hormone concentrations for which dose-dependent responses remain possible.[5] A well-documented example is the dual activation mechanism of extracellular-signal-regulated kinases (ERK) by GPCRs: G protein-mediated ERK activation is rapid, transient, and translocates to the nucleus. In contrast, the ERK activated via β-arrestins are slower but persistent and are sequestered in the cytosol (Figure 17.2).[6] Different hormones have been reported to activate ERK via this type of spatially and temporally encoded mechanism: angiotensin, vasopressin, parathyroid hormone and follicule-stimulating hormone.[7-11] Recently, intracellular G_s-receptor coupling and cyclic adenosine monophosphate (cAMP) production has also gained a lot of attention.[12-15] Interestingly, two hormone-activated receptors (i.e., thyroid-stimulating hormone and parathyroid hormone receptors) have been reported to induce transient cAMP production while at the plasma membrane,

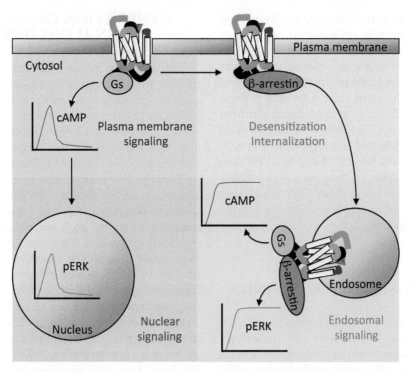

FIGURE 17.2 **Compartmentalization of signals.** Signals can be generated with different kinetics from different cellular compartments. In this example, cyclic adenosine monophosphate (cAMP) is transiently produced at proximity of the plasma membrane upon activation and functional coupling of a receptor present at the cell surface. This cAMP response leads to a fast but transient accumulation of protein kinase RNA-like endoplasmic reticulum kinase (pERK) in the nucleus. The activated receptor is rapidly turned off through desensitization and internalization, both processes involving β-arrestins. However, once in the endosomes, the receptors still signals via two distinct mechanisms: extracellular signal-regulated kinases are phosphorylated in a β-arrestin-dependent manner whereas cAMP is generated through a G_s-dependent process. Both signals are now sustained.

whereas they lead to sustained cAMP accumulation via an endosomal complex involving the internalized receptor, β-arrestin and G_s (Figure 17.2). Such spatial and temporal differences in intracellular signals substantially enhance the complexity of signaling systems, which also increase their processing capabilities.

Feedback and cross-talk also play key roles in the processing of signals through networks. For instance, negative feedback allows pathways to adapt or desensitize to persistent stimuli. On the other hand, cross-inhibition limits cross-talk between pathways.[16,17] In addition to

transmitting qualitative information, signaling pathways must also convey quantitative information about the strength of the stimulus. For instance, it has been recently shown that signaling pathways can take advantage of their nonlinear nature to convert stimulus intensity into signal duration.[18] Moreover, complex interactions among the proteins and metabolites involved in signaling cascades can lead to oscillations,[19,20] for example, in: receptor, cAMP and PKA;[21,22] calcium and IP3;[23] NF-κB, IκB and IKK;[24,25] p53 and MDM2 or Msn2;[26–28] adenylate cyclase, cAMP and PKA.[29,30] These examples clearly demonstrate that encoding within signaling networks can, under certain circumstances, operate on a digital rather than analog mode.

Another major factor generating complexity comes from the fact that multiple hormones hit a target cell at the same time. Obviously, these hormonal cues have to be processed simultaneously by the recipient cells and transformed into the most adapted biological response.[31] Another important feature that has to be considered in endocrine systems is the pulsatility that some hormones present and which can certainly affect the activation dynamics of the intracellular networks.[32,33]

PATHWAY-SELECTIVE LIGANDS

There is clear evidence that some GPCR ligands (including hormones) can selectively modulate a subset of the signaling events triggered by the full agonist while having no or even negative effects on the other signaling events. This concept, which is generally referred to as "biased agonism," resonates with the accumulating structural data demonstrating the existence of multiple inactive and active conformations of the receptor, a subset of which may be stabilized by biased ligands.[34] Interestingly, biased ligands also seem to exist in other classes of receptors including receptor tyrosine kinases (RTKs).[35,36] This new concept opens new research avenues in drug discovery, as it raises the possibility of developing new classes of drugs with more selective actions and hence fewer side effects. A growing number of examples illustrate the therapeutic potential of biased agonists at different GPCRs.[37] The complexity of signaling mechanisms underlying the integration of hormonal cues is such that computational modeling is necessary in order to rationalize the discovery of new ligands exhibiting biased properties at hormone receptors.

GENERATION OF ADAPTED SIGNALING DATA

The first step when building computational models of signaling networks is the collection and/or generation of informative and reliable biological data. Signaling events are propagated within the cell by protein—protein interactions and enzymatic activities, primarily reversible protein phosphorylation.[38] Therefore, the comprehensive and quantitative analysis of protein phosphorylation profiles in cells undergoing different hormonal exposure, is a vital piece of information to be acquired. Classically, phosphorylation dynamics are assessed with phosphospecific antibodies combined to Western blotting (WB). However, this approach requires large amounts of biological material and, due to its poor throughput, does not allow large-scale analysis. Lately, the isolation of phosphorylated peptides from complex samples and their analysis by mass spectrometry (MS)[39,40] has come to the fore, typically allowing the identification of thousands of phosphopeptides in a single sample.[39–41] Despite this unparalleled analytical power, these methods are often limited to the generation of snapshots statically comparing a limited number of biological conditions. Protein microarraying, which consists in the automated spotting of concentrated and complex protein extracts, has emerged as a means to efficiently exploring signaling mechanisms using phosphospecific antibodies. In particular, reverse-phase protein array (RPPA) allows the simultaneous analysis of thousands of samples. Moreover, as it requires only very small quantities of biological material; microarrays fabrication can easily be parallelized, allowing their subsequent screening with many different antibodies.[7,42–45] Thanks to this substantial throughput, kinetics, dose—responses, and targeted perturbations can be systematically carried out and analyzed. Importantly, RPPA typically allows simultaneous quantification of thousands of samples in a single assay and this drastically reduces data heterogeneity and variability, which hinder modeling. In addition, with known amounts of purified proteins arrayed on the same slide, molar quantities of the assayed protein present in the cell lysates can easily be calculated, thereby greatly favoring quantitative modeling. The main limitation is that the specificity of signals is difficult to estimate in this format and the technique therefore heavily relies on the availability of antibodies validated for their use in RPPA. To overcome this limitation, microwestern arrays have been proposed. This technology enables quantitative, sensitive and high-throughput assessment of protein abundance and modifications after electrophoretic separation of microarrayed cell lysates.[46] An emerging technology, based on micro-isoelectrofocusing, allows post-translational analyses at medium throughput from minute amounts of biological material. Importantly, all the isoforms of a given protein as well as their relative quantities can be measured using a pan-antibody.[47] Cell-sorting-based approaches have also proven to be a powerful approach to acquire cell-signaling data with high-throughput and reliability.[48]

Most cellular signaling processes are carried out by multiproteic complexes. The identification and analysis of these complexes provide insight on the architecture of signaling networks. High-throughput approaches have long been developed to tackle this problem: yeast two-hybrid screens,[49] tandem-affinity purification and mass spectrometry,[50] and luminescence-based mammalian interactome mapping[51] have been successfully used on a genome-wide scale.

As previously pointed out, intracellular signals are encoded in time and space. Therefore, monitoring of signaling events in living cells is of great interest for modeling. In particular, imaging approaches that use fluorescent sensors of signaling activities have been developed that combine unmatched time and spatial resolutions. Protein−protein interactions and changes in protein conformation can also be analyzed in real time in living cells using similar approaches. Genetically encoded fluorescence resonance energy transfer (FRET)-based sensors have helped to capture the spatiotemporal patterns of second messengers, kinase activities and GPCR activation.[52−54] As FRET data are generally acquired using fluorescence microscopy, they allow the measurement of signaling events in single cells, a feature amenable to stochastic modeling.[55] Bioluminescence resonance energy transfer (BRET) has also been extensively used to monitor signaling, protein−protein interactions, and conformational changes in living cells.[56,57] BRET is typically carried out in multi-well plate format and ensures the production of huge amounts of dynamic data from cell populations that are very well suited for deterministic dynamical modeling.[55]

ITERATIVE DIALOG BETWEEN MODELING AND EXPERIMENTATION

The aptitude to precisely control hormone exposure, in both time and concentration, is also a very important aspect of the implementation of computational modeling in the context of cellular endocrinology. In fact, physiological hormonal exposure is rarely monotonous — pulsatile release of some hormones being an extreme illustration. Moreover, *in vivo*, hormonal cues are combinatorial, and simultaneously activate multiple receptors linked to distinct transduction mechanisms. It is therefore important to be able to reproduce these physiological situations as closely as possible *in vitro*. Exposing cells to tightly controlled and complex hormonal patterns is also a means to challenging model predictions with experimental validations. The use of microfluidic devices might help negate the limitations found with conventional approaches, as microfluidics offers the possibility of applying well-defined stimulation.[58,59]

Another important aspect is the ability to apply targeted perturbations and to measure the consequences on the network behavior. There are many different options to perturb signaling networks. Schematically, they can be classified into three categories: i) the perturbations acting at the hormonal level (biased hormones[60] or application of non-monotonous stimulation patterns); ii) those acting at the receptor level (biased mutant receptors);[10] and iii) targeted perturbations hitting specific nodes within the signaling network (Figure 17.3). For the third category, in addition to classical approaches (kinase inhibitors, dominant negative constructs, etc.), interfering RNAs offer the unique opportunity to easily and specifically achieve gene knockdowns. Moreover, genome-wide siRNA screenings are now available either in multi-well liquid phase or in transfected cell array format.[61-63]

Finally, the process of parameter optimization, which will be explained later, also offers a very nice opportunity for establishing an iterative dialog between experimental biology and modeling. Indeed, the endpoint of parameter optimization is the fitting of the simulated model to the experimental curves. When the fitting is not satisfactory (i.e., the distance between simulations and experimental data is too large), the architecture of the signaling network needs to be revised.[64]

BUILDING THE STATIC MODEL

The static model is a mathematical formalization of the molecular species involved in the studied process and the relationships they share: catalysis, physical interaction, etc. Building this model is what a biologist implicitly does when constructing a project and designing a series of experiments.

These last years, a number of databases gathering knowledge essentially extracted from literature have been published (an extensive list can be found on the Pathguide website: http://www.pathguide.org). Two types of databases can be distinguished: community efforts that aim at collecting the knowledge of multiple experts, and model repositories. Within the first category, the largest and most used is probably Reactome.[65] Pathways presented in Reactome are manually curated and peer-reviewed. Two other pathways are

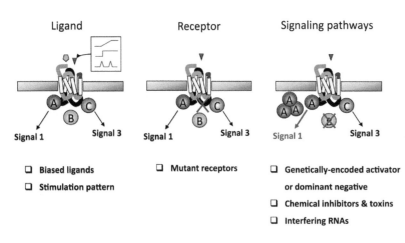

FIGURE 17.3 Targeted perturbations. When building computational models and/or when validating their accuracy, the ability to selectively inhibit discrete nodes within the predicted network plays a crucial role. Targeted perturbations can be achieved by operating at the ligand, the receptor or the signaling pathways levels. Upon targeted inhibition, functional read outs are recorded and can serve as a basis to reconstruct the network architecture or can be compared to the model's predictions.

very promising: WikiPathways,[66] which offers an easy-to-use platform for the deposition and community curation of biological pathways; and NetPath,[67] which is still of small size, but presents the decisive advantage of giving the references that led to the presented pathway. Within the second category, Biomodels[68] is becoming the reference.

One very important point is that all these models rely on manual analysis of experimental data, which has two major drawbacks. First, the models rely on the biologist's interpretation of experimental results. As these models are often built using published data and interpretations, there are multiple levels of human analysis that can lead to what are considered as true facts (and what were hypotheses to begin with). Secondly, high-throughput data are not fully taken into account, and only the genes or proteins that were suspected to be involved are generally considered, leaving out potentially important results. The resulting model is thus necessarily incomplete and biased, although the community efforts greatly damp these effects.

To overcome these difficulties, various automated methods have been developed. As transcriptomics was the first high-throughput experimental method, it has received considerable attention from computer scientists. Many methods are available to build transcription regulation networks from microarray data.[69] However, these networks alone cannot explain the link between stimulus and biological outcome, which justifies inferring the signaling networks. Further methods have been designed, working from RNAi data,[70] phosphoproteomic data[71] or micro Western array data.[46] However, all these methods are based on the exploitation of a single experimental method, whereas experimental exploration of a signaling system requires multiple methods. Since the biologist is able to manually integrate data coming from different experimental methods, the authors have implemented an automated method that could reproduce this reasoning, and extend it to high-throughput data. To this aim, the elementary reasoning elements have been formalized using first order logic rules. To infer a given signaling network, these rules, together with experimental data present in literature, are fed to an inference engine.[72] The authors have demonstrated that they could build the follicle-stimulating hormone receptor (FSHR), epidermal growth factor receptor (EGFR) and angiotensin II receptor signaling pathways using this method.

BUILDING AND PARAMETERIZING THE DYNAMICAL MODEL

Formalisms

Most models use the ordinary differential equations (ODE) formalism. In this formalism, the concentration variation of a given molecule is expressed as a function of the concentration of other molecules (Figure 17.4). At this step, a rate law has to be chosen. The simplest kinetic law is derived from mass action law. For enzymatic reactions, many authors still prefer Michaëlis−Menten kinetic law. However, using the right parameters, the simulations of both kinetic laws are completely superimposable (Figure 17.4). Similarly, although more-accurate kinetic laws exist for other processes such as transcription or translation, mass action law can still be a good approximation, especially since the differences in simulations arising from these different kinetic laws are in all cases far lower than the experimental variability.

Another popular formalism is the Bayesian networks. In this formalism the time course is discrete, and at each time point each molecule has a probability to undergo a state transition. If considering the simplest example in Figure 17.4A, each A molecule has a probability k_0 to become a B molecule, and each

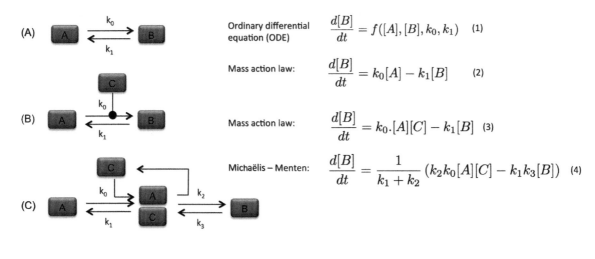

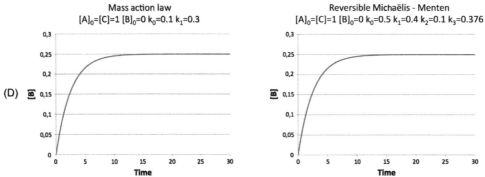

FIGURE 17.4　**Formalism for dynamic models.** (A) Simple state transition, general differential expression of the variation of B's concentration (1) and expression when considering a mass action law (2). (B) State transition depending on a third molecule, for example state transition mediated by an enzyme, and expression of the variation of B's concentration when considering mass action law (3). (C) Michaëlis–Menten model for a reversible state transition mediated by an enzyme and corresponding expression of B's concentration (4). (D) Parameter values can be found for which models B and C give superimposable simulations.

pre-existing B molecule has a probability k_1 to become an A molecule. At a given time point t, and for each molecule in the system, a random number is generated, if this number is lower than the transition probability, this individual molecule undergoes state transition. The molecules in the different species are then counted, and the process is repeated for time point $t + 1$. This method has the great advantage to represent better the reality of molecular reactions, and unlike ODE allows simulating stochastic events.

However, it can become very computationally inefficient when addressing large number of molecules.

Simulation

Once the model is precisely defined, and if all the parameters are known (kinetic constants, affinities, initial concentrations, etc.), the system can be simulated to show the evolution of the concentrations of the different molecular species as a function of time. Different software tools allow these simulations to be performed. CellDesigner,[73] which is valued by biologists for its clear and simple, yet rigorous, formalism, includes a simulation module. Moreover, the individual parameters can be interactively modified, the simulation being recomputed on the fly, which gives an easy way to understand how each parameter modifies the evolution of the different molecular species. Copasi[74] is another major Systems Biology Markup Language (SBML)-compliant software tool for dynamic systems. The user interface is less user-friendly than that of CellDesigner, however it offers a number of advanced tools for dynamic systems, such as parameter estimation or sensitivity analysis. Both software tools use the SBML language[75] and consequently the models can be exported to most system's biology tools.

Model reduction

When building a static model, the aim is usually to obtain a picture as detailed as possible of a given cellular process. This can lead to models involving hundreds of molecules, which is not compatible with dynamic modeling, and especially with parameter estimation (see below). Consequently, some molecules have to be eliminated, a process known as "model reduction," to obtain a model that retains the essential properties of the complete system, but with a reduced number of variables.

Various automated methods have been implemented to perform model reduction.[76,77] However, the reduction is often performed manually because it should not be only based on graph properties. Indeed, it is essential to keep in the model molecules that are known to play crucial roles (for example because they exist in limiting quantities) and molecules that can be measured experimentally.

An example of model reduction, from our model of the angiotensin signaling pathway,[64] is illustrated on Figure 17.5. In the pathway going from receptor activation to protein kinase C (PKC) activation, many steps have been omitted. Although very simplified, this pathway retains the major characteristics of the complete pathway, and leads to accurate estimation of PKC activation half-life — 25 seconds — whereas a value of 30 seconds has been experimentally measured.[78]

Parameter estimation

As stated above, if all the parameters of the system are known, the evolution of the concentrations of molecular species as a function of time can be simulated. However, this situation is only exceptional. Indeed, measuring rate constants, for example, is a difficult task. Moreover, the systems are often simplified, and successive reactions are aggregated. In this case, the kinetic constants do not have biological significance anymore, and cannot be measured. Another point is that very often the evolutions of molecular species are measured relative to a maximum, and not as absolute quantities. Some authors choose to pick values from publications, often obtained in different biological conditions, or to use an "educated guess." The other option is to search for a parameter set that gives a simulation reproducing experimental data (Figure 17.6), a process called parameter estimation.

Parameter estimation has become an important challenge in systems biology (see ref [79] for a review). Some methods are easily available through Copasi. These methods rely on classical continuous optimization methods: genetic algorithms,[80] evolutionary programming,[81] Hooke–Jeeves,[82] Levenberg–Marquardt,[83] PRAXIS,[84] particle swarm,[85] or steepest descent. These methods have the great advantage of providing an easy-to-use environment, but are not very efficient for difficult systems.

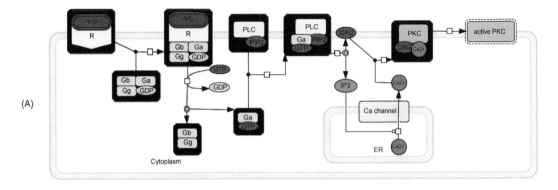

(A)

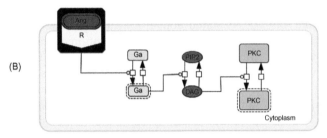

(B)

FIGURE 17.5   **Model reduction.** (A) Retailed pathway leading from angiotensin-bound receptor to activated protein kinase C. (B) Simplified model for the same process.[64]

More efficient methods have been developed, such as SSmGo[79] or BIOCHAM.[86,87]

The goal in biological systems modeling is to be able to predict the behavior of the system in biological conditions different from those of the original experimental measures. The authors have shown on the angiotensin signaling pathway that a good model, and accurate parameter estimation can lead to accurate predictions.[64] In this work they were able to make multiple predictions using this model that were experimentally validated at a later stage. Moreover, the simulations enabled the computation of half-life times for the different molecules in the system that corresponded amazingly well with those measured in biological systems.

Parameter estimation is a central processing unit (CPU)-intensive process, and some models are more difficult to process than others. Unexpectedly, the number of parameters is not predominant in this. Two main features drive the increase of the CPU time: the number of differential equations (tightly related to the number of molecules in the system) and the stiffness of the system (if some molecules vary very rapidly). Finally, some systems give very few converging estimations, which is the case for the angiotensin signaling pathway. Factors that decrease convergence rate are mainly the ratio between the number of observables and the number of unknown parameters, and the range between highest and lowest parameters.

Sensitivity and indentifiability

Quite naturally, the question arises as to whether the obtained parameters are unique or if unrelated parameter sets can lead to simulations fitting the experimental data. This is the identifiability issue. A correlated, but

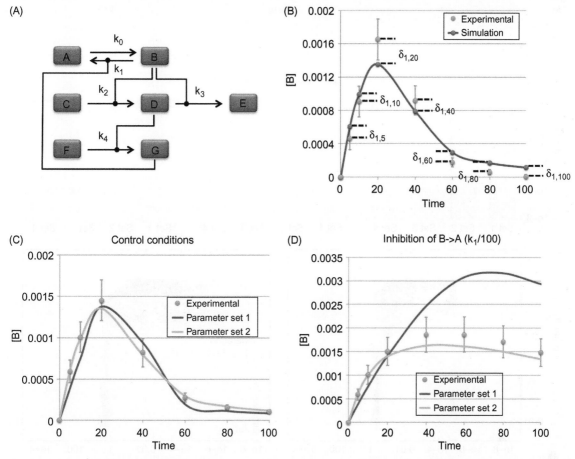

FIGURE 17.6 **Parameter estimation.** (A) Toy model. The values of the parameters governing the system are fixed, and the system is simulated. A set of (time, concentration) values are chosen for the different molecules. Variability is added to these values to simulate an experimental standard deviation of 20%. These "experimental" values are then used to estimate the parameters. (B) The difference between each experimental point and its simulated counterpart is measured. The error is a function of these differences, usually obtained by summing the squares. (C) Comparison of experimental values (blue) and simulation of the toy model in control conditions using two parameter sets, obtained through parameter estimation, giving low errors, 0.088 for set 1 and 0.0054 for set 2. (D) Same comparison for a perturbed condition. The parameter sets used are those giving the simulations in C, the blue points show the values that should be obtained.

not equivalent, question is the sensitivity (opposite of the robustness): when one individual parameter is varied, how does that change the behavior of the system?

To evaluate the sensitivity of a parameter, the usual way is to change its value and compute the error between simulation and experimental data (that was used to estimate the parameters), a process called "local parameter sensitivity analysis" (LPSA). Most authors vary these parameters by between 1 and 10%. However, this does not allow for identifying another minimum for this parameter. In the authors' study of the angiotensin signaling system they proposed, instead, to vary the value of each individual parameter in

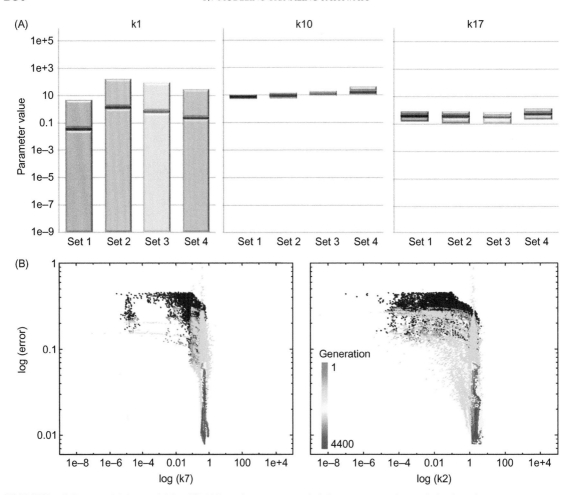

FIGURE 17.7   **Sensitivity and identifiability of parameters.** (A) Sensitivity analysis of the four low error parameter sets obtained for the angiotensin signaling system.[64] The darker bar corresponds to the estimated value of the parameter; the clear region shows the value region for each parameter giving errors lower than three-fold the minimal error. Two different behaviors can be observed: parameter k1 can be set to 0 in the four sets without changing much the error. The estimated values are different but the "tolerance" regions are very similar. Parameters k10 and k17 have close estimated values in the four sets. The tolerance regions for parameter k17 are larger than those for parameter k10, showing that k17 is more robust (less sensitive) than k10. (B) Identifiability analysis for parameters k2 and k7 of the toy system (Figure 17.6). The log of the error is plotted against the log of parameters k2 and k7 values for 4.4e + 6 parameter sets explored during parameter estimation.

a very large range: from $1e - 9$ to $1e + 5$. They then determined the region within which the error between simulation and experimental data remained within three-fold of the minimum obtained through parameter estimation (Figure 17.7). This methodology shows that there is never a second minimum. Two distinct patterns are found: parameters having the same value in all the parameter sets, and parameters that could be set to 0 without changing to overall behavior of the system. Finally, the width of the region where the error is less than three-fold of the minimum gives a measure of the sensitivity of the parameter: sensible parameters will give small regions, whereas robust parameters give large regions.

However, this type of sensitivity analysis only shows that there is a single minimum for each parameter when the other parameters remain constant. This does not prove that another unrelated parameter set does not exist. Ideally, this process, called "global parameter sensitivity analysis" (GPSA), would require computing the error for all the possible parameter sets, which is not feasible. Different methods have been proposed to address this question, the guiding principle being the generation of many, unrelated parameter sets, covering as well as possible the solution space.[88,89] Rather than applying one of these methods, the authors have exploited a peculiar property of their in-house parameter estimation method. Indeed, because they use evolutionary machine learning methods, a very large number of parameter sets are evaluated during optimization. Moreover, their method makes it possible to start with random parameter sets, and uses genetic algorithms, guaranteeing that the whole parameter space is reachable. During one optimization, the algorithm estimates 200 parameter sets at each generation for 4400 generations. In Figure 17.7B the values of parameters k7 and k2 have been plotted against the error within best parameter set of each generation, for 1000 optimizations of the toy model shown Figure 17.6. For each parameter there is a single minimum showing that they are all identifiable in practice. Again, the width of the funnel can be used to estimate the robustness of the parameter.

CONCLUSION

Modeling signaling pathways is a difficult but rewarding task. Experimental methods now exist to acquire dynamic data with precision and throughput compatible with mathematical modeling. This data acquisition step is essential and has to be carefully planned. Both the nature of the data to acquire, but also the frequency of measurements will condition the quality of the model. Computational tools for model building, reduction, and parameter estimation are now reaching a level where models become predictive, leading to unprecedented understanding of the cellular machinery and how the integration of multiple signals are processed in adapted biological response. Although a lot still has to be done, especially concerning the software tools and formal aspects of modeling, the models can already be used to explore the possible behaviors of the cell as a function of external stimuli, which opens the way to rational design of a new generation of drugs with more targeted action.

References

1. Preisinger C, von Kriegsheim A, Matallanas D, Kolch W. Proteomics and phosphoproteomics for the mapping of cellular signalling networks. *Proteomics* 2008;**8** (21):4402–15.
2. Kell DB. Theodor Bucher Lecture. Metabolomics, modelling and machine learning in systems biology — towards an understanding of the languages of cells. Delivered on 3 July 2005 at the 30th FEBS Congress and the 9th IUBMB conference in Budapest. *Febs J.* 2006;**273** (5):873–94.
3. Kholodenko B, Yaffe MB, Kolch W. Computational approaches for analyzing information flow in biological networks. *Sci Signal.* 2012;**5**(220):re1.
4. Gesty-Palmer D, Chen M, Reiter E, Ahn S, Nelson CD, Wang S, et al. Distinct beta-arrestin- and G protein-dependent pathways for parathyroid hormone receptor-stimulated ERK1/2 activation. *J Biol Chem* 2006;**281** (16):10856–64.
5. Kholodenko BN, Hancock JF, Kolch W. Signalling ballet in space and time. *Nat Rev Mol Cell Biol* 2010; **11**(6):414–26.
6. Reiter E, Lefkowitz RJ. GRKs and beta-arrestins: roles in receptor silencing, trafficking and signaling. *Trends Endocrinol Metab* 2006;**17**(4):159–65.

7. Ahn S, Shenoy SK, Wei H, Lefkowitz RJ. Differential kinetic and spatial patterns of beta-arrestin and G protein-mediated ERK activation by the angiotensin II receptor. *J Biol Chem* 2004;**279**(34):35518−25.

8. Kara E, Crepieux P, Gauthier C, Martinat N, Piketty V, Guillou F, et al. A phosphorylation cluster of five serine and threonine residues in the C-terminus of the follicle-stimulating hormone receptor is important for desensitization but not for beta-arrestin-mediated ERK activation. *Mol Endocrinol* 2006;**20**(11): 3014−26.

9. Barnes WG, Reiter E, Violin JD, Ren XR, Milligan G, Lefkowitz RJ. β-Arrestin 1 and Gαq/11 coordinately activate RhoA and stress fiber formation following receptor stimulation. *J Biol Chem* 2005;**280**(9): 8041−50.

10. Shenoy SK, Drake MT, Nelson CD, Houtz DA, Xiao K, Madabushi S, et al. beta-arrestin-dependent, G protein-independent ERK1/2 activation by the beta2 adrenergic receptor. *J Biol Chem* 2006;**281**(2):1261−73.

11. Lyons-Weiler J, Patel S, Bhattacharya S. A classification-based machine learning approach for the analysis of genome-wide expression data. *Genome Res* 2003; **13**(3):503−12.

12. Wehbi VL, Stevenson HP, Feinstein TN, Calero G, Romero G, Vilardaga JP. Noncanonical GPCR signaling arising from a PTH receptor-arrestin-Gbetagamma complex. *Proc Natl Acad Sci U S A* 2013;**110**(4):1530−5.

13. Vilardaga JP, Gardella TJ, Wehbi VL, Feinstein TN. Non-canonical signaling of the PTH receptor. *Trends Pharmacol Sci* 2012;**33**(8):423−31.

14. Calebiro D, Nikolaev VO, Gagliani MC, de Filippis T, Dees C, Tacchetti C, et al. Persistent cAMP-signals triggered by internalized G-protein-coupled receptors. *PLoS Biol* 2009;**7**(8):e1000172.

15. Irannejad R, Tomshine JC, Tomshine JR, Chevalier M, Mahoney JP, Steyaert J, et al. Conformational biosensors reveal GPCR signalling from endosomes. *Nature* 2013;**495**(7442):534−8.

16. Komarova NL, Zou X, Nie Q, Bardwell L. A theoretical framework for specificity in cell signaling. *Mol Syst Biol*. 2005;**1**:0023.

17. McClean MN, Mody A, Broach JR, Ramanathan S. Cross-talk and decision making in MAP kinase pathways. *Nat Genet* 2007;**39**(3):409−14.

18. Behar M, Hao N, Dohlman HG, Elston TC. Dose-to-duration encoding and signaling beyond saturation in intracellular signaling networks. *PLoS Comput Biol* 2008;**4**(10):e1000197.

19. Novak B, Tyson JJ. Design principles of biochemical oscillators. *Nat Rev Mol Cell Biol* 2008;**9**(12):981−91.

20. Goldbeter A. Computational approaches to cellular rhythms. *Nature* 2002;**420**(6912):238−45.

21. Martiel JL, Goldbeter AA. Model based on receptor desensitization for cyclic AMP signaling in dictyostelium cells. *Biophys J*. 1987;**52**(5):807−28.

22. Goldbeter A. Mechanism for oscillatory synthesis of cyclic AMP in Dictyostelium discoideum. *Nature* 1975;**253**(5492):540−2.

23. Meyer T, Stryer L. Molecular model for receptor-stimulated calcium spiking. *Proc Natl Acad Sci USA* 1988;**85**(14):5051−5.

24. Nelson DE, Ihekwaba AE, Elliott M, Johnson JR, Gibney CA, Foreman BE, et al. Oscillations in NF-kappaB signaling control the dynamics of gene expression. *Science* 2004;**306**(5696):704−8.

25. Hoffmann A, Levchenko A, Scott ML, Baltimore D. The IkappaB-NF-kappaB signaling module: temporal control and selective gene activation. *Science* 2002;**298** (5596):1241−5.

26. Ma L, Wagner J, Rice JJ, Hu W, Levine AJ, Stolovitzky GA. A plausible model for the digital response of p53 to DNA damage. *Proc Natl Acad Sci USA* 2005; **102**(40):14266−71.

27. Monk NA. Oscillatory expression of Hes1, p53, and NF-kappaB driven by transcriptional time delays. *Curr Biol* 2003;**13**(16):1409−13.

28. Ciliberto A, Novak B, Tyson JJ. Steady states and oscillations in the p53/Mdm2 network. *Cell Cycle* 2005;**4** (3):488−93.

29. Jacquet M, Renault G, Lallet S, De Mey J, Goldbeter A. Oscillatory nucleocytoplasmic shuttling of the general stress response transcriptional activators Msn2 and Msn4 in Saccharomyces cerevisiae. *J Cell Biol* 2003;**161**(3):497−505.

30. Garmendia-Torres C, Goldbeter A, Jacquet M. Nucleocytoplasmic oscillations of the yeast transcription factor Msn2: evidence for periodic PKA activation. *Curr Biol* 2007;**17**(12):1044−9.

31. Janes KA, Albeck JG, Gaudet S, Sorger PK, Lauffenburger DA, Yaffe MB. A systems model of signaling identifies a molecular basis set for cytokine-induced apoptosis. *Science* 2005;**310**(5754):1646−53.

32. Gan EH, Quinton R. Physiological significance of the rhythmic secretion of hypothalamic and pituitary hormones. *Prog Brain Res* 2010;**181**:111−26.

33. Bonnefont X. Circadian timekeeping and multiple timescale neuroendocrine rhythms. *J Neuroendocrinol* 2010;**22**(3):209−16.

34. Reiter E, Ahn S, Shukla AK, Lefkowitz RJ. Molecular mechanism of beta-arrestin-biased agonism at seven-transmembrane receptors. *Annu Rev Pharmacol Toxicol* 2012;**52**:179−97.

35. Zheng H, Shen H, Oprea I, Worrall C, Stefanescu R, Girnita A, et al. β-Arrestin-biased agonism as the central mechanism of action for insulin-like growth factor

1 receptor-targeting antibodies in Ewing's sarcoma. *Proc Natl Acad Sci USA* 2012;**109**(50):20620−5.

36. Girnita A, Zheng H, Gronberg A, Girnita L, Stahle M. Identification of the cathelicidin peptide LL-37 as agonist for the type I insulin-like growth factor receptor. *Oncogene* 2012;**31**(3):352−65.

37. Whalen EJ, Rajagopal S, Lefkowitz RJ. Therapeutic potential of beta-arrestin- and G protein-biased agonists. *Trends Mol Med* 2011;**17**(3):126−39.

38. Mann M, Jensen ON. Proteomic analysis of post-translational modifications. *Nat Biotechnol* 2003;**21**(3):255−61.

39. Linding R, Jensen LJ, Ostheimer GJ, van Vugt MA, Jorgensen C, Miron IM, et al. Systematic discovery of in vivo phosphorylation networks. *Cell* 2007;**129**(7):1415−26.

40. Olsen JV, Blagoev B, Gnad F, Macek B, Kumar C, Mortensen P, et al. Global, in vivo, and site-specific phosphorylation dynamics in signaling networks. *Cell* 2006;**127**(3):635−48.

41. Witze ES, Old WM, Resing KA, Ahn NG. Mapping protein post-translational modifications with mass spectrometry. *Nat Methods* 2007;**4**(10):798−806.

42. Gembitsky DS, Lawlor K, Jacovina A, Yaneva M, Tempst P. A prototype antibody microarray platform to monitor changes in protein tyrosine phosphorylation. *Mol Cell Proteomics* 2004;**3**(11):1102−18.

43. Sheehan KM, Calvert VS, Kay EW, Lu Y, Fishman D, Espina V, et al. Use of reverse phase protein microarrays and reference standard development for molecular network analysis of metastatic ovarian carcinoma. *Mol Cell Proteomics* 2005;**4**(4):346−55.

44. Spurrier B, Ramalingam S, Nishizuka S. Reverse-phase protein lysate microarrays for cell signaling analysis. *Nat Protoc* 2008;**3**(11):1796−808.

45. Dupuy L, Gauthier C, Durand G, Musnier A, Heitzler D, Herledan A, et al. A highly sensitive near-infrared fluorescent detection method to analyze signalling pathways by reverse-phase protein array. *Proteomics* 2009;**9**(24):5446−54.

46. Ciaccio MF, Wagner JP, Chuu CP, Lauffenburger DA, Jones RB. Systems analysis of EGF receptor signaling dynamics with microwestern arrays. *Nat Methods* 2010;**7**(2):148−55.

47. Fan AC, Deb-Basu D, Orban MW, Gotlib JR, Natkunam Y, O'Neill R, et al. Nanofluidic proteomic assay for serial analysis of oncoprotein activation in clinical specimens. *Nat Med* 2009;**15**(5):566−71.

48. Bowen WP, Wylie PG. Application of laser-scanning fluorescence microplate cytometry in high content screening. *Assay Drug Dev Technol* 2006;**4**(2):209−21.

49. Stelzl U, Worm U, Lalowski M, Haenig C, Brembeck FH, Goehler H, et al. A human protein-protein interaction network: a resource for annotating the proteome. *Cell* 2005;**122**(6):957−68.

50. Gavin AC, Bosche M, Krause R, Grandi P, Marzioch M, Bauer A, et al. Functional organization of the yeast proteome by systematic analysis of protein complexes. *Nature* 2002;**415**(6868):141−7.

51. Barrios-Rodiles M, Brown KR, Ozdamar B, Bose R, Liu Z, Donovan RS, et al. High-throughput mapping of a dynamic signaling network in mammalian cells. *Science* 2005;**307**(5715):1621−5.

52. Allen MD, DiPilato LM, Ananthanarayanan B, Newman RH, Ni Q, Zhang J. Dynamic visualization of signaling activities in living cells. *Sci Signal* 2008;**1**(37):pt6.

53. Giepmans BN, Adams SR, Ellisman MH, Tsien RY. The fluorescent toolbox for assessing protein location and function. *Science* 2006;**312**(5771):217−24.

54. Lohse MJ, Nikolaev VO, Hein P, Hoffmann C, Vilardaga JP, Bunemann M. Optical techniques to analyze real-time activation and signaling of G-protein-coupled receptors. *Trends Pharmacol Sci* 2008;**29**(3):159−65.

55. Radhakrishnan K, Halasz A, Vlachos D, Edwards JS. Quantitative understanding of cell signaling: the importance of membrane organization. *Curr Opin Biotechnol* 2010;**21**(5):677−82.

56. Charest PG, Terrillon S, Bouvier M. Monitoring agonist-promoted conformational changes of beta-arrestin in living cells by intramolecular BRET. *EMBO Rep* 2005;**6**(4):334−40.

57. Gales C, Rebois RV, Hogue M, Trieu P, Breit A, Hebert TE, et al. Real-time monitoring of receptor and G-protein interactions in living cells. *Nat Methods* 2005;**2**(3):177−84.

58. Azizi F, Mastrangelo CH. Generation of dynamic chemical signals with pulse code modulators. *Lab Chip* 2008;**8**(6):907−12.

59. Hung PJ, Lee PJ, Sabounchi P, Lin R, Lee LP. Continuous perfusion microfluidic cell culture array for high-throughput cell-based assays. *Biotechnol Bioeng* 2005;**89**(1):1−8.

60. Wehbi V, Tranchant T, Durand G, Musnier A, Decourtye J, Piketty V, et al. Partially deglycosylated equine LH preferentially activates beta-arrestin-dependent signaling at the follicle-stimulating hormone receptor. *Mol Endocrinol* 2010;**24**(3):561−73.

61. Castel D, Debily MA, Pitaval A, Gidrol X. Cell microarray for functional exploration of genomes. *Methods Mol Biol* 2007;**381**:375−84.

62. Erfle H, Neumann B, Rogers P, Bulkescher J, Ellenberg J, Pepperkok R. Work flow for multiplexing siRNA assays

by solid-phase reverse transfection in multiwell plates. *J Biomol Screen* 2008;**13**(7):575–80.

63. Starkuviene V, Pepperkok R, Erfle H. Transfected cell microarrays: an efficient tool for high-throughput functional analysis. *Expert Rev Proteomics* 2007;**4**(4):479–89.

64. Heitzler D, Durand G, Gallay N, Rizk A, Ahn S, Kim J, et al. Competing G protein-coupled receptor kinases balance G protein and beta-arrestin signaling. *Mol Syst Biol* 2012;**8**:590.

65. Matthews L, Gopinath G, Gillespie M, Caudy M, Croft D, de Bono B, et al. Reactome knowledgebase of human biological pathways and processes. *Nucleic Acids Res* 2009;**37**(Database issue):D619–22.

66. Kelder T, van Iersel MP, Hanspers K, Kutmon M, Conklin BR, Evelo CT, et al. WikiPathways: building research communities on biological pathways. *Nucleic Acids Res* 2012;**40**(Database issue):D1301–7.

67. Kandasamy K, Mohan SS, Raju R, Keerthikumar S, Kumar GS, Venugopal AK, et al. NetPath: a public resource of curated signal transduction pathways. *Genome Biol* 2010;**11**(1):R3.

68. Li C, Donizelli M, Rodriguez N, Dharuri H, Endler L, Chelliah V, et al. BioModels database: an enhanced, curated and annotated resource for published quantitative kinetic models. *BMC Syst Biol* 2010;**4**:92.

69. De Smet R, Marchal K. Advantages and limitations of current network inference methods. *Nat Rev Microbiol* 2010;**8**(10):717–29.

70. Kaderali L, Dazert E, Zeuge U, Frese M, Bartenschlager R. Reconstructing signaling pathways from RNAi data using probabilistic Boolean threshold networks. *Bioinformatics* 2009;**25**(17):2229–35.

71. Mitsos A, Melas IN, Siminelakis P, Chairakaki AD, Saez-Rodriguez J, Alexopoulos LG. Identifying drug effects via pathway alterations using an integer linear programming optimization formulation on phosphoproteomic data. *PLoS Comput Biol* 2009;**5**(12):e1000591.

72. Aslaoui-Errafi Z, Cohen-Boulakia S, Froidevaux C, Gloaguen P, Poupon A, Rougny A, et al. Towards a logic-based method to infer provenance-aware molecular networks. 1st ECML/PKDD International workshop on Learning and Discovery in Symbolic Systems Biology (LDSSB). 2012.

73. Funahashi A, Matsuoka Y, Jouraku A, Morohashi M, Kikuchi N, Kitano H. CellDesigner 3.5: A versatile modeling tool for biochemical networks. *IEEE* 2008; 1254–65.

74. Mendes P, Hoops S, Sahle S, Gauges R, Dada J, Kummer U. Computational modeling of biochemical networks using COPASI. *Methods Mol Biol* 2009;**500**:17–59.

75. Hucka M, Finney A, Sauro HM, Bolouri H, Doyle JC, Kitano H, et al. The systems biology markup language (SBML): a medium for representation and exchange of biochemical network models. *Bioinformatics* 2003;**19**(4):524–31.

76. Anderson J, Chang YC, Papachristodoulou A. Model decomposition and reduction tools for large-scale networks in systems biology. *Automatica* 2011;**47**(6):1165–74.

77. Gay S, Soliman S, Fages F. A graphical method for reducing and relating models in systems biology. *Bioinformatics* 2010;**26**(18):i575–81.

78. Violin JD, Dewire SM, Barnes WG, Lefkowitz RJ. G protein-coupled receptor kinase and beta-arrestin-mediated desensitization of the angiotensin II type 1A receptor elucidated by diacylglycerol dynamics. *J Biol Chem* 2006;**281**(47):36411–9.

79. Moles CG, Mendes P, Banga JR. Parameter estimation in biochemical pathways: a comparison of global optimization methods. *Genome Res* 2003;**13**(11):2467–74.

80. Michalewicz Z. *Genetic algorithms + data structures = evolution programs*. Berlin, Germany: Springer-Verlag; 1994.

81. Fogel D, Fogel L, Atmar J. Meta-evolutionary programming. Signals, Systems and Computers. Asilomar; 1991: p. 540–45.

82. Hooke R, Jeeves T. "Direct search" solution of numerical and statistical problems. *JACM* 1961;**8**(2):212–29.

83. Goldfeld S, Quandt R, Trotter H. Maximization by quadratic hill-climbing. *Econometrica* 1966;**34**:541–51.

84. Brent P. *New algorithm for minimizing a function of several variables without calculating derivatives. Algorithms for minimization without derivatives.* Englewood Cliffs, NJ, USA: Courier Dover Publications; 1973. p. 117–67.

85. Kennedy J, Eberhart R. Particle swarm optimization. Fourth IEEE International Conference on Neural Networks; 1995: p. 1942–48.

86. Calzone L, Fages F, Soliman S. BIOCHAM: an environment for modeling biological systems and formalizing experimental knowledge. *Bioinformatics* 2006;**22**(14):1805–7.

87. Hansen N, Müller S, Koumoutsakos P. Reducing the time complexity of the derandomized evolution strategy with covariance matrix adaptation (cma-es). *Evol Comput* 2003;**11**(1):1–18.

88. Rodriguez-Fernandez M, Banga JR, Doyle JC. Novel global sensitivity analysis methodology accounting for the crucial role of the distribution of input parameters: application to systems biology models. *Int J Robust Nonlinear Control* 2011;**0**:1–18.

89. van Riel NA. Dynamic modelling and analysis of biochemical networks: mechanism-based models and model-based experiments. *Brief Bioinform* 2006;**7**(4):364–74.

Defects in Ovarian Steroid Hormone Biosynthesis

Jerome F. Strauss III*, Bhavi Modi* and Jan M. McAllister†

Virginia Commonwealth University, Richmond, VA, USA, †Pennsylvania State University College of Medicine, Hershey, PA, USA

INTRODUCTION

The follicle and its derivative structure, the corpus luteum, are the main steroidogenic structures in the ovary during reproductive life. Ovarian hilar cells also produce steroid hormones and may be a significant site of hormone production in the post-menopausal ovary. These compartments elaborate three classes of steroid hormones: progestins, androgens and estrogens. The amount of each type of steroid hormone produced at a given time largely depends upon the number of follicles undergoing maturation and the stage of maturation, which is programmed either by endogenous pituitary gonadotropins or, in the case of assisted reproduction, exogenous gonadotropins; the adequacy of gonadotropic support; the number of follicles that ovulate and/or luteinize to form corpora lutea; and during the luteal phase, the availability of the steroid hormone precursor, cholesterol.

Pathological conditions, including certain ovarian tumors, and benign conditions that disturb follicular development lead to abnormal patterns of steroid hormone production. The most common benign disorder associated with arrested follicular development and abnormal steroid production is polycystic ovary syndrome (PCOS). It is the leading cause of anovulatory infertility. Rare mutations in genes involved in steroid hormone synthesis and its regulation are also associated with disordered patterns of pituitary gonadotropin release that prevent ovulation and lead to the formation of large follicular cysts due to the lack of an appropriately programmed ovulatory signal.

Fundamentals of Steroidogenesis

Steroid hormone synthesis involves a series of sequential modifications of cholesterol, the precursor molecule, that clip off the side chain,

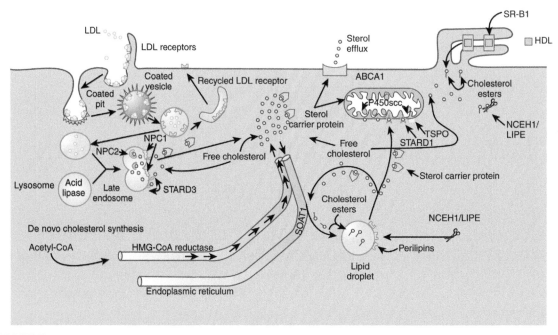

FIGURE 18.1　**The acquisition, storage, and trafficking of cholesterol in steroidogenic cells.** ABCA1, ATP-binding cassette transporter A1; FFA, free fatty acid; HDL, high-density lipoprotein; HMG-CoA, 3-hydroxy-3-methyl-glutaryl-coenzyme A; LDL, low-density lipoprotein; LIPE, hormone-sensitive lipase; NCEH1, neutral pH cholesterol ester hydrolase; SR-B1, scavenger receptor type B; STARD1, steroidogenic acute regulatory protein; STARD3, (steroidogenic acute regulatory protein)-related lipid transfer domain 3; sterol carrier proteins include sterol carrier protein$_2$, STARD4, and STARD5; SOAT1, sterol-O-acyltransferase-1; TSPO, translocator protein.

alter the location of olefinic bonds, and add hydroxyl groups. The process involves conversion of cholesterol (a cholestane, 27 carbons) to a pregnane (21 carbons), then to an androstane (19 carbons), and finally, to an estrane (18 carbons) backbone.

The machinery for steroidogenesis is compartmentalized at the organ, cellular, and subcellular levels, which has important implications for the control of steroid hormone production.[1-4] Specific cell types can accomplish several of the sequential steps in metabolism of cholesterol, but rarely can they generate an estrogen from cholesterol. The requirement for cooperative efforts by two different tissues or cell types is a characteristic of estrogen biosynthesis. This joint effort enables independent control of the cells involved in androgen precursor synthesis and aromatization.[5]

Steroidogenic cells have ultrastructural features that facilitate the uptake of cholesterol from blood lipoproteins, *de novo* synthesis of cholesterol, or its storage in cytoplasmic lipid droplets for future use in steroidogenesis. A number of proteins are involved in intracellular trafficking of cholesterol (Figure 18.1).[3,4] Unlike protein hormone-producing endocrine cells, steroid hormone-producing cells do not store prefabricated hormone; they synthesize the hormones on demand from cholesterol that has been acquired from the plasma, synthesized *de novo*, or stored in membranes or as sterol esters in lipid droplets.

The manufacture of bioactive steroid hormones requires the activity of several classes of enzymes: the cytochrome P450s (named because of their distinctive absorption peak at 450 nm when reduced in the presence of carbon monoxide) and the hydroxysteroid dehydrogenases.[3,6] Cytochrome P450s catalyze the major alterations in the sterol backbone: cleavage of the side chain, hydroxylations, and aromatization. These heme-containing proteins require molecular oxygen and a source of electrons to complete a catalytic cycle. The hydroxysteroid dehydrogenases reduce ketone groups or oxidize hydroxyl functions, employing pyridine nucleotide cofactors, usually with a stereospecific substrate preference and reaction direction.

The rate-limiting step in steroidogenesis is the movement of cholesterol into the mitochondria, a process mediated by the steroidogenic acute regulatory protein, encoded by the *STARD1* gene (Figure 18.1).[7] The first enzymatic step in steroid hormone synthesis takes place in the inner mitochondrial membranes, catalyzed by the P450 cholesterol side chain cleavage enzyme (P450scc), encoded by the *CYP11A1* gene, in which cholesterol is converted into pregnenolone (P5). Metabolism of P5 in the ovary can occur along two different pathways: the delta 4 pathway in which P5 is converted into progesterone (P4) by 3β-hydroxysteroid dehydrogenase 2 (HSD3B2); or the delta 5 pathway in which further metabolism of P5 occurs before the product is acted on by HSD3B2. The significance of these two pathways is that P5 is the preferred substrate for the human enzyme that converts pregnanes into androstanes, cytochrome P450 17α-hydroxylase (P450c17), encoded by the *CYP17A1* gene, which has both 17α-hydroxylase and 17,20-lyase activity.

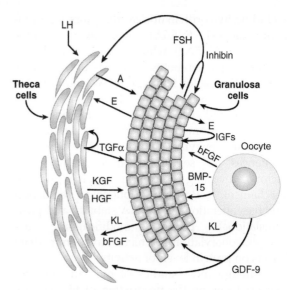

FIGURE 18.2 **Paracrine and autocrine interactions between theca cells, granulosa cells, and the oocyte.** A, androstenedione; BMP, bone morphogenetic protein; E, estradiol; FGF, fibroblast growth factor; FSH, follicle-stimulating hormone; GDF, growth differentiation factor; HGF, hepatocyte growth factor; IGF, insulin-like growth factors; KGF, keratinocyte-derived growth factor; KL, kit ligand; LH, luteinizing hormone; TGF, transforming growth factor.

The expression of genes encoding proteins involved in steroidogenesis and their activity are governed by the gonadotropins, luteinizing hormone (LH) and follicle stimulating hormone (FSH), that trigger intracellular signaling cascades, including those involving protein kinase A and Akt (protein kinase B), through activation of their respective receptors. Gonadotropin signaling is also modulated (either dampened or enhanced) by growth factors produced locally in the ovary (Figure 18.2). Gonadotropins and their modulators influence the steroidogenic capacity of ovarian cells through transcriptional and post-transcriptional mechanisms. In transcriptional regulation, the signaling pathways converge to activate several key transcription factors, including two orphan nuclear receptors, steroidogenic factor-1 (NR5A1) and liver receptor homolog-1 (NR5A2), which recognize similar *cis* element sequence motifs, members of the GATA transcription factor family, GATA4 and GATA6, sterol response-element binding proteins (SREBP1) and CCAT/enhancer binding protein.[8,9] The promoters of a number of the steroidogenic enzyme genes contain binding sequences for several of these transcription factors, revealing a combinatorial regulatory mechanism. Post-transcriptional regulation involves alterations in mRNA stability and post-translational protein modifications (e.g., phosphorylation), which have immediate effects on protein levels or activity.[10,11]

Although most studies of ovarian steroidogenesis focus on the production of biologically active hormones, ovarian steroidogenic cells secrete metabolites of these hormones that lack the classical hormone activity. For example, the human corpus luteum secretes large quantities of 5α-dihydroprogesterone, a molecule that has no progestational activity, but can modulate neuronal γ-aminobutyric acid (GABA) receptors.[12] The corpus luteum can metabolize estradiol into 2-methoxyestradiol, a steroid that regulates angiogenesis.[13]

Steroidogenic Cells of the Ovary

Granulosa and Granulosa Lutein Cells

Granulosa cells are believed to originate from the ovarian surface epithelium.[14] These pregranulosa cells surround primordial oocytes. The cohort of granulosa cells that surrounds each oocyte has an oligoclonal origin.[15] Three to five parent cells are estimated to give rise to the full complement of granulosa cells in a mature follicle.[15] In the follicle, granulosa cells receive no direct blood supply. They rest on a basal lamina that separates them from the vascularized theca interna. Consequently, there is a relative

blood–follicle barrier that restricts access of high-molecular-weight substances, such as low-density lipoproteins (LDL), to the granulosa cell compartment.

The key steroid hormone produced by granulosa cells is estradiol. The synthesis of this hormone requires a collaborative relationship between theca cells surrounding the follicle, which produce androgens (i.e., dehydroepiandrosterone (DHEA), androstenediol, androstenedione, testosterone) in response to LH, which then diffuse into granulosa cells and are converted to estrogens (i.e., estrone, estradiol), by cytochrome P450 aromatase (CYP19A1) in granulosa cells, in response to FSH (Figure 18.3).

One of the major actions of FSH is the induction of aromatase expression. FSH also induces expression of NADPH cytochrome P450 reductase, which transfers electrons to aromatase; HSD3B2 which converts DHEA to androstenedione, and type 1 17β-hydroxysteroid dehydrogenase (HSD17B1), the "estrogenic" 17β-HSD that reduces estrone to estradiol. Studies of isolated granulosa cells from preovulatory follicles have shown that FSH, but not LH, stimulates estrogen production when the cells are provided with an aromatizable substrate.[16] Isolated human theca cells do not produce any appreciable amounts of estrogen. Aromatase activity in granulosa cells is estimated to be at least 700 times greater in large preovulatory follicles than in the associated theca cells.

Granulosa cells display different phenotypes depending on their location in the follicle.[17–19] The mural granulosa cells, antral granulosa cells, and cumulus cells surrounding the oocyte each have distinguishing features that are likely determined by their proximity to the oocyte and theca cells, and by the paracrine substances that they produce. Mural granulosa cells in the antral follicle have the greatest steroidogenic activity. In addition, mural granulosa cells in the preovulatory follicle have the highest level of LH

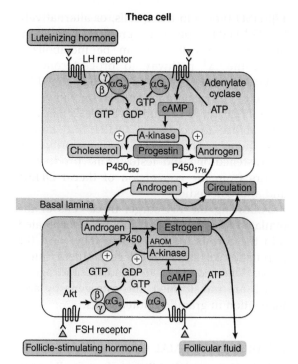

Theca cell

Granulosa cell

FIGURE 18.3   The two-cell–two-gonadotropin system for estradiol synthesis in the follicle. Luteinizing hormone (LH) and follicle-stimulating hormone (FSH) are shown to stimulate adenylate cyclase via G protein-coupled receptors. The cyclic adenosine monophosphate (cAMP) generated from adenosine triphosphate (ATP) activates protein kinase A to stimulate expression of the respective steroidogenic enzymes in theca and granulosa cells. In addition, in granulosa cells, FSH binding to the FSH receptor leads to activation of AKT, also known as protein kinase B, probably via a phosphatidyl inositol second messenger, which augments aromatase expression. GDP, guanosine diphosphate; GTP, guanosine triphosphate.

receptors. Cumulus cells express low amounts of aromatase, and their LH receptor content and level of LH responsiveness are substantially lower than those of their mural counterparts.

FSH induces LH receptors in the granulosa cells of the preovulatory follicle. Luteinizing

hormone/choriogonadotropin receptor (LHCGR) mRNA is detectable in antral follicles, reaching maximal levels in granulosa cells in preovulatory follicles. In contrast, FSH receptor (FSHR) mRNA levels decline in granulosa cells as follicular diameter increases. Consequently, in the late stages of follicular maturation, LH can take over FSH's function in propelling follicular maturation.[20]

The follicle "chosen" to ovulate synthesizes estradiol in significant amounts resulting in passage of this hormone into the systemic circulation. Asymmetry in ovarian estrogen secretion related to the emergence of the dominant follicle is detectable as early as day 5 to 7 of the cycle.[21,22] In the late follicular phase, the intrafollicular concentrations of estradiol are directly correlated with follicular size, and achieve concentrations of approximately $1\,\mu g/mL$ at a time when circulating estradiol levels reach their peak concentrations.[22–25]

With the ovulatory LH surge, granulosa cells undergo terminal differentiation into granulosa lutein cells, representing the large luteal cell population of the corpus luteum. Their major steroid secretory product is P4. Luteinization of granulosa cells is associated with an upregulation of the machinery to acquire cholesterol from circulating lipoproteins and convert it into progestins. There is an increase in LDL receptors, and expression of the *STARD1*, *CYP11A1* and *HSD3B2* genes. In addition, the granulosa lutein cells also retain the capacity to synthesize estrogens from androgen precursors produced by theca lutein cells. The steroidogenesis of both granulosa lutein and theca lutein cells is dependent upon LH stimulation, and suppression of LH secretion markedly reduces progestin and estrogen output from the corpus luteum.

Theca and Theca Lutein Cells

The theca and theca lutein cells are the primary contributors to androgen production in the normal ovary. The fate of androgen producing cells in the ovary appears to be regulated by WNT4, a member of the wingless transcription factor family. When WNT4 is inactivated in humans as a consequence missense mutations, ovarian androgen production is increased, in part due to reduced repression of expression of steroidogenic enzymes involved in androgen biosynthesis.[26,27]

The theca and interstitial cells are believed to arise from fibroblast-like mesenchymal cells in the stromal compartment. They first appear around follicles that have two or more layers of granulosa cells.[28,29] The Kit and Kit ligand system is thought to play a prominent role in early development of the androgen producing cells of the ovary and testis.[29] Theca cells express Kit, the receptor for Kit ligand, and researchers have postulated that Kit ligand produced by granulosa cells is important in organizing the theca layer around the developing follicle. Production of growth differentiation factor-9 (GDF-9) by the oocyte is required for development of the thecal layer.[30] In the absence of GDF-9, theca cells are recruited to surround the follicle but do not undergo proper functional differentiation.

Theca cells engage in a bidirectional dialog with granulosa cells through the production of growth factors, including keratinocyte-derived growth factor (KGF) and hepatocyte growth factor (HGF). KGF and HGF stimulate granulosa cells to produce Kit ligand, whereas Kit ligand acts on the theca cells to promote expression of KGF and HGF in a positive feedback loop.

Androgen biosynthesis in theca cells is stimulated by LH, which increases cholesterol movement to the mitochondria and its conversion into P5. In human theca cells, androgen biosynthesis occurs predominately via the Δ^5 pathway in which P5 is converted to 17α-hydroxypregnenolone and DHEA by P450c17, then further converted to androstenediol by 17β-HSD type 5 in theca cells, or alternatively by 17β-HSD type 1 in granulosa cells. In human theca cells, P5 can be metabolized down the Δ^4 pathway to progesterone by HSD3B2, and further converted to 17α-hydroxy progesterone (17OHP4); however, the C17,20 lyase of human P450c17 has limited ability to convert 17OHP4 into androstenedione. Therefore, the majority of androstenedione and testosterone production occurs from the conversion of thecal DHEA and androstenediol.

Following the LH surge, theca cells of the ovulated follicle luteinize and are incorporated into the corpus luteum, where they form the population of small luteal cells (theca lutein cells). They continue to produce androgens, which are transformed into estrogens by aromatization in granulosa lutein cells.

HILAR AND STROMAL CELLS

Hilar interstitial cells are large cells with structural and functional characteristics indistinguishable from those of differentiated Leydig cells. The endocrine activity of these cells is thought to be most prominent at the time of puberty, during pregnancy, and around menopause.

The ovarian stroma contains fibroblastic cells that express some steroidogenic enzymes and may also harbor precursors of androgen producing cells. Ovarian stroma cells express androgen receptors, and may proliferate under the influence of androgens, contributing to the increased stromal density characteristic of hyperandrogenemia of ovarian origin (i.e., PCOS).

Stromal cells serve as insulators in the ovary, separating follicles and corpora lutea from adjacent structures physically, as well as biochemically. The biochemical insulator function of stromal cells is mediated, in part, by the production of growth factors and growth factor-binding proteins.

MUTATIONS AND POLYMORPHISMS IN GENES AFFECTING OVARIAN STEROIDOGENESIS

Mutations in Gonadotropin Receptor Genes

FSHR

FSH is required for the transition of secondary preantral follicles to the antral stage. It also is a survival factor for antral follicles, and its withdrawal triggers programmed cell death in the absence of local factors that sensitize the follicle to FSH action or amplify its effects.[31,32] Follicular maturation initiated at the start of a new menstrual cycle is driven by an increase in FSH levels in the late luteal phase as progesterone, estradiol, and inhibin A levels fall. Preantral follicles apparently require a threshold FSH concentration to sustain growth, and this threshold level is reached during the late luteal phase. Remarkably, the threshold can be crossed with as little as a 10 to 30% increase in FSH, indicating that granulosa cells have a highly sensitive detection system that interprets circulating FSH levels. FSH can induce follicular growth to a preovulatory size of at least 17 mm in the virtual absence of LH. Although estradiol production is greatly impaired under these circumstances, inhibin production is induced, reflecting the normal response of granulosa cells to FSH.

The importance of FSH in follicular development has been documented by the discovery of mutations that inactivate the FSH β subunit and the FSH receptor in humans, and by targeted deletions of these genes in mice.[33,34] Women with inactivating mutations of the FSH receptor gene have the features of hypergonadotropic hypogonadism, with absent or poor development of secondary sexual characteristics, and high FSH and LH levels. In the absence of the functional FSH β subunit or FSH receptor, the ovaries are small and follicular development generally goes no further than the preantral stage.

Conversely, mutations in the transmembrane helices of the FSH receptor and the extracellular domain are associated with spontaneous ovarian hyperstimulation syndrome.[35] These mutations in the transmembrane helices result in ligand promiscuity, allowing the receptor to respond to endogenous human chorionic gonadotropin (hCG) and thyroid stimulating hormone (TSH). Mutations in the extracellular domain display increased specificity and sensitivity to hCG.

Polymorphisms in the FSHR gene have been associated with response of the gonads to gonadotropins, secretion and with PCOS, as described later in this chapter.

LHCGR

F Genetic females with inactivating LHCGR mutations have female external genitalia, and spontaneous breast development and pubic hair at puberty. They have normal or late menarche, followed by oligo/amenorrhea and infertility. Levels of estradiol and P4 are similar to those in the early to mid-follicular phase, but do not reach levels characteristic of the preovulatory or luteal phases. LH levels are elevated, but FSH levels are either normal or slightly increased. Enlarged ovaries with multiple cysts reflect the disordered follicular development resulting from the abnormal pattern of gonadotropin secretion.

Women with activating mutations of the LH receptor have no obvious reproductive phenotype, in contrast to the precocious puberty found in males resulting from Leydig cell elaboration of testosterone. The absence of hyperandrogenemia in female with these activating mutations probably reflects the fact that theca cell function is governed by factors other than LH. Moreover, premature luteinization of granulosa cells does not evidently occur. Genome-wide association studies (GWAS) and candidate gene studies have found associations with variants in the *LHCGR* and PCOS as described below.

Mutations and Polymorphisms in Genes Affecting Cholesterol Availability

Cholesterol, the precursor of steroid hormones, is acquired by ovarian steroidogenic cells from *de novo* synthesis and uptake of lipoprotein-carried cholesterol from the circulation. The quantitative importance of circulating cholesterol carried by LDL, high-density lipoproteins (HDL), and other lipoproteins as steroid hormone precursor as opposed to *de novo* cholesterol synthesis, at least in structures that produce large quantities of hormone (corpus luteum), is demonstrated by the fact that radiolabeled plasma cholesterol in humans is almost fully equilibrated with the steroidogenic pool of cholesterol. Evidence for an important role of circulating lipoprotein cholesterol in steroidogenesis in the corpus luteum comes from the study of hypobetalipoproteinemia, a disorder in which there is virtually no circulating LDL.[36,37] This rare metabolic disease is associated with diminished P4 levels in the luteal phase and in pregnancy. However, individuals with familial hypercholesterolemia due to inactivating mutations in the LDL receptor have only modest impairment of steroidogenic gland function, reflecting the capacity of alternative sterol uptake mechanisms to compensate for LDL receptor deficiency.

HDL also provides cholesterol for hormone synthesis through uptake via the scavenger receptor type B, class (SCARB1).[5,38] Human heterozygote carriers of a functional P297S mutation have an attenuated adrenal glucocorticoid output. The impact on ovarian function has not been studied.

The commonly used statins (which inhibit 3-hydroxy-3-methylglutaryl coenzyme A reductase, the rate-limiting enzyme in *de novo* cholesterol synthesis) do not impair luteal steroidogenesis in adult humans despite the lowering of plasma LDL levels.[39,40]

Steroidogenic Acute Regulatory Protein (STARD1)

Translocation of cholesterol from the outer mitochondrial membranes to the relatively sterol-poor inner membranes is the critical step in steroidogenesis.[5,6] This translocation process occurs at modest rates in the absence of specific effectors. It is markedly enhanced by the steroidogenic acute regulatory protein (StAR or STARD1). The expression of STARD1 is directly correlated with steroidogenesis, and coexpression of STARD1 and the cholesterol side-chain cleavage enzyme system in cells that are not normally steroidogenic results in substantial P5 synthesis above that produced by cells expressing the cholesterol side-chain cleavage enzyme system alone. Definitive evidence of the importance of STARD1 was provided by the discovery that mutations in the *STARD1* gene cause congenital lipoid adrenal hyperplasia, a rare autosomal recessive disorder in which the synthesis of all adrenal and gonadal steroid hormones is severely impaired before the cholesterol side-chain cleavage step.

Human STARD1 is synthesized as a 285-amino acid protein. The *N*-terminus of STARD1 is characteristic of proteins synthesized in the cytoplasm and then imported into mitochondria. Newly synthesized STARD1 preprotein (37 kDa) is rapidly imported into mitochondria and processed to the mature 30-kDa form. The preprotein has a very short half-life (minutes), but the mature form is longer lived (hours).

STARD1 contains two consensus sequences for cyclic adenosine monophosphate (cAMP)-dependent protein kinase phosphorylation at serine 57 and serine 195. Serine 195 of human STARD1 must be phosphorylated for maximal steroidogenic activity in model systems.

The abundance of STARD1 protein in steroidogenic cells is determined primarily by the rate of *STARD1* gene transcription, which is influenced by transcription factors involved in the control of other genes involved in cholesterol metabolism (e.g., SREBPs, NR5A1, NR5A2, LXRs), although translational mechanisms may also contribute. In differentiated cells, the *STARD1* gene is activated by the cAMP signal transduction cascade within 15 to 30 min. In differentiating cells (e.g., luteinizing granulosa cells), the induction of *STARD1* transcription takes hours and requires ongoing protein synthesis.

STARD1 was initially believed to stimulate cholesterol movement from the outer to the inner mitochondrial membrane as it was imported into the mitochondria. The importation process was proposed to create contact sites between the two membranes, allowing cholesterol to flow down a chemical gradient. However, a STARD1 protein lacking the N-terminal 62 amino acids (N-62 STARD1), which contain the mitochondrial targeting sequence, was found to be as effective as STARD1 in stimulating steroidogenesis. Other STARD1 constructs engineered for prolonged tethering to the surface of the mitochondria were very active in stimulating P5 production, suggesting that the residency time of the protein on the mitochondrial surface determines the duration of the steroidogenic stimulus. Recombinant human N-62 STARD1 in nanomolar concentrations enhanced P5 production by isolated ovarian mitochondria in a dose- and time-dependent fashion, with significant increases in steroid production observed within minutes. Collectively, these findings suggest that STARD1 acts on the outer mitochondrial membrane to promote cholesterol translocation. This model implies that import of the protein into the mitochondrial matrix, rather than being the trigger to

steroid production, is the "off" mechanism, because it removes STARD1 from its site of action.

Existing evidence is most consistent with the idea that STARD1 enhances desorption of cholesterol from the sterol-rich outer mitochondrial membrane to the relatively sterol-poor inner membranes. The desorption process may involve a pH-dependent conformational change (molten globule transition). Even though STARD1 contains a hydrophobic pocket that binds cholesterol, sterol binding is not required for steroidogenic activity.

The molecule or structure on the mitochondrial outer membrane that STARD1 acts on has not yet been identified. It could be a lipid configuration or a protein. One protein candidate is translocator protein (TSPO), also known as the peripheral-type benzodiazepine receptor (PBR), an outer mitochondrial membrane protein that also binds cholesterol.[3,4] Knockdown of TSPO expression blocks steroidogenesis in cultured cells, even in the presence of STARD1, suggesting that TSPO is required for STARD1 action and that it may serve as a pore through which cholesterol could flow to the inner mitochondrial membrane in the presence of STARD1. This notion has been recently challenged by the creation of tissue-specific TSPO knock-out mice which have normal steroidogenesis. Thus, the role of TSPO in STARD1 action remains to be elucidated.

Mutations found in the *STARD1* gene in subjects with congenital lipoid adrenal hyperplasia include frameshifts caused by deletions or insertions, splicing errors, and nonsense and missense mutations. All of these mutations lead to the absence of STARD1 protein or the production of functionally inactive protein. Several nonsense mutations were shown to result in C-terminus truncations of STARD1. One of these mutations, Q258X, results in the deletion of the C-terminal 28 amino acids of the STARD1 protein and accounts for 80%

of the known mutant alleles in the affected Japanese population. Known point mutations that produce amino acid substitutions occur in exons 5 to 7 of the gene, the exons that encode the *C*-terminus. Mutations that cause partial loss of STARD1 activity are associated with a milder disease phenotype.[3,4,41]

The pathophysiology of lipoid congenital adrenal hyperplasia entails a two-step process in which impaired use of cholesterol for steroidogenesis leads to accumulation of sterol esters in lipid droplets. These droplets ultimately compress cellular organelles, causing damage through the formation of lipid peroxides. This damage occurs prominently in the adrenal cortex and Leydig cells.

Although affected XY subjects are pseudohermaphrodites because of an inability to generate sufficient fetal testicular testosterone to masculinize the external genitalia, XX subjects have normal external genitalia, develop female secondary sexual characteristics, and experience menarche. They are, however, anovulatory, develop follicular cysts, and are unable to produce large amounts of estradiol and P4 in a cyclic fashion. The fact that some ovarian estradiol synthesis occurs reflects the existence of STARD1-independent substrate movement to the cholesterol side-chain cleavage system.

MUTATIONS AND POLYMORPHISMS IN GENES ENCODING STEROIDOGENIC ENZYMES

The Cholesterol Side-Chain Cleavage Enzyme (CYP11A1)

Cholesterol side-chain cleavage is catalyzed by cytochrome P450scc (CYP11A1) and its associated electron transport system, consisting of a flavoprotein reductase (ferredoxin or adrenodoxin reductase) and an iron sulfoprotein (ferredoxin or adrenodoxin), which shuttles electrons to cytochrome P450scc.[3,42,43] The side-chain cleavage reaction involves three catalytic cycles: the first two lead to the introduction of hydroxyl groups at positions C-22 and C-20, and the third results in scission of the side chain between these carbons. Each catalytic cycle requires one molecule of reduced nicotinamide adenine dinucleotide phosphate (NADPH) and one molecule of oxygen so that the formation of one mole of the cleavage products, P5 and isocaproaldehyde, uses three moles of NADPH and three moles of oxygen.

The rate of formation of P5 is determined by: access of cholesterol to the inner mitochondrial membranes; the quantity of cholesterol side-chain cleavage enzyme and, secondarily, its flavoprotein and iron−sulfur protein electron transport chain; and catalytic activity of P450scc, which can be influenced by posttranslational modification. Acute alterations in steroidogenesis generally result from changes in the delivery of cholesterol to P450scc, whereas long-term alterations involve changes in the quantity of enzyme proteins as well as cholesterol delivery.

Mutations in the *CYP11A1* gene that result in significantly diminished cholesterol side-chain cleavage activity (12% of normal to undetectable enzymatic activity) have been reported in association with adrenal insufficiency and 46, XY disorder of sex development, phenotypes that are similar to those associated with inactivating mutations in the *STARD1* gene.[3,44−46] Partial enzyme defects with normal sexual development, but late onset adrenal insufficiency have been reported. The ovarian phenotype of genetic females with *CYP11A1* mutations has not been described, but it is expected to be similar to XX individuals with *STARD1* mutations (spontaneous pubertal development, large follicular cysts resulting from elevated but disordered gonadotropin secretion, anovulation), especially if there is residual enzyme activity.

17α-Hydroxylase/17,20-lyase (CYP17A1)

P450c17 (CYP17A1) is an endoplasmic reticulum enzyme that catalyzes two reactions: hydroxylation of P5 and P4 at carbon 17, and conversion of pregnenolone into C19 steroids (in the case of human CYP17A1, P4 is also converted to C19 products, but to a much lesser extent).[3] The 17α-hydroxylation reaction requires one pair of electrons and molecular oxygen. The lyase reaction requires a second electron pair and molecular oxygen. The reducing equivalents are transferred to the P450c17 heme iron from NADPH by NADPH cytochrome P450 reductase (POR). The hydroxylase and lyase reactions are both believed to proceed through a ferryl oxene mechanism, with the substrate bound to the enzyme in the catalytic pocket in the same orientation.

The importance of POR in steroid metabolism catalyzed by endoplasmic reticulum cytochrome P450 enzymes has been illuminated by the phenotype of individuals with POR deficiency.[38,39,47,48] The autosomal recessive disorder results in a steroid profile that is suggestive of combined 21-hydroxylase and 17α-hydroxylase/17-20 lyase deficiency, which presents as a range of phenotypes, including adrenal insufficiency, and Antley–Bixler skeletal malformation syndrome. Males with POR deficiency are often under-virilized, while females can be virilized.

Several factors determine whether substrates only undergo 17α-hydroxylation or subsequent scission of the 17,20 bond, including: the nature of the substrate; the amount POR and flux of reducing equivalents; allosteric effectors; and post-translational modification of P450c17. Collectively, these factors determine the nature of the products produced by the enzyme, which in the gonads and zona reticularis favor androgens through augmentation of lyase activity.

Human P450c17 preferentially uses Δ5 substrates for 17,20 bond cleavage. Cytochrome b5 (CYB5A) promotes electron transfer for the lyase reaction acting as an allosteric effector, and not in an electron donor *per se*, because the apo b5 protein is also effective.[49] Cytochrome b5 also increases the use of 17OHP4 as a substrate for androstenedione synthesis. The distribution and regulation of cytochrome b5 expression in the adrenal cortex, being greatest in the adult zona reticularis, supports a role for this protein in the regulation of lyase activity. The product of a second cytochrome b5 gene (type 2 cytochrome b5, CYB5B), found in human testis and expressed in the adrenals, also increases lyase activity. Recently, missense mutations in *CYB5B* were found to be a cause of 17,20 lyase deficiency.

Phosphorylation of P450c17 at serine and threonine residues by the kinase p38α, appears to be necessary for maximal 17,20-lyase activity.[10] The phosphorylated P450c17 protein is evidently a substrate for protein phosphatase 2A. Inhibitors of PP2A enhance lyase activity.

Mutations in the *CYP17A1* gene cause combined or isolated deficiency states for each activity of P450c17.[3,48] Individuals with combined deficiency have marked diminution in production of C19 and C18 steroids, low levels of cortisol result in elevated adrenocorticotropic hormone (ACTH) secretion and excess production of steroids proximal to the P40c17 reaction. The inability to produce sex steroid hormones prevents adrenarche and puberty in females, and results in incomplete or absent development of male genitalia (XY disorder of sexual development).

Four recurrent mutations have been described that cause combined 17α-hydroxylase and lyase deficiency in different populations: a 4-bp insertion resulting in a frame shift and altered C-terminal sequence; in-frame deletions of amino acid residues 487−489; a deletion of phenylalanine in the N-terminus of the protein; and non-synonymous mutations (W406R and R362C). Isolated 17,20-lyase deficiency is very rare.[3,48] The documented cases result from point mutations that permit P5 or P4 to be bound and undergo 17α-hydroxylation, but prohibit the efficient receipt or usage of a second pair of electrons provided by POR to support the 17,20-lyase reaction.

Insight into the importance of estrogens in ovarian function in women comes from the study of subjects in whom estrogen synthesis is impaired. Only a few studies have been performed on women with 17α-hydroxylase/17,20 lyase deficiency who are incapable of producing thecal androgens to support granulosa cell estradiol synthesis. Promotion of follicular growth to the preovulatory stage in an estrogen-impoverished environment is possible in these individuals with exogenous gonadotropins after pituitary desensitization. The same is true in severely hypogonadotropic women given exogenous FSH. Follicles grow, but in the absence of exogenous LH, estradiol synthesis is minimal. Moreover, the development of large follicular cysts in association with low estrogen levels is common in women with steroidogenic acute regulatory protein (StARD1), 17α-hydroxylase/17,20 lyase, and aromatase deficiency.[50]

Aromatase (CYP19A1)

Aromatase, an endoplasmic reticulum enzyme, catalyzes three sequential hydroxylations of a C19 substrate by using 3 moles of NADPH and 3 moles of molecular oxygen to produce C18 steroids with a phenolic A ring.[3,51] The aromatase protein is encoded by a single large gene, CYP19A1, which gives rise to cell-specific transcripts from different promoters.[6,44] The promoter driving ovarian aromatase expression lies adjacent to the exon encoding the translation start site (promoter II). In granulosa cells, FSH stimulates transcription of the genes encoding both aromatase and POR, which provides its reducing equivalents.

Cases of aromatase deficiency due to mutations in the CYP19A1 gene have been described.[3,52−54] Pregnancies in which the fetus is affected with aromatase deficiency are characterized by low maternal urinary estrogen excretion, maternal virilization, and ambiguous genitalia or female 46,XX disorder of sexual development in affected females. The maternal and fetal virilization in the absence of placental aromatase activity highlights the importance and efficiency of the placenta in converting maternal and fetal androgens into estrogens.

In prepubertal girls with aromatase deficiency, basal FSH levels are elevated, but basal LH levels are normal. After puberty, both basal FSH and LH are elevated, as are gonadotropin responses to gonadotropin-releasing hormone (GnRH) stimulation. Estradiol levels are extremely low or not detectable. High levels of gonadotropins promote the formation of large hemorrhagic ovarian cysts. Sexual development does not take place in pubertal girls, and symptoms of androgen excess develop due to elevated androgen levels. Bone mineralization is affected and bone aging is delayed. The phenotypes are, however, variable depending upon the level of residual aromatase activity.

Among the mutations identified in the CYP19A1 gene are missense mutations, mostly in exons 9 and 10, insertions and deletions causing frameshifts resulting in nonsense codons, and an 87-bp insertion at the splice junction between exon 6 and intron 6, causing the addition of 29 in-frame amino acid residues. The mutant protein with the 29 in-frame amino acid residues displayed less than 3% of normal aromatase activity. Expression of the mutant complementary DNA confirmed that the protein had only a trace of aromatase activity. Other mutations in coding sequences found in patients with aromatase deficiency have been shown to have minimal enzyme activity (<2% of normal activity).

Hydroxysteroid Dehydrogenases

Hydroxysteroid dehydrogenases (HSDs) or oxidoreductases catalyze the interconversion of alcohol and carbonyl functions in a position- and stereospecific manner on the steroid nucleus and side chain, using oxidized (+) or reduced (H) nicotinamide adenine dinucleotide (NAD(H)), or nicotinamide adenine dinucleotide phosphate NADP(H) as cofactors.[3,6,55] In some instances, HSDs display bifunctionality (e.g., they oxidize or reduce 17β- and 20α- oxy functions). Although the HSDs can catalyze both the oxidation and reduction reactions *in vitro* under different conditions (e.g., substrate, pH and cofactor), *in vivo* they catalyze reactions in one direction and can be classified as dehydrogenases or reductases. These enzymes are members of the short chain dehydrogenase or aldo-keto reductase superfamilies.

3β-Hydroxysteroid Dehydrogenase/Δ^{5-4} Isomerases

The 3β-HSD/Δ^{5-4} isomerases are membrane-bound enzymes localized to the endoplasmic reticulum and mitochondria that use nicotinamide adenine dinucleotide (NAD$^+$) as a cofactor. These enzymes catalyze dehydrogenation of the 3β-hydroxyl group and the subsequent isomerization of the Δ^5 olefinic bond to yield a Δ^4 three-ketone structure. They convert P5 into P4, 17α-hydroxypregnenolone into 17OHP4, and DHEA into androstenedione.[3,56]

The dehydrogenase and isomerase reactions are performed at a single bifunctional catalytic site that adopts different conformations for each activity. The 3β-hydroxysteroid dehydrogenase step is rate-limiting in the overall reaction sequence, and the NADH formed in this reaction is believed to alter the enzyme conformation to promote the isomerase reaction.

The human genome has two active 3β-HSD/Δ^{5-4} isomerase genes: one (*HSD3B1*) encodes a protein predominantly expressed in the placenta, liver, breast and brain; the other (*HSD3B2*) is a protein expressed in the gonads and adrenal cortex. These two genes, each consisting of four exons, lie 100 kb apart on band 1p13.1. The DNA sequences of the exons of the two active genes are very similar, and the encoded proteins differ in only 23 amino acid residues. HSD3B1, however, has a lower K$_m$ for substrate than HSD3B2, which facilitates metabolism of lower concentrations of Δ^5 substrate. Electron microscope cytochemistry localized HSD3B2 activity to the perimitochondrial endoplasmic reticulum and in subcellular fractions containing STARD1 and P450scc. In some cell types HSD3B2 appears to be localized in the inner mitochondrial membrane. Thus, these enzymes are positioned to act on P5 produced by the cholesterol side-chain cleavage system.

Mutations causing deficiency of HSD3B2 produce a form of congenital adrenal hyperplasia characterized by impaired adrenal and gonadal steroidogenesis with accumulation of Δ^5 steroids in the circulation. Males with classical 3β-HSD deficiency present with either perineal hypospadias or perineoscrotal hypospadias. However, complete or partial inhibition of 3β-HSD activity in the adrenals and ovaries does not affect the differentiation of the external genitalia of female patients.

A so-called "attenuated" or "late-onset" form of 3β-HSD deficiency, diagnosed by steroid measurements, has been described in the literature. However, no mutations have yet been discovered in the genes encoding *HSD3B1* and *HSD3B2* in subjects with this clinical diagnosis.

17β-Hydroxysteroid Dehydrogenases

The type 1 enzyme (HSD17B1) is referred to as the "estrogenic" 17β-HSD because it catalyzes the final step in estrogen biosynthesis by preferentially reducing the weak estrogen estrone to yield the potent estrogen 17β-estradiol. The enzyme is a cytosolic protein that uses either NADH or NADPH as a cofactor.[57] It has 100-fold higher affinity for C18 steroids than for C19 steroids. The *HSD17B1* structural gene consists of six exons. It is located on bands 17q11-12 in tandem with a highly homologous pseudogene.[57] The structural gene is expressed in granulosa cells of the ovary and the placental syncytiotrophoblast.

HSD17B3 is referred to as the "androgenic" 17β-HSD because it catalyzes the final step in androgen biosynthesis in the Leydig cells, reducing androstenedione to testosterone using NADPH as a cofactor. It also can reduce estrone to estradiol. HSD17B3 is not expressed in the ovary, requiring androgen-producing cells of the ovary to employ another enzyme, probably HSD17B5, to synthesize testosterone.

HSD17B5, located on chromosome 10p15-14, is a member of the aldo-keto reductase family (AKR1C3). It has 3β-HSD and 20α-HSD activity in addition to 17β-HSD activity that produces testosterone from androstenedione. It is thought to serve as the "androgenic" 17β-HSD of ovarian theca cells.[58,59]

HSD17B6 (also known as RODH), encoded by a gene on chromosome 12q13.3 has 3α-HSD activity and catalyzes the conversion of androstanediol to dihydrotestosterone in the prostate. Polymorphisms in the *HSD17B6* gene have been associated with polycystic ovary syndrome phenotypes in some but not all studies.

$\Delta^{4\text{-}5}$ Reductases

The $\Delta^{4\text{-}5}$ reductases are membrane-associated enzymes that reduce the double bond between carbons 4 and 5 in steroid hormones by catalyzing direct hydride transfer from NADPH to the carbon 5 position of the steroid substrate.[3,60,61] They produce either 5α or 5β-dihydrosteroids.

Two different human 5α-reductases sharing 50% similarity in amino acid sequence and a molecular weight of approximately 29 kDa have been identified: type 1 (SRD5A1) and 2 (SRD5A2).[60,61] SRDA2 is predominantly expressed in male genital structures, including genital skin and prostate, where it reduces testosterone to yield the more potent androgen, 5α-dihydrotestosterone. SRD5A1, which catalyzes a similar reaction on both C21 and C19 steroid hormones, is expressed in the ovary, liver, kidneys, skin, and brain. Although this enzyme can also make 5α-dihydrotestosterone, its tissue distribution suggests that its predominant function is to inactivate steroid hormones.

Inactivating mutations in *SRD5A2* cause male pseudohermaphrodism. The enzyme defect is characterized by abnormal testosterone: 5α-dihydrotestosterone ratios. Affected males have varying degrees of abnormal development of the external genitalia, ranging from mild hypospadias to severe defects in which the external genitalia are essentially female.

Females carrying mutations in the *SRD5A2* gene have a normal sex phenotype and normal menstrual cycles. They have a low incidence of hirsutism and acne, and like males with the disease, have low ratios of 5α: 5β-dihydrosteroid metabolites in urine.

Among the mutations reported are deletions that inactivate SRD5A2, and missense mutations that impair enzyme activity by affecting substrate or cofactor binding. A *SRD5A2* variant (Ala49Thr) that has increased catalytic activity has been associated with increased risk of prostate cancer. Mutations have not been described in the human *SRA5A1* gene.

PATHOPHYSIOLOGY AND GENETICS OF POLYCYSTIC OVARY SYNDROME

Polycystic ovary syndrome (PCOS) is a common endocrine disorder that affects ~5–7% of reproductive aged women, characterized by hyperandrogenemia, menstrual irregularities, chronic anovulation, "cystic" ovaries, and infertility.[62,63] Other reproductive and endocrine abnormalities include disordered gonadotropin secretion, oligomenorrhea, enlarged polycystic ovaries, endometrial hyperplasia, hirsutism, and acne. PCOS is also associated with metabolic consequences, including an increased risk of obesity,[64] insulin resistance leading to type 2 diabetes mellitus (T2DM)[65] and premature arteriosclerosis,[66] and increased cardiovascular risk. PCOS is a heterogeneous disorder that shows genetic predisposition among affected individuals.[67,68]

Although there has been debate about the diagnostic criteria for PCOS, hyperandrogenemia/hyperandrogenism not explained by other causes is a hallmark of the disorder, and it is included as an essential element in all "consensus" diagnosis schemes.[62,63] The ovaries are the primary source of excess androgen production in PCOS. However, in a significant number of women there is excessive secretion of adrenal DHEA. The fact that destruction of follicle cysts or removal of a section of the PCOS ovary can reduce androgen production, and the tendency of hyperandrogenemia to lessen and ovulatory cycles return with depletion of follicles and the approach of menopause, highlights the important underlying role of the follicles in androgen excess in PCOS.

The ovarian phenotype of PCOS is characterized by an increased number of small antral follicles, with a hypertrophied thecal wall, under a thickened capsule.[69] These follicles are arrested in development at the stage where selection of a dominant follicle would normally occur. Compared with follicles of the same size in cycling ovaries, these growth arrested follicles contain an increased number of steroidogenic cells in the theca interna, with increased CYP17A1 and CYP11A1 gene expression and androgen biosynthesis,[70] and a decreased number of granulosa cells, with lowered CYP19A1 expression. During normal follicle development, the granulosa cells begin to express CYP19A1 mRNA when the follicle grows to approximately 6–7 mm in diameter.[71] In the PCOS ovary, follicles arrest at 5–8 mm in diameter, and granulosa cells express lower levels of CYP19A1 mRNA independent of the follicle size, even though there are normal concentrations of FSH in the follicular fluid.[71] Thus, follicular fluid from PCOS follicles contain elevated levels of androgens and lower amounts of estradiol,[72] resulting from an increase in androgen biosynthesis in theca cells and a decrease in the conversion of androgens to estrogens in granulosa cells.[71,73] In addition, both theca and granulosa cells appear to be luteinized as they express LH receptors and have increased CYP11A1 expression and production of progestins.[70] 5α-reductase activity has also been reported to be elevated in the granulosa cells of PCOS ovaries, leading to markedly increased metabolism of androstenedione to 5α-androstanedione, a competitive inhibitor of aromatase activity.[74] Therefore, it is clear that there are abnormalities of proliferation and differentiation in both the theca and granulosa compartments of the polycystic ovary in vivo. A reduction in circulating FSH relative to elevated LH, has been proposed to contribute to reduced follicular development and elevated thecal androgen production.[75] Alterations in anti-Müllerian hormone in granulosa cells and primordial and transitional follicles in ovaries from anovulatory women with polycystic ovaries is also thought to contribute to disordered early follicle development in PCOS.[76] Granulosa cells isolated from follicles of women with PCOS have decreased expression of apoptotic effectors and increased expression of cell survival factors, which may contribute to the PCOS ovarian phenotype.[77]

Women with PCOS also exhibit an exaggerated serum estradiol response to recombinant human FSH compared with similarly treated normal women.[75] This is consistent with excessive follicular development following gonadotropin therapy in PCOS women, and the corresponding risk of ovarian hyperstimulation syndrome. Reports also suggest that inhibin B and estradiol induction by FSH is significantly enhanced in PCOS women, occurring more rapidly and robustly than in normal women. In contrast, the induction of inhibin A was similar in normal and PCOS women. In PCOS, the increased response to FSH does not appear to be the result of elevated androgens, since treatment with the anti-androgen flutamide does not alter FSH responsiveness.[78]

Studies on freshly isolated ovarian tissue, thecal tissue from normal and PCOS ovaries, and cultures of human theca cells derived from normal and PCOS women have demonstrated that PCOS theca cells have increased expression of steroidogenic enzymes and secrete greater amounts of androgen than thecal tissue or cells from regularly ovulating women.[70,79-87] Immunohistochemical studies have demonstrated that there is greater immunoreactive CYP17A1 protein in the theca interna cell layer of small, large, and atretic follicles from PCOS ovaries that is independent of the thickness of the theca interna layer. Gilling-Smith et al. published the first studies demonstrating that androgen and progestin production are elevated on a per cell basis in PCOS theca cells in primary culture.[80] In women with PCOS, androgen biosynthesis was found to be inhibited following suppression of pituitary LH production in patients with PCOS, consistent with an ovarian thecal source of excess androgen biosynthesis in PCOS.[80] A comparison of the steroid biosynthetic capacity of normal and PCOS theca cells grown for multiple population doublings verified that androgen and progestin production were increased on a per cell basis in PCOS theca cells compared to cells propagated from normal cycling women.[81] Increased theca cell steroidogenesis has been attributed to increased CYP17A1, CYP11A1, HSD3B2 gene expression and mRNA stability in PCOS theca cells.[11,73,83,88,89] In contrast, STARD1 gene expression, which controls the rate-limiting step in steroidogenesis, and HSD17B5 (AKR1C3) gene expression, which is the theca cell enzyme thought to be responsible for the reduction of androstenedione into testosterone, are not different between normal and PCOS theca cells.[81,82,90] Thus, these collective studies have defined a stable steroidogenic phenotype for the PCOS theca cell that includes altered expression of a subset of proteins that are important for androgen synthesis.[90]

The molecular characterization of PCOS theca cells and normal theca cells from multiple individuals by microarray analysis and quantitative polymerase chain reaction (PCR) established that normal and PCOS theca cells have distinctive molecular signatures.[90-92] Several gene networks that are involved in steroid hormone biosynthesis, cell signaling, retinoic acid biosynthesis, insulin and glucose homeostasis, scaffolding, cell structure, many growth factors, transcription factors and receptors are differentially expressed in normal and PCOS theca cells. Of note, the promoters of the CYP11A1 and CYP17A1 genes are more active in the context of PCOS theca cells compared to normal theca cells. The increased transcription of these and other genes involved in steroidogenesis is presumed to contribute to the increase in thecal androgen secretion. Elevated levels of GATA6 in PCOS theca cells may explain, in part, the increased *CYP11A1* and *CYP17A1* promoter activity in PCOS theca cells.[93]

Genetics of PCOS

PCOS is a complex genetic disorder influenced by environmental factors. Familial aggregation of PCOS has been recognized for many years, supporting a genetic contribution to its etiology, and its incidence appears to be similar across racial/ethnic groups. Twin studies and family-based association studies carried out on a large Dutch population including more than 1300 monozygotic twins and 1800 dizygotic twins/singleton sisters of twins provided strong evidence that PCOS is a heritable condition.[94] The multiple reproductive and metabolic features that define PCOS as a disorder reflect underlying genetic heterogeneity. Male and female first-degree relatives of women with PCOS are also at increased risk of developing obesity, insulin resistance and T2DM.[95,96] Instead of a classic Mendelian pattern, an oligogenic/polygenic model likely contributes to the underlying pathophysiology. Incomplete penetrance, epigenetic modification, and environmental contributions have hindered attempts to clarify the underlying model of inheritance.

Given the evidence for excessive thecal androgen production, early candidate gene studies focused on variation in genes encoding steroidogenic enzymes or genes that regulate androgen synthesis as a risk factor for PCOS. These included *CYP11A1, CYP17A1, CYP19A1, HSD3B2, HSD17B5, HSD17B6, SRD5A2, GATA6,* and *WNT4*. Association studies involving these genes were carried out mostly on small populations. Although several reports described significant associations, these have not been confirmed in GWAS carried out on large populations. Numerous candidate gene association studies were also conducted on genes associated with obesity, insulin resistance and type 2 diabetes, including sex hormone binding globulin (*SHBG*), fibrillin-3 (*FBN3*), insulin (*INS*), insulin receptor substrate 1 (*IRS1*), transcription factor 7-like 2 (*TCF7L2*), calpain 10 (*CAPN10*), and fat and obesity associated gene (*FTO*).[97] Findings for these and many more candidate gene studies have been mostly inconclusive, but few have yielded statistically significant associations that have been replicated.[98] Moreover, none of the studies were found to be significant at the genome-wide level ($P < 1 \times 10^{-8}$), the gold standard for genetic studies.[97]

Genetic heterogeneity, combined with broad variability in the diagnostic criteria, has contributed to the complexity of candidate gene approaches to the genetics of PCOS.[97] A further problem with the candidate gene approach is that such studies rely upon some prior understanding of the pathogenesis of PCOS to determine the candidacy of the gene chosen. As a consequence, despite advances in genetic technologies, very few PCOS susceptibility genes have been validated.[98]

This limitation is overcome through the use of GWAS. GWAS do not rely upon any *a priori* understanding of the pathogenesis, and thus have the potential to reveal previously unexpected or even unknown pathogenic pathways that may be involved in the development of PCOS. Data obtained from these studies will transform our understanding of PCOS pathogenesis, and also provide a basis to develop future novel therapeutic strategies.

A major milestone was achieved with the publication of a GWAS by Chen *et al.*, who reported their findings on a Han Chinese population.[99] Chen and colleagues identified three loci, two on chromosome 2 (2p16.3 and 2p21), and one on chromosome 9 (9q33.3) that had significant associations with PCOS at levels exceeding the threshold statistical significance for genome-wide associations. A subsequent GWAS with additional subjects identified eight new putative PCOS loci on chromosomes 2p16.3, *LHCGR*; 9q22.32, *DENND1A*; 11q22.1, *YAP1*; 12q13.2, *RAB5B*; 12q14.3, *HMGA2*; 16q21.1, *TOX1*; 19p13.3, *INSR*; and 20q13.2, *ZNF217*.[100]

While loci on 2p16.3 are in or near the *FSHR* and *LHCGR* genes, and the 19p13.3 signal is near the insulin receptor (*INSR*) gene, and are plausible PCOS candidates, the pathophysiological links of the other GWAS loci to reproduction or ovarian function are less obvious. As noted earlier, significant associations were not found in genes encoding steroidogenic enzymes.

Among the GWAS PCOS candidate genes, the *DENND1A* locus at 9q22.32 assumed significance among these candidates as a result of several studies confirming the association of *DENND1A* single nucleotide polymorphisms (SNPs) with PCOS in European populations.[30,101−104] Due to the success of replication attempts in both Asian and European populations, *DENND1A* has gained recognition as a true PCOS susceptibility gene.[97] Interestingly, none of the genes encoding proteins involved in the paracrine regulation of theca and granulosa cell function were identified as PCOS risk loci in the published GWAS.

The *DENND1A* gene encodes a protein named connecdenn 1, which interacts with members of the Rab family of small GTPases, like RAB5B, which are involved in membrane trafficking.[105] Connecdenn 1 has a clathrin-binding domain and is thought to facilitate endocytosis.[105] *DENND1A* encodes two transcripts as a result of alternative splicing.[105] One of these transcripts, DENND1A variant 1, codes for a 1009-amino acid protein with *C*-terminal proline-rich domain; the other, DENND1A variant 2, codes for a truncated 559-amino acid protein that contains the DENN domain and the clathrin-binding domain, but lacks the proline-rich domain, and includes a *C*-terminal 33-amino acid sequence that differs from the larger connecdenn 1 variant. Up until the recently, little has been known about DENND1A expression in cells and tissues related to reproduction

with the exception that it is expressed in testes, theca cells, and H295 adrenal carcinoma cells, cells that make androgens.[64]

Among the loci associated with PCOS in Han Chinese, several reside in or near genes that potentially define a network, including the *FSHR*, *LHCGR*, and I*NSR*, which encode receptors that reside on the plasma membrane, and which are internalized by coated pits where DENND1A protein is located.[98,100] RAB5B is thought to be involved in endocytosis and could, therefore, be a molecule interacting with the DENN domain. YAP1, TOX3, HMGA2, and ZNF217 are all involved in transcriptional regulation, although none of them have been specifically implicated in the expression of genes involved in steroidogenesis. However, TOX3 (transcriptional coactivator of the p300/CBP-mediated transcription complex) transactivates through cAMP response element (CRE) sites, which are present in genes encoding steroidogenic proteins. These genes can be assembled into a signaling network beginning at the receptor level, receptor coupling, or recycling, and downstream molecules that ultimately regulate gene transcription, either of steroidogenic genes directly, or possibly through the upregulation of other transcription factors that directly influence steroidogenic gene promoter function (Figure 18.4).

Five of the GWAS candidates, *DENND1A*, *LHCGR*, *RAB5B*, *C9orf3*, and *INSR*, have been reported to be associated with similar and overlapping signaling pathways.[106] The DENN/MADD domain is reported to be involved with MAPK/ERK signaling.[105] DENND1A has also been shown to regulate phosphotidyl inositol kinase (PI3K) and AKT- protein kinase B (PKB)-dependent signaling cascades.[105,107,108] RAB5B signaling has also been reported to involve PI3K, PKB, and MAPK/ERK components.[107−110] In addition to activating cAMP-dependent PKA-mediated signaling pathways, LH-dependent

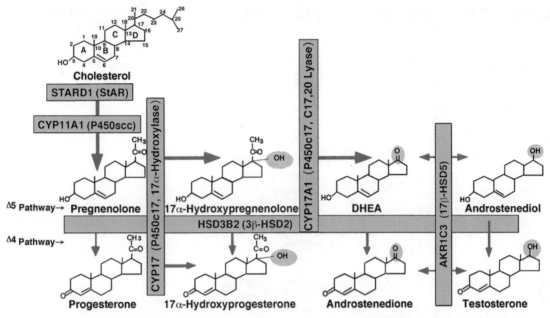

FIGURE 18.4 **Pathway of thecal androgen biosynthesis.** Thickness of red arrows indicates relative flux through the pathways.

LHCGR activation has been reported to be associated with PI3K-, PKB-, and MAPK-signaling cascades.[111–116] YAP1, the key protein in the Hippo tumor suppressor pathway, is translocated from the cytosolic compartment to the nucleus, where it acts as a transcription factor. YAP1 has been reported to be associated with regulation of upstream PI3K, PKB, and MAPK signaling.[117–121] The expression of C9orf3, or amino-pepidase O, is regulated by PI3K, PKB, and MAPK signaling.[122–124] C9orf3 shuttles between both the cytoplasm and the nucleus, and functions as a metalloprotease.[125] Combined, these data support the hypothesis that GWAS candidates form a hierarchical signaling pathway that converges downstream on the transcriptional apparatus to potentiate the PCOS phenotype of increased *CYP17A1* expression and androgen biosynthesis in theca cells (Figure 18.5).

This framework provides a road map for the identification of genetic variation/mutations that predispose to PCOS and the molecular basis for the action of the identified risk alleles. However, based on the relatively low odds ratios for *DENND1A* and other GWAS loci risk alleles, it is likely that variation in genes other than those disclosed from the Chinese GWAS contribute to PCOS risk, perhaps in a population-specific manner. With the remarkable advancements in gene sequencing technology and development of new analytical methods, the ability to examine the role of functional GWAS candidates and their variants in the PCOS ovary, and determine their role within the signaling networks that affect androgen biosynthesis, will provide a clearer understanding of the complex genetic architecture of PCOS. This will not only help elucidate the ovarian pathophysiology of PCOS, but most likely the other PCOS-associated phenotypes.

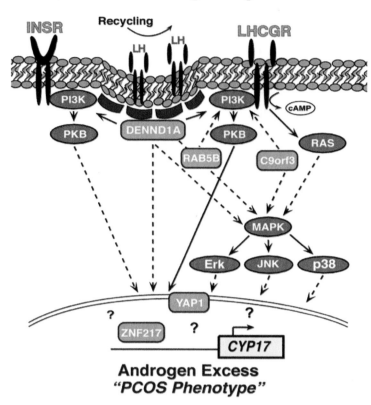

GWAS Candidate Signaling Cascades?

Androgen Excess "PCOS Phenotype"

FIGURE 18.5 **A network of genes postulated to contribute to the hyperandrogenemia associated with PCOS based on loci identified in published GWAS and replication studies.**

References

1. Hu J, Zhang Z, Shen WJ, et al. Cellular cholesterol delivery, intracellular processing and utilization for biosynthesis of steroid hormones. *Nutr Metab (Lond)* 2010;**7**:47.
2. Rainey WE. Adrenal zonation: clues from 11beta-hydroxylase and aldosterone synthase. *Mol Cell Endocrinol* 1999;**151**:151−60.
3. Miller WL, Auchus RJ. The molecular biology, biochemistry, and physiology of human steroidogenesis and its disorders. *Endocr Rev* 2011;**32**:81−151.
4. Miller WL, Bose HS. Early steps in steroidogenesis: intracellular cholesterol trafficking. *J Lipid Res* 2011;**52**:2111−35.
5. Conley AJ, Corbin CJ, Thomas JL, et al. Costs and consequences of cellular compartmentalization and substrate competition among human enzymes involved in androgen and estrogen synthesis. *Biol Reprod* 2012;**86**:1−8.
6. Penning TM. Human hydroxysteroid dehydrogenases and pre-receptor regulation: insights into inhibitor design and evaluation. *J Steroid Biochem Mol Biol* 2011;**125**:46−56.
7. Christenson LK, Strauss III JF. Steroidogenic acute regulatory protein: an update on its regulation and mechanism of action. *Arch Med Res* 2001;**32**:576−86.
8. Fayard E, Auwerx J, Schoonjans K. LRH-1: an orphan nuclear receptor involved in development, metabolism and steroidogenesis. *Trends Cell Biol* 2004;**14**:250−60.
9. Martin LJ, Taniguchi H, Robert NM, et al. GATA factors and the nuclear receptors, steroidogenic factor 1/liver receptor homolog 1, are key mutual partners in the regulation of the human 3beta-hydroxysteroid dehydrogenase type 2 promoter. *Mol Endocrinol* 2005;**19**:2358−70.
10. Zhang LH, Rodriguez H, Ohno S, et al. Serine phosphorylation of human P450c17 increases 17,20-lyase

activity: implications for adrenarche and the polycystic ovary syndrome. *Proc Natl Acad Sci U S A* 1995;**92**:10619−23.

11. Wickenheisser JK, Nelson-Degrave VL, McAllister JM. Dysregulation of cytochrome P450 17alpha-hydroxylase messenger ribonucleic acid stability in theca cells isolated from women with polycystic ovary syndrome. *J Clin Endocrinol Metab* 2005;**90**:1720−7.

12. Backstrom T, Andersson A, Baird DT, et al. The human corpus luteum secretes 5 alpha-pregnane-3,20-dione. *Acta Endocrinol (Copenh)* 1986;**111**:116−21.

13. Kohen P, Henriquez S, Rojas C, et al. 2-Methoxyestradiol in the human corpus luteum throughout the luteal phase and its influence on lutein cell steroidogenesis and angiogenic activity. *Fertil Steril* 2013. in press

14. Mork L, Maatouk DM, McMahon JA, et al. Temporal differences in granulosa cell specification in the ovary reflect distinct follicle fates in mice. *Biol Reprod* 2012;**86**:37.

15. Van Deerlin PG, Cekleniak N, Coutifaris C, et al. Evidence for the oligoclonal origin of the granulosa cell population of the mature human follicle. *J Clin Endocrinol Metab* 1997;**82**:3019−24.

16. McNatty KP, Makris A, DeGrazia C, et al. The production of progesterone, androgens, and estrogens by granulosa cells, thecal tissue, and stromal tissue from human ovaries in vitro. *J Clin Endocrinol Metab* 1979;**49**:687−99.

17. Zoller LC, Weisz J. A quantitative cytochemical study of glucose-6-phosphate dehydrogenase and delta 5-3 beta-hydroxysteroid dehydrogenase activity in the membrana granulosa of the ovulable type of follicle of the rat. *Histochemistry* 1979;**62**:125−35.

18. Turner KJ, Macpherson S, Millar MR, et al. Development and validation of a new monoclonal antibody to mammalian aromatase. *J Endocrinol* 2002;**172**:21−30.

19. Lawrence TS, Dekel N, Beers WH. Binding of human chorionic gonadotropin by rat cumuli oophori and granulosa cells: a comparative study. *Endocrinology* 1980;**106**:1114−8.

20. Sullivan MW, Stewart-Akers A, Krasnow JS, et al. Ovarian responses in women to recombinant follicle-stimulating hormone and luteinizing hormone (LH): a role for LH in the final stages of follicular maturation. *J Clin Endocrinol Metab* 1999;**84**:228−32.

21. McNatty KP, Smith DM, Makris A, et al. The microenvironment of the human antral follicle: interrelationships among the steroid levels in antral fluid, the population of granulosa cells, and the status of the oocyte in vivo and in vitro. *J Clin Endocrinol Metab* 1979;**49**:851−60.

22. van Dessel HJ, Schipper I, Pache TD, et al. Normal human follicle development: an evaluation of correlations with oestradiol, androstenedione and progesterone levels in individual follicles. *Clin Endocrinol (Oxf)* 1996;**44**:191−8.

23. Westergaard L, Christensen IJ, McNatty KP. Steroid levels in ovarian follicular fluid related to follicle size and health status during the normal menstrual cycle in women. *Hum Reprod* 1986;**1**:227−32.

24. Fujiwara T, Sidis Y, Welt C, et al. Dynamics of inhibin subunit and follistatin mRNA during development of normal and polycystic ovary syndrome follicles. *J Clin Endocrinol Metab* 2001;**86**:4206−15.

25. McNatty KP, Hunter WM, MacNeilly AS, et al. Changes in the concentration of pituitary and steroid hormones in the follicular fluid of human graafian follicles throughout the menstrual cycle. *J Endocrinol* 1975;**64**:555−71.

26. Vainio S, Heikkila M, Kispert A, et al. Female development in mammals is regulated by Wnt-4 signalling. *Nature* 1999;**397**:405−9.

27. Naillat F, Prunskaite-Hyyrylainen R, Pietila I, et al. Wnt4/5a signalling coordinates cell adhesion and entry into meiosis during presumptive ovarian follicle development. *Hum Mol Genet* 2010;**19**:1539−50.

28. Young JM, McNeilly AS. Theca: the forgotten cell of the ovarian follicle. *Reproduction* 2010;**140**:489−504.

29. Merkwitz C, Lochhead P, Tsikolia N, et al. Expression of KIT in the ovary, and the role of somatic precursor cells. *Prog Histochem Cytochem* 2011;**46**:131−84.

30. Yamamoto N, Christenson LK, McAllister JM, et al. Growth differentiation factor-9 inhibits 3'5'-adenosine monophosphate-stimulated steroidogenesis in human granulosa and theca cells. *J Clin Endocrinol Metab* 2002;**87**:2849−56.

31. Chun SY, Billig H, Tilly JL, et al. Gonadotropin suppression of apoptosis in cultured preovulatory follicles: mediatory role of endogenous insulin-like growth factor I. *Endocrinology* 1994;**135**:1845−53.

32. Chun SY, Eisenhauer KM, Minami S, et al. Hormonal regulation of apoptosis in early antral follicles: follicle-stimulating hormone as a major survival factor. *Endocrinology* 1996;**137**:1447−56.

33. Danilovich N, Javeshghani D, Xing W, et al. Endocrine alterations and signaling changes associated with declining ovarian function and advanced biological aging in follicle-stimulating hormone receptor haploinsufficient mice. *Biol Reprod* 2002;**67**:370−8.

34. Themmen APN, Huhtaniemi IT. Mutations of gonadotropins and gonadotropin receptors: elucidating the physiology and pathophysiology of pituitary-gonadal function. *Endocr Rev* 2000;**21**:551−83.

35. Aittomaki K, Lucena JL, Pakarinen P, et al. Mutation in the follicle-stimulating hormone receptor gene causes hereditary hypergonadotropic ovarian failure. *Cell* 1995;**82**:959−68.

36. Illingworth DR, Corbin DK, Kemp ED, et al. Hormone changes during the menstrual cycle in abetalipoproteinemia: reduced luteal phase progesterone in a patient with homozygous hypobetalipoproteinemia. *Proc Natl Acad Sci USA* 1982;**79**:6685−9.

37. Parker Jr. CR, Illingworth DR, Bissonnette J, et al. Endocrine changes during pregnancy in a patient with homozygous familial hypobetalipoproteinemia. *N Engl J Med* 1986;**314**:557−60.

38. Hoekstra M, Van Eck M, Korporaal SJ. Genetic studies in mice and humans reveal new physiological roles for the high-density lipoprotein receptor scavenger receptor class B type I. *Curr Opin Lipidol* 2012;**23**:127−32.

39. Plotkin D, Miller S, Nakajima S, et al. Lowering low density lipoprotein cholesterol with simvastatin, a hydroxy-3-methylglutaryl-coenzyme a reductase inhibitor, does not affect luteal function in premenopausal women. *J Clin Endocrinol Metab* 2002;**87**:3155−61.

40. Laue L, Hoeg JM, Barnes K, et al. The effect of mevinolin on steroidogenesis in patients with defects in the low density lipoprotein receptor pathway. *J Clin Endocrinol Metab* 1987;**64**:531−5.

41. Bose HS, Sugawara T, Strauss III JF, et al. The pathophysiology and genetics of congenital lipoid adrenal hyperplasia. *N Engl J Med* 1996;**335**:1870−8.

42. Heyl BL, Tyrrell DJ, Lambeth JD. Cytochrome P-450scc-substrate interactions. Role of the 3 beta- and side chain hydroxyls in binding to oxidized and reduced forms of the enzyme. *J Biol Chem* 1986;**261**:2743−9.

43. Lambeth JD, Seybert DW, Kamin H. Phospholipid vesicle-reconstituted cytochrome P-450SCC. Mutually facilitated binding of cholesterol and adrenodoxin. *J Biol Chem* 1980;**255**:138−43.

44. Katsumata N, Ohtake M, Hojo T, et al. Compound heterozygous mutations in the cholesterol side-chain cleavage enzyme gene (CYP11A) cause congenital adrenal insufficiency in humans. *J Clin Endocrinol Metab* 2002;**87**:3808−13.

45. Hiort O, Holterhus PM, Werner R, et al. Homozygous disruption of P450 side-chain cleavage (CYP11A1) is associated with prematurity, complete 46,XY sex reversal, and severe adrenal failure. *J Clin Endocrinol Metab* 2005;**90**:538−41.

46. Hauffa B, Hiort O. P450 side-chain cleavage deficiency--a rare cause of congenital adrenal hyperplasia. *Endocr Dev* 2011;**20**:54−62.

47. Krone N, Reisch N, Idkowiak J, et al. Genotype-phenotype analysis in congenital adrenal hyperplasia due to P450 oxidoreductase deficiency. *J Clin Endocrinol Metab* 2012;**97**:E257−67.

48. Miller WL. The syndrome of 17,20 lyase deficiency. *J Clin Endocrinol Metab* 2012;**97**:59−67.

49. Auchus RJ, Lee TC, Miller WL. Cytochrome b5 augments the 17,20-lyase activity of human P450c17 without direct electron transfer. *J Biol Chem* 1998;**273**:3158−65.

50. Rabinovici J, Blankstein J, Goldman B, et al. In vitro fertilization and primary embryonic cleavage are possible in 17 alpha-hydroxylase deficiency despite extremely low intrafollicular 17 beta-estradiol. *J Clin Endocrinol Metab* 1989;**68**:693−7.

51. Sohl CD, Guengerich FP. Kinetic analysis of the three-step steroid aromatase reaction of human cytochrome P450 19A1. *J Biol Chem* 2010;**285**:17734−43.

52. Bulun SE. Clinical review 78: aromatase deficiency in women and men: would you have predicted the phenotypes? *J Clin Endocrinol Metab* 1996;**81**:867−71.

53. Belgorosky A, Guercio G, Pepe C, et al. Genetic and clinical spectrum of aromatase deficiency in infancy, childhood and adolescence. *Horm Res* 2009;**72**:321−30.

54. Rochira V, Carani C. Aromatase deficiency in men: a clinical perspective. *Nat Rev Endocrinol* 2009;**5**:559−68.

55. Saloniemi T, Jokela H, Strauss L, et al. The diversity of sex steroid action: novel functions of hydroxysteroid (17beta) dehydrogenases as revealed by genetically modified mouse models. *J Endocrinol* 2012;**212**:27−40.

56. Simard J, Ricketts ML, Gingras S, et al. Molecular biology of the 3beta-hydroxysteroid dehydrogenase/delta5-delta4 isomerase gene family. *Endocr Rev* 2005;**26**:525−82.

57. Hong Y, Chen S. Aromatase, estrone sulfatase, and 17beta-hydroxysteroid dehydrogenase: structure-function studies and inhibitor development. *Mol Cell Endocrinol* 2011;**340**:120−6.

58. Luu-The V, Dufort I, Pelletier G, et al. Type 5 17beta-hydroxysteroid dehydrogenase: its role in the formation of androgens in women. *Mol Cell Endocrinol* 2001;**171**:77−82.

59. Nelson LR, Bulun SE. Estrogen production and action. *J Am Acad Dermatol* 2001;**45**:S116−24.

60. Russell DW, Wilson JD. Steroid 5 alpha-reductase: two genes/two enzymes. *Annu Rev Biochem* 1994;**63**:25−61.

61. Wilson JD, Griffin JE, Russell DW. Steroid 5 alpha-reductase 2 deficiency. *Endocr Rev* 1993;**14**:577−93.

62. Azziz R, Carmina E, Dewailly D, et al. Positions statement: criteria for defining polycystic ovary syndrome as a predominantly hyperandrogenic syndrome: an Androgen Excess Society guideline. *J Clin Endocrinol Metab* 2006;**91**:4237−45.

63. The Rotterdam ESHRE/ASRM-Sponsored PCOS Consensus Workshop Group Revised 2003 consensus

on diagnostic criteria and long-term health risks related to polycystic ovary syndrome (PCOS). *Hum Reprod* 2004;**19**:41–7.

64. Hoeger K. Obesity and weight loss in polycystic ovary syndrome. *Obstet Gynecol Clin North Am* 2001;**28**:85–97 vi–vii

65. Legro RS, McAllister JM. Heirarchical clustering and beyond in PCOS endometrium: brave new world. *J Clin Endocrinol Metab* 2009;**94**:1084–5.

66. Talbott EO, Guzick DS, Sutton-Tyrrell K, et al. Evidence for association between polycystic ovary syndrome and premature carotid atherosclerosis in middle-aged women. *Arterioscler Thromb Vasc Biol* 2000;**20**:2414–21.

67. Legro R, Driscoll D, Strauss III J, et al. Evidence for a genetic basis for hyperandrogenemia in polycystic ovary syndrome. *Proc Natl Acad Sci USA* 1998;**95**:14956–60.

68. Legro RS, Strauss JF. Molecular progress in infertility: polycystic ovary syndrome. *Fertil Steril* 2002;**78**:569–76.

69. Mason H. Function of the polycystic ovary. *Hum Fertil (Camb)* 2000;**3**:80–5.

70. Magoffin DA. Ovarian enzyme activities in women with polycystic ovary syndrome. *Fertil Steril* 2006;**86** (Suppl. 1):S9–11.

71. Jakimiuk AJ, Weitsman SR, Brzechffa PR, et al. Aromatase mRNA expression in individual follicles from polycystic ovaries. *Mol Hum Reprod* 1998;**4**:1–8.

72. Welt CK, Taylor AE, Fox J, et al. Follicular arrest in polycystic ovary syndrome is associated with deficient inhibin A and B biosynthesis. *J Clin Endocrinol Metab* 2005;**10**:5582–7.

73. Wickenheisser JK, Nelson-DeGrave VL, McAllister JM. Human ovarian theca cells in culture. *Trends Endocrinol Metab* 2006;**17**:65–71.

74. Jakimiuk AJ, Weitsman SR, Magoffin DA. 5alpha-reductase activity in women with polycystic ovary syndrome. *J Clin Endocrinol Metabol* 1999;**84**:2414–8.

75. Chang RJ, Cook-Andersen H. Disordered follicle development. *Mol Cell Endocrinol* 2013;**373**:51–60.

76. Stubbs SA, Hardy K, Da Silva-Buttkus P, et al. Anti-mullerian hormone protein expression is reduced during the initial stages of follicle development in human polycystic ovaries. *J Clin Endocrinol Metab* 2005;**90**:5536–43.

77. Das M, Djahanbakhch O, Hacihanefioglu B, et al. Granulosa cell survival and proliferation are altered in polycystic ovary syndrome. *J Clin Endocrinol Metab* 2008;**93**:881–7.

78. Mehta RV, Malcom PJ, Chang RJ. The effect of androgen blockade on granulosa cell estradiol production after follicle-stimulating hormone stimulation in women with polycystic ovary syndrome. *J Clin Endocrinol Metab* 2006;**91**:3503–6.

79. Gilling-Smith C, Storey H, Rogers V, et al. Evidence for a primary abnormality in theca cell steroidogenesis in the polycystic ovarian syndrome. *Clin Endocrinol* 1997;**47**:1158–65.

80. Gilling-Smith C, Willis DS, Beard RW, et al. Hypersecretion of androstenedione by isolated thecal cells from polycystic ovaries. *J Clin Endocrinol Metab* 1994;**79**:1158–65.

81. Nelson VL, Legro RS, Strauss III JF, et al. Augmented androgen production is a stable steroidogenic phenotype of propagated theca cells from polycystic ovaries. *Mol Endocrinol* 1999;**13**:946–57.

82. Nelson VL, Qin KN, Rosenfield RL, et al. The biochemical basis for increased testosterone production in theca cells propagated from patients with polycystic ovary syndrome. *J Clin Endocrinol Metab* 2001;**86**:5925–33.

83. Wickenheisser JK, Quinn PG, Nelson VL, et al. Differential activity of the cytochrome P450 17alpha-hydroxylase and steroidogenic acute regulatory protein gene promoters in normal and polycystic ovary syndrome theca cells. *J Clin Endocrinol Metab* 2000;**85**:2304–11.

84. Jakimiuk AJ, Weitsman SR, Navab A, et al. Luteinizing hormone receptor, steroidogenesis acute regulatory protein, and steroidogenic enzyme messenger ribonucleic acids are overexpressed in thecal and granulosa cells from polycystic ovaries. *J Clin Endocrinol Metabol* 2001;**86**:1318–23.

85. Barnes R, Rosenfield R, Burnstein S, et al. Pituitary-ovarian responses to nafarelin testing in the polycystic ovary syndrome. *N Engl J Med* 1989;**320**:559–65.

86. Rosenfield R, Barnes R, Carr J, et al. Dysregulation of cytochrome P450c17 as the cause of polycystic ovarian syndrome. *Fert Steril* 1990;**53**:785–91.

87. Rosenfield RL. Ovarian and adrenal function in polycystic ovary syndrome. *Endocrinol Metab Clin North Am* 1999;**28**:265–93.

88. Wickenheisser JK, Nelson-DeGrave VL, Quinn PG, et al. Increased cytochrome P450 17alpha-hydroxylase promoter function in theca cells isolated from patients with polycystic ovary syndrome involves nuclear factor-1. *Mol Endocrinol* 2004;**18**:588–605.

89. Wickenheisser JK, Biegler JM, Nelson-Degrave VL, et al. Cholesterol side-chain cleavage gene expression in theca cells: augmented transcriptional regulation and mRNA stability in polycystic ovary syndrome. *PLoS One* 2012;**7**:e48963.

90. Strauss III JF. Some new thoughts on the pathophysiology and genetics of polycystic ovary syndrome. *Ann N Y Acad Sci* 2003;**997**:42−8.

91. Wood JR, Ho CK, Nelson-Degrave VL, et al. The molecular signature of polycystic ovary syndrome (PCOS) theca cells defined by gene expression profiling. *J Reprod Immunol* 2004;**63**:51−60.

92. Wood JR, Nelson VL, Ho C, et al. The molecular phenotype of polycystic ovary syndrome (PCOS) theca cells and new candidate PCOS genes defined by microarray analysis. *J Biol Chem* 2003;**278**:26380−90.

93. Ho CK, Wood JR, Stewart DR, et al. Increased transcription and increased messenger ribonucleic acid (mRNA) stability contribute to increased GATA6 mRNA abundance in polycystic ovary syndrome theca cells. *J Clin Endocrinol Metab* 2005;**90**:6596−602.

94. Vink JM, Sadrzadeh S, Lambalk CB, et al. Heritability of polycystic ovary syndrome in a dutch twin-family study. *J Clin Endocrinol Metab* 2006;**91**:2100−4.

95. Ehrmann DA, Kasza K, Azziz R, et al. Effects of race and family history of type 2 diabetes on metabolic status of women with polycystic ovary syndrome. *J Clin Endocrinol Metab* 2005;**90**:66−71.

96. Legro RS, Bentley-Lewis R, Driscoll D, et al. Insulin resistance in the sisters of women with polycystic ovary syndrome: association with hyperandrogenemia rather than menstrual irregularity. *J Clin Endocrinol Metab* 2002;**87**:2128−33.

97. Kosova G, Urbanek M. Genetics of the polycystic ovary syndrome. *Mol Cell Endocrinol* 2013;**373**:29−38.

98. Strauss III JF, McAllister JM, Urbanek M. Persistence pays off for PCOS gene prospectors. *J Clin Endocrinol Metab* 2012;**97**:2286−8.

99. Chen ZJ, Zhao H, He L, et al. Genome-wide association study identifies susceptibility loci for polycystic ovary syndrome on chromosome 2p16.3, 2p21 and 9q33.3. *Nat Genet* 2011;**43**:55−9.

100. Shi Y, Zhao H, Shi Y, et al. Genome-wide association study identifies eight new risk loci for polycystic ovary syndrome. *Nat Genet* 2012;**44**:1020−5.

101. Goodarzi MO, Jones MR, Li X, et al. Replication of association of DENND1A and THADA variants with polycystic ovary syndrome in European cohorts. *J Med Genet* 2012;**49**:90−5.

102. Welt CK, Styrkarsdottir U, Ehrmann DA, et al. Variants in DENND1A are associated with polycystic ovary syndrome in women of European ancestry. *J Clin Endocrinol Metab* 2012;**97**:E1342−7.

103. Lerchbaum E, Trummer O, Giuliani A, et al. Susceptibility loci for polycystic ovary syndrome on chromosome 2p16.3, 2p21, and 9q33.3 in a cohort of Caucasian women. *Horm Metab Res* 2011;**43**:743−7.

104. Eriksen MB, Brusgaard K, Andersen M, et al. Association of polycystic ovary syndrome susceptibility single nucleotide polymorphism rs2479106 and PCOS in Caucasian patients with PCOS or hirsutism as referral diagnosis. *Eur J Obstet Gynecol Reprod Biol* 2012;**163**:39−42.

105. Marat AL, Dokainish H, McPherson PS. DENN domain proteins: regulators of Rab GTPases. *J Biol Chem* 2011;**286**:13791−800.

106. Munir I, Yen HW, Geller DH, et al. Insulin augmentation of 17alpha-hydroxylase activity is mediated by phosphatidyl inositol 3-kinase but not extracellular signal-regulated kinase-1/2 in human ovarian theca cells. *Endocrinology* 2004;**145**:175−83.

107. Stenmark H. Rab GTPases as coordinators of vesicle traffic. *Nat Rev Mol Cell Biol* 2009;**10**:513−25.

108. Stenmark H, Olkkonen VM. The rab GTPase family. *Genome Biol* 2001;2 REVIEWS3007

109. Chiariello M, Bruni CB, Bucci C. The small GTPases Rab5a, Rab5b and Rab5c are differentially phosphorylated in vitro. *FEBS Lett* 1999;**453**:20−4.

110. Allaire PD, Marat AL, Dall'Armi C, et al. The Connecdenn DENN domain: a GEF for Rab35 mediating cargo-specific exit from early endosomes. *Mol Cell* 2010;**37**:370−82.

111. Carvalho CR, Carvalheira JB, Lima MH, et al. Novel signal transduction pathway for luteinizing hormone and its interaction with insulin: activation of Janus kinase/signal transducer and activator of transcription and phosphoinositol 3-kinase/Akt pathways. *Endocrinology* 2003;**144**:638−47.

112. Menon KM, Menon B. Structure, function and regulation of gonadotropin receptors - a perspective. *Mol Cell Endocrinol* 2012;**356**:88−97.

113. Zeleznik AJ, Saxena D, Little-Ihrig L. Protein kinase B is obligatory for follicle-stimulating hormone-induced granulosa cell differentiation. *Endocrinology* 2003;**144**:3985−94.

114. Palaniappan M, Menon KM. Luteinizing hormone/human chorionic gonadotropin-mediated activation of mTORC1 signaling is required for androgen synthesis by theca-interstitial cells. *Mol Endocrinol* 2012;**26**:1732−42.

115. Tai P, Ascoli M. Reactive oxygen species (ROS) play a critical role in the cAMP-induced activation of Ras and the phosphorylation of ERK1/2 in Leydig cells. *Mol Endocrinol* 2011;**25**:885−93.

116. Stocco DM, Wang X, Jo Y, et al. Multiple signaling pathways regulating steroidogenesis and steroidogenic acute regulatory protein expression: more complicated than we thought. *Mol Endocrinol* 2005;**19**:2647−59.

117. Kang W, Tong JH, Chan AW, et al. Yes-associated protein 1 exhibits oncogenic property in gastric cancer

and its nuclear accumulation associates with poor prognosis. *Clin Cancer Res* 2011;**17**:2130−9.

118. Lapi E, Di Agostino S, Donzelli S, et al. PML, YAP, and p73 are components of a proapoptotic autoregulatory feedback loop. *Mol Cell* 2008;**32**:803−14.

119. Cyert MS. Regulation of nuclear localization during signaling. *J Biol Chem* 2001;**276**:20805−8.

120. Basu S, Totty NF, Irwin MS, et al. Akt phosphorylates the Yes-associated protein, YAP, to induce interaction with 14-3-3 and attenuation of p73-mediated apoptosis. *Mol Cell* 2003;**11**:11−23.

121. Gulshan K, Rovinsky SA, Moye-Rowley WS. YBP1 and its homologue YBP2/YBH1 influence oxidative-stress tolerance by nonidentical mechanisms in *Saccharomyces cerevisiae*. *Eukaryot Cell* 2004;**3**:318−30.

122. Bloethner S, Chen B, Hemminki K, et al. Effect of common B-RAF and N-RAS mutations on global gene expression in melanoma cell lines. *Carcinogenesis* 2005;**26**:1224−32.

123. Tsuruta T, Kozaki K, Uesugi A, et al. miR-152 is a tumor suppressor microRNA that is silenced by DNA hypermethylation in endometrial cancer. *Cancer Res* 2011;**71**:6450−62.

124. Andarawewa KL, Erickson AC, Chou WS, et al. Ionizing radiation predisposes nonmalignant human mammary epithelial cells to undergo transforming growth factor beta induced epithelial to mesenchymal transition. *Cancer Res* 2007;**67**:8662−70.

125. Diaz-Perales A, Quesada V, Sanchez LM, et al. Identification of human aminopeptidase O, a novel metalloprotease with structural similarity to aminopeptidase B and leukotriene A4 hydrolase. *J Biol Chem* 2005;**280**:14310−7.

Control of the GnRH Pulse Generator

Manuel Tena-Sempere

University of Córdoba, Córdoba, Spain, Instituto de Salud Carlos III, Córdoba, Spain, Instituto
Maimónides de Investigaciones Biomédicas (IMIBIC)/Hospital Universitario Reina Sofia, Córdoba,
Spain

INTRODUCTION: THE GnRH PULSE GENERATOR

Reproduction, as the complex series of processes that enable organisms to generate new organisms of the same kind, is mandatory for the perpetuation of species. Accordingly, this body function is precisely controlled; the development, maturation, and dynamic regulation of the reproductive system are under the regulation of sophisticated mechanisms, which impinge and integrate at different levels of the hypothalamic–pituitary–gonadal (HPG) axis.[1–3] This neurohormonal system is primarily composed by the hypothalamic decapeptide, gonadotropin-releasing hormone (GnRH), the pituitary gonadotropins – luteinizing hormone (LH) and follicle-stimulating hormone (FSH), and gonadal hormones of steroid and peptidergic nature.[4] These factors are connected by feedforward and feedback regulatory loops, which are organized at early developmental periods and are responsible for the dynamic regulation of the reproductive system during the lifespan.[4]

Because of their pivotal position in this neuroendocrine network, GnRH neurons are considered a major hierarchical element of this system. Hence, GnRH operates as the final output signal for the hypothalamic regulation of the downstream elements of the HPG axis, by virtue of its ability to drive the pulsatile secretion of LH and FSH. As is the case for other hypophysitropic hormones, in order to conduct its gonadotropin-releasing effects, adequate pulsatile secretion of GnRH is mandatory;[5] thus, conditions of deregulated secretion of GnRH are incompatible with proper attainment and maintenance of reproductive function. The pulsatile release of GnRH is driven by the activity of the so-called *GnRH pulse generator*. The concept of a hypothalamic system involving GnRH neurons and key afferents responsible for the precise generation of GnRH secretory pulses was originally suggested by Knobil more than three decades ago.[5]

311

Yet, the morphological and functional basis of such a *generator* is still under active investigation. Nonetheless, experimental evidence from different species, including rodents and primates, has documented that the secretory profiles of GnRH are the result of the combination of the intrinsic oscillatory activity of GnRH neurons, which can generate *per se* secretory effects,[6,7] and the essential contribution of various hypothalamic afferents, including those originating from the mediobasal hypothalamus (MBH).[8]

The GnRH pulse generator dictates the function of the HPG axis and, accordingly, is the target of the regulatory actions of numerous peripheral and central regulators that permit the functional coupling between reproduction and other key body systems.[4] However, the precise central mechanisms for the tight regulation of the GnRH neuronal system, and how these are targeted by peripheral signals and environmental cues, are yet to be fully exposed. In this sense, while it is known that a substantial component of the control of the HPG axis ultimately occurs at the level of GnRH neurons,[4,9] it has become clear in recent years that many key regulatory signals, such as sex steroids and various metabolic factors, conduct their actions via indirect modification of GnRH neuronal function. Indeed, experimental evidence for the relevance of indirect actions on GnRH neurons has fueled specific neuroanatomical and functional studies aimed at deciphering the nature and mode of action of the major regulatory afferents responsible for the control of the GnRH neurosecretory activity. In this chapter, we will focus our attention on a recently recognized player of the central networks governing reproduction, namely, the Kiss1/NKB/dynorphin system. The major elements of the system will be presented in the following sections, with special emphasis in the description of the interplay between these ligands and their receptors in the precise control of pulsatile GnRH secretion.

ROLES OF THE Kiss1/Kiss1R SYSTEM IN THE CONTROL OF GnRH NEURONS

Among the pathways and afferent signals responsible for transmitting the regulatory actions of different signals to GnRH neurons, a great deal of effort has been devoted in recent years to elucidate the putative roles of kisspeptins in the metabolic control of puberty and fertility.[4,10] The *Kiss1* gene and their products, kisspeptins, were originally identified as metastasis suppressors; the canonical kisspeptin receptor, the G protein-coupled receptor, Kiss1R (also termed Gpr54), was initially a orphan in 1999, and later linked to kisspeptin signaling in 2001. In mammals, kisspeptins exist in different lengths: kisspeptin-54 (kp-54), kp-14, kp-13 and kp-10; the latter sequence is shared by all active forms of kisspeptin and is sufficient to bind and activate Kiss1R.[4]

The connection between the Kiss1 system and reproductive function was disclosed in 2003, when patients with null mutations of the *KISS1R* gene were found to display hypogonadotropic hypogonadism (HH), i.e., reproductive failure of central origin.[11,12] These initial findings have been recently complemented by the first demonstration of an inactivating mutation of the *KISS1* gene in humans with HH.[13] A grossly similar phenotype is observed in mice lacking functional *Kiss1* or *Kiss1R* genes.[14] These findings boosted an extraordinary interest that has allowed, by a combination of molecular, physiological, and pharmacological studies, to rapidly define the major features of kisspeptins as essential regulators of GnRH neurons.[4,15]

Indeed, compelling evidence has conclusively documented that the primary site of action for the *reproductive* effects of kisspeptins is the hypothalamus, where they potently and directly activate GnRH neurons. This mode of action is supported by the facts that in mammals GnRH neurons express *Kiss1R* and respond to kisspeptins with *C-fos* induction.[16] Moreover, GnRH antagonists abrogate gonadotropin responses to kisspeptin administration, while kisspeptins potently increases the firing rate of GnRH neurons and elicit GnRH release *ex vivo* and *in vivo*. In addition, kisspeptin antagonism decreases the electrical activity of GnRH neurons and pulsatile release of GnRH.[4,16] This evidence clearly demonstrates a major primary action of kisspeptins on GnRH neurons, which does not preclude the possibility of additional, indirect effects, via modulation of other excitatory afferents to GnRH neurons, and/or direct actions at the pituitary level,[17] presumably of less physiological relevance.

The source of kisspeptins, as trans-synaptic regulators of GnRH neurons, has been also actively investigated in recent years. Indeed, neuroanatomical studies in various mammalian species have allowed the identification of discrete neuronal populations in the hypothalamus that are (presumably) responsible for the stimulatory control of the secretory activity of GnRH neurons.[4,16,18] The location and functional features of such populations has been thoroughly characterized in rodents, in which two major groups of Kiss1 neurons are sited in the arcuate nucleus (ARC) and anteroventral periventricular nucleus (AVPV).[4,16,18] Of note, ARC and AVPV neurons diverge not only in their neuroanatomical location but also in their patterns of co-transmitters and molecular responses to key regulators, such as sex steroids.[4,16] On the former, ARC (but not AVPV) Kiss1 neurons have been shown to co-express neurokinin-B (NKB) and dynorphin that, as described in later sections of this chapter, play important roles in the dynamic regulation of this population of Kiss1 neurons.[16] On the latter, it has been well documented that while ARC Kiss1 neurons respond to sex steroid with a decrease in *Kiss1* (and *NKB*) mRNA expression in various mammalian species,[19]

Kiss1 neurons in the AVPV display completely opposite responses, so that estrogen enhances *Kiss1* expression levels in this area.[19] Accordingly, ARC Kiss1 neurons have been proposed to play a substantial role in mediating the negative feedback effects of gonadal steroids, whereas AVPV Kiss1 neurons appear to be essential for the positive feedback of estradiol and the generation of the pre-ovulatory surge of gonadotropins,[19] as supported also by functional genomics and pharmacological studies.[20,21]

In fact, solid anatomical, genetic, electrophysiological, pharmacological, and hormonal data have collectively illustrated in recent years the essential physiological roles of kisspeptins in the regulation of all key aspects of reproductive maturation and function, including brain sex differentiation, timing of puberty onset, tonic control of gonadotropin secretion, generation of the ovulatory surge of gonadotropins, seasonal regulation of reproduction, and modulation of the reproductive axis by metabolic cues. While extensive recapitulation of these important features exceeds the scope of this chapter and can be found elsewhere, it is important to stress that these functions appear to be primarily mediated by the ability of kisspeptins to regulate the pulsatile secretion of GnRH. In this context, direct appositions between kisspeptin fibers and GnRH neurons have been demonstrated in rodents;[22] yet, such kisspeptin projections seem to stem mainly from the AVPV, while ARC projections are less numerous and have been difficult to document.[23] Interestingly, neuroanatomical studies in mammals have shown kisspeptin fibers in the vicinity not only of GnRH perikarya but also nerve terminals.[24] Nonetheless, the number of synaptic contacts between Kiss1 and GnRH neurons appears to be low, thus suggesting some inter-neuronal or even non-synaptic communication between these two neuronal populations.

PARTNERS OF Kiss1/Kiss1R IN THE CONTROL OF GnRH NEURONS: NKB AND DYNORPHIN

As mentioned above, the recognition of the different patterns of co-transmitters expressed by the different populations of Kiss1 neurons prompted anatomical and functional studies aimed at defining the physiological relevance of such kisspeptin partners. While evidence for the synthesis of other neuropeptides, such as met-enkephalin and galanin, has been presented for AVPV Kiss1,[25] the co-expression of NKB and dinorphyn selectively in the ARC population of Kiss1 neurons has been, by far, the one that has attracted more interest.[26] Such co-localization was first documented in the sheep in 2007, and later confirmed in the mouse, goat and monkey, among other species;[26] hence, the term KNDy (standing for kisspeptin, NKB and dynorphin) has been coined for this population. The physiological interest of the above anatomical finding was reinforced by previous literature on the effects of NKB and dynorphin in the control of gonadotropin secretion,[27] and (especially) by the demonstration that inactivating mutations of TAC3 or TACR3 (the genes encoding NKB and its canonical receptor, NK3) were associated with central hypogonadism,[28] just as previously reported for KISS1R mutations.

NKB belongs to the family of tachykinins, which also includes substance P, neurokinin-A (NKA), hemokinin-1, and neuropeptides K and γ.[29] In addition, three tachykinin receptors have so far been reported: NK1R (which is preferentially bound by substance P), NK2R (that binds NKA) and NK3R; NKB preferentially activates NK3R, which is considered the canonical NKB receptor.[27] The gene encoding NKB is named TAC3 in humans and Tac2 in rodents, whereas the NK3R is encoded by the TACR3/Tacr3 gene.[27] In turn, dynorphin is a member of the endogenous opioid peptide (EOP) family, which is encoded by the prodynorphin gene

and acts via the kappa opioid receptor (κOR).[30] Of note, a number of pharmacological studies had addressed the roles of NKB and other tachykinins in the regulation of gonadotropin secretion in various species.[27,31,32] Yet, before the initial human TAC3/TACR3 mutation studies,[28] evidence regarding the role of NKB in the control of the gonadotropic axis remained scarce, and rodent experiments had suggested the possibility of inhibitory effects of NKB agonist, senktide, on LH secretion, an effect that was later found at odds with the reported failure of gonadal function in patients with genetic inactivation of NKB signaling. On the other hand, dynorphin, as well as other EOPs, had been long cataloged as a proven inhibitor of gonadotropin secretion.[30,33] Yet, the ultimate mechanisms for the central regulatory effects of NKB and dynorphin on the HPG axis remained partially unknown.

The co-expression of NKB and dynorphin in ARC Kiss1 neurons was first documented in the ewe,[26] and was later confirmed in different species, such as mouse and goat, where in situ hybridization has demonstrated the co-localization of Kiss1, Tac2 and prodynorphin mRNAs in the ARC.[16] Likewise, indirect evidence has suggested the existence of ARC KNDy neurons in rats, and co-expression of NKB and kisspeptins has been shown in the monkey.[16] Moreover, the presence of NKB has been demonstrated in as much as 77% of all Kiss1 neurons of the infundibular nucleus in women.[34] Yet, some species and/or sex differences in the proportion of KNDy neurons might exist, since a recent study suggested limited overlapping between kisspeptin-, NKB-, and dynorphin-positive neurons in young men;[35] however, in a later report, as much as 75% of Kiss1 neuron perikarya were also labeled for NKB. In any event, while the co-expression of the three neuropeptides may be regarded as an important functional signature of this set of neurons, it is plausible that ARC Kiss1 neurons express other neuropeptides and transmitters,[26] which may cooperate in the regulation of GnRH neurons or other functions. Likewise, based on neuroanatomical studies, it is likely that NKB neurons not expressing *Kiss1*/kisspeptins may exist in the ARC, at least of humans and rodents[35] (and Bentsen & Mikkelsen, personal communication).

Immunohistochemical and *in situ* hybridization studies have suggested that KNDy neurons express also the NKB receptor, NK3R, thus providing the basis for auto-regulatory effects of NKB on KNDy neurons.[16] The presence of the dynorphin receptor, κOR (encoded by *Oprk1*), in this population was initially suggested,[16] but has been recently questioned on the basis of *in situ* hybridization data in mice.[36] Of note, initial analyses in the rat suggested the presence of NK3R in GnRH axons, and *Tacr3* mRNA was detected in mouse GnRH neurons.[37] However, immunohistochemical analyses documented that only a minute percentage of GnRH cells expressed NK3R in the rat.[38] Likewise, dual-label studies in the sheep failed to demonstrate any detectable co-expression of GnRH and NK3R, neither at the level of cell bodies or the median eminence.[39] In the same vein, recent *in situ* hybridization studies in the mouse have failed to demonstrate expression of *NK3R* mRNA in GnRH neurons, which did not show detectable electrophysiological responses to the NKB agonist, senktide.[36] These data would suggest a preferential, if not exclusive, action of NKB on KNDy neurons for the indirect control of GnRH secretion.

Another interesting feature of KNDy neurons in the ARC of potential functional relevance is that they are profusely interconnected, a fact that has been documented in sheep and rodents.[26,40,41] Thus, ARC KNDy neurons seem to conform to a tightly connected network that is well suited to operate in a synchronized manner to control key neurosecretory events, such as GnRH pulsatile release. It is also interesting

to note that virtually all KNDy neurons in the sheep seem to express estrogen receptor (ER)α, as well as progesterone and androgen receptors, therefore providing the basis for their key role in the feedback control of GnRH/gonadotropin secretion.[26] Likewise, the presence of ERα has been documented in neurons expressing kisspeptin, NKB and/or dynorphin in other species, including rodents, monkeys and humans.[27,40,42] Given the conspicuous lack of ERα in GnRH neurons, this feature strongly suggests an important role of ARC KNDy neurons in transmitting the effects of estrogen in the dynamic control of pulsatile GnRH secretion.

INTERPLAY BETWEEN KISSPEPTINS, NKB AND DYNORPHIN IN THE CONTROL OF GnRH: THE KNDy PARADIGM

The data from human genetics, as well as pharmacological and neuroanatomical analyses reviewed in previous sections strongly suggested functional interplay between kisspeptins, NKB and dynorphin in the central control of the HPG axis. Hence, the effects of NKB, and to a lesser extent dynorphin, in the central regulation of GnRH/gonadotropin secretion have been actively investigated in recent years.[16] Indeed, initial hormonal studies addressing the gonadotropic actions of NKB revealed inhibitory effects of this tachykinin on LH secretion,[32] findings that were later considered discordant with the observed phenotype of central hypogonadism of humans with inactivating mutations of the NKB system.

In the last 3 years, systematic analyses of the effects of NKB on the gonadotropic axis have been conducted in different species, including primates (monkey), sheep, goats, rats, mice, and non-mammalian (fish) vertebrates, using mainly LH levels as surrogate marker of GnRH activation. These studies have documented different (even contradictory) findings regarding the effects of NKB, or the agonist of NK3R, senktide, on LH secretion, although there is a predominance of reports showing stimulatory actions in physiological conditions.[16] Thus, potent LH responses have been reported in male mice following central injection of senktide.[36] Similarly, robust LH-releasing responses to intracerebroventricular (icv) administration of senktide have been documented in cyclic female rats and ovariectomized (OVX) rats with physiological supplementation of estradiol,[43] as well as in prepubertal male and female rats.[44,45] In addition, potent stimulatory effects of NKB and/or senktide on LH secretion have been detected in the adult ewe (at the follicular phase of the ovarian cycle) and in the juvenile monkey.[46,47] The effects of senktide in the monkey were blocked by pre-administration of a GnRH antagonist, therefore suggesting a central site of action.[46] Very recently, stimulatory effects of two Tac2-derived peptides, namely NKB and NKF, have been reported in zebrafish.[48]

Despite the above evidence for conserved stimulatory effects of NKB on LH secretion, null or even inhibitory actions have been also reported. In fact, senktide administration to OVX mice or rats, without hormonal replacement or very low (or ineffective) estrogen substitution, has been shown to suppress (rather than stimulate) LH secretion.[32,49,50] In addition, effective doses of senktide in adult female rats were unable to elicit LH secretion in adult male rats,[44] in which NKB even partially suppressed LH secretory responses to kisspeptin-10.[51] While the latter findings do not question the predominant stimulatory effect of NKB in the control of the gonadotropic axis, they nicely illustrate that the net effects of NKB might be tightly dependent on different physiological parameters, including the endogenous sex steroid levels, the prevailing GnRH neurosecretory activity, and the stage of development. In addition, there might exist also species differences in

the roles of NKB in reproductive control; this was initially suggested by the fact that, in contrast to humans, mice with inactivating mutations of Tacr3 gene appeared to have preserved fertility.[52] Yet, different reproductive deficits have been recently reported in a novel Tacr3 null mouse line,[53] and, as mentioned above, evidence for a (stimulatory) role of the NKB system in the control of the HPG axis in non-mammalian vertebrates has been recently provided.[48]

The presence of dynorphin in KNDy neurons justified also the re-evaluation of the effects of this neuropeptide on gonadotropin secretion. It must be noted, however, that contrary to NKB, for which opposite actions have been reported (see above), unanimous evidence had set the contention that dynorphin is an inhibitor of gonadotropin secretion,[30,33] as is also the case for other endogenous opioids. In good agreement, dynorphin has been recently shown to suppress LH secretion in mice and to inhibit multiunit activity (MUA) volleys, as markers of GnRH pulses, and pulsatile LH secretion in goats.[49,54] Similarly, studies in cyclic female rats and young male rats have shown that dynorphin (moderately) inhibits basal LH secretion and LH responses to kisspeptin-10.[51]

The integration of the neuroanatomical and functional data reviewed above led to the proposal of a hypothetical model that we may regard as the KNDy paradigm, which aimed to incorporate the dynamic effects of NKB and dynorphin to the pre-existing Kiss1 neuronal model for the fine control of pulsatile GnRH secretion. Thus, according to this model (tentatively depicted in Figure 19.1), NKB and dynorphin operate, in a reciprocal manner, as major positive and negative regulators of the pulsatile release of kisspeptins by KNDy neurons in the ARC,[49,54] kisspeptin being the output effector of this neuronal population for the stimulation of the neurosecretory activity of GnRH neurons. This model is supported not only by the pharmacological data summarized

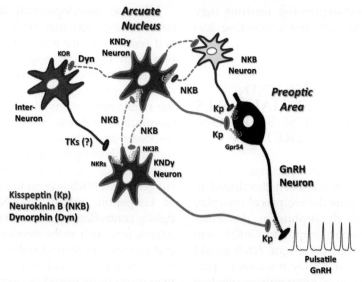

FIGURE 19.1   Tentative model for the control of pulsatile GnRH secretion by KNDy neurons. Neurons in the arcuate nucleus (ARC) that co-express kisspeptins (Kp), neurokinin B (NKB) and dynorphin (Dyn), also termed "KNDy neurons," have been proposed to play an essential role in the control of the pulsatile secretory activity of GnRH neurons. While the original model suggested that NKB and Dyn actions on Kiss1-expressing neurons in the ARC would operate as switch-on and -off mechanism (respectively) for kisspeptin release, recent evidence suggest a more complex mode of action in which the net effect of NKB may depend, among other factors, on the developmental stage, sex steroid milieu, and/or the prevailing GnRH secretory activity, and may result in stimulatory (predominant) or inhibitory effects. Further refinement of this model comes from the suggestion that Dyn may preferentially act on inter-neuronal afferents to KNDy neurons, and the possibility that NKB (or Dyn) neurons not expressing Kp may also integrate in this network. Similarly, the participation of other tachykinins (TKs) in the control of KNDy neurons has been recently proposed. In addition, via as yet unknown mechanisms, NKB and Dyn may also modulate GnRH responses to kisspeptins (rather than kisspeptin release). Moreover, for some of its effects, NKB would require a preserved Dyn signaling.

above, with predominant stimulatory effects of NKB in numerous species and the invariant inhibitory action of dynorphin, but also by the neuroanatomical features of ARC KNDy neurons, which are profusely interconnected, thus permitting NKB and dynorphin to operate auto-synaptically within the network to modulate kisspeptin outflow to GnRH perikarya and/or nerve terminals. Further support for this mode of action comes from the proven ability of senktide to activate Kiss1 neurons, as measured by c-fos induction or electrophysiological recordings in rats and mice;[36,43] conversely, dynorphin has been recently shown to suppress the firing rate of ARC KNDy neurons.[55,56] In addition, senktide was unable to

stimulate LH secretion (as proxy marker of GnRH activation) in mice without kisspeptin signaling (Kiss1R knockout mice).[20] In the same vein, recent evidence suggests that the stimulatory actions of NKB on GnRH secretion take place upstream of kisspeptin signaling, and are blocked after Kiss1R desensitization in the monkey.[57] Of note, this KNDy paradigm is not invalidated by the possibility that the percentage of actual KNDy neurons (i.e., neurons co-expressing the three peptides) might be low in some species or physiological conditions. In fact, it is not clear whether the dynorphin receptor is actually expressed in KNDy neurons or in upstream afferents (see above). Similarly, it might be possible that some of the

NKB input on Kiss1-expressing neurons may stem from NKB neurons not expressing kisspeptins and/or dynorphin.

RE-SHAPING THE KNDy PARADIGM: DIFFERENTIAL ROLES OF KISSPEPTINS VERSUS NKB ON LH AND FSH SECRETION

As described in previous sections, a hypothetical (KNDy) model has been developed in recent years to integrate the reciprocal interplay between NKB and dynorphin in the auto-regulation of kisspeptin output to GnRH neurons. According to this paradigm, NKB would operate as activator of KNDy neurons to promote GnRH pulses in a kisspeptin-dependent manner. Experimental evidence suggests, however, that the above model might be an over-simplification of a complex neuronal circuit, so that NKB activation is not always associated to potent stimulation of kisspeptin release and, hence, GnRH/LH secretion. This is nicely illustrated by findings in OVX rodents (as discussed in previous sections), showing that NKB activation is linked to clear-cut inhibitory responses in conditions of null/low sex steroid levels and prevailing activation of GnRH secretory activity. The mechanisms for such a switch from stimulatory (in conditions of adequate sex steroid milieu) to inhibitory (in conditions of null or insufficient sex steroid levels) responses remain unclear. One possibility is that in OVX conditions the endogenous NKB tone is supposedly high, thus making the system prone to display desensitization responses after pharmacological doses of a NKB agonist (that would further enhance the NKB signal). As an additional note, studies in young male rats have shown that co-stimulation with senktide might partially suppress LH responses to exogenous kisspeptin-10, therefore suggesting another lever of potential interplay, namely a step beyond the regulation of kisspeptin release.[51]

Likewise, developmental studies in the rat have documented that, while prepubertal male and female rats are responsive to NKB stimulation with clear LH responses, male (but not female) rats become non-responsive to senktide after puberty, suggesting sex-specific changes in the regulatory actions of NKB during postnatal maturation.[44] Of note, very potent gonadotropin responses are detected after exogenous administration to kisspeptins in male and female rats, regardless of the developmental stage.[58] This suggests that NKB actions in terms of regulation of kisspeptin output by KNDy neurons are tightly controlled by a number of physiological parameters, such as the developmental stage and endogenous sex steroid milieu, so that, contrary to kisspeptin stimulation, the net effect of NKB may result in increased or decreased GnRH pulsatility depending on the prevailing physiological conditions.

In the same vein, the specific roles of NKB in the control of FSH secretion have remained largely unexplored. In fact, while kisspeptins have been demonstrated to stimulate the secretion of both gonadotropins in a variety of species and conditions,[4] virtually all the studies published to date have focused on the effects of NKB on LH secretion, with little attention being paid to elucidate the influence of NKB signaling in the regulation of FSH secretion. However, our recent preliminary work in male and female rats suggests that senktide can induce significant LH and FSH secretory responses, but only in pre-pubertal rats (author's unpublished data). In contrast, this NKB agonist failed to evoke FSH release in pubertal or adult rats, either gonadal-intact or OVX followed by sham or estradiol replacement (author's unpublished data). This is in clear contrast with the ability of senktide to robustly stimulate LH release in pubertal and adult female rats.[44,45] Of note, adult male rats, which are highly responsive to kisspeptins, did not display FSH (or LH) responses to senktide.[44] The above dissociations between kisspeptin and NKB responses, in terms of LH versus FSH release, clearly illustrate that the net effects of NKB in the control of kisspeptin/GnRH/gonadotropin secretion are complex and diverse, and may help to explain the divergences in the secretory profiles between the two gonadotropins observed in certain physiological or pathological conditions.

Finally, the relative importance of the endogenous dynorphin tone in the KNDy control of GnRH release is yet to be fully characterized. Again, recent (partially unpublished) work studying the effects of the antagonist of dynorphin receptor, nor-BNI, on gonadotropin responses to kisspeptin-10 and senktide administration has revealed that pre-treatment of adult rats with the antagonist increased basal LH and FSH levels. Furthermore, blockade of endogenous dynorphin tone enhanced gonadotropin responses to kisspeptin-10,[51] and allowed the manifestation of senktide-induced LH release in adult males (author's unpublished data), which are non-responsive to NKB agonist in basal conditions. Moreover, adult female rats, which do not display detectable FSH responses to senktide (see above), become responsive to senktide after blockade of dynorphin receptors (author's unpublished data). These data strongly suggest a role of dynorphin signaling in inhibiting basal gonadotropin secretion and their responses to NKB and kisspeptin. All in all, the available evidence documents that, rather than simple switch-on and -off mechanisms, the delicate balance between NKB and dynorphin signaling, in conjunction with a number of physiological variables, is determinant for the fine-tuning of a range of kisspeptin secretory responses and actions, which in turn are essential for the dynamic regulation of the HPG axis.

FUTURE PERSPECTIVES AND CONCLUDING REMARKS

Identification of the essential roles of Kiss1 neurons in the control of reproduction, and the later refinement of our knowledge of their mode of action, with the disclosure of the interplay between kisspeptins, NKB, dynorphin, and their receptors, have revolutionized our understanding of the neuro-endocrine mechanisms by which essential facets of the development and function of the HPG axis are precisely controlled. In this context, formulation of the KNDy paradigm, as to define the interactions of kisspeptins/NKB/dynorphin in the regulation of the ARC population of Kiss1 neurons, has been extremely useful and has provided a tenable model to explain how the elements of the KNDY system operate to finely modulate the pulsatile secretion of GnRH.

While we believe this model remains valid, recent developments will likely help to re-shape and refine the KNDy paradigm, in order to provide a more realistic view of the operational mode of action of this complex neuropeptide network. As illustrative example, recent immunohistochemical analyses have questioned the existence of a large population of KNDy neurons, at least in men.[35] While these observations raise the interesting issue as to whether NKB or dynorphin neurons not expressing kisspeptins, share with KNDy neurons their role in the control of GnRH secretion, it must be stressed that the KNDy model of regulation of kisspeptin output to GnRH neurons would not be invalidated even if the number of ARC neurons actually co-expressing the three peptides turns out to be low. In fact, by the presence of colateral projections, NKB or dynorphin neurons may modulate kisspeptin release by KNDy or Kiss1-only neurons in the ARC.

One interesting feature that has been recently recognized is that tachykinin control of kisspeptin/GnRH secretion is likely not restricted to NKB/NK3R pathways. In fact, in a very recent electrophysiological study, de Croft and co-workers documented that it was not only NKB and its agonist, senktide, that were able to activate ARC Kiss1 neurons, but also agonists of NK1R and NK2R (substance-P and NKA, respectively).[56] Interestingly, in that study, the effects of NKB were not suppressed by the blockade of individual tachykinin receptor subtypes; on the contrary, NKB effects were only prevented by a cocktail of NK1R, NK2R, and NK3R antagonists.[56] These observations suggest that there is a complex mode of action of NKB, acting through multiple receptors, in the control of KNDy neurons, and raise the possibility of a role for other tachykinins in the control of pulsatile GnRH secretion. Whether, as proposed for NKB, the effects of NK1R and/or NK2R pathways would operate via modulation of kisspeptin secretion and/or actions is presently under investigation.

Similarly, the interplay between dynorphin and NKB (or other tachykinins) in the control of kisspeptin output by KNDy neurons needs further analyses. As an example of the complexity of these interactions, it has been recently suggested that the inhibitory effects of NKB on pulsatile GnRH secretion (i.e., in OVX models of insufficient sex steroid replacement) require a preserved κOR signaling pathway, and do not manifest in the presence of a dynorphin antagonist.[50] Conversely, the author's unpublished data strongly suggest that changes in the endogenous dynorphin tone are partially responsible for the differences in gonadotropin responses between adult male and female rats. Similarly, endogenous dynorphin seems to play an important role in the negative feedback effects of estradiol on LH secretion,[59] whose physiological relevance needs further investigation.

In the same vein, while the specific roles of kisspeptins in the control of key facets of reproductive maturation and function have been thoroughly investigated,[4] the contribution of the KNDy network, and specifically of NKB and/or dynorphin, in these phenomena warrants additional investigation. Yet, preliminary evidence coming from rodent and sheep studies strongly suggests that NKB may cooperate with kisspeptins in the control of specific aspects of brain sex

differentiation, the timing of puberty, the feedback control of gonadotropin secretion and the metabolic regulation of puberty and fertility. These facets of the physiology of KNDy neurons need further clarification, and additional studies along these lines are needed as they will help to better delineate the integral role of kisspeptins, NKB, dynorphin, and their receptors in the central control of the HPG axis in normal and, eventually, pathological conditions.

Acknowledgements

The author is indebted to E. Aguilar, L. Pinilla, V.M. Navarro, J. Roa, J.M. Castellano, D. García-Galiano, and other the members of the research team in the Physiology Section of the University of Córdoba, who actively participated in the generation of experimental data discussed herein. The work from the author's laboratory summarized in this article was supported by grants BFU 2008-00984 and BFU2011-25021 (Ministerio de Economía and Competitividad, Spain; covered in part by EU-FEDER funds), grant P08-CVI-03788 (Junta de Andalucía, Spain), grants from Instituto de Salud Carlos III (Red de Centros RCMN C03/08 and Project PI042082; Ministerio de Sanidad, Spain), and EU research contract DEER FP7-ENV-2007-1. CIBER is an initiative of Instituto de Salud Carlos III (Ministerio de Sanidad, Spain).

References

1. Fink G. Neuroendocrine regulation of pituitary function: general principles. In: Conn PM, Freeman ME, editors. *Neuroendocrinology in physiology and medicine.* Totowa, New Jersey: Humana Press; 2000. p. 107–34.
2. Tena-Sempere M, Huhtaniemi I. Gonadotropins and gonadotropin receptors. In: Fauser BCJM, editor. *Reproductive medicine – molecular, cellular and genetic fundamentals.* New York: Parthenon Publishing; 2003. p. 225–44.
3. Schwartz NB. Neuroendocrine regulation of reproductive cyclicity. In: Conn PM, Freeman ME, editors. *Neuroendocrinology in physiology and medicine.* Totowa, New Jersey: Humana Press; 2000. p. 135–46.
4. Pinilla L, Aguilar E, Dieguez C, Millar RP, Tena-Sempere M. Kisspeptins and reproduction: physiological roles and regulatory mechanisms. *Physiol Rev* 2012;**92**(3):1235–316.
5. Knobil E. The neuroendocrine control of the menstrual cycle. *Recent Prog Horm Res* 1980;**36**:53–88.

6. Maeda K, Ohkura S, Uenoyama Y, Wakabayashi Y, Oka Y, Tsukamura H, et al. Neurobiological mechanisms underlying GnRH pulse generation by the hypothalamus. *Brain Res* 2010;**1364**:103−15.

7. Terasawa E, Kurian JR, Guerriero KA, Kenealy BP, Hutz ED, Keen KL. Recent discoveries on the control of gonadotrophin-releasing hormone neurones in non-human primates. *J Neuroendocrinol* 2010;**22**(7):630−8.

8. Halasz B, Pupp L. Hormone secretion of the anterior pituitary gland after physical interruption of all nervous pathways to the hypophysiotrophic area. *Endocrinology* 1965;**77**(3):553−62.

9. Elias CF, Purohit D. Leptin signaling and circuits in puberty and fertility. *Cell Mol Life Sci* 2012;**70**:841−62.

10. Castellano JM, Bentsen AH, Mikkelsen JD, Tena-Sempere M. Kisspeptins: bridging energy homeostasis and reproduction. *Brain Res* 2010;**1364**:129−38.

11. de Roux N, Genin E, Carel JC, Matsuda F, Chaussain JL, Milgrom E. Hypogonadotropic hypogonadism due to loss of function of the KiSS1-derived peptide receptor GPR54. *Proc Natl Acad Sci U S A* 2003;**100**(19): 10972−6.

12. Seminara SB, Messager S, Chatzidaki EE, Thresher RR, Acierno Jr. JS, Shagoury JK, et al. The GPR54 gene as a regulator of puberty. *N Engl J Med* 2003;**349**(17): 1614−27.

13. Topaloglu AK, Tello JA, Kotan LD, Ozbek MN, Yilmaz MB, Erdogan S, et al. Inactivating KISS1 mutation and hypogonadotropic hypogonadism. *N Engl J Med* 2012; **366**(7):629−35.

14. Colledge WH. Transgenic mouse models to study Gpr54/kisspeptin physiology. *Peptides* 2009;**30**(1): 34−41.

15. Oakley AE, Clifton DK, Steiner RA. Kisspeptin signaling in the brain. *Endocr Rev* 2009;**30**(6):713−43.

16. Navarro VM, Tena-Sempere M. Neuroendocrine control by kisspeptins: role in metabolic regulation of fertility. *Nat Rev Endocrinol* 2012;**8**(1):40−53.

17. Luque RM, Cordoba-Chacon J, Gahete MD, Navarro VM, Tena-Sempere M, Kineman RD, et al. Kisspeptin regulates gonadotroph and somatotroph function in nonhuman primate pituitary via common and distinct signaling mechanisms. *Endocrinology* 2011;**152**(3): 957−66.

18. Mikkelsen JD, Simonneaux V. The neuroanatomy of the kisspeptin system in the mammalian brain. *Peptides* 2009;**30**(1):26−33.

19. Garcia-Galiano D, Pinilla L, Tena-Sempere M. Sex steroids and the control of the Kiss1 system: developmental roles and major regulatory actions. *J Neuroendocrinol* 2012;**24**(1):22−33.

20. Garcia-Galiano D, van Ingen Schenau D, Leon S, Krajnc-Franken MA, Manfredi-Lozano M, Romero-Ruiz A, et al. Kisspeptin signaling is indispensable for neurokinin B, but not glutamate, stimulation of gonadotropin secretion in mice. *Endocrinology* 2012;**153** (1):316−28.

21. Pineda R, Garcia-Galiano D, Roseweir A, Romero M, Sanchez-Garrido MA, Ruiz-Pino F, et al. Critical roles of kisspeptins in female puberty and preovulatory gonadotropin surges as revealed by a novel antagonist. *Endocrinology* 2010;**151**(2):722−30.

22. Clarkson J, d'Anglemont de Tassigny X, Colledge WH, Caraty A, Herbison AE. Distribution of kisspeptin neurones in the adult female mouse brain. *J Neuroendocrinol* 2009;**21**(8):673−82.

23. Yeo SH, Herbison AE. Projections of arcuate nucleus and rostral periventricular kisspeptin neurons in the adult female mouse brain. *Endocrinology* 2011;**152**(6): 2387−99.

24. Ramaswamy S, Guerriero KA, Gibbs RB, Plant TM. Structural interactions between kisspeptin and GnRH neurons in the mediobasal hypothalamus of the male rhesus monkey (Macaca mulatta) as revealed by double immunofluorescence and confocal microscopy. *Endocrinology* 2008;**149**(9):4387−95.

25. Porteous R, Petersen SL, Yeo SH, Bhattarai JP, Ciofi P, de Tassigny XD, et al. Kisspeptin neurons co-express met-enkephalin and galanin in the rostral periventricular region of the female mouse hypothalamus. *J Comp Neurol* 2011;**519**(17):3456−69.

26. Lehman MN, Coolen LM, Goodman RL. Minireview: kisspeptin/neurokinin B/dynorphin (KNDy) cells of the arcuate nucleus: a central node in the control of gonadotropin-releasing hormone secretion. *Endocrinology* 2010;**151**(8):3479−89.

27. Rance NE, Krajewski SJ, Smith MA, Cholanian M, Dacks PA. Neurokinin B and the hypothalamic regulation of reproduction. *Brain Res* 2010;**1364**:116−28.

28. Topaloglu AK, Reimann F, Guclu M, Yalin AS, Kotan LD, Porter KM, et al. TAC3 and TACR3 mutations in familial hypogonadotropic hypogonadism reveal a key role for Neurokinin B in the central control of reproduction. *Nat Genet* 2009;**41**(3):354−8.

29. Almeida TA, Rojo J, Nieto PM, Pinto FM, Hernandez M, Martin JD, et al. Tachykinins and tachykinin receptors: structure and activity relationships. *Curr Med Chem* 2004;**11**(15):2045−81.

30. Yen SS, Quigley ME, Reid RL, Ropert JF, Cetel NS. Neuroendocrinology of opioid peptides and their role in the control of gonadotropin and prolactin secretion. *Am J Obstet Gynecol* 1985;**152**(4):485−93.

31. Kalra SP, Sahu A, Dube G, Kalra PS. Effects of various tachykinins on pituitary LH secretion, feeding, and sexual behavior in the rat. *Ann N Y Acad Sci* 1991; **632**:332−8.

32. Sandoval-Guzman T, Rance NE. Central injection of senktide, an NK3 receptor agonist, or neuropeptide Y inhibits LH secretion and induces different patterns of Fos expression in the rat hypothalamus. *Brain Res* 2004;**1026**(2):307–12.

33. Goodman RL, Coolen LM, Anderson GM, Hardy SL, Valent M, Connors JM, et al. Evidence that dynorphin plays a major role in mediating progesterone negative feedback on gonadotropin-releasing hormone neurons in sheep. *Endocrinology* 2004;**145**(6):2959–67.

34. Hrabovszky E, Ciofi P, Vida B, Horvath MC, Keller E, Caraty A, et al. The kisspeptin system of the human hypothalamus: sexual dimorphism and relationship with gonadotropin-releasing hormone and neurokinin B neurons. *Eur J Neurosci* 2010;**31**(11):1984–98.

35. Hrabovszky E, Sipos MT, Molnar CS, Ciofi P, Borsay BA, Gergely P, et al. Low degree of overlap between kisspeptin, neurokinin B, and dynorphin immunoreactivities in the infundibular nucleus of young male human subjects challenges the KNDy neuron concept. *Endocrinology* 2012;**153**(10):4978–89.

36. Navarro VM, Gottsch ML, Wu M, Garcia-Galiano D, Hobbs SJ, Bosch MA, et al. Regulation of NKB pathways and their roles in the control of Kiss1 neurons in the arcuate nucleus of the male mouse. *Endocrinology* 2011;**152**(11):4265–75.

37. Todman MG, Han SK, Herbison AE. Profiling neurotransmitter receptor expression in mouse gonadotropin-releasing hormone neurons using green fluorescent protein-promoter transgenics and microarrays. *Neuroscience* 2005;**132**(3):703–12.

38. Krajewski SJ, Anderson MJ, Iles-Shih L, Chen KJ, Urbanski HF, Rance NE. Morphologic evidence that neurokinin B modulates gonadotropin-releasing hormone secretion via neurokinin 3 receptors in the rat median eminence. *J Comp Neurol* 2005;**489**(3):372–86.

39. Amstalden M, Coolen LM, Hemmerle AM, Billings HJ, Connors JM, Goodman RL, et al. Neurokinin 3 receptor immunoreactivity in the septal region, preoptic area and hypothalamus of the female sheep: colocalisation in neurokinin B cells of the arcuate nucleus but not in gonadotrophin-releasing hormone neurones. *J Neuroendocrinol* 2010;**22**(1):1–12.

40. Burke MC, Letts PA, Krajewski SJ, Rance NE. Coexpression of dynorphin and neurokinin B immunoreactivity in the rat hypothalamus: morphologic evidence of interrelated function within the arcuate nucleus. *J Comp Neurol* 2006;**498**(5):712–26.

41. Krajewski SJ, Burke MC, Anderson MJ, McMullen NT, Rance NE. Forebrain projections of arcuate neurokinin B neurons demonstrated by anterograde tract-tracing and monosodium glutamate lesions in the rat. *Neuroscience* 2010;**166**(2):680–97.

42. Rance NE. Menopause and the human hypothalamus: evidence for the role of kisspeptin/neurokinin B neurons in the regulation of estrogen negative feedback. *Peptides* 2009;**30**(1):111–22.

43. Navarro VM, Castellano JM, McConkey SM, Pineda R, Ruiz-Pino F, Pinilla L, et al. Interactions between kisspeptin and neurokinin B in the control of GnRH secretion in the female rat. *Am J Physiol Endocrinol Metab* 2011;**300**(1):E202–10.

44. Ruiz-Pino F, Navarro VM, Bentsen AH, Garcia-Galiano D, Sanchez-Garrido MA, Ciofi P, et al. Neurokinin B and the control of the gonadotropic axis in the rat: developmental changes, sexual dimorphism, and regulation by gonadal steroids. *Endocrinology* 2012;**153**(10):4818–929.

45. Navarro VM, Ruiz-Pino F, Sanchez-Garrido MA, Garcia-Galiano D, Hobbs SJ, Manfredi-Lozano M, et al. Role of neurokinin B in the control of female puberty and its modulation by metabolic status. *J Neurosci* 2012;**32**:2388–97.

46. Ramaswamy S, Seminara SB, Ali B, Ciofi P, Amin NA, Plant TM. Neurokinin B stimulates GnRH release in the male monkey (*Macaca mulatta*) and is colocalized with kisspeptin in the arcuate nucleus. *Endocrinology* 2010;**151**:4494–503.

47. Billings HJ, Connors JM, Altman SN, Hileman SM, Holaskova I, Lehman MN, et al. Neurokinin B acts via the neurokinin-3 receptor in the retrochiasmatic area to stimulate luteinizing hormone secretion in sheep. *Endocrinology* 2010;**151**(8):3836–46.

48. Biran J, Palevitch O, Ben-Dor S, Levavi-Sivan B. Neurokinin Bs and neurokinin B receptors in zebrafish-potential role in controlling fish reproduction. *Proc Natl Acad Sci USA* 2012;**109**(26):10269–74.

49. Navarro VM, Gottsch ML, Chavkin C, Okamura H, Clifton DK, Steiner RA. Regulation of gonadotropin-releasing hormone secretion by kisspeptin/dynorphin/neurokinin B neurons in the arcuate nucleus of the mouse. *J Neurosci* 2009;**29**(38):11859–66.

50. Kinsey-Jones JS, Grachev P, Li XF, Lin YS, Milligan SR, Lightman SL, et al. The inhibitory effects of neurokinin B on GnRH pulse generator frequency in the female rat. *Endocrinology* 2012;**153**(1):307–15.

51. Garcia-Galiano D, Pineda R, Roa J, Ruiz-Pino F, Sanchez-Garrido MA, Castellano JM, et al. Differential modulation of gonadotropin responses to kisspeptin by aminoacidergic, peptidergic, and nitric oxide neurotransmission. *Am J Physiol Endocrinol Metab* 2012;**303**(10):E1252–63.

52. Siuciak JA, McCarthy SA, Martin AN, Chapin DS, Stock J, Nadeau DM, et al. Disruption of the neurokinin-3 receptor (NK3) in mice leads to cognitive deficits. *Psychopharmacology (Berl)* 2007;**194**(2):185–95.

53. Yang JJ, Caligioni CS, Chan YM, Seminara SB. Uncovering novel reproductive defects in neurokinin B receptor null mice: closing the gap between mice and men. *Endocrinology* 2012;**153**(3):1498−508.

54. Wakabayashi Y, Nakada T, Murata K, Ohkura S, Mogi K, Navarro VM, et al. Neurokinin B and dynorphin A in kisspeptin neurons of the arcuate nucleus participate in generation of periodic oscillation of neural activity driving pulsatile gonadotropin-releasing hormone secretion in the goat. *J Neurosci* 2010;**30**(8): 3124−32.

55. Ruka KA, Burger LL, Moenter SM. Regulation of arcuate neurons coexpressing kisspeptin, neurokinin B, and dynorphin by modulators of neurokinin 3 and kappa-opioid receptors in adult male mice. *Endocrinology* 2013;**154**(8):2761−71.

56. de Croft S, Boehm U, Herbison AE. Neurokinin B activates arcuate kisspeptin neurons through multiple tachykinin receptors in the male mouse. *Endocrinology* 2013;**154**(8):2750−60.

57. Ramaswamy S, Seminara SB, Plant TM. Evidence from the agonadal juvenile male rhesus monkey (*Macaca mulatta*) for the view that the action of neurokinin B to trigger gonadotropin-releasing hormone release is upstream from the kisspeptin receptor. *Neuroendocrinology* 2011;**94**(3):237−45.

58. Castellano JM, Navarro VM, Fernandez-Fernandez R, Castano JP, Malagon MM, Aguilar E, et al. Ontogeny and mechanisms of action for the stimulatory effect of kisspeptin on gonadotropin-releasing hormone system of the rat. *Mol Cell Endocrinol* 2006;**257−258**:75−83.

59. Mostari MP, Ieda N, Deura C, Minabe S, Yamada S, Uenoyama Y, et al. Dynorphin-kappa opioid receptor signaling partly mediates estrogen negative feedback effect on LH pulses in female rats. *J Reprod Dev* 2013;**59**:266−72.

Endocrinology of the Single Cell: Tools and Insights

Judith L. Turgeon and Dennis W. Waring

University of California, Davis, USA

INTRODUCTION

The emerging field of single-cell analysis is providing remarkable, novel insights into cell function and heterogeneity. Sparked by the availability of an expanding repertoire of innovative, sensitive tools, research has begun to address long-standing questions such as: How does a cell integrate or restrict input in space and time across a cell? What is the basis of cellular heterogeneity? And how do whole genome sequences differ between two single cancer cells? Recently the National Institutes of Health (NIH) unveiled its Single Cell Analysis Program, the long-term goal of which "...is to accelerate the move towards personalizing health to the cellular level by understanding the link between cell heterogeneity, tissue function and emergence of disease through the discovery, development and translation of innovative approaches which will dramatically change the way cells are characterized" (http://commonfund.nih.gov/singlecell/).

Single-cell analysis has a long history in the study of endocrine cells, including, for example, electrophysiological measurement of membrane activity, assessment of intracellular calcium changes, single-cell secretion, and imaging of intracellular associations using immunofluorescence technology. In this chapter we review a sampling of new tools that allow selective manipulation and data capture in single cells and discuss examples of output from these studies and how the results might lead to new concepts or a changed understanding. Included in the discussion is an exploration of the constraints and advantages of population-based and single-cell studies as mechanistic models. The chapter is not meant to be a comprehensive review of the single-cell analysis field but rather the aim is to focus on its potential for new discoveries and resolution of old questions in endocrinology. Finally, and to this aim, the pituitary gonadotrope and the pancreatic β-cell will be used as parallel examples for the application of single-cell analysis in the examination of the secretory process.

Cellular Endocrinology in Health and Disease.
DOI: http://dx.doi.org/10.1016/B978-0-12-408134-5.00020-2

SINGLE-CELL STUDIES: BENEFITS AND PITFALLS

A fundamental issue in biological studies is whether all cells within a group ascribed to the same type are indeed identical. Do all cells in primary cultures of parathyroid chief cells respond to manipulation identically? Is a transformed endocrine cell line without cell-to-cell variation? Are differences in responses between cells in a population due only to stochastic noise or do the variations provide valuable information about, for example, pathway and/or expression capacity or microenvironmental responses that are regulated by active mechanisms?

A wealth of information in the literature suggests that, while some of the cell-to-cell variation in response to stimuli in cell populations can be attributed to intrinsic stochastic fluctuations of the component reactions, extrinsic deterministic factors underlie the variability to a significant extent (see for example refs[1–6]). Examples of extrinsic factors include cell population context (cell size, local cell density, location within the population) and cell history (cell cycle stage, internal and external metabolite concentration, spatiotemporal response to previous stimulus). The behavioral cell-specific response to these deterministic factors may be changes in transcription, metabolism, signaling pathway intermediates, endocytosis, etc., resulting in a cell population with multiple phenotypes. Given the potential complexity, diversity, and non-linearity of adaptive multicellular responses to these factors, the assumptions that all cells in a population are identical and that the averaged readout provides unambiguous information can obscure crucial differences and important biological functions.[4] Cell-to-cell variability is recognized as a core biological principle, and technological advances have contributed to an explosion of single-cell studies to parse out the dynamic and adaptive heterogeneous responses to stochastic and regulated noise — detrimental as well as beneficial noise. The larger, next-step questions are if, when, and how are the individual cell responses integrated into a population response? What are the mechanisms by which populations are able to selectively dampen or amplify individual cell perturbations and responses, and what happens when those mechanisms are dysfunctional?[4]

As a logical extension of these studies, single-cell analysis is now being used to understand the dynamics of the multiple components in regulatory networks for genes, intracellular signaling molecules, and extracellular autocrine/paracrine factors. As described in the following section, the capability to interrogate, manipulate with spatiotemporal specificity, and acquire data on multiple activities in the same cell can be powerful for understanding normal function, dysfunction, and drug targeting.

Single-cell studies are not without potential pitfalls, one of which is the possible difference in the behavior of single cells isolated from their usual contacts and microenvironment. Ideally, variables such as the impact of intercellular matrix and paracrine factors should be reflected in the single-cell experimental design, but the fix for this may not be feasible or the variable unknown. Ultimately, results from single cells in isolation are usually considered to be first-stage information, subsequently tested in other models. To some extent, this problem can be bypassed with the use of current tools, which allow for single-cell manipulation and real-time data sampling from intact tissue, for example single β-cells within pancreatic islets or single neurons within hypothalamic slices.[7,8] Another challenge that is the byproduct of current innovative single-cell technology can be a mass of multidimensional data generated that, in some cases, may require mathematical modeling and new analytical tools.[9] Finally, the combination of single-cell tools required for experimental manipulation, data acquisition, and multiplex analysis may be costly and require skills that come with a steep learning curve.

SINGLE-CELL TOOLS AND OUTCOMES

In the last few years significant progress has been made in the development and refinement of tools for the measurement of cellular function at the single-cell level. Many of these advances will be important for increasing our understanding of endocrine cell function in health and disease, from genomics and transcriptomics to receptor activation and signaling, to hormone secretion. As discussed below (and reviewed in depth in refs[3,6,9−14]), the methodologies used rely on continued development in the established technologies of electrophysiology, microscopy and imaging, advances in genomics and transcriptomics, and newly developed tools, e.g., optogenetics and microfluidics.

Access to single cells and their contents is achieved through various, experimental objective-dependent means. For example, patch clamp micropipettes used in electrophysiological studies also are being used to aspirate the contents of single cells in culture, in tissue slices and in whole tissue for genomic and transcriptomic analysis.[10,15] Individual single cells can also be harvested from fixed tissue using laser capture microdissection (LCM). Fluorescence-activated cell sorting (FACS) is used to segregate large numbers of cells of a particular type, and a much newer approach, microfluidics, can be used to separate large numbers of selected cells individually into microchambers for experimental treatment and analysis, allowing collection of data from a large number of individual cells simultaneously.[10,16,17] The

following is an overview sample of methodologies and tools used in single-cell studies (Table 20.1). Their categorization by subcellular function is artificial in that many of the tools can be used for assessment and even manipulation of multiple cell function targets.

Secretion

Electrophysiological methods are of continuing importance in assessing endocrine cell function at the single-cell level. They were initially used to establish the electrical excitability of endocrine cells, the ion channels that underlie it, and its role in stimulus−secretion coupling.[23] With introduction of the patch clamp technique, patch clamp micropipettes were used for studies of ion channels and, subsequently, for measurement of hormone secretion as reflected in changes in plasma membrane capacitance (C_m) associated with secretory granule−plasma membrane fusion, exocytosis. This methodology now is being used to explore, for example, stimulus−secretion coupling, the initial step in the release of vesicle contents, fusion pore formation, and vesicle dynamics. Combinations of single-cell methodologies have been useful in studies of exocytosis in endocrine cells. For example, simultaneous measurement of C_m and intracellular Ca^{2+} with the fluorescent indicator fura-2 showed that the ability of estradiol to increase gonadotropin-releasing hormone (GnRH)-induced secretion is demonstrable in single gonadotropes, but that estradiol does not affect the simultaneously determined GnRH-induced rise in intracellular Ca^{2+}, demonstrating that an action of estradiol in the gonadotrope is to increase the sensitivity of the exocytotic pathway to Ca^{2+}.[27] Caged compounds are bound, inactive forms of molecules that can be introduced into cells; subsequently, the active molecule is released by a pulse of ultraviolet (UV)

TABLE 20.1   Examples of Single-Cell Methodologies

Name	Methods/Technologies	[a]Single-Cell Selection/Isolation Techniques
Genomics	Single-cell genomic sequencing[17,18]	Fluorescence-activated cell sorting (FACS)
		Microfluidics
Transcriptomics	Single-cell transcriptomics[10]	Patch pipette harvest
	Select/targeted transcripts in single cells[10]	FACS
		Laser capture microdissection (LCM)
Receptor activation/signaling	Optogenetics[13,19]	Patch pipette
	[b]TIRF microscopy[7,20−22]	Fluorescence microscopy
	Electrophysiology[23]	
	Quantitative immunofluorescence[15]	
	[c]FRET reporters[11,19,24]	
	Caged compounds[25]	
	Nanobodies[20]	
	SNAP-tag[14,26]	
Secretion	Capacitance measurement[25,27,28]	Patch pipette
	Amperometry [29]	Carbon fiber electrode
	TIRF microscopy[7,12,22]	Fluorescence microscopy
	Microfluidics[9,16,30]	Microfluidics

[a]Single cell: a cell in isolation (for example, dispersed, cultured or sorted) or an individual cell in a tissue slice or whole tissue.
[b]Total internal reflection fluorescence microscopy (TIRFM; also, evanescent wave fluorescence microscopy).
[c]Förster resonance energy transfer (FRET, also fluorescence resonance energy transfer; BRET, bioluminescence resonance energy transfer)

light, producing a step change in the concentration of the active form with precise temporal control. Renström *et al.* used two caged molecules, cyclic adenosine monophosphate (cAMP) and Ca^{2+}, for temporal control of their intracellular concentrations to demonstrate that cAMP stimulates Ca^{2+}-dependent exocytosis by both protein kinase A-dependent and -independent pathways in single mouse pancreatic β-cells.[25]

The final step in exocytosis is fusion of the secretory vesicle and plasma membranes and release of the vesicle contents. This process starts with formation of a narrow fusion pore that can reversibly widen and proceed to full-fusion exocytosis. Calejo *et al.* used high-resolution C_m measurements of single exocytotic events to examine cAMP's modulation of this process in isolated rat lactotropes.[28] They showed that cAMP increases the frequency of transient changes in fusion pore diameter and that the openings are wider with a longer duration open time, demonstrating that this late stage of exocytosis, after secretory vesicle—plasma membrane fusion, is still subject to regulation.

While C_m measurements provide data for calculating the geometry and kinetics of the final step(s) in exocytosis, this methodology does not provide information on the secreted product. Electrochemical methods to directly measure secretory products from single cells are well established, and they complement and inform data obtained from C_m measurements. Detection is through oxidation of the secreted molecules on the surface of a carbon microelectrode positioned close to the cell.[29] In theory, this method is limited to secreted molecules that can be oxidized on the electrode surface; however, the application has been extended to cells whose secretory product(s) do not meet this criterion by loading reporter molecules into the cell's secretory vesicles that are oxidized at the electrode upon exocytosis. Amperometry, one of the two main electrochemical techniques used, has a time resolution of milliseconds and can detect single exocytotic events; thus, the number and frequency of events in response to a stimulus can be determined. In addition, from analysis of the amperometric traces associated with single exocytotic events, the kinetics of the release of secreted molecules can be obtained, e.g., the maximum oxidative current directly translates to the maximum flux of secreted product.[29] Microfluidics technology is now being developed to determine single-cell secretion, and its potential has been demonstrated using immune cell cytokine secretion. In one approach, the single cells, along with microspheres with attached fluorescently labeled antibodies to detect the secreted product, are captured in microfluidic droplets of pico- to nanoliter volume; fluorescence images of the droplets are captured and analyzed for secretion.[30] Another application of microfluidics captures single cells in small wells, nanowells, on arrays on glass slides, each containing thousands of wells; fluorescence images of the arrays are collected and analyzed. Thus, thousands of individual cells can be tested at one time.[16]

Receptor Activation/Signaling

Established and more-recently developed methods of microscopy and imaging, usually incorporating fluorescence-based tools, are major components of studies of single-cell function. The methodologies range from the simple visualization of individual cells to aid electrophysiological studies to laser scanning microscopy and the use of total internal reflection fluorescence microscopy (TIRFM) to study, for example, the mechanisms involved in secretory granule dynamics[12] or the sub-plasma membrane oscillations in concentration of Ca^{2+} and cAMP (Figure 20.1).[7] Fluorescence microscopy also is an essential tool in optogenetics studies and in the examination of molecular interactions using Förster resonance energy transfer (FRET; also fluorescence resonance energy transfer) and bioluminescence resonance energy transfer (BRET).[11]

Light-activated molecules can provide precise temporal and spatial control and, recently, optogenetics has provided a wealth of tools for studying single cell dynamics.[13] The general approach is to use activation of light-sensitive proteins, opsins, to initiate or control functions at the single-cell level and, more recently, for spatiotemporal control in selected regions within a single cell. Optogenetics began with the use of type I opsins (microbial opsins), which are light-sensitive ion channels expressed in neurons. Light activation of these channels was used to depolarize or hyperpolarize neurons to initiate or alter neuronal electrical activity and function. More recently, type II opsins (animal opsins), which are G protein-coupled receptors (GPCRs), have been engineered for control of GPCR signaling pathways. This has led to development of light-activated chimeric receptors for the temporal and spatial control of GPCR activity within single cells. Targeted expression of these tools has been shown from cell culture through transgenic animals.[13]

In a recent study from Karunarathne and colleagues, $G_{i/o}$, G_q, or G_s signaling was activated in selected regions of single cells with optical inputs to control the

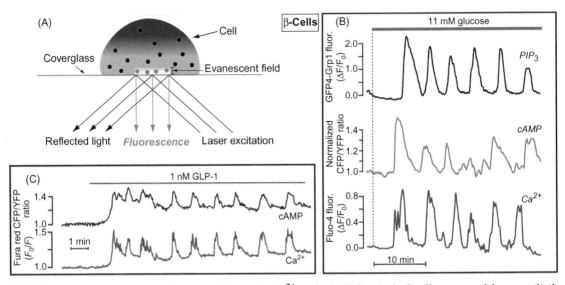

FIGURE 20.1 **Sub-plasma membrane oscillations in Ca^{2+} and cAMP in single β-cells measured by quantitative real-time imaging with confocal and total internal reflection fluorescence (TIRF) microscopy.** (A) Principle for TIRF imaging of a cell or tissue attached to a coverslip. Reflection of the laser excitation light at the coverglass—medium interface generates an evanescent field within an ∼100 nm zone above the interface, which will excite fluorescent molecules near the plasma membrane of a cell adhering to the coverslip. (B) Elevated glucose-triggered oscillations underlying pulsatile insulin secretion. TIRF microscopy recordings of insulin secretion kinetics indicated by the surrogate plasma membrane phosphatidylinositol trisphosphate (PIP_3; formed by autocrine insulin receptor activation) and sub-plasma membrane space cAMP and Ca^{2+} concentrations for three separate single MIN6 cells (pancreatic β-cell line). *Adapted with permission from The American Society for Biochemistry and Molecular Biology.*[22] (C) GLP-1-triggered oscillations in sub-plasma membrane cAMP and Ca^{2+} concentrations simultaneously recorded in an INS-1β cell (β-cell line). *Adapted with permission from Macmillan Publishers Ltd.*[21]

concentrations of phosphatidylinositol-(3,4,5)-trisphosphate (PIP_3), inositol-(1,4,5)-trisphosphate (IP_3), or cAMP.[19] The optical triggers based on non-rhodopsin opsins developed by these investigators allow for reversible, rapidly switching patterns of spatiotemporally restricted GPCR activity in a single cell. Optically evoked molecular and cellular response dynamics were monitored by spectrally selective imaging with confocal microscopy. An application model in this elegant work was GPCR-regulated polarized cell behavior; the question was whether the optical triggers could be used to induce symmetry-breaking cellular events – can neurite outgrowth be spatially controlled by optically localizing GPCR signaling within an individual rat hypothalamic neuron?[19] They showed

that localized triggers of $G_{i/o}$ signaling in single neurons induced PIP_3 increase, neurite initiation, and actin remodeling. This is an example of the extent to which spatial and temporal control of GPCR activity can be programmed, monitored, and applied to a confined region of a single cell, an entire cell, or a portion of a tissue.[19]

GPCRs serve as a prototype of how understanding of function at the single-cell level – from receptor activation to downstream signaling – has benefited from ongoing methodological and technological developments. The analysis of conformational changes within a GPCR or the monitoring of protein–protein interactions of the receptor and its G-protein subunits has been made possible by FRET, which allows the monitoring of distances between two fluorescent labels on a nanometer scale. Such studies have led to understanding the activation and signaling of GPCRs from purified proteins to intact cells (reviewed in ref[11]).

Studies of parathyroid hormone (PTH) and thyroid stimulating hormone (TSH) GPCRs using a FRET reporter to measure intracellular cAMP showed that these receptors continue to signal to cAMP after internalization into endosomes.[24] More recently, nanobodies (conformation-specific single-domain antibodies) and TIRF microscopy have been used to show that agonist activation of β2-adrenergic receptor (β2AR) and its associated G protein, G_s, can occur in early endosome membrane after receptor internalization. The internalized receptor activation contributes a delayed cellular cAMP response.[20] These results are consistent with, and support, a revised generalized model of GPCR–cAMP signaling that incorporates signaling at the endosome as proposed by Calebrio et al.[24] Another method for labeling GPCRs, SNAP-tags, has been used together with single-molecule TIRF microscopy to dynamically monitor single receptors on intact single cells. The receptors examined, β1AR, β2AR and GABA$_B$, varied in their mobility and degree of dimerization/oligomerization, but all were highly dynamic in their interactions, including with scaffolding proteins to maintain cell-surface location.[26]

One of the most intriguing questions of cellular function is: How are ubiquitous signals managed within a single cell to provide unique temporal and spatial outcomes? Methodologies for single-cell analysis continue to evolve with ever-increasing spatiotemporal resolution, sensitivity, and specificity, and, thus, are providing progressively clearer pictures of how this core function is accomplished.

Genomics and Transcriptomics

The genome and transcriptome can now be studied at the level of the single cell, providing understanding of the contribution of heterogeneity in genetic composition and transcriptional expression between cells, to population or tissue variation, even within a "homogeneous" cell population. For example, in the genome of a single cell copy number can be determined, which allows for a gene copy number comparison between cells and an assessment of its contribution to phenotypic variation. Baslan, et al. developed a single nucleus sequencing (SNS) approach involving flow sorting of single nuclei, whole genome amplification, and next-generation sequencing.[18] Applied to tumor cells, comparisons based on SNS are particularly useful for understanding the evolutionary processes occurring in tumor cells and tumor progression.[18] Another potentially powerful application of single-cell genome-wide analysis is in the study of gametes. During the meiotic recombination shuffle, the gamete genome undergoes programmed and spontaneous changes, and accumulated replication errors and associated point mutations may affect gamete function. The upshot of this process is that a variety of new genomes are created in the gametes, contributing to genetic diversity.[17] But exactly how is this critically important editing process occurring? To begin to address this in human sperm, Wang, et al. applied microfluidics to single-sperm whole-genome amplification and analysis.[17] The group also created a personal recombination map and was able to measure the rate and de novo mutations in an individual's germline. Single-cell genome analysis is early in its development, but advances in the molecular tools and high-throughput data analysis and bioinformatics should facilitate the move of this approach into the endocrine sphere, for example, in studies of the progression of pancreatic β-cell malfunction or thyroid tumorigenesis.

Determination of levels for select RNAs at the single cell level has been part of the endocrine research armamentarium for some time (for example, reviewed in refs[6,10]). Now, in addition to methodological advances for assessment of changes in select transcripts in individual cells, a variety of methodologies are being developed and refined that allow for transcriptome-wide profiling of the identities and abundances of RNA molecules in a single cell (as reviewed in ref[10]). The capability to monitor single-cell transcriptomics adds a new dimension to studies of endocrine signaling networks and drug target discovery.

PARALLEL APPLICATIONS IN SINGLE-CELL ANALYSIS: THE SECRETORY PROCESS

A capstone event in an endocrine cell is stimulus-induced release of hormone. For exocytotic secretion in particular, single-cell models have been extremely useful for examining the complex regulatory apparatus supporting this endocrine activity. Although quantification of the secretory product from single cells has been challenging, the capability to simultaneously monitor multiple components, for example in the pathway between activated membrane receptors and exocytotic events, has been transformative. Illustrative of this are the advances in our understanding of the regulation of insulin secretion from the pancreatic β-cell. The following is a brief overview of one component of regulated insulin secretory process, including novel insights generated by recent single-cell studies and their potential for application to another secretory cell, the pituitary gonadotrope.

Pancreatic β-cells

Glucose is a primary driver of insulin secretion, but other signals reflecting metabolic or energy status can modify the signaling pathway in the pancreatic β-cell. An example is glucagon-like peptide 1 (GLP-1), a gastrointestinal incretin hormone produced in response to nutrients, which potentiates or primes glucose-stimulated insulin secretion. In the presence of basal or zero levels of glucose, GLP-1 has little or no effect on insulin secretion, but when glucose levels are increased, GLP-1-mediated stimulation of glucose-induced insulin secretion exceeds that seen with glucose alone. In type 2 diabetes mellitus, the first phase of insulin secretion in response to elevated glucose characteristically is lost and the second phase impaired. Because of the insulinotropic action of GLP-1, notable particularly in the first-phase response, GLP-1 receptor agonists have become an important component of therapy for type 2 diabetes mellitus (as reviewed in ref[31]).

The overall relationship of glucose, GLP-1, and insulin secretion is shown in the simplified schematic (see Figure 20.4): Glucose enters the β-cell via glucose transporters and is metabolized, resulting in an increase in the cytosolic adenosine triphosphate/adenosine diphosphate (ATP/ADP) ratio and closure of ATP-sensitive K^+ (K_{ATP}) channels, leading to membrane depolarization and opening of voltage-dependent Ca^{2+} channels (VDCCs) and Ca^{2+} influx.[32] The resulting increase in cytosolic Ca^{2+} concentration triggers exocytosis of the readily releasable pool of secretory granules docked at the cell membrane (as reviewed in ref[33]). Signaling by GLP-1 is through activation of its membrane receptor, a GPCR, and release of G_s, which activates adenylyl cyclase leading to a rapid increase in cAMP production.[34] The two known downstream effectors of cAMP in β-cells are protein kinase A (PKA) and the cAMP-regulated guanine nucleotide exchange factor, Epac (exchange protein directly activated by cAMP).[33] The role of cAMP's downstream effectors is complex, and for this discussion the focus primarily will be on PKA-independent actions.

Epac2, the predominant isoform in β-cells, has been implicated as having a major role in the potentiation of glucose-stimulated insulin secretion through effects on secretory granule scaffold protein dynamics, and also on increased cytoplasmic Ca^{2+}, most likely through interactions with the K_{ATP} channel and also the endoplasmic reticulum ryanodine channel (as reviewed in refs[34−36]). Although not all of Epac's downstream effectors have been established, an obligatory role for Rap1 as a signaling mediator is clear.[37] Rap1, a Ras-related small GTPase, cycles between an inactive guanosine diphosphate (GDP)-bound and an active guanosine triphosphate (GTP)-bound state; cAMP-bound Epac catalyzes guanine nucleotide exchange on Rap1 thereby activating it. One of several intriguing targets of activated Rap1 in β-cells is phospholipase C-ε (PLC-ε). The hydrolysis of phosphatidylinositol-4,5-bisphosphate (PIP_2) by Rap1-activated PLC-ε is suggested to increase K_{ATP}-channel responsiveness to ATP and also to have a role in Ca^{2+}-induced Ca^{2+} release thereby contributing to Ca^{2+}-dependent insulin granule exocytosis.[36,38]

In addition to GLP-1-induced increase in β-cell cAMP, earlier studies established that elevated glucose can result in a modest increase in total cAMP content of β-cells but that cAMP by itself does not lead to insulin secretion (as reviewed in refs[8,31]). Questions still linger as to the mechanism for this glucose effect on increasing cAMP with most attention directed to a role for increased ATP and possibly Ca^{2+}-dependent calmodulin on a soluble adenylyl cyclase, which is structurally distinct from the transmembrane forms.[8] The cAMP query that has become the primary focus is whether spatial and temporal confinement dictates the action of cAMP in β-cells. Compartmentalization by scaffolding of secretory elements, anchoring mechanisms for critical kinases and other enzymes, and microdomains of elevated cAMP and Ca^{2+} are

hallmarks of regulated secretion and now have been shown to operate in β-cells (as reviewed in refs[33,36,39]). Of particular interest are compartments in the sub-plasma membrane region. For example, Epac2 has been found at plasma membranes, and Rap1 localizes to various compartments, including vesicle membranes and plasma membrane.

Recent studies in single cells utilizing quantitative real-time imaging with confocal and TIRF microscopy to assess spatiotemporal activity below the plasma membrane (Figure 20.1A) have led to new concepts incorporating a glucose-stimulated cAMP signaling mode into the insulin secretion picture.[7,21,22,40] Oscillatory insulin secretion is well established, and the subcellular events generating this pattern became a bit more clear with the report in single cells that glucose triggers coordinated oscillations of the sub-plasma membrane Ca^{2+} and cAMP concentrations, synergizing to generate pulsatile insulin secretion (Figure 20.1B).[22] Submembrane oscillations of cAMP and Ca^{2+} have also been demonstrated in response to GLP-1 in single β-cells (Figure 20.1C).[21] The amplitude and duration of the GLP-1-induced cAMP oscillations was concentration-dependent with sustained cAMP elevations at high GLP-1 concentrations that rapidly reversed with GLP-1 removal.[40] Dyachok and colleagues suggest that brief transients of cAMP are important for conferring specificity for regulation of rapid local events,[21] for example ion channel activity and exocytosis. As would be expected from the insulin secretory response to combined elevated glucose and GLP-1, the submembrane cAMP elevation induced by a combination of elevated glucose and GLP-1 is larger than the effect of either stimulus alone.[40] What happens to localization of Epac during these oscillations? Using real-time confocal and TIRF microscopy in single cells, Idevall-Hagren and colleagues report that an increase in submembrane cAMP recruits Epac translocation to the plasma membrane, and this translocation is cyclic in the presence of elevated glucose (Figure 20.2A).[7] Further, the Epac2 oscillations are accompanied by oscillations of Rap GTPase activity in the submembrane area, positioning Rap1 at potential sites of exocytosis. In cells in which Epac2 is knocked down by siRNA or in cells expressing a dominant-negative Rap1b mutant (Figure 20.2B), the magnitude of the pulsatile insulin secretion indicator in response to either glucose stimulation or K^+ stimulation in single cells is significantly reduced.[7,22]

It appears, then, that cAMP-activated Epac and subsequently Rap1 participate in the β-cell response to elevated glucose and that activation of the GLP-1 receptor kicks Epac/Rap1 activity up several notches − potentiating β-cell exocytotic activity as a response to incoming nutrients. Many issues remain to be addressed for insulin-secretion regulation, such as the identity of other Rap1 targets and also non-Rap1 Epac effectors, regulation of the proteins that are involved in the spatial control of submembranous activity, decoding of the submembrane oscillations, particularly related to Ca^{2+}, the mechanism by which glucose metabolism leads to elevated cAMP, and the roles for pancreatic islet paracrine factors in synergizing the response. But acknowledging the work yet to be done, it is clear that single-cell analyses of β-cell spatiotemporal dynamics and discrete signaling complexes are contributing to an emergent model for understanding normal function and, crucially, impaired insulin secretory function and therapeutic targets. An additional important contribution of the β-cell work is to our understanding of common modular components and secretory tactics that are shared with other endocrine cells.

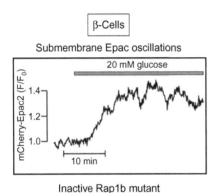

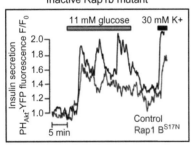

FIGURE 20.2 (A) Epac2 translocation to the sub-plasma membrane space as determined by TIRF microscopy recording in a primary cell within an intact pancreatic islet stimulated by an increase in glucose concentration. (B) Representative single-cell recordings by TIRF microscopy in MIN6 β-cells expressing a dominant-negative Rap1B[S17N] mutant or a control cell. Insulin secretion indicated by plasma membrane PIP_3 concentration in response to elevated glucose or K^+ depolarization was significantly reduced in the mutant expressing cells compared to control cells (n = 30 and 52 cells). *Adapted from ref[7], with permission from the American Association for the Advancement of Science.*

Pituitary Gonadotropes

Signal amplification is an efficient strategy in endocrine systems to conserve resources and to provide mechanisms for responding to specific physiological demands, as demonstrated by GLP-1-enhanced glucose-dependent insulin secretion. In the reproductive endocrine system, the periodic explosive release of luteinizing hormone (LH) leading to ovulation is the product of a fail-safe system of several overlapping mechanisms, including increased amplitude and optimal frequency of the GnRH signal and amplification of the signaling potential of GnRH at the gonadotrope. An overview of GnRH-mediated LH secretion with an emphasis on the amplification of the GnRH signaling potential and a comparison with the pancreatic β-cell is presented in a simplified schematic (see Figure 20.4).

One GnRH signal amplification device is self-priming or potentiation that manifests as enhanced LH secretion in response to a subsequent, identical GnRH stimulation, and has been demonstrated *in vivo* in humans[41–43] and *in vivo* and *in vitro* in the rat (for example, in refs[44–46]) and the mouse.[47,48] Potentiation requires a background of estradiol, and the measurable physiological endpoint is an augmented LH secretory response seen within 40 min of initial GnRH exposure (as reviewed in ref[45]). Potentiation depends on early gene induction, requiring transcription but only for the first hour of induction, and is not dependent on changes in LH synthesis or in the GnRH receptor. Once the process is activated by an initial GnRH pulse, it can be elicited by secretagogues such as a depolarizing pulse of K^+ (as reviewed in ref[45]). The signaling pathway mediating potentiation is not fully established, but we have shown that elevation in intracellular cAMP, which by itself does not stimulate LH secretion, can substitute for the first GnRH pulse to potentiate subsequent responses to secretagogues.[49,50] This is similar to observations in pancreatic β-cells as noted above: Elevated intracellular cAMP by itself does not stimulate insulin secretion, but can potentiate glucose-stimulated insulin exocytosis.

Because time-dependent transcription is required for GnRH priming, in collaboration with Stuart Sealfon and colleagues the authors examined early genomic events following GnRH receptor activation using high-density oligonucleotide microarray analysis and real-time polymerase chain reaction (PCR) confirmation. In rat primary cultures enriched for gonadotropes, we found that the small

GTPase, rap1b, was among a small cohort of transcripts upregulated 40 min after GnRH stimulation.[15] Because of the potential cellular heterogeneity of the gonadotrope-enriched pituitary cultures, we examined GnRH activation of rap1b gene expression in single rat gonadotropes cultured in the presence of estradiol.[15] Figure 20.3A depicts the harvesting of cell contents by aspiration into a patch pipette for assay by single-cell quantitative real-time PCR. At 40 min after a GnRH stimulus, the rap1b transcript level increased six-fold compared to control; the increase was two-fold after stimulation with a cAMP analogue (Figure 20.3B). Rap1 protein following GnRH stimulation was evaluated by dual immunofluorescence and confocal microscopy of single gonadotropes. By 60 min after a GnRH stimulus, Rap1 complexes appeared as focal points of intense immunofluorescence with a distribution suggesting positioning of the complexes at the periphery of the gonadotrope in areas compact with LH secretory granules (Figure 20.3C). Object volume quantification and analysis for multiple single cells 1-hour post GnRH stimulation showed Rap1 immunofluorescence to be significantly increased (Figure 20.3D).

These results in single gonadotropes place Rap1 as candidate player in the GnRH-stimulated priming pathway, similar to that in the GLP-1 priming pathway in β-cells. Epac2 is present in the gonadotrope, but future studies must establish that Epac-specific cAMP analogs directly affect Epac and subsequently Rap1 activation, and that Epac/Rap1 activation is causally linked to potentiated GnRH-stimulated LH secretion in gonadotropes. Also to be considered is whether pituitary adenylyl cyclase activating peptide (PACAP), an autocrine component of the gonadotrope exosignaling circuit, has a modulatory role in this GnRH action.[51,52] GnRH receptor activation results in increased global cAMP concentration in rat gonadotropes[50] and in the gonadotrope

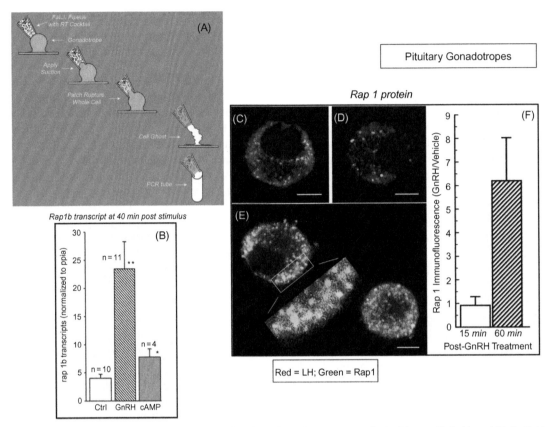

FIGURE 20.3   Female rat anterior pituitary cells cultured in the presence of 0.2 nM estradiol. (A and B) Individual gonadotropes were perfused with a 15-min pulse of extracellular medium (Ctrl, control) or with medium containing either 10 nM GnRH or 1 mM 8-bromo-cAMP followed by perfusion with control medium for 25 min and harvested (shown in panel (A) by aspiration into a patch pipette containing reverse transcriptase (RT) cocktail, expelled into a polymerase chain reaction (PCR) tube, and immediately processed for reverse transcription. Single-cell DNA samples were assayed by quantitative real-time PCR. Rap1b transcript data (panel B) were normalized to peptidylprolyl isomerase A (ppia) transcripts determined in the same gonadotrope. Number of individual gonadotropes per treatment group is indicated. Compared to control: **$P < 0.001$; *$P < 0.02$. C−F) Rat pituitary cells were incubated in the presence of vehicle for 60 min (C) or stimulated with a 15-min pulse of 1 nM GnRH and either fixed immediately (D) or 45 min later (E). Fixed cells were immunofluorescently stained for LH (red) and Rap1 (green) and analyzed by confocal microscopy; panels show single 2 μm optical sections. Inset of (E): expanded view of immunofluorescent punctate Rap1-associated objects. Scale bar = 5 μm. Rap1 associated object volume analysis data are shown (F) as fold-change in Rap1 immunofluorescence in gonadotropes (GnRH-treated/vehicle-treated) fixed at either 15 min (open bar) or 60 min (hatched bar). Number of individual cells analyzed/group: 16−20; *$P < 0.03$ compared to cells fixed at 15 min. (B−F) from ref[15] with permission from Elsevier.

cell line, LβT2 cells (as reviewed in ref[53]), but does it lead to compartmentalized cAMP oscillations? In single LβT2 cells, GnRH pulse treatment induces a general cellular pulsatile cAMP response pattern, and, interestingly, the cAMP response did not desensitize with a rapid GnRH pulse frequency,[53] a GnRH delivery pattern associated *in vivo* with the preovulatory LH surge. No information on compartmentalized cAMP oscillations in single gonadotropes in primary culture is available. Very useful observations, however, are that

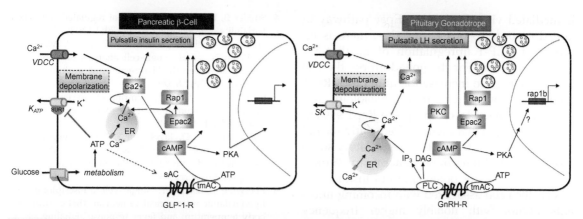

FIGURE 20.4 **Schematic comparison of proposed roles for Epac2 and Rap1 in β-cell insulin secretion and gonado-trope LH secretion.** In the pancreatic β-cell, glucose metabolism results in an increase in the ATP/ADP ratio and closure of ATP-sensitive K$^+$ (K$_{ATP}$) channels, leading to membrane depolarization and opening of voltage-dependent Ca^{2+} channels (VDCC) and Ca^{2+} influx. Elevated cytosolic Ca^{2+} concentration triggers exocytosis of insulin granules docked at the cell membrane. Glucose metabolism also leads to small increases in cytosolic cAMP via activation of soluble adenylyl cyclases (sAC), which may be mediated via ATP and possibly Ca^{2+}-dependent calmodulin (not shown). Stimulation of the GLP-1 receptor (GLP-1-R) activates membrane associated adenylyl cyclase (tmAC) leading to increased cAMP levels resulting in potentiation of glucose-stimulated insulin secretion via both protein kinase A (PKA)-dependent and Epac-dependent mechanisms. Activated Epac2 has several targets, for example: specific exocytotic scaffolding and fusion proteins involved in insulin granule recruitment and fusion with the plasma membrane; the regulatory subunit of the K$_{ATP}$ channel, SUR1 (sulfonylurea receptor 1); endoplasmic reticulum (ER) ryanodine and inositol trisphosphate (IP$_3$) receptors; and increased Ca^{2+} mobilization from stores. Epac2 actions are mediated through its activation of the GTPase, Rap1, or potentially through direct protein interaction with the target protein. In the pituitary gonadotrope, activation of the GnRH receptor (GnRH-R) stimulates phospholipase C (PLC) leading to formation of IP$_3$ and diacylglycerol (DAG) and ultimately activation of protein kinase C (PKC). The consequences of the PLC arm of GnRH-R activation are multiple, and the emphasis in this simplified scheme is on calcium signaling and exocytotic events. GnRH-R stimulation also activates adenylyl cyclase leading to increased cAMP levels with proposed roles in potentiating the LH secretory response to a subsequent stimulus with mediation by both PKA and Epac2. The lag time following GnRH-R activation and potentiation (~40 min) may be partially explained by the requirement for increased GnRH-induced rap1b transcription; whether this cAMP-associated transcription event is mediated by PKA or Epac2 is unknown. Following the increase in rap1b transcript, the protein increases in the sub-plasma membrane region in the vicinity of LH granules. SK, small conductance calcium-activated potassium channels.

GnRH self-potentiation cannot be demonstrated functionally in LβT2 cells and, consistent with this, the rap1b gene has not been found to be activated by GnRH in this cell line (for example, see refs[54,55]). Using manipulations similar to that reported for the β-cell model of potentiated exocytosis, a comparison of single-cell outcomes in primary gonadotropes and LβT2 cells should be particularly informative.

The gonadotrope brings two novel twists to the β-cell Rap1 story. First, instead of potentiating a different signal, which is what GLP-1 does for glucose-stimulated insulin secretion, GnRH is priming itself for stimulation of LH secretion; and secondly, GnRH receptor activation results in a rapid increase in rap1b transcripts as well as potentially activating Rap1 protein. Regarding the change in transcript number, the signaling pathway has not been established, although cAMP stimulation was found to result in a rapid increase in rap1b transcripts (Figure 20.3B); it will be of interest to determine whether this cAMP effect is mediated via a PKA or an Epac pathway in gonadotropes. To our knowledge, there is no report of GLP-1 upregulation of rap1b transcripts in β-cells. For the gonadotrope, the use of this strategy to increase the amount of Rap1 mediator may be related to physiology and timing. The need to ramp up the Rap1 system to amplify the signal occurs infrequently for LH secretion associated with the preovulatory surge (once every 4−5 days in rats; once every 28 days in humans), but for potentiation of insulin exocytosis the need associated with incoming nutrients occurs with notably higher frequency (usually daily). The gonadotrope appears to be using the pattern of increasing estradiol concentration as one of its permissive signals reflecting the reproductive endocrine status and enabling the increased rap1b expression response to GnRH. The strategies used by the β-cell to regulate replenishment of Rap1 are unknown, but one might speculate that it involves information on nutrient status and may occur on a more long-term than acute basis.

Application of single-cell analysis has revealed similarities in the secretion priming pathways in β-cells and gonadotropes (Figure 20.4). Exemplifying biological parsimony, two endocrine cell types utilize a basic pathway blueprint, and adaptive strategies and nuances are added on to address the cell-specific endocrine function required for metabolism and reproduction.

References

1. Colman-Lerner A, Gordon A, Serra E, Chin T, Resnekov O, Endy D, et al. Regulated cell-to-cell variation in a cell-fate decision system. *Nature* 2005;**437**:699−706.
2. Hsieh Y-Y, Hung PH, Leu JY. Hsp90 regulates nongenetic variation in response to environmental stress. *Mol Cell* 2013;**50**:82−92.
3. Raj A, van Oudenaarden A. Single-molecule approaches to stochastic gene expression. *Annu Rev Biophys* 2009;**38**:255−70.
4. Snijder B, Pelkmans L. Origins of regulated cell-to-cell variability. *Nat Rev Mol Cell Biol* 2011;**12**:119−25.
5. Snijder B, Sacher R, Ramo P, Liberali P, Mench K, Wolfrum N, et al. Single-cell analysis of population context advances RNAi screening at multiple levels. *Mol Syst Biol* 2012;**8**:579.
6. Selimkhanov J, Hasty J, Tsimring LS. Recent advances in single-cell studies of gene regulation. *Curr Opin Biotechnol* 2012;**23**:34−40.
7. Idevall-Hagren O, Jakobsson I, Xu Y, Tengholm A. Spatial control of Epac2 activity by cAMP and Ca^{2+}-mediated activation of Ras in pancreatic β cells. *Sci Signal* 2013;**6** ra29-6
8. Eberwine J, Bartfai T. Single cell transcriptomics of hypothalamic warm sensitive neurons that control core body temperature and fever response signaling asymmetry and an extension of chemical neuroanatomy. *Pharmacol Ther* 2011;**129**:241−59.
9. Spiller DG, Wood CD, Rand DA, White MRH. Measurement of single-cell dynamics. *Nature* 2010;**465**:736−45.
10. Spaethling JM, Eberwine JH. Single-cell transcriptomics for drug target discovery. *Curr Opin Pharmacol* 2013.
11. Lohse MJ, Nuber S, Hoffmann C. Fluorescence/bioluminescence resonance energy transfer techniques to study G-protein-coupled receptor activation and signaling. *Pharmacol Rev* 2012;**64**:299−336.
12. Steyer JA, Almers W. A real-time view of life within 100 nm of the plasma membrane. *Nat Rev Mol Cell Biol* 2001;**2**:268−75.
13. Fenno L, Yizhar O, Deisseroth K. The development and application of optogenetics. *Annu Rev Neurosci* 2011;**34**:389−412.
14. Crivat G, Taraska JW. Imaging proteins inside cells with fluorescent tags. *Trends Biotechnol* 2012;**30**:8−16.
15. Yuen T, Choi SG, Pincas H, Waring DW, Sealfon SC, Turgeon JL. Optimized amplification and single-cell analysis identify GnRH-mediated activation of Rap1b in primary rat gonadotropes. *Mol Cell Endocrinol* 2012;**350**:10−9.
16. Yamanaka YJ, Szeto GL, Gierahn TM, Forcier TL, Benedict KF, Brefo MS, et al. Cellular barcodes for efficiently profiling single-cell secretory responses by microengraving. *Anal Chem* 2012;**84**:10531−6.
17. Wang J, Fan HC, Behr B, Quake SR. Genome-wide single-cell analysis of recombination activity and de novo mutation rates in human sperm. *Cell* 2012;**150**:402−12.
18. Baslan T, Kendall J, Rodgers L, Cox H, Riggs M, Stepansky A, et al. Genome-wide copy number analysis of single cells. *Nat Protoc* 2012;**7**:1024−41.
19. Karunarathne WKA, Giri L, Kalyanaraman V, Gautam N. Optically triggering spatiotemporally confined GPCR activity in a cell and programming neurite

initiation and extension. *Proc Natl Acad Sci USA* 2013;**110**:E1565–74.

20. Irannejad R, Tomshine JC, Tomshine JR, Chevalier M, Mahoney JP, Steyaert J, et al. Conformational biosensors reveal GPCR signalling from endosomes. *Nature* 2013;**495**:534–8.

21. Dyachok O, Isakov Y, Sagetorp J, Tengholm A. Oscillations of cyclic AMP in hormone-stimulated insulin-secreting β-cells. *Nature* 2006;**439**:349–52.

22. Idevall-Hagren O, Barg S, Gylfe E, Tengholm A. cAMP mediators of pulsatile insulin secretion from glucose-stimulated single β-cells. *J Biol Chem* 2010;**285**: 23007–18.

23. Stojilkovic SS, Tabak J, Bertram R. Ion channels and signaling in the pituitary gland. *Endocr Rev* 2010; **31**:845–915.

24. Calebiro D, Nikolaev VO, Lohse MJ. Imaging of persistent cAMP signaling by internalized G protein-coupled receptors. *J Mol Endocrinol* 2010;**45**:1–8.

25. Renström E, Eliasson L, Rorsman P. Protein kinase A-dependent and -independent stimulation of exocytosis by cAMP in mouse pancreatic B-cells. *J Physiol* 1997;**502**:105–18.

26. Calebiro D, Rieken F, Wagner J, Sungkaworn T, Zabel U, Borzi A, et al. Single-molecule analysis of fluorescently labeled G-protein-coupled receptors reveals complexes with distinct dynamics and organization. *Proc Natl Acad Sci USA* 2013;**110**:743–8.

27. Thomas P, Waring DW. Modulation of stimulus-secretion coupling in single rat gonadotrophs. *J Physiol* 1997;**504**:705–19.

28. Calejo AI, Jorgacevski J, Kucka M, Kreft M, Goncalves PP, Stojilkovic SS, et al. cAMP-mediated stabilization of fusion pores in cultured rat pituitary lactotrophs. *J Neurosci* 2013;**33**:8068–78.

29. Amatore C, Arbault S, Guille M, Lemaître F. Electrochemical monitoring of single cell secretion: vesicular exocytosis and oxidative stress. *Chem Rev* 2008;**108**:2585–621.

30. Konry T, Dominguez-Villar M, Baecher-Allan C, Hafler DA, Yarmush ML. Droplet-based microfluidic platforms for single T cell secretion analysis of IL-10 cytokine. *Biosens Bioelectron* 2011;**26**:2707–10.

31. Meloni AR, DeYoung MB, Lowe C, Parkes DG. GLP-1 receptor activated insulin secretion from pancreatic β-cells: mechanism and glucose dependence. *Diabetes Obes Metab* 2013;**15**:15–27.

32. Ashcroft FM. ATP-sensitive potassium channelopathies: focus on insulin secretion. *J Clin Invest* 2005; **115**:2047–58.

33. Gloerich M, Bos JL. Epac: defining a new mechanism for cAMP action. *Annu Rev Pharmacol Toxicol* 2010; **50**:355–75.

34. Doyle ME, Egan JM. Mechanisms of action of glucagon-like peptide 1 in the pancreas. *Pharmacol Ther* 2007;**113**:546–93.

35. Cabot MC, Welsh CJ, Zhang Z-C, Cao H-T. Evidence for a protein kinase C-directed mechanism in the phorbol diester-induced phospholipase D pathway of diacylglycerol generation from phosphatidylcholine. *FEBS Lett* 1989;**245**:85–90.

36. Schmidt M, Dekker FJ, Maarsingh H. Exchange protein directly activated by cAMP (epac): a multidomain cAMP mediator in the regulation of diverse biological functions. *Pharmacol Rev* 2013;**65**:670–709.

37. Shibasaki T, Takahashi H, Miki T, Sunaga Y, Matsumura K, Yamanaka M, et al. Essential role of Epac2/Rap1 signaling in regulation of insulin granule dynamics by cAMP. *Proc Natl Acad Sci U S A* 2007; **104**:19333–8.

38. Dzhura I, Chepurny OG, Kelley GG, Leech CA, Roe MW, Dzhura E, et al. Epac2-dependent mobilization of intracellular Ca^{2+} by glucagon-like peptide-1 receptor agonist exendin-4 is disrupted in {beta}-cells of phospholipase C- knockout mice. *J Physiol* 2010;**588**: 4871–89.

39. Zaccolo M. Spatial control of cAMP signalling in health and disease. *Curr Opin Pharmacol* 2011;**11**:649–55.

40. Tian G, Sandler S, Gylfe E, Tengholm A. Glucose- and hormone-induced cAMP oscillations in α- and β-cells within intact pancreatic islets. *Diabetes* 2011;**60**: 1535–43.

41. Hoff JD, Lasley BL, Yen SSC. The functional relationship between priming and releasing actions of luteinizing hormone-releasing hormone. *J Clin Endocrinol Metab* 1979;**49**:8–11.

42. Rommler A. Short-term regulation of LH and FSH secretion in cyclic women. I. Altered pituitary response to a second of two LH-RH injections at short intervals. *Acta Endocrinol (Copenh)* 1978;**87**:248–58.

43. Sollenberger MJ, Carlsen EC, Booth Jr. RA, Johnson ML, Veldhuis JD, Evans WS. Nature of gonadotropin-releasing hormone self-priming of luteinizing hormone secretion during the normal menstrual cycle. *Am J Obstet Gynecol* 1990;**163**:1529–34.

44. Fink G. Oestrogen and progesterone interactions in the control of gonadotrophin and prolactin secretion. *J Steroid Biochem* 1988;**30**:169–78.

45. Turgeon JL, Waring DW. Functional cross-talk between receptors for peptide and steroid hormones. *Trends Endocrinol Metab* 1992;**3**:360–5.

46. Fink G. The self-priming effect of LHRH: a unique servomechanism and possible cellular model for memory. *Front Neuroendocrinol* 1995;**16**:183–90.

47. Chappell PE, Schneider JS, Kim P, Xu M, Lydon JP, O'Malley BW, et al. Absence of gonadotropin surges

and gonadotropin-releasing hormone self-priming in
ovariectomized (OVX), estrogen (E2)-treated, progesterone receptor knockout (PRKO) mice. *Endocrinology*
1999;**140**:3653–8.

48. Turgeon JL, Waring DW. Luteinizing hormone secretion from wild-type and progesterone receptor knockout mouse anterior pituitary cells. *Endocrinology*
2001;**142**:3108–15.

49. Turgeon JL, Waring DW. Activation of the progesterone
receptor by the gonadotropin-releasing hormone self-
priming signaling pathway. *Mol Endocrinol* 1994;**8**:860–9.

50. Waring DW, Turgeon JL. A pathway for luteinizing
hormone releasing-hormone self- potentiation: cross-
talk with the progesterone receptor. *Endocrinology*
1992;**130**:3275–82.

51. McArdle CA, Counis R. GnRH and PACAP action in
gonadotropes. Cross-talk between phosphoinositidase
C and adenylyl cyclase mediated signaling pathways.
Trends Endocrinol Metab 1996;**7**:168–75.

52. Pincas H, Choi SG, Wang Q, Jia J, Turgeon JL, Sealfon
SC. Outside the box signaling: secreted factors modulate
GnRH receptor-mediated gonadotropin regulation. *Mol
Cell Endocrinol* 2013. Available from: http://dx.doi.org/
10.1016/j.mce.2013.08.015. [Epub ahead of print]
PubMed PMID: 23994024.

53. Tsutsumi R, Mistry D, Webster NJ. Signaling responses
to pulsatile gonadotropin-releasing hormone in LbetaT2
gonadotrope cells. *J Biol Chem* 2010;**285**:20262–72.

54. Yuen T, Wurmbach E, Ebersole BJ, Ruf F, Pfeffer RL,
Sealfon SC. Coupling of GnRH concentration and the
GnRH receptor-activated gene program. *Mol Endocrinol*
2002;**16**:1145–53.

55. Turgeon JL, Kimura Y, Waring DW, Mellon PL. Steroid
and pulsatile gonadotropin-releasing hormone (GnRH)
regulation of luteinizing hormone and GnRH receptor
in a novel gonadotrope cell line. *Mol Endocrinol*
1996;**10**:439–50.

Intracellular Trafficking of G Protein-Coupled Receptors to the Plasma Membrane in Health and Disease

Alfredo Ulloa-Aguirre,† and P. Michael Conn†*

*Universidad Nacional Autónoma de México and Instituto Nacional de Salud Pública, Mexico D.F., Mexico, †Departments of Internal Medicine and Cell Biology-Biochemistry, Texas Tech University Health Services Center, Lubbock, TX, USA

INTRODUCTION

Proteins are essential macromolecule components of the organism that play important roles in the structure, function, and regulation of all the body's tissues and organs. The synthesis and initial processing of proteins occur in the endoplasmic reticulum (ER) and both are tightly regulated events controlled at transcriptional, translational, and post-translational levels by diverse signaling pathways. The ER has the daunting task of synthesis and assembly of thousands of proteins and provides the specialized environment necessary for folding, glycosylation, oxidation, and dimeric/oligomeric assembly of proteins, prior to their export to other organelles and/or domains of the cell, including the cell surface plasma membrane (PM).[1,2] As proteins are synthesized in the ER cell compartment, they fold (that is, the protein acquires its native structure, departing from a completely or partially unfolded state) and adopt distinct, low free-energy conformations that lead to a stable, tightly compact structure compatible with ER export (Figure 21.1).[4−6]

Given the proximity and marked diversity of proteins that accumulate in the cytosol (to a nominal concentration of 300−400 mg/ml), protein folding is, by necessity, a complex process that should yield unique structures that allow selectivity and diversity in cell function.[7] Current models of protein folding[5,8] indicate that the dihedral angles of the protein backbone fold synchronously in groups of four or more ("protein wriggling"),[9] thereby avoiding dramatic steric clashes. Since the number of possible conformations of a newly synthesized polypeptide chain is tremendously large, protein folding cannot involve an ordered series of mandatory steps between particular folding

Cellular Endocrinology in Health and Disease.
DOI: http://dx.doi.org/10.1016/B978-0-12-408134-5.00021-4

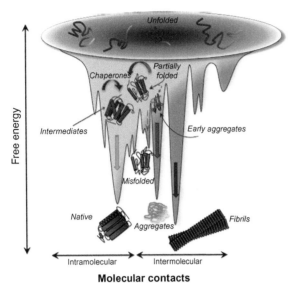

Free energy (y-axis)

Unfolded

Chaperones / Partially folded

Intermediates

Early aggregates

Misfolded

Native

Aggregates

Fibrils

← Intramolecular → ← Intermolecular →

Molecular contacts

FIGURE 21.1 **Funnel-shape model of folding kinetics.** Schematic of the funnel-shaped free-energy surface (*y*-axis) of newly synthesized proteins as they fold and approach the native state (see text for details). Kinetically trapped conformations and intermediate structures traverse free energy barriers (helped by molecular chaperones) in order to reach a *favorable*, low energy downhill funnel. Potentially toxic aggregates or ordered amyloid fibrils may be formed as a result of overlapping between free-energy surface of folding and intermolecular aggregation (*x*-axis), when intermediates are folding simultaneously in the same compartment. *Modified from ref[3], with permission from Nature Publishing Group.*

intermediates (i.e., partially structured states along the folding pathway), which would take considerable time before reaching the correct, native-like structure. Rather, folding follows a stochastic search of the many conformations accessible to a polypeptide chain based, essentially, on searching through inherent fluctuations of the protein-stable interactions between residues, which will progressively lead to lower and lower energy conformational ensembles, until finding the lowest. According to this protein-folding landscape, funnel-shaped model (Figure 21.1), the number of possible conformations progressively decreases so that only a reduced number needs to be sampled during this search process.[10] Although this mechanism is quite fast (microseconds for some proteins), errors in folding are common given the enormous complexity of the system, and about half of the structures synthesized may contain folding errors, which are eventually recognized by an efficient cellular quality control system (see below). In fact, the three-dimensional and complex protein structure is stabilized by a number of physico-chemical reactions, including several noncovalent interactions involving hydrogen

bonds, van der Waals and electrostatic interactions, backbone-angle preferences, chain entropy, and hydrophobic interactions, as well as by covalent bonds between cysteine side chains that form disulfide bridges.

Although the intimate mechanisms subserving protein folding are still unknown (and thus we are still unable to accurately predict the native, low-energy, three-dimensional structure of the protein based on its amino acid sequence and, hence, its propensity to aggregate, for example), proteins appear to fold in units of secondary structures, their stability increases with the growing partial structure as they fold, and a protein appears to initially develop local structures in the chain (e.g., helices and turns) forming "basic" folding nuclei, followed by growth in more global structures.[11]

It has been recognized that mutations resulting in protein sequence variations may lead to misfolding (i.e., an error in folding due to a sufficient number of persistent non-native interactions that affect the overall architecture of the protein and/or its properties in a significant manner)[12] and disease-causing proteins that, not rarely, are transcribed and translated at normal levels, but are unable to reach their functional destinations in the cell or to engage the secretory pathway.[5,12] Misfolding can also be triggered by other factors including protein overexpression, temperature, oxidative stress, and activation of signaling pathways associated with protein folding and quality control.[2] In particular cases, misfolding results in *loss-of-function* of the conformationally defective protein.[5,13] Although normally some non-native interactions with other molecules may occur during the folding process (to bury highly aggregation-prone regions such as exposed hydrophobic surfaces), misfolded proteins may aggregate, leading to potentially toxic intracellular accumulation, or even to protein accumulation in the plasma with extracellular amyloid deposition.[5,14,15] In the past decade, extraordinary efforts have been made to understand how abnormal folding occurs and to design therapeutic interventions that could prevent or correct the structural abnormality of disease-causing misfolded proteins. It has also been increasingly evident that some mutations that lead to misfolding do not compromise domains involved in the function of the protein, as is the case of some G protein-coupled receptors or ion channel mutants;[2,16] in this case, strategies to correct folding and promote trafficking of the defective receptor to its final destination (i.e., the cell surface PM) may restore function of the mutant protein and eventually cure disease.[17,18]

This chapter is focused on protein misfolding and misrouting in disease states. Special emphasis is placed on G protein-coupled receptors (GPCRs), given the role played by misfolding of these particular membrane proteins in a variety of endocrine disorders (Table 21.1). G protein-coupled receptors are the largest family of membrane proteins in the human genome, involving nearly 800 genes; GPCRs are activated by highly diverse ligands, including photons, odorants, pheromones, hormones, lipids, and neurotransmitters, that vary in size, from small biogenic amines to peptides to large proteins.[39] Therefore, GPCRs are among the essential nodes of communication between the internal and external environments of cells, transducing the information provided by extracellular stimuli into intracellular signals. In accordance with the large variety of endogenous and exogenous agonists, GPCRs are involved in the regulation of an array of aspects of normal physiology and pathophysiology. Consequently, they have enormous potential as drug targets; in fact, nearly 50% of all approved drugs derive their benefits by selective targeting to these proteins.[40–42]

Although GPCRs may vary considerably in molecular size, all share a common molecular topology that consists of a single polypeptide chain of variable lengths that traverses the lipid bilayer seven times, forming characteristic transmembrane (TM) hydrophobic alpha-helices connected by alternating extracellular and intracellular sequences or loops (EL and IL, respectively), with an extracellular NH_2-terminus and an intracellular COOH-terminal tail.[39,43] Upon activation by agonist, GPCRs undergo conformational changes that facilitate activation of heterotrimeric guanine nucleotide-binding proteins (G proteins) and other membrane-associated intracellular proteins, which, in turn, regulate activation of a variety of G protein-dependent and independent signaling cascades. Activated GPCRs are rapidly desensitized and internalized via formation of endosomes, where receptor-mediated signaling usually terminates and the fate of the internalized receptor is determined (Figure 21.2). Thus, the net amount of a given GPCR at the PM will depend on several factors, including: a) the dynamics of intracellular export from its site of synthesis (the ER) to its final destination, the cell surface PM (which is one of the main subjects of this chapter); b) the fate of the receptor following ligand-stimulated internalization, either to the degradative or to the recycling pathway; and c) the normal membrane turnover (Figure 21.2).

TABLE 21.1 Loss-of-Function Abnormalities Caused by GPCR Misfolding and Examples of Pharmacoperones Tested *in vitro* and/or *in vivo*

Disease or Anomaly	GPCR Involved	Pharmacoperones	References
Retinitis pigmentosa	Rhodopsin	Retinoids (9-*cis*-retinal, 11-*cis*-retinal, 11-*cis*-7-ring retinal, vitamin A palmitate)	19−24
Obesity	MC3R, MC4R	ML00253764 (and compounds described in ref 170)	25−27
Red head color phenotype and propensity to skin cancer	MC1R	NBA-A	28
Nephrogenic diabetes insipidus	V2R	Satavaptan, relcovaptan, VPA-985, YM087, tolvaptan, OPC31260	29−35
Hypogonadotropic hypogonadism	GnRHR	IN3, IN30, Q89, A177775, TAK-013	Reviewed in ref 13
Familial hypocalciuric hypercalcemia	CaR	NPS R-568	Reviewed in ref 36
Premature ovarian failure	FSHR	Org 41841	37
Leydig cell hypoplasia	LHR	Org 42599	38
Familial glucocorticoid deficiency	MC2R	−	−
Congenital hypothyroidism	TSHR	−	−
Hirschsprung's disease	E-BR	−	−
Resistance to HIV-1 infection	CCR5	−	−

Abbreviations: CaR, calcium-sensing receptor; CCR5, chemokine receptor-5; E-BR, endothelin-B receptor; FSHR, follicle-stimulating hormone receptor; GnRHR, human gonadotropin-releasing hormone receptor; HIV, human immunodeficiency virus; LHR, luteinizing hormone receptor; MC1R, melanocortin-1 receptor; MC2R, melanocortin-2 receptor (or adrenocorticotropin (ACTH) receptor); MC3R, melanocortin-3 receptor; MC4R, melanocortin-4 receptor; TSHR, thyrotropin receptor; V2R, vasopressin V2 receptor.

Synthesis and intracellular trafficking of GPCRs is a complex process that involves a number of co-translational and post-translational mechanisms and events that subserve PM targeting of these key proteins.[44] Let us briefly review some of the mechanisms that regulate intracellular trafficking and PM targeting of GPCRs from the ER.

THE ER QUALITY CONTROL SYSTEM

As with any other protein, GPCRs are synthesized in the ER. Here, they fold and adopt distinct conformations (determined by their amino acid sequence), that eventually provide a three-dimensional structure recognized by the quality control system (QCS) machinery of the cell as a structure compatible with ER export.[45] The ER QCS guards against aberrant protein structures and non-productive protein aggregation, and checks for adequate folding, processing, and structural integrity of nascent proteins, ensuring proper intracellular trafficking of the newly synthesized protein to the Golgi apparatus (where processing is completed), and eventually, to its final destination within the cell (e.g., the plasma membrane). By monitoring the structural correctness of newly synthesized proteins, the QCS prevents accumulation of defective proteins that may

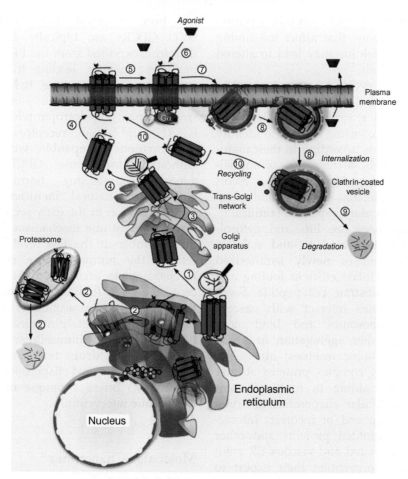

FIGURE 21.2 **Anterograde and retrograde trafficking of GPCRs.** Newly synthesized proteins fold in the endoplasmic reticulum (ER). Correctly folded proteins are then translocated to the Golgi apparatus to initiate/complete processing such as glycosylation (magnifiers) (steps 1 and 3). Misfolded and misassembled products are retained in the ER and exposed to resident chaperones, which attempt to correct folding and stabilize the protein in a conformation compatible with ER export. When correct folding fails, the misfolded protein is dislocated into the cytoplasm for proteosomal degradation (step 2). Mature receptors are then exported to the plasma membrane (step 4) where they interact with cognate ligands (steps 5 and 6). Agonist-provoked activation of the receptor (step 6) is followed by phosphorylation of the receptor and recruitment of β-arrestins (purple circles), which promote endocytosis (step 7) and internalization of the receptor (step 8). Agonist dissociates from the internalized receptor and may be recycled or degraded in lysosomes. The receptor embedded in clathrin-coated vesicles may be either targeted to lysosomes for degradation (step 9) or recycled to the plasma membrane (step 10) to interact again with cognate agonist (step 5).

aggregate and interfere with normal cell function. Thus, the ER QCS provides the means for exporting functional proteins to other cell compartments or outside the cell.[45,46] Since the scrutiny by the ER quality control system relies on conformational rather than on functional criteria, even minor alterations in the secondary or tertiary structure of a protein may lead to intracellular retention and degradation (Figure 21.2). Correct folding and trafficking are delicately balanced and even minor synonymous polymorphisms that affect the timing of co-translational folding may lead to altered function of the protein.[47]

The ER QCS has evolved different strategies that operate at several levels and that require the participation of a variety of mechanisms, including a complex sorting system, to identify and separate proteins according to their maturation status as well as the action of specialized folding factors and catalysts, escort proteins, retention factors, enzymes, and members of major molecular chaperone families.[6,29] Molecular chaperones are ER- and cytosol-resident proteins that bind to and stabilize unstable conformers of newly synthesized polypeptides to facilitate efficient folding and assembly of the substrate polypeptide. Some molecular chaperones interact with nascent chains at the ribosomes and bind non-specifically to prevent aggregation of prone regions in hydrophobic residues; others are involved in guiding complex proteins at later stages of folding, aiding in multimolecular assemblies.[3,5] Molecular chaperones not only prevent aggregation and/or incorrect interactions between misfolded proteins and other molecules in a crowded and viscous ER environment (thereby preventing their export to other cellular compartments), but also may act to disassemble protein aggregates.[48] Thus, molecular chaperones guard nascent polypeptide chains against potentially unproductive and even toxic interactions that may occur during the different stages of the folding process (i.e., while still attached to the ribosome, just after release from the ribosome, as a folding intermediate with exposed hydrophobic surfaces, and even as a misfolded protein). Proteins that do not fulfill the criteria of the ER QCS are degraded in proteasomes or lysosomes.[2]

Similar to other proteins, GPCRs have to be correctly folded in order to pass through the ER QCS (see above). Nevertheless, it has been observed that certain wild-type (WT) GPCRs are typically misfolded and, therefore, exported from the ER in a relatively inefficient manner, leading to restricted PM expression.[49] For example, it has been found that only a fraction (40–60%) of newly synthesized human gonadotropin-releasing hormone (GnRH)[2] and δ-opioid receptors[50] reach mature conformations compatible with ER export. Although, for some GPCRs (e.g., the gonadotropin-releasing hormone receptor, GnRHR), this natural "inefficiency" in folding and maturation in the early secretory pathway may represent one mechanism for controlling their number at the PM level,[49] in others it reflects the normal failure of the folding mechanisms to accomplish the complex process of protein folding in a timely way. The possibility that intracellularly retained misfolded or incompletely mature receptors may be rescued from intracellular trapping and degradation by drugs that act as chaperones (i.e., pharmacological chaperones, or "pharmacoperones") offers a unique opportunity for therapeutic interventions.

Molecular Chaperones

Molecular chaperones are endogenous proteins that serve as an essential control mechanism for recognizing, retaining, and targeting misfolded proteins for their eventual degradation. They also assist other proteins to fold or refold, in case they have lost their native conformation, and to travel to their destination in the cell. As mentioned above, molecular chaperones may additionally participate in the dissolution of protein aggregates and in ushering damaged peptides toward degradation.[48] Although the steric character of the protein backbone restricts the spectrum of protein shapes that are recognized by the stringent quality control mechanisms, some features displayed by client proteins, including exposure of hydrophobic shapes, unpaired cysteines, immature glycans, and particular sequence motifs, have been identified as important for chaperone recognition and association.[44] In fact, molecular chaperones possess the ability to recognize misfolded proteins when they expose hidden hydrophobic domains or specific sequences.[51,52] Through this association, chaperones may stabilize unstable conformers of nascent polypeptides to prevent aggregation and facilitate correct folding or assembly of the substrate via binding and release cycles. Molecular chaperones may form functional complexes with other molecules of the same chaperone group or of another group, or even with molecules that are not typical chaperones themselves (co-chaperones and co-factors) (see below). Thus, these multi-molecular protein complexes actually represent machines that carry out the chaperoning process.

Several GPCR interacting proteins and chaperone complexes that support trafficking of the receptor molecule to the cell surface plasma membrane have been identified. Some examples follow. Nina A (neither inactivation nor after potential A), a photoreceptor-specific integral membrane glycoprotein, is a molecular chaperone that facilitates cell surface membrane expression of the sensory GPCR *Drosophila melanogaster* rhodopsin 1; its absence leads to rhodopsin 1 accumulation in the ER and eventually to its degradation. The mammalian homolog of Nina A is RanBP2, which specifically binds red/green opsin molecules that act as a chaperone aiding folding, transportation to and localization of the mature receptors in the PM.[53] Another chaperone is ODR4, a molecular chaperone that assists in folding, ER exit and/or targeting of olfactory GPCRs (e.g., ODR10 in *Caenorhabditis elegans*) to olfactory cilia.[54] RAMPs (receptor activity modifying proteins), are proteins that interact with several GPCRs (e.g., the calcitonin receptor-like receptor, the vasoactive intestinal polypeptide/pituitary adenylate cyclase-activating peptide receptor, the glucagon receptor, and the parathyroid hormone receptor), fostering the transport of the associated receptor to, and regulating its signaling function at the PM,[55] whereas gC1q-R (receptor for globular heads of C1q) interacts with the COOH-terminus of the α_{1B} adrenergic receptor (AR), regulating its maturation and PM expression.[56]

A broad range of glycoproteins including several GPCRs (e.g., the GnRHR, the vasopressin (V)-2 receptor (V_2R), and the glycoprotein hormone receptors) bind the chaperones calnexin and calreticulin.[57–60] The action of these latter chaperones is predominantly centered on *N*-glycans as substrates in newly synthesized proteins, adding hydrophobicity to the folding protein. Protein disulfide isomerase (PDI),[61] which is an ER-resident folding catalyst involved in disulfide bond formation and reorganization of folding intermediates, probably acts as a co-chaperone with calnexin and calreticulin during their association with the glycoprotein hormone receptors. Another chaperone thought to physically associate with calnexin and calreticulin is Erp57, a thiol-disufide oxido-reductase that catalyzes disulfide bond formation and isomerization.[62] When *N*-linked glycosylation or early glycan processing fails, glycoproteins may misfold, fail the QCS, and limit their traffic to the PM.[57] BiP/Grp78 is a chaperone that maintains proteins in a state competent for subsequent folding and oligomerization, and that mediates retrograde translocation of misfolded conformers for proteosomal degradation. This chaperone is involved in the protective unfolded protein response, which is a cell-stress program activated when misfolded proteins accumulate and/or aggregate in the ER.[63] Finally, DriP78 is an ER-membrane-associated protein that binds to the $F(x)_3F(x)_3F$ motif of the dopamine receptor (and presumably other GPCRs bearing this motif; see below), facilitating its maturation and export to the PM.[64] Identification of these particular molecular chaperones is important to identify differences between misfolded species of the same protein. For example, the inactivating A593P and S616Y mutants of the human luteinizing hormone receptor (LHR), which cause different forms of Leydig cell hypoplasia and male pseudohermaphroditism, exhibit absent or severely impaired response to agonist due to misfolding and intracellular retention of the mutant receptors.[61,65] These LHR mutants are conformationally distinct and display particular folding conformations during their maturation process at the ER as suggested by their differential association with molecular chaperones.[61] Whereas, similar to the WT LHR, the A593P and S616Y mutants were found associated with calnexin, only S616Y was associated with PDI. Further, although both LHR mutants appeared to interact with BiP/Grp78, only the former was found to interact with Grp94.[61]

Molecular chaperones are also important as potential targets to manipulate ER retention and/or export mechanisms, and hence, as potential therapeutic means for exogenously influencing protein trafficking and secretion. In fact, it has been shown that reducing the intracellular levels of BiP/Grp78 may increase the secretion of proteins that otherwise might be retained by the quality control machinery within the ER.[66]

As mentioned above, surveillance of the ER QCS for correct folding and assembly of newly synthesized proteins that ensure correct intracellular trafficking, relies more on some general structural features of the unfolded or partially folded client protein rather than on particular sequences. Nevertheless, transmembrane cargo proteins (such as GPCRs) also bear particular motifs, post-translational modifications, and structural features that regulate export trafficking of the protein from the ER through the Golgi to the PM. Let us briefly examine some of these structural features.

GPCR MOTIFS, POST-TRANSLATIONAL MODIFICATIONS AND STRUCTURAL FEATURES THAT PROMOTE/LIMIT ER-GOLGI EXPORT AND ANTEROGRADE TRAFFICKING TO THE PM

GPCRs ER Export and Retention Motifs

Properly folded and assembled secretory proteins are packed into ER-derived COPII-coated vesicles for exporting to the ER—Golgi intermediate complex and then to the Golgi apparatus and the trans-Golgi network. During their transport in this latter compartment, proteins undergo post-translational modifications (e.g., glycosylation), which are essential for reaching mature status before moving to their final destination.[67] Several mechanisms of bulk-flow, ER-retention and receptor-mediated export have been identified to operate during this critical vesicle-mediated transport step. One mechanism is the *association* of particular sequences with COPII transport vesicles, which are formed from three components: the small GTPase Sar1, which recruits the Sec23/24 heterodimer and then the Sec13/31 complex. Deactivation of Sar1 leads to the release of the COPII vesicles carrying cargo proteins from the ER membrane.[68] This mechanism, however, has only been demonstrated for a few cargoes bearing the diacidic (DxE) and dihydrophobic (FF) signals, which are present in the COOH-terminus of the vesicular stomatitis viral glycoprotein,[69] the cystic fibrosis transmembrane conductance regulator protein,[70] proteins belonging to the p24 family,[71] and angiotensin II receptors.[72] Nevertheless, additional motifs have been shown to be involved in exit of GPCRs from the ER and the Golgi, and some of these sequences might well also associate with components of the COPII transport machinery and small GTPases, such as members of Rab (e.g., Rab1), and Sar1/ARF subfamilies (which are involved in almost every step of vesicle-mediated transport, particularly the targeting, tethering, and fusion of transport vesicles) to exit the ER.[73] These include the dileucin motif E $(x)_3$LL (identified in the human V_2R),[74] the FN $(x)_2$LL(x)3L motif (found in the human V_3R),[75] and F$(x)_6$LL motif identified in the COOH-terminus of several GPCRs,[76] as well as the triple phenylalanine F$(x)_3$F$(x)_3$F sequence (identified in the COOH-terminus of the dopamine D_1 receptor, the M_2-muscarinic receptor and the angiotensin II AT$_1$ receptor).[77] Mutation at these motifs markedly reduces receptor expression at the PM due to intracellular retention of the altered receptor. Interestingly, in addition to the regular exit from the ER to the Golgi, some receptors might use these motifs to be exported from the trans-Golgi network to the PM via binding to the small GTPase Rab8.[73]

Although studies on export motifs present in the NH$_2$-terminal domain of GPCRs are scarce, a distinct YS motif in α_2-ARs has been identified as important for receptor export from the Golgi. Residues or sequences embedded in regions other than the termini of GPCRs and involved in receptor trafficking from the ER to the PM have also been identified. These include the IL1, where a single leucine residue located in the center of this loop (and which is a highly conserved residue among family A members of GPCRs) appears to play an essential role in ER export in a number of adrenergic receptors, including the α_{2B}- and α_{1B}-ARs, the β_2-AR, and the angiotensin II AT$_1$ receptor.[78] More recently a three-arginine (3R) motif in the IL3 has been identified as important in mediating interaction of the α_{2B}-AR with Sec24C/D isoforms.[79]

Two highly conserved motifs, the E/DRY motif (at the boundary of the TM3 and the IL2) and the N/DPxxY motif (at the TM7 near the cytoplasmic face of the PM) are important structural determinants in many GPCRs.[80] In some receptors, mutations in these motifs (such as the E/DRY motif in the V_2R and the GnRHR, and the N/DpxxY motif in the V_2R, GnRHR, endothelin-B receptor, melanocortin-4 receptor and the chemokine receptor 5 (CCR5))[81–83] may lead to different functional outcomes, including defective intracellular trafficking, depending on the particular receptor. On the other hand, the amino acid sequence of GPCRs belonging to family A, predicts formation of a disulfide bridge between the first and second extracellular loops; this structural feature is associated with the stabilization of the heptahelical structure of many GPCRs, and mutations at the vicinity of this bridge usually result in a complete loss of activity, due to retention in the ER of the mutant receptor. In the case of the human GnRHR and rhodopsin, receptors bearing mutations at this location are recalcitrant to or cannot be easily stabilized with pharmacological chaperones.[81,84]

In some transmembrane cargo proteins, it has been possible to identify specific ER *retention* sequences that restrict trafficking of the protein to the ER for further processing. When some of these sequences are removed by mutagenesis, the trafficking of the cargo protein markedly improves. Among these retention signals are the RXR sequence identified in the COOH-terminus of the Kir6.2 potassium channel,[85] the RSRR sequence present in the γ-aminobutyric acid (GABA)-B1 receptor,[86] the penta-arginine (RRRRR) sequence identified in the α_{2C}-AR,[87] and the conserved ALAAALAAAAA hydrophobic sequence present in the NH$_2$-termini of α_2-ARs.[88] Although, in general, the regulatory role of these retention signals is still unclear, they might restrict trafficking of receptors that have failed to properly heterodimerize or fold, leading to exposure of such retention signals. The former is the case of GABA-B1 and GABA-B2 receptors, in which heterodimerization masks the RSRR retention signal present in the COOH-terminus of GABA-B1 receptor species, preventing retrograde transport from the Golgi to the ER via COP-I vesicles, and allowing forward transport and trafficking of the obligatory heterodimer to the PM.[89] Thus, the presence of export and/or retention signals within the structure of the protein, ensures that only correctly folded and assembled, functional molecules are exported to the PM.

Post-Translational Modifications Regulating GPCRs Targeting to the PM

Post-translational modifications are also important for GPCR targeting to the cell surface. Two modifications are particularly important: palmitoylation and N-linked glycosylation. In some GPCRs, S-acylation with palmitic acid of conserved cysteine residues in the COOH-terminus provide an additional site for anchoring of the receptor to the PM, creating a fourth intracellular loop.[39] Palmitoylation occurs at the ER–Golgi intermediate compartment and plays a significant role in receptor export to the PM, as abrogation of this modification limits to variable extents receptor export to the PM.[90] It has also been shown that, depending on the GPCR, palmitoylation is important for internalization and efficiency of recycling; in some GPCRs, palmitoylation promotes β-arrestin recruitment, endocytosis, and degradation, whereas in others, it has opposite effects. Another common post-translational modification is N-linked glycosylation at the consensus sequence NxS/T.[39] This post-translational modification facilitates folding by increasing protein solubility and stabilizing protein conformation.[91] For several GPCRs, glycosylation is absolutely required for cell-surface expression of the receptor as mutation of the glycosylation sites lead to intracellular accumulation of mutated receptors. This is the case, for example, of the gonadotropin receptors in which mutations at the AFNGT sequence of the NH_2-terminus alters glycosylation and causes intracellular

sequestration of the mutant receptor protein.[92] In contrast, in other GPCRs this modification is not absolutely required for receptor transport to the cell surface.

Oligomerization and Anterograde Receptor Trafficking

Biochemical and pharmacological studies support the concept that association (dimerization or oligomerization) among cell surface membrane-expressed receptors is a fundamental process for receptor activity. It appears that GPCRs approach this issue differently. Some receptors are monomeric in the membrane and oligomerize upon ligand binding and activation, whereas others constitutively form multiunit complexes as they are synthesized in the ER or processed in the Golgi, an apparent requisite for correct targeting to the cell surface.[93] Constitutive oligomerization has been demonstrated for a number of GPCRs, including the GABA-B receptors,[94] melatonin receptors,[95] some dopamine receptors (the dopamine D_2 receptor),[96] the vasopressin[97] and serotonin receptors,[98] as well as the δ-opioid receptor,[99] the $β_2$-AR,[100–102] and the gonadotropin receptors.[103,104] Functions of homo- and hetero-oligomerization at the ER include effective quality control of protein folding prior to export to the PM.[93] For example, inhibiting homodimerization of the $β_2$-AR, leads to ER retention and perturbation of cell-surface targeting.[100] In the case of the GABA-B receptor, heterodimerization between GABA-B receptor 1 and GABA-B receptor 2 is an obligatory prerequisite for cell-surface expression of a functional receptor. Apparently, formation of a coil–coil domain between the COOH-terminus of the GABA-B receptor subtypes masks an ER retention signal (RxR) located in the COOH-terminus of the GABA-B receptor 1, thus promoting the ER export of the heterodimer to the PM (see above).[105] A similar role in receptor outward trafficking has been shown for the α_{1D}- and α_{1B}-ARs[106] and the β_2-AR.[107]

Introduction of ER retention motifs (e.g., RxR motif) into GPCRs may lead not only to retention of the mutant receptor but also to hindering of the cell-surface delivery of a homologous unmodified receptor.[100] This dominant-negative effect of mutant receptors on wild-type anterograde trafficking has been demonstrated for a number of GPCRs, including the human GnRHR,[108,109] the gonadotropin receptors,[110] the V_2R,[111] the D_2- and D_3-dopamine receptors,[112,113] and the CCR5.[114] The dominant-negative action that the mutant GnRH receptors have on the WT receptor appears to be due to ER retention of aggregates formed by WT and mutant proteins. The WT and mutant receptors appear to form oligomers in the ER and those oligomers are retained and presumably degraded. The dominant-negative effect of mutant receptors on WT receptor expression might play a role in the phenotypic expression of diseases in individuals bearing simple heterozygous mutations. Nevertheless, the fact that individuals who are heterozygous for misfolded mutations in the gonadotropin receptors do not exhibit detectable reproductive abnormalities suggests that the decrease in PM expression of the WT receptor that results from the dominant-negative effect of the misfolded mutant is not low enough to impact on cell function, given that occupancy of only a low fraction of gonadotropin receptors per cell is sufficient to elicit normal responses. In this regard, it is interesting that heterozygous subjects bearing misfolded thyroid-stimulating hormone receptor (TSHR) mutants (e.g., C41S, L467P, and C600R) express clinical phenotypes of partial thyrotropin resistance presumably due to the dominant-negative effect of the mutant receptors on WT receptor PM expression,[115] underlining the potential effects of misfolded receptors on the dominant transmission of partial GPCR resistance in heterozygous individuals.

Oligomerization in the ER, as part of the intracellular transport of GPCRs, might be a more general mechanism also applicable to other membrane proteins. Although the intrinsic mechanism(s) that serve these protein–protein associations as well as the mechanistic basis for the general need for oligomerization of membrane proteins are not well understood, studies suggest that oligomerization chaperones 14-3-3 epsilon- and zeta, which, in turn, probe for the valency and spatial arrangement of recognition domains (i.e., the carboxyl tails), function as a checkpoint for forward trafficking of maturing multimeric proteins.[116]

In summary, anterograde trafficking of GPCRs and other proteins from the ER to their final destination, relies on several factors, including: a) the QCS of the cell (e.g., molecular chaperones) that check for proper folding; b) particular motifs embedded within the protein molecule that regulate retention in or export from the ER to the Golgi and from the trans-Golgi network to the PM; c) post-translational modifications (e.g., palmitoylation, glycosylation); d) interaction of the GPCRs and other secretory proteins with the microtubule networks, to control their cell-surface movement;[117] and e) a well-organized association (dimerization/oligomerization) between GPCR, which promotes proper folding-assembly. Let's move now to the consequences of protein misfolding and how defective folding may be corrected, thereby preventing unproductive associations (aggregation) between incompletely folded/misfolded proteins and/or their degradation and, thus, failure to express at the correct destination in the cell.

PROTEIN MISFOLDING AND DISEASE

As mentioned previously, defective protein conformation due to misfolding can lead to disease, and conformational diseases are among the most debilitating, socially disruptive and costly diseases these days. They include Alzheimer's and Parkinson's diseases, spongiform encephalopathies, late-onset diabetes, cystic fibrosis, and familial amyloidoses, to mention a few. In the case of GPCRs, mutations may cause misrouting of otherwise-functional proteins and provoke disease (Table 21.1). For example, in retinitis pigmentosa (an inherited, degenerative eye disease that causes severe vision impairment and often blindness), trapping of misfolded mutant rhodopsin occurs resulting in aggregates at the ER. This event ultimately leads to rod photoreceptor degeneration followed by cone degeneration.[84] Mutations in the V_2R gene cause X-linked nephrogenic diabetes insipidus, a disease in which, despite normal or elevated plasma concentrations of the antidiuretic hormone arginine vasopressin, the kidney is unable to concentrate urine due to failure of the V_2R mutants to express at the PM and thus respond to agonists.[118] Trafficking-defective mutants of the glycoprotein hormone receptors (LHR, follicle-stimulating hormone receptor (FSHR), and TSHR) have been described as a cause of ovarian failure, Leydig cell hypoplasia, and congenital hypothyroidism,[119–121] respectively. Mutations in the melanocortin-1 receptor, may lead to skin and hair abnormalities, as well as to increased susceptibility to certain skin cancers; among the ~ 60 mutants described, at least four display decreased PM expression, presumably due to intracellular retention of the abnormal receptor.[122] Misfolding and intracellular retention of mutants from the melanocortin-3 and melanocortin-4 receptors, which are associated with regulation of fat deposition and energy homeostasis, have been detected in patients with morbid obesity.[25] Mutations that lead to misfolding and trapping of the endothelin-B receptor have been detected in some patients with Hirschsprung's disease or aganglionic megacolon, while mutations in the calcium-sensing receptor leading to intracellular retention of the receptor have been identified in patients with familial hypocalciuric hypercalcemia.[123–125] Intracellular retention of the mutant CCR5 at the ER has been observed in a subset of subjects with resistance to human immunodeficiency virus (HIV) infection,[126] whereas mutations leading to receptor misfolding of the human GnRHR frequently lead to congenital hypogonadotropic hypogonadism (HH).[127]

As discussed above, variable (but sometimes significant) amounts of WT GPCRs are inefficiently expressed at the PM (i.e. retained within the ER), apparently as a result of a combination of factors, including inefficient folding and poor coupling to the ER export machinery, aggregation, and/or increased degradation provoked by particular structural features of the receptor.[45,49] The importance of this phenomenon is that this relative "inefficiency" for PM expression may impact on intracellular information networking. For example, recent understanding of the unfolded protein response suggests that there is the potential for information transfer between the ER and the transcriptional machinery as a result of protein accumulation at the ER. In principle, the chaperoning system could mediate the fraction of the newly synthesized receptor that traverses to the PM and, thus, contributes to the changes in net levels of GPCRs that are observed in different physiological conditions. This might be the case for the GnRHR, whose level of expression at the PM remarkably fluctuates throughout the estrous cycle, which is important for differential regulation of certain agonist-mediated effects.[128,129] This level of post-translational control might represent another level of potential therapeutic intervention.

Correcting Protein Misfolding

Physical, Genetic and Chemical Approaches to Rescue Function of Misfolded Proteins

Several *in vitro* approaches to correct folding and promote trafficking of the protein from the ER to the PM have been studied. These approaches include physical, chemical, genetic, and pharmacological strategies. Studies on the biosynthesis of the CFTR F508 deletion mutant (which is the most common mutation that leads to cystic fibrosis) showed that although the mutation leads to subtle misfolding (that does not grossly impact on the function of the chloride ion channel) it also provokes marked intracellular retention of the mutant protein.[130] Incubation of cells expressing this particular mutant at reduced temperatures (20–30°C), reverted processing of the CFTR mutant toward the WT species, promoting PM expression of the ion channel and thus normal function. Similarly, increased PM expression of several conformationally defective human GnRHRs bearing different point mutations (see below) resulted from incubating cells bearing the mutant receptor at lower temperatures (32°C).[2] It seems that for certain misfolded proteins that are temperature sensitive, the use of physical methods may counteract their retention at the ER by the QCS and facilitate trafficking of the defective protein to their physiological site of action. Another strategy for enhancing PM expression of misfolded proteins is by introducing or deleting specific sequences into the conformationally abnormal protein ("genetic rescue").[131] This approach either overexpresses or stabilizes molecules rendered unstable by genetic defects and, in theory, does not provoke global changes in the ER secretory activity. Examples of genetic rescue include addition of carbohydrates, NH_2-terminal or COOH-terminal sequences to expression-deficient GPCRs, maneuvers that have been shown to markedly enhance ER export of inefficiently expressed receptors. In the case of the mammalian GnRHR, which is unique among members of the GPCR superfamily in that it lacks the COOH-terminal extension, addition of this domain from other species (e.g., fish) or deletion of K191 (which, when present, restricts PM expression)[132]

dramatically increase PM expression in both cases.[133] Genetic approaches, albeit effective, are impractical as therapeutic intervention because, if it were possible to access the gene sequence, the primary error could be directly corrected. Rescue of misfolded membrane receptors can also be achieved by manipulating ER and/or post-ER mechanisms that regulate GPCR export. For example, in the GnRHR the molecular chaperone calnexin seems to act as a quality control protein by retaining misfolded receptors and steering properly folded receptors to the PM. Overexpression in COS-7 cells of WT human GnRHR (whose PM expression is normally inefficient)[2] with calnexin, decreased receptor expression and receptor-mediated second messenger production, an effect that was counteracted by using siRNA to knock down overexpressed calnexin.[58] In the case of the P23H mutation in the rhodopsin gene, which causes rhodopsin misfolding and autosomal dominant retinitis pigmentosa, overexpression of the chaperone BiP/Grp78 (see above) led to reduction of photoreceptor apoptosis and retinal degeneration, and allowed recovery of retinal activity in rats expressing this particular rhodopsin mutant.[134] Another strategy to manipulate the cellular QCS to enhance misfolded receptor PM expression is by employing cell-penetrating peptides that modify cytosolic Ca^{2+} stores, thereby affecting function of Ca^{2+}-regulated chaperones as those involved in post-ER quality control.[135] Nevertheless, as in the case of chemical chaperones (see below), the major drawback of these maneuvers is their lack of specificity for the target protein.

In chemical rescue, PM expression and function of misfolded proteins are achieved by incubating cells expressing the mutant protein with non-specific stabilizing agents (e.g., polyols and sugars).[136] Chemical chaperones, which are small molecular weight compounds, promote protein folding by stabilizing their conformation *without interacting* with them or interfering with their function. Osmolites stabilize proteins by reducing the free movement of proteins as well as by increasing their hydration,[136] preventing aggregation of partially folded conformers and promoting protein stabilization through modifying the free-energy difference between partially folded and more compact native structures (Figure 21.1). For example, incubation of stable CFTR delF508 transfectants with glycerol or trimethylamine N-oxide (TMAO) resulted in accumulation of functional F508 deletion molecules and an increase in whole-cell chloride conductance.[130] In the case of temperature-sensitive mutants, such as the A135V mutant of the tumor suppressor protein p53,[137] when cells expressing this mutant were cultured at non-permissive temperatures, the mutant was localized in the cytoplasm and was biologically inactive unless cells were exposed to chemical chaperones, such as glycerol, TMAO or deuterated water.[138] Given that chemical chaperones require high concentrations for effective folding of mutant proteins, they are too toxic for in vivo applications. Although chemical chaperones can rescue some misfolded proteins, their effects are nonspecific and might potentially lead to increased secretion or intracellular retention of many different proteins in various cellular compartments, provoking inappropriate changes in the local levels and/or secretion of many proteins, thereby compromising cell function.[139] As an exception, it has been observed that glycerol, 4-phenylbutyric acid and TMAO may selectively increase the secretion efficiency of α1-antitrypsin without influencing that of other proteins or decreasing proteasomal degradation.[140]

Pharmacological Approaches

Pharmacochaperones or pharmacoperones are small, PM-permeable molecules (frequently agonists or antagonists of the natural ligand) that enter cells and serve as a molecular scaffold to promote correct folding and prevent aggregation of intermediate, incompletely folded or misfolded proteins within the cell (Figures 21.1 and 21.3).[13] In contrast with chemical chaperones, these small molecules have the advantage of selective binding to the conformationally defective protein, which allows the beneficial (normal) degradation of other misfolded proteins that require elimination from the cell as part of the normal process in protein biosynthesis. For example, in transthyretin amyloidogenesis several small molecules bind with high affinity to binding sites within the transthyretin molecule leading to stabilization of the native state of the protein, decreasing the concentration of the intermediate species and hence amyloid formation.[141,142] Short β-sheet breaker peptides have been designed for blocking the aggregation undergone by β-amyloid;[143] these peptides have a structure homologous to the central hydrophobic region of the fibril aggregate, and inhibit and dissolve β-amyloid aggregates in vitro and in vivo.[144] Another example is the competitive inhibitor 1-deoxy-galactonojirimycin, which increases the activity of the R301Q mutant form of α-galactosidase A (whose accumulation in the ER leads to the lysosomal storage disease, Fabry's disease in humans), thereby promoting its transport from the ER to lysosomes.[145] More recently, it has been reported that cyclosporin A (a substrate of the multidrug-resistant protein ATP-binding cassette transporter ABCB1) improved maturation and PM expression of the misfolded I541F mutant ABCB4 transporter, which is linked to progressive familial intrahepatic cholestasis type 3, a disease featured by early onset of persistent cholestasis that usually progresses to cirrhosis and liver failure before adulthood.[146]

The efficiency of pharmacoperones to rescue PM expression and function will depend on the particular structure of the pharmacoperone (either agonist or antagonist), which determines selectivity toward the target protein, the severity of the folding defect present in the

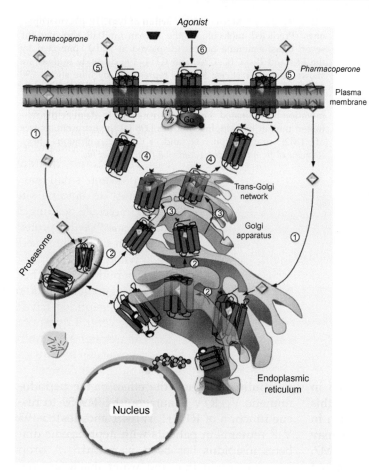

FIGURE 21.3 **Rescue of misfolded GPCRs by pharmacoperones.** Pharmacoperones (diamond-like structures) traverse the plasma membrane into the cytoplasm (step 1) and interact specifically with the misfolded protein (step 2). Misfolded proteins sensitive to pharmacoperones are then stabilized in a conformation compatible with ER export (step 2). The "rescued" receptor enters into and is processed in the Golgi apparatus (step 3) and then exported to the plasma membrane (step 4), where the pharmacoperone dissociates from the receptor (step 5) to allow interaction with agonist (step 6). Dissociation from the receptor is necessary whenever the pharmacoperone is an agonist or antagonist that competes with the natural ligand for the binding site at the receptor or, in the case of allosteric molecules, that keeps the receptor in inactive conformation.

target protein, and the particular location of the mutation (i.e., the mutation *should not* include critical residues involved in agonist binding, receptor activation, or coupling to effectors).[118] For example, mutant human V_2Rs displaying amino acid exchanges at the interface of the TM2 and TM4 (H80R, W164R, and S167L mutants), are resistant to pharmacoperone-mediated cell surface delivery, probably because the replacing residues lead to a severe folding defect.[147] In the case of the S168R and S217R human GnRHR mutants, replacement of any of these serine residues (which in the three-dimensional structure of the receptor are located in the lipid

membrane-contact phase of the TMs 4 and 5, respectively) by the highly hydrophilic arginine involves a thermodynamically unfavorable exchange that rotates the TM4 and TM5, moving the EL2, and making formation of the C14–C200 bridge improbable (Figure 21.4).[148] In this particular receptor, the presence of a C14–C200 disulfide bridge is a key structural feature since this bridge stabilizes the receptor in a conformation that is compatible with ER export.[148] Both GnRHR mutants are completely resistant to pharmacoperone treatment *in vitro*.[81] In the case of the P320L GnRHR mutant, the abnormal protein is unrescuable by genetic approaches because the

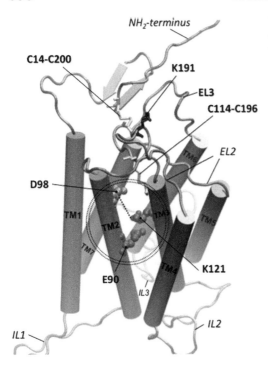

FIGURE 21.4 **Mechanism of action of GnRHR pharmacoperones.** Predicted molecule of the human GnRHR showing the seven transmembrane helices (displayed as rods) connected by the ELs and ILs. C14–C200 and C114–196 disulfide bridges are shown as yellow sticks; K191 is represented by blue sticks. E90 (at the TM2; red spheres) forms a salt bridge with K121 (at the TM3; purple spheres), which in the case of the inactivating E90K mutation is eliminated. Pharmacoperones act to stabilize the misfolded mutant by bridging residues D98 (at the extracellular face of TM2; orange spheres) and K121 (discontinuous line). *Reproduced from ref [13] with permission from Elsevier Inc.*

peptide backbone of proline is constrained in a ring structure; in fact, occurrence of this amino acid is associated with a forced turn in the protein sequence and its replacement may severely disturb the structure of the TM7. Mutational defects that interfere with ligand binding would also be expected to show minimal or no functional rescue of the misfolded receptor in response to pharmacoperones, despite promoting receptor PM expression.

The efficacy of pharmacoperones to rescue function and prevent abnormal intracellular accumulation has been demonstrated for a number of misfolded GPCRs (Table 21.1). In the case of the V_2R, it has been shown that distinct cell membrane-permeable antagonists effectively rescue *in vitro* function of several misfolded, traffic-defective mutants that cause diabetes insipidus in humans.[83] These observations are important since the majority of V_2R mutations leading to nephrogenic diabetes insipidus are provoked by receptor

misrouting. Further, the effect of the peptidomimetic $V_{1A}R/V_2R$ antagonist SR49059 to rescue function of R137H, W164S, and des185-193 V_2R mutants in patients with nephrogenic diabetes insipidus has been examined.[83] A drop in urine production and water intake as well as a significant increase in urine osmolarity in response to this particular compound was observed. In retinitis pigmentosa, the vast majority of mutations in rhodopsin affect the folding of the receptor protein, leading to decreased PM expression, intracellular retention, aggregation, and eventually cell death (see above). Rescue of the T17M and P23H mutants associated with retinitis pigmentosa has been achieved by *in vitro* exposure of the cells to 11-*cis*-retinal or *11-cis*-7-ring-retinal (a seven-membered ring variant of 11-*cis*-retinal, the chromophore of rhodopsin that plays a central role in the photoactivation process).[19] Further, in transgenic mice bearing the T17M opsin mutation, administration of vitamin A palmitate was followed by a significant decrease in retinal degeneration.[20] In the majority of melanocortin-4 receptor mutants, which may lead to monogenic obesity, the mutant receptors are retained intracellularly due to misfolding.[25] Recent studies on a palette of misfolded mutants of this receptor demonstrated that the pharmacoperone ML00253764 and 4-phenylbutyric acid rescued PM expression and function of the mutant receptors to different extents.[25] In the case of PM expression-deficient μ-opioid receptors and melanin concentrating hormone receptor-1 mutants, different cell-permeable agonists and antagonists also have been shown to effectively enhance cell-surface expression of the mutant receptors.[28] Finally, with the exception of the pharmacoperone-resistant mutant human GnRHRs described above, all the remaining misfolded GnRHR mutants leading to HH and tested to date are sensitive (to a greater or lesser extent) to functional rescue by a variety of small molecules of diverse chemical structure (e.g., quinolones, indoles, and erythromycin-related macrolides).[13]

Small molecules that rescue misfolded human GnRHR mutants are one of few pharmacoperones whose mechanism of action at the molecular level has been elucidated with some detail.[13,149] Using site-directed mutagenesis, confocal microscopy, computer modeling and ligand docking, it has been found that different chemical classes of compounds stabilize the misfolded E90K mutant (whose functional rescue is complete when genetic and pharmacological approaches are applied) in a conformation compatible with ER export. Stabilization of the misfolded GnRHR by pharmacoperones is achieved by associating D98 (at the extracellular face of the TM1) and K121 (at the TM3) by creating a surrogate bond for the highly conserved E90-K121 salt bridge (Figure 21.4), leading to stabilization of the TM2-TM3 conformation disturbed by the E90K substitution. Apparently, a stable TM2—TM3

configuration is a structural requirement for passing the GnRHR through the ER QCS. Interestingly, the observation that all GnRHR pharmacoperones rescued most of the human GnRHR mutants tested despite their different distribution along the receptor, indicates that the E90—K121 bridge stabilizes the orientation of, and relation between, the TM2 and TM3 and thus might be a specific structural domain subjected to scrutiny by the ER QCS.

Other pharmacoperones act via allosteric interactions with the misfolded receptor. The advantage of this class of drugs is that allosteric compounds *do not compete* with the natural agonist for the binding site. For example, in the case of the misfolded A189V human FSHR, *in vitro* exposure of COS-7 cells expressing this mutant receptor to the thienopyr(im)idine Org41841 (which is a molecule that binds a conserved region of the human LHR without competing for the natural agonist binding site or the site for interaction with effector)[150] resulted in almost a two-fold increase in PM expression and FSH-stimulated cAMP production, without significantly altering mRNA expression of the receptor nor its ligand binding affinity.[37] Similarly, incubation of cells expressing the misfolded human LHR mutants A593P and S616Y (which lead to Leydig cell hypoplasia and varying degrees of genital ambiguity; Table 21.1) with the cell-permeant, allosterically binding small-molecule agonist Org 42599, led to rescue of PM expression and signaling of the two LHR misfolded mutants.[38] Another example is the allosteric agonist NPS R-568, which binds the calcium-sensing receptor promoting PM expression and function of misfolded loss-of-function mutants of this receptor.[36]

Current challenges in this fascinating area include identification of novel small molecules that may adequately function *in vivo*, and thus cure disease caused by GPCR misfolding. In addition, the identification of molecules that may bind (inefficiently expressed) WT receptors and promote not only increased PM expression but also conformational bias at the ER that can favor GPCR-signalosome assembly and/or signaling in response to agonist[151] may be useful. Although some assays for identification of pharmacoperones that lack antagonistic activity are available,[152,153] more are required to obviate the need to deal with the complex pharmacology that occurs when both activities are present. As we learn more about the anterograde transfer of GPCRs from the ER and Golgi apparatus to the PM, we anticipate that new sites for therapeutic intervention will be presented. The development of pharmacoperone drugs will benefit by the creation of animal models as proof-of-principle test models.

Acknowledgements

The authors of this chapter are supported by CONACyT Grant 86881 (to A.U.-A) and the National Institutes of Health Grants OD012220, DK85040, DK99090, OD011092 and TW/HD-00668 (to P.M.C.).

References

1. Ellgaard L, Helenius A. Quality control in the endoplasmic reticulum. *Nat Rev Mol Cell Biol* 2003;**4**:181–91.
2. Ulloa-Aguirre A, Janovick JA, Brothers SP, Conn PM. Pharmacologic rescue of conformationally-defective proteins: implications for the treatment of human disease. *Traffic* 2004;**5**:821–37.
3. Hartl FU, Bracher A, Hayer-Hartl M. Molecular chaperones in protein folding and proteostasis. *Nature* 2011;**475**:324–32.
4. Aridor M. Visiting the ER: the endoplasmic reticulum as a target for therapeutics in traffic related diseases. *Adv Drug Deliv Rev* 2007;**59**:759–81.
5. Dobson CM. Principles of protein folding, misfolding and aggregation. *Semin Cell Dev Biol* 2004;**15**:3–16.
6. Hartl FU, Hayer-Hartl M. Molecular chaperones in the cytosol: from nascent chain to folded protein. *Science* 2002;**295**:1852–8.
7. Ellis RJ, Minton AP. Protein aggregation in crowded environments. *Biol Chem* 2006;**387**:485–97.
8. Hartl FU, Hayer-Hartl M. Converging concepts of protein folding in vitro and in vivo. *Nat Struct Mol Biol* 2009;**16**:574–81.
9. Cahill M, Cahill S, Cahill K. Proteins wriggle. *Biophys J* 2002;**82**:2665–70.
10. Dinner AR, Sali A, Smith LJ, Dobson CM, Karplus M. Understanding protein folding via free-energy surfaces from theory and experiment. *Trends Biochem Sci* 2000;**25**:331–9.
11. Dill KA, MacCallum JL. The protein-folding problem, 50 years on. *Science* 2012;**338**:1042–6.
12. Dobson CM. Protein folding and misfolding. *Nature* 2003;**426**:884–90.
13. Conn PM, Ulloa-Aguirre A. Pharmacological chaperones for misfolded gonadotropin-releasing hormone receptors. *Adv Pharmacol* 2011;**62**:109–41.
14. Horwich A. Protein aggregation in disease: a role for folding intermediates forming specific multimeric interactions. *J Clin Invest* 2002;**110**:1221–32.
15. Shastry BS. Neurodegenerative disorders of protein aggregation. *Neurochem Int* 2003;**43**:1–7.
16. Bernier V, Lagace M, Bichet DG, Bouvier M. Pharmacological chaperones: potential treatment for conformational diseases. *Trends Endocrinol Metab* 2004;**15**:222–8.
17. Conn PM, Ulloa-Aguirre A. Trafficking of G-protein-coupled receptors to the plasma membrane: insights for pharmacoperone drugs. *Trends Endocrinol Metab* 2010;**21**:190–7.
18. Ulloa-Aguirre A, Conn PM. Pharmacoperones: a new therapeutic approach for diseases caused by misfolded G protein-coupled receptors. *Recent Pat Endocr Metab Immune Drug Discov* 2011;**5**:13–24.
19. Noorwez SM, Malhotra R, McDowell JH, Smith KA, Krebs MP, Kaushal S. Retinoids assist the cellular folding of the autosomal dominant retinitis pigmentosa opsin mutant P23H. *J Biol Chem* 2004;**279**:16278–84.
20. Li T, Sandberg MA, Pawlyk BS, Rosner B, Hayes KC, Dryja TP, et al. Effect of vitamin A supplementation on rhodopsin mutants threonine-17 --> methionine and proline-347 --> serine in transgenic mice and in cell cultures. *Proc Natl Acad Sci USA* 1998;**95**:11933–8.
21. Krebs MP, Holden DC, Joshi P, Clark III CL, Lee AH, Kaushal S. Molecular mechanisms of rhodopsin retinitis pigmentosa and the efficacy of pharmacological rescue. *J Mol Biol* 2010;**395**:1063–78.
22. Noorwez SM, Kuksa V, Imanishi Y, Zhu L, Filipek S, Palczewski K, et al. Pharmacological chaperone-mediated in vivo folding and stabilization of the P23H-opsin mutant associated with autosomal dominant retinitis pigmentosa. *J Biol Chem* 2003;**278**:14442–50.
23. Ostrov DA, Kaushai S, Noorwez SM. Opsin stabilizing compounds and methods of use. US Patent application number 20090286808 (2009).
24. Sung CH, Schneider BG, Agarwal N, Papermaster DS, Nathans J. Functional heterogeneity of mutant rhodopsins

responsible for autosomal dominant retinitis pigmentosa. *Proc Natl Acad Sci U S A* 1991;**88**:8840−4.

25. Tao YX. The melanocortin-4 receptor: physiology, pharmacology, and pathophysiology. *Endocr Rev* 2010;**31**:506−43.

26. Fan J-Q, Valenzano K, Lee G, Bouvier M. Pharmacological chaperones for treating obesity. US Patent application number 20090312345 (2009).

27. Fan ZC, Tao YX. Functional characterization and pharmacological rescue of melanocortin-4 receptor mutations identified from obese patients. *J Cell Mol Med* 2009;**13**:3268−82.

28. Fan J, Perry SJ, Gao Y, Schwarz DA, Maki RA. A point mutation in the human melanin concentrating hormone receptor 1 reveals an important domain for cellular trafficking. *Mol Endocrinol* 2005;**19**:2579−90.

29. Morello JP, Petaja-Repo UE, Bichet DG, Bouvier M. Pharmacological chaperones: a new twist on receptor folding. *Trends Pharmacol Sci* 2000;**21**:466−9.

30. Albright JD, Reich MF, Delos Santos EG, Dusza JP, Sum FW, Venkatesan AM, et al. 5-Fluoro-2-methyl-N-[4-(5H-pyrrolo[2,1-c]-[1, 4]benzodiazepin-10(11H)-ylcarbonyl)-3-chlorophenyl]benzamide (VPA-985): an orally active arginine vasopressin antagonist with selectivity for V2 receptors. *J Med Chem* 1998;**41**:2442−4.

31. Hawtin SR. Pharmacological chaperone activity of SR49059 to functionally recover misfolded mutations of the vasopressin V1a receptor. *J Biol Chem* 2006;**281**:14604−14.

32. Robben JH, Sze M, Knoers NV, Deen PM. Rescue of vasopressin V2 receptor mutants by chemical chaperones: specificity and mechanism. *Mol Biol Cell* 2006;**17**:379−86.

33. Robben JH, Sze M, Knoers NV, Deen PM. Functional rescue of vasopressin V2 receptor mutants in MDCK cells by pharmacochaperones: relevance to therapy of nephrogenic diabetes insipidus. *Am J Physiol Renal Physiol* 2007;**292**:F253−60.

34. Serradeil-Le Gal C, Lacour C, Valette G, Garcia G, Foulon L, Galindo G, et al. Characterization of SR 121463A, a highly potent and selective, orally active vasopressin V2 receptor antagonist. *J Clin Invest* 1996;**98**:2729−38.

35. Tahara A, Tomura Y, Wada KI, Kusayama T, Tsukada J, Takanashi M, et al. Pharmacological profile of YM087, a novel potent nonpeptide vasopressin V1A and V2 receptor antagonist, in vitro and in vivo. *J Pharmacol Exp Ther* 1997;**282**:301−8.

36. Huang Y, Cavanaugh A, Breitwieser GE. Regulation of stability and trafficking of calcium-sensing receptors by pharmacologic chaperones. *Adv Pharmacol* 2011;**62**:143−73.

37. Janovick JA, Maya-Nunez G, Ulloa-Aguirre A, Huhtaniemi IT, Dias JA, Verbost P, et al. Increased plasma membrane expression of human follicle-stimulating hormone receptor by a small molecule thienopyr(im)idine. *Mol Cell Endocrinol* 2009;**298**:84−8.

38. Newton CL, Whay AM, McArdle CA, Zhang M, van Koppen CJ, van de Lagemaat R, et al. Rescue of expression and signaling of human luteinizing hormone G protein-coupled receptor mutants with an allosterically binding small-molecule agonist. *Proc Natl Acad Sci USA* 2011;**108**:7172−6.

39. Ulloa-Aguirre A, Conn PM. G Protein-coupled receptors and the G protein family. In: Conn PM, editor. *Handbook of physiology. Section 7: the endocrine system*. New York: Oxford University Press; 1998. . p. 87−124..

40. Lagerstrom MC, Schioth HB. Structural diversity of G protein-coupled receptors and significance for drug discovery. *Nat Rev Drug Discov* 2008;**7**:339−57.

41. Overington JP, Al-Lazikani B, Hopkins AL. How many drug targets are there? *Nat Rev Drug Discov* 2006;**5**:993−6.

42. Schlyer S, Horuk R. I want a new drug: G-protein-coupled receptors in drug development. *Drug Discov Today* 2006;**11**:481−93.

43. Palczewski K, Kumasaka T, Hori T, Behnke CA, Motoshima H, Fox BA, et al. Crystal structure of rhodopsin: a G protein-coupled receptor. *Science* 2000;**289**:739−45.

44. Ulloa-Aguirre A, Conn PM. Targeting of G protein-coupled receptors to the plasma membrane in health and disease. *Front Biosci* 2009;**14**:973−94.

45. Sitia R, Braakman I. Quality control in the endoplasmic reticulum protein factory. *Nature* 2003;**426**:891−4.

46. Ellgaard L, Helenius A. ER quality control: towards an understanding at the molecular level. *Curr Opin Cell Biol* 2001;**13**:431−7.

47. Kimchi-Sarfaty C, Oh JM, Kim IW, Sauna ZE, Calcagno AM, Ambudkar SV, et al. A "silent" polymorphism in the MDR1 gene changes substrate specificity. *Science* 2007;**315**:525−8.

48. Duennwald ML, Echeverria A, Shorter J. Small heat shock proteins potentiate amyloid dissolution by protein disaggregases from yeast and humans. *PLoS Biol* 2012;**10**:e1001346.

49. Conn PM, Janovick JA, Brothers SP, Knollman PE. 'Effective inefficiency': cellular control of protein trafficking as a mechanism of post-translational regulation. *J Endocrinol* 2006;**190**:13−6.

50. Petaja-Repo UE, Hogue M, Laperriere A, Walker P, Bouvier M. Export from the endoplasmic reticulum represents the limiting step in the maturation and cell surface expression of the human delta opioid receptor. *J Biol Chem* 2000;**275**:13727−36.

51. Tan CM, Brady AE, Nickols HH, Wang Q, Limbird LE. Membrane trafficking of G protein-coupled receptors. *Annu Rev Pharmacol Toxicol* 2004;**44**:559–609.

52. Dong C, Filipeanu CM, Duvernay MT, Wu G. Regulation of G protein-coupled receptor export trafficking. *Biochim Biophys Acta* 2007;**1768**:853–70.

53. Ferreira PA, Nakayama TA, Pak WL, Travis GH. Cyclophilin-related protein RanBP2 acts as chaperone for red/green opsin. *Nature* 1996;**383**:637–40.

54. Gimelbrant AA, Haley SL, McClintock TS. Olfactory receptor trafficking involves conserved regulatory steps. *J Biol Chem* 2001;**276**:7285–90.

55. Christopoulos A, Christopoulos G, Morfis M, Udawela M, Laburthe M, Couvineau A, et al. Novel receptor partners and function of receptor activity-modifying proteins. *J Biol Chem* 2003;**278**:3293–7.

56. Xu Z, Hirasawa A, Shinoura H, Tsujimoto G. Interaction of the alpha(1B)-adrenergic receptor with gC1q-R, a multifunctional protein. *J Biol Chem* 1999;**274**:21149–54.

57. Morello JP, Salahpour A, Petaja-Repo UE, Laperriere A, Lonergan M, Arthus MF, et al. Association of calnexin with wild type and mutant AVPR2 that causes nephrogenic diabetes insipidus. *Biochemistry* 2001;**40**:6766–75.

58. Brothers SP, Janovick JA, Conn PM. Calnexin regulated gonadotropin-releasing hormone receptor plasma membrane expression. *J Mol Endocrinol* 2006;**37**:479–88.

59. Ayala Yanez R, Conn PM. Protein disulfide isomerase chaperone ERP-57 decreases plasma membrane expression of the human GnRH receptor. *Cell Biochem Funct* 2010;**28**:66–73.

60. Lucca-Junior W, Janovick JA, Conn PM. Participation of the endoplasmic reticulum protein chaperone thiooxidoreductase in gonadotropin-releasing hormone receptor expression at the plasma membrane. *Braz J Med Biol Res* 2009;**42**:164–7.

61. Mizrachi D, Segaloff DL. Intracellularly located misfolded glycoprotein hormone receptors associate with different chaperone proteins than their cognate wild-type receptors. *Mol Endocrinol* 2004;**18**:1768–77.

62. Gething MJ. Role and regulation of the ER chaperone BiP. *Semin Cell Dev Biol* 1999;**10**:465–72.

63. Schroder M, Kaufman RJ. The mammalian unfolded protein response. *Annu Rev Biochem* 2005;**74**:739–89.

64. Bermak JC, Li M, Bullock C, Zhou QY. Regulation of transport of the dopamine D1 receptor by a new membrane-associated ER protein. *Nat Cell Biol* 2001;**3**:492–8.

65. Martens JW, Lumbroso S, Verhoef-Post M, Georget V, Richter-Unruh A, Szarras-Czapnik M, et al. Mutant luteinizing hormone receptors in a compound heterozygous patient with complete Leydig cell hypoplasia:

66. abnormal processing causes signaling deficiency. *J Clin Endocrinol Metab* 2002;**87**:2506–13.

66. Dorner AJ, Krane MG, Kaufman RJ. Reduction of endogenous GRP78 levels improves secretion of a heterologous protein in CHO cells. *Mol Cell Biol* 1988;**8**:4063–70.

67. Tang BL, Wang Y, Ong YS, Hong W. COPII and exit from the endoplasmic reticulum. *Biochim Biophys Acta* 2005;**1744**:293–303.

68. Gurkan C, Stagg SM, Lapointe P, Balch WE. The COPII cage: unifying principles of vesicle coat assembly. *Nat Rev Mol Cell Biol* 2006;**7**:727–38.

69. Malkus P, Jiang F, Schekman R. Concentrative sorting of secretory cargo proteins into COPII-coated vesicles. *J Cell Biol* 2002;**159**:915–21.

70. Wang X, Matteson J, An Y, Moyer B, Yoo JS, Bannykh S, et al. COPII-dependent export of cystic fibrosis transmembrane conductance regulator from the ER uses a di-acidic exit code. *J Cell Biol* 2004;**167**:65–74.

71. Miller EA, Beilharz TH, Malkus PN, Lee MC, Hamamoto S, Orci L, et al. Multiple cargo binding sites on the COPII subunit Sec24p ensure capture of diverse membrane proteins into transport vesicles. *Cell* 2003;**114**:497–509.

72. Dong C, Zhou F, Fugetta EK, Filipeanu CM, Wu G. Endoplasmic reticulum export of adrenergic and angiotensin II receptors is differentially regulated by Sar1 GTPase. *Cell Signal* 2008;**20**:1035–43.

73. Wang G, Wu G. Small GTPase regulation of GPCR anterograde trafficking. *Trends Pharmacol Sci* 2012;**33**:28–34.

74. Thielen A, Oueslati M, Hermosilla R, Krause G, Oksche A, Rosenthal W, et al. The hydrophobic amino acid residues in the membrane-proximal C tail of the G protein-coupled vasopressin V2 receptor are necessary for transport-competent receptor folding. *FEBS Lett* 2005;**579**:5227–35.

75. Robert J, Clauser E, Petit PX, Ventura MA. A novel C-terminal motif is necessary for the export of the vasopressin V1b/V3 receptor to the plasma membrane. *J Biol Chem* 2005;**280**:2300–8.

76. Duvernay MT, Zhou F, Wu G. A conserved motif for the transport of G protein-coupled receptors from the endoplasmic reticulum to the cell surface. *J Biol Chem* 2004;**279**:30741–50.

77. Leclerc PC, Auger-Messier M, Lanctot PM, Escher E, Leduc R, Guillemette G. A polyaromatic caveolin-binding-like motif in the cytoplasmic tail of the type 1 receptor for angiotensin II plays an important role in receptor trafficking and signaling. *Endocrinology* 2002;**143**:4702–10.

78. Duvernay MT, Dong C, Zhang X, Robitaille M, Hebert TE, Wu G. A single conserved leucine residue on the

first intracellular loop regulates ER export of G protein-coupled receptors. *Traffic* 2009;**10**:552–66.

79. Dong C, Nichols CD, Guo J, Huang W, Lambert NA, Wu G. A triple arg motif mediates alpha(2B)-adrenergic receptor interaction with Sec24C/D and export. *Traffic* 2012;**13**:857–68.

80. Gershengorn MC, Osman R. Minireview: Insights into G protein-coupled receptor function using molecular models. *Endocrinology* 2001;**142**:2–10.

81. Leanos-Miranda A, Janovick JA, Conn PM. Receptor-misrouting: an unexpectedly prevalent and rescuable etiology in gonadotropin-releasing hormone receptor-mediated hypogonadotropic hypogonadism. *J Clin Endocrinol Metab* 2002;**87**:4825–8.

82. Topaloglu AK, Lu ZL, Farooqi IS, Mungan NO, Yuksel B, O'Rahilly S, et al. Molecular genetic analysis of normosmic hypogonadotropic hypogonadism in a Turkish population: identification and detailed functional characterization of a novel mutation in the gonadotropin-releasing hormone receptor gene. *Neuroendocrinology* 2006;**84**:301–8.

83. Bernier V, Morello JP, Zarruk A, Debrand N, Salahpour A, Lonergan M, et al. Pharmacologic chaperones as a potential treatment for X-linked nephrogenic diabetes insipidus. *J Am Soc Nephrol* 2006;**17**:232–43.

84. Mendes HF, van der Spuy J, Chapple JP, Cheetham ME. Mechanisms of cell death in rhodopsin retinitis pigmentosa: implications for therapy. *Trends Mol Med* 2005;**11**:177–85.

85. Zerangue N, Schwappach B, Jan YN, Jan LY. A new ER trafficking signal regulates the subunit stoichiometry of plasma membrane K(ATP) channels. *Neuron* 1999;**22**:537–48.

86. Pagano A, Rovelli G, Mosbacher J, Lohmann T, Duthey B, Stauffer D, et al. C-terminal interaction is essential for surface trafficking but not for heteromeric assembly of GABA(b) receptors. *J Neurosci* 2001;**21**:1189–202.

87. Ma D, Zerangue N, Lin YF, Collins A, Yu M, Jan YN, et al. Role of ER export signals in controlling surface potassium channel numbers. *Science* 2001;**291**:316–9.

88. Angelotti T, Daunt D, Shcherbakova OG, Kobilka B, Hurt CM. Regulation of G-protein coupled receptor traffic by an evolutionary conserved hydrophobic signal. *Traffic* 2010;**11**:560–78.

89. Benke D, Zemoura K, Maier PJ. Modulation of cell surface GABA(B) receptors by desensitization, trafficking and regulated degradation. *World J Biol Chem* 2012;**3**:61–72.

90. Uribe A, Zarinan T, Perez-Solis MA, Gutierrez-Sagal R, Jardon-Valadez E, Pineiro A, et al. Functional and structural roles of conserved cysteine residues in the carboxyl-terminal domain of the follicle-stimulating hormone receptor in human embryonic kidney 293 cells. *Biol Reprod* 2008;**78**:869–82.

91. Helenius A, Aebi M. Roles of N-linked glycans in the endoplasmic reticulum. *Annu Rev Biochem* 2004;**73**:1019–49.

92. Huhtaniemi IT, Themmen AP. Mutations in human gonadotropin and gonadotropin-receptor genes. *Endocrine* 2005;**26**:207–17.

93. Milligan G. G protein-coupled receptor dimerisation: molecular basis and relevance to function. *Biochim Biophys Acta* 2007;**1768**:825–35.

94. Margeta-Mitrovic M. Assembly-dependent trafficking assays in the detection of receptor-receptor interactions. *Methods* 2002;**27**:311–7.

95. Ayoub MA, Couturier C, Lucas-Meunier E, Angers S, Fossier P, Bouvier M, et al. Monitoring of ligand-independent dimerization and ligand-induced conformational changes of melatonin receptors in living cells by bioluminescence resonance energy transfer. *J Biol Chem* 2002;**277**:21522–8.

96. Guo W, Shi L, Javitch JA. The fourth transmembrane segment forms the interface of the dopamine D2 receptor homodimer. *J Biol Chem* 2003;**278**:4385–8.

97. Terrillon S, Barberis C, Bouvier M. Heterodimerization of V1a and V2 vasopressin receptors determines the interaction with beta-arrestin and their trafficking patterns. *Proc Natl Acad Sci USA* 2004;**101**:1548–53.

98. Herrick-Davis K, Grinde E, Mazurkiewicz JE. Biochemical and biophysical characterization of serotonin 5-HT2C receptor homodimers on the plasma membrane of living cells. *Biochemistry* 2004;**43**:13963–71.

99. McVey M, Ramsay D, Kellett E, Rees S, Wilson S, Pope AJ, et al. Monitoring receptor oligomerization using time-resolved fluorescence resonance energy transfer and bioluminescence resonance energy transfer. The human delta -opioid receptor displays constitutive oligomerization at the cell surface, which is not regulated by receptor occupancy. *J Biol Chem* 2001;**276**:14092–9.

100. Salahpour A, Angers S, Mercier JF, Lagace M, Marullo S, Bouvier M. Homodimerization of the beta2-adrenergic receptor as a prerequisite for cell surface targeting. *J Biol Chem* 2004;**279**:33390–7.

101. Mercier JF, Salahpour A, Angers S, Breit A, Bouvier M. Quantitative assessment of beta 1- and beta 2-adrenergic receptor homo- and heterodimerization by bioluminescence resonance energy transfer. *J Biol Chem* 2002;**277**:44925–31.

102. Angers S, Salahpour A, Joly E, Hilairet S, Chelsky D, Dennis M, et al. Detection of beta 2-adrenergic

receptor dimerization in living cells using biolumines-cence resonance energy transfer (BRET). *Proc Natl Acad Sci USA* 2000;**97**:3684—9.

103. Thomas RM, Nechamen CA, Mazurkiewicz JE, Muda M, Palmer S, Dias JA. Follicle-stimulating hormone receptor forms oligomers and shows evidence of carboxyl-terminal proteolytic processing. *Endocrinology* 2007;**148**:1987—95.

104. Guan R, Feng X, Wu X, Zhang M, Zhang X, Hebert TE, et al. Bioluminescence resonance energy transfer studies reveal constitutive dimerization of the human lutropin receptor and a lack of correlation between receptor activation and the propensity for dimeriza-tion. *J Biol Chem* 2009;**284**:7483—94.

105. Margeta-Mitrovic M, Jan YN, Jan LY. A trafficking checkpoint controls GABA(B) receptor heterodimeri-zation. *Neuron* 2000;**27**:97—106.

106. Hague C, Uberti MA, Chen Z, Hall RA, Minneman KP. Cell surface expression of alpha1D-adrenergic receptors is controlled by heterodimerization with alpha1B-adrenergic receptors. *J Biol Chem* 2004;**279**:15541—9.

107. Uberti MA, Hague C, Oller H, Minneman KP, Hall RA. Heterodimerization with beta2-adrenergic receptors promotes surface expression and functional activity of alpha1D-adrenergic receptors. *J Pharmacol Exp Ther* 2005;**313**:16—23.

108. Leanos-Miranda A, Ulloa-Aguirre A, Ji TH, Janovick JA, Conn PM. Dominant-negative action of disease-causing gonadotropin-releasing hormone receptor (GnRHR) mutants: a trait that potentially coevolved with decreased plasma membrane expression of GnRHR in humans. *J Clin Endocrinol Metab* 2003;**88**:3360—7.

109. Brothers SP, Cornea A, Janovick JA, Conn PM. Human loss-of-function gonadotropin-releasing hor-mone receptor mutants retain wild-type receptors in the endoplasmic reticulum: molecular basis of the dominant-negative effect. *Mol Endocrinol* 2004;**18**:1787—97.

110. Zarinan T, Perez-Solis MA, Maya-Nunez G, Casas-Gonzalez P, Conn PM, Dias JA, et al. Dominant nega-tive effects of human follicle-stimulating hormone receptor expression-deficient mutants on wild-type receptor cell surface expression. Rescue of oligomerization-dependent defective receptor expres-sion by using cognate decoys. *Mol Cell Endocrinol* 2010;**321**:112—22.

111. Zhu X, Wess J. Truncated V2 vasopressin receptors as negative regulators of wild-type V2 receptor function. *Biochemistry* 1998;**37**:15773—84.

112. Karpa KD, Lin R, Kabbani N, Levenson R. The dopa-mine D3 receptor interacts with itself and the truncated D3 splice variant d3nf: D3-D3nf interaction causes mislocalization of D3 receptors. *Mol Pharmacol* 2000;**58**:677—83.

113. Lee SP, O'Dowd BF, Ng GY, Varghese G, Akil H, Mansour A, et al. Inhibition of cell surface expression by mutant receptors demonstrates that D2 dopamine receptors exist as oligomers in the cell. *Mol Pharmacol* 2000;**58**:120—8.

114. Benkirane M, Jin DY, Chun RF, Koup RA, Jeang KT. Mechanism of transdominant inhibition of CCR5-mediated HIV-1 infection by ccr5delta32. *J Biol Chem* 1997;**272**:30603—6.

115. Calebiro D, de Filippis T, Lucchi S, Covino C, Panigone S, Beck-Peccoz P, et al. Intracellular entrap-ment of wild-type TSH receptor by oligomerization with mutants linked to dominant TSH resistance. *Hum Mol Genet* 2005;**14**:2991—3002.

116. Yuan H, Michelsen K, Schwappach B. 14-3-3 dimers probe the assembly status of multimeric membrane proteins. *Curr Biol* 2003;**13**:638—46.

117. Duvernay MT, Wang H, Dong C, Guidry JJ, Sackett DL, Wu G. Alpha2B-adrenergic receptor interaction with tubulin controls its transport from the endoplas-mic reticulum to the cell surface. *J Biol Chem* 2011;**286**:14080—9.

118. Conn PM, Ulloa-Aguirre A, Ito J, Janovick JA. G protein-coupled receptor trafficking in health and dis-ease: lessons learned to prepare for therapeutic mutant rescue in vivo. *Pharmacol Rev* 2007;**59**:225—50.

119. Aittomaki K, Lucena JL, Pakarinen P, Sistonen P, Tapanainen J, Gromoll J, et al. Mutation in the follicle-stimulating hormone receptor gene causes hereditary hypergonadotropic ovarian failure. *Cell* 1995;**82**:959—68.

120. Biebermann H, Schoneberg T, Krude H, Schultz G, Gudermann T, Gruters A. Mutations of the human thyrotropin receptor gene causing thyroid hypoplasia and persistent congenital hypothyroidism. *J Clin Endocrinol Metab* 1997;**82**:3471—80.

121. Gromoll J, Schulz A, Borta H, Gudermann T, Teerds KJ, Greschniok A, et al. Homozygous mutation within the conserved Ala-Phe-Asn-Glu-Thr motif of exon 7 of the LH receptor causes male pseudohermaphrodit-ism. *Eur J Endocrinol* 2002;**147**:597—608.

122. Beaumont KA, Shekar SN, Newton RA, James MR, Stow JL, Duffy DL, et al. Receptor function, dominant negative activity and phenotype correlations for MC1R variant alleles. *Hum Mol Genet* 2007;**16**:2249—60.

123. D'Souza-Li L, Yang B, Canaff L, Bai M, Hanley DA, Bastepe M, et al. Identification and functional characterization of novel calcium-sensing receptor mutations in familial hypocalciuric hypercalcemia

and autosomal dominant hypocalcemia. *J Clin Endocrinol Metab* 2002;**87**:1309–18.

124. Fuchs S, Amiel J, Claudel S, Lyonnet S, Corvol P, Pinet F. Functional characterization of three mutations of the endothelin B receptor gene in patients with Hirschsprung's disease: evidence for selective loss of Gi coupling. *Mol Med* 2001;**7**:115–24.

125. Tanaka H, Moroi K, Iwai J, Takahashi H, Ohnuma N, Hori S, et al. Novel mutations of the endothelin B receptor gene in patients with Hirschsprung's disease and their characterization. *J Biol Chem* 1998;**273**:11378–83.

126. Rana S, Besson G, Cook DG, Rucker J, Smyth RJ, Yi Y, et al. Role of CCR5 in infection of primary macrophages and lymphocytes by macrophage-tropic strains of human immunodeficiency virus: resistance to patient-derived and prototype isolates resulting from the delta ccr5 mutation. *J Virol* 1997;**71**:3219–27.

127. Ulloa-Aguirre A, Janovick JA, Leanos-Miranda A, Conn PM. Misrouted cell surface GnRH receptors as a disease aetiology for congenital isolated hypogonadotrophic hypogonadism. *Hum Reprod Update* 2004;**10**:177–92.

128. Bedecarrats GY, Kaiser UB. Differential regulation of gonadotropin subunit gene promoter activity by pulsatile gonadotropin-releasing hormone (GnRH) in perifused L beta T2 cells: role of GnRH receptor concentration. *Endocrinology* 2003;**144**:1802–11.

129. Ferris HA, Shupnik MA. Mechanisms for pulsatile regulation of the gonadotropin subunit genes by GNRH1. *Biol Reprod* 2006;**74**:993–8.

130. Brown CR, Hong-Brown LQ, Welch WJ. Strategies for correcting the delta F508 CFTR protein-folding defect. *J Bioenerg Biomembr* 1997;**29**:491–502.

131. Maya-Nunez G, Janovick JA, Ulloa-Aguirre A, Soderlund D, Conn PM, Mendez JP. Molecular basis of hypogonadotropic hypogonadism: restoration of mutant (E(90)K) GnRH receptor function by a deletion at a distant site. *J Clin Endocrinol Metab* 2002;**87**:2144–9.

132. Arora KK, Chung HO, Catt KJ. Influence of a species-specific extracellular amino acid on expression and function of the human gonadotropin-releasing hormone receptor. *Mol Endocrinol* 1999;**13**:890–6.

133. Lin X, Janovick JA, Brothers S, Blomenrohr M, Bogerd J, Conn PM. Addition of catfish gonadotropin-releasing hormone (GnRH) receptor intracellular carboxyl-terminal tail to rat GnRH receptor alters receptor expression and regulation. *Mol Endocrinol* 1998;**12**:161–71.

134. Gorbatyuk MS, Knox T, LaVail MM, Gorbatyuk OS, Noorwez SM, Hauswirth WW, et al. Restoration of visual function in P23H rhodopsin transgenic rats by

gene delivery of BiP/Grp78. *Proc Natl Acad Sci U S A* 2010;**107**:5961–6.

135. Oueslati M, Hermosilla R, Schonenberger E, Oorschot V, Beyermann M, Wiesner B, et al. Rescue of a nephrogenic diabetes insipidus-causing vasopressin V2 receptor mutant by cell-penetrating peptides. *J Biol Chem* 2007.

136. Arakawa T, Ejima D, Kita Y, Tsumoto K. Small molecule pharmacological chaperones: from thermodynamic stabilization to pharmaceutical drugs. *Biochim Biophys Acta* 2006;**1764**:1677–87.

137. Michalovitz D, Halevy O, Oren M. Conditional inhibition of transformation and of cell proliferation by a temperature-sensitive mutant of p53. *Cell* 1990;**62**:671–80.

138. Brown CR, Hong-Brown LQ, Welch WJ. Correcting temperature-sensitive protein folding defects. *J Clin Invest* 1997;**99**:1432–44.

139. Castro-Fernandez C, Maya-Nunez G, Conn PM. Beyond the signal sequence: protein routing in health and disease. *Endocr Rev* 2005;**26**:479–503.

140. Perlmutter DH. Chemical chaperones: a pharmacological strategy for disorders of protein folding and trafficking. *Pediatr Res* 2002;**52**:832–6.

141. Cohen FE, Kelly JW. Therapeutic approaches to protein-misfolding diseases. *Nature* 2003;**426**:905–9.

142. Hammarstrom P, Wiseman RL, Powers ET, Kelly JW. Prevention of transthyretin amyloid disease by changing protein misfolding energetics. *Science* 2003;**299**:713–6.

143. Soto C. Protein misfolding and disease; protein refolding and therapy. *FEBS Lett* 2001;**498**:204–7.

144. Estrada LD, Soto C. Disrupting beta-amyloid aggregation for Alzheimer disease treatment. *Curr Top Med Chem* 2007;**7**:115–26.

145. Ishii S, Chang HH, Kawasaki K, Yasuda K, Wu HL, Garman SC, et al. Mutant alpha-galactosidase A enzymes identified in Fabry disease patients with residual enzyme activity: biochemical characterization and restoration of normal intracellular processing by 1-deoxygalactonojirimycin. *Biochem J* 2007;**406**:285–95.

146. Gautherot J, Durand-Schneider AM, Delautier D, Delaunay JL, Rada A, Gabillet J, et al. Effects of cellular, chemical, and pharmacological chaperones on the rescue of a trafficking-defective mutant of the ATP-binding cassette transporter proteins ABCB1/ABCB4. *J Biol Chem* 2012;**287**:5070–8.

147. Wuller S, Wiesner B, Loffler A, Furkert J, Krause G, Hermosilla R, et al. Pharmacochaperones post-translationally enhance cell surface expression by increasing conformational stability of wild-type and mutant vasopressin V2 receptors. *J Biol Chem* 2004;**279**:47254–63.

148. Ulloa-Aguirre A, Janovick JA, Miranda AL, Conn PM. G-protein-coupled receptor trafficking: understanding the chemical basis of health and disease. *ACS Chem Biol* 2006;**1**:631−8.

149. Janovick JA, Patney A, Mosely R, Goulet M, Altman M, Rush T, et al. Molecular mechanism of action of pharmacoperone rescue of misrouted GPCR mutants: the GnRH receptor. *Mol Endocrinol* 2009;**23**:157−68.

150. van Straten NC, Schoonus-Gerritsma GG, van Someren RG, Draaijer J, Adang AE, Timmers CM, et al. The first orally active low molecular weight agonists for the LH receptor: thienopyr(im)idines with therapeutic potential for ovulation induction. *Chembiochem* 2002;**3**:1023−6.

151. Ulloa-Aguirre A, Crepieux P, Poupon A, Maurel MC, Reiter E. Novel pathways in gonadotropin receptor signaling and biased agonism. *Rev Endocr Metab Disord* 2011;**12**:259−74.

152. Conn PM, Smith E, Hodder P, Janovick JA, Smithson DC. High-throughput screen for pharmacoperones of the vasopressin type 2 Receptor. *J Biomol Screen*, in press.

153. Smithson DC, Janovick JA, Conn PM. Therapeutic rescue of misfolded/mistrafficked mutants: automation-friendly high-throughput assays for identification of pharmacoperone drugs of GPCRs. *Methods Enzymol* 2013;**521**:3−16.

Iodothyronine Deiodinases: Emerging Clinical Crossroads

Carlos Valverde-R*, Aurea Orozco*,
Juan Carlos Solís-S† and Ludivina Robles-Osorio†
*Instituto de Neurobiologia, Universidad Nacional Autónoma de México (UNAM), Querétaro, México,
†Facultad de Medicina, Universidad Autónoma de Querétaro, Querétaro, México

INTRODUCTION

In the past decades thyroid physiology has experienced a rapid increase in knowledge via different interrelated fields that impact on clinical practice. This is the case for iodothyronine deiodinases, the set of thioredoxin fold-containing selenoenzymes that in a tissue- and physiologically specific manner customize thyroid hormone (TH) intracellular requirements. The aim of this chapter is to summarize current developments regarding the finely tuned regulation of the intracellular supply of active or inactive THs, highlighting the association between deiodination and several clinical entities, including metabolic syndrome, type 2 diabetes, lipid metabolism, mood disorders, and cancer. Recent publications have been preferentially included.

THs are a singular family of iodine-containing endocrine messengers that derive from the aromatic amino acid tyrosine (Tyr) and are exclusively synthesized by the thyroid gland. In humans, the gland synthesizes, in an approximate ratio of 13:1, two main THs: 3,5,3′,5′-tetraiodothyronine (also known as thyroxine or T_4) and 3,5,3′-triiodothyronine or T_3. The other THs, both circulating and intracellular, are not the product of thyroid hormonogenesis; instead, they are metabolized peripherally by the sequential and stereospecific removal of the iodine atoms contained in T_4 or T_3. Indeed, the modern paradigm of thyroid hormone action recognizes that, through specific membrane TH transporters, they enter target cells and are deiodinated before accessing the cell nucleus where, by binding to thyroid hormone receptors (TR), they activate or repress the expression of specific TH-responsive genes. Thus, by differentially expressing a specific combination of deiodinases, T_3-responsive cells can customize intracellular TH bioavailability to meet their particular requirements, since the net amount of T_3 eventually occupying the TR defines the TH transcriptional footprint. Therefore, in all vertebrates and at precise

Cellular Endocrinology in Health and Disease.
DOI: http://dx.doi.org/10.1016/B978-0-12-408134-5.00022-6

species-specific ontogenetic stages, the appropriate supply and local concentrations of THs are of paramount importance for the development and differentiation of most organs and systems, including the nervous system. Concomitantly, THs are the major endocrine regulators of energy expenditure during adulthood.[1–3]

DEIODINASES

The deiodinase family comprises three enzymes, D1, D2, and D3, each with different catalytic properties and tissue- and developmentally-specific expression. All are homodimers of ~ 60 kDa anchored to cellular membranes through a single transmembrane segment. This membrane linkage determines, in part, the specific subcellular topology that characterizes the family members: D1 and D3 reside at the plasma membrane and D2 at the endoplasmic reticulum (ER). This allocation, together with their catalytic properties, allows deiodinases to fine-tune the intracellular availability of active and inactive TH on the basis of tissue-specific and functional demand. D1 and D3 have a long half-life (~ 8 h and ~ 12 h respectively), while D2 has the shortest (~ 45 min). All deiodinases share $\sim 50\%$ sequence similarity and, at their catalytic center, contain the rare amino acid selenocysteine (SeCys) as the key residue. As in other selenoproteins, the transcription and incorporation of SeCys requires an ancient and very complex *trans*-acting machinery; the mechanisms involved in this process have been recently reviewed.[4,5]

Deiodinase-catalyzed iodine removal is highly selective. While D1 and D2 serve the activating or outer ring-deiodinating pathway (ORD) by converting T_4 to T_3, and T_3 to 3,5-diiodothyronine, the inactivating or inner ring-deiodinating pathway (IRD) is catalyzed primarily by D3, which produces the inactive iodothyronines reverse T_3 (rT_3) and the 3′,5′- and 3,3′-diiodothyronines (Figure 22.1). The expression and activity of the three deiodinases

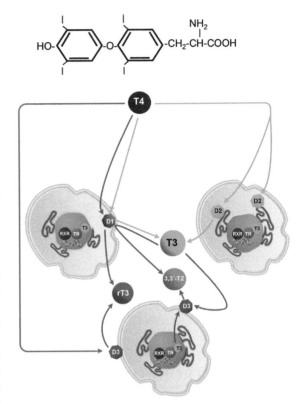

FIGURE 22.1　**Schematic representation of the subcellular topology and function of deiodinases.** The chemical structure of the prohormone T_4 is depicted at the top. Green and red arrows show the activating and inactivating pathway, respectively. See text for details.

is closely regulated, primarily by the thyroid status, although other factors such as nutritional balance and environmental cues also participate. Thus, a rise in intracellular T_4 and/or T_3 increases the expression and activity of D1 and D3 and reduces that of D2.[4,6,7]

Deiodinase Type 1 (D1)

The human D1 gene (*hDio1*), localized on chromosome 1 (locus p32.3), spans 17 kb, encompasses three introns and four exons, and has two half sites of thyroid hormone response elements (TRE) in its promoter region (see below). The expression and activity of D1 is greater in organs with a high metabolic rate, i.e., liver, kidney, thyroid, and lactating gland. The *Dio1* gene is highly sensitive to T_3, with a ~ 175-fold induction during the transition from hypo- to hyperthyroid states, thus functioning as an indicator of systemic thyroid status. Furthermore, new insights from two transgenic mouse models (conditional inactivation of hepatic selenoprotein synthesis and D1-KO) indicate that D1 plays a scavenger role, recycling iodine from sulfated iodothyronines in the process of being eliminated in the bile and urine.[4,6,8]

Deiodinase Type 2 (D2)

The human D2 gene (*hDio2*) is located in the long arm of chromosome 14 (14q24.2–q24.3), spans 7.5 kb, and encompasses two exons separated by one intron. In the rat, mouse, and human, the promoter region of *Dio2* contains a cyclic adenosine monophosphate (cAMP) response element, but only the *hDio2* gene contains response elements for the thyroid transcription factors TTF-1 and PAX8, as well as for Nfkβ. The transmembrane domain of D2 attaches the enzyme to the ER membrane, while the rest of the protein, including the active site, is oriented towards the cytosol. Residency in the ER is critical for D2, since D2-generated T_3 is transferred and ligated to the TR contained in the nuclear compartment. D2 exclusively catalyzes the ORD pathway, producing active hormones for both local consumption within the cell expressing the enzyme, as well as contributing to plasma T_3. In humans, D2 is highly expressed in the pituitary, brain (particularly in glial cells, astrocytes, and tanycytes), endothelial cells, placenta, brown adipose tissue, thyroid gland, skeletal muscle, skin, and osteoblasts, but not in liver or kidney.[4,7,9]

D2 has a singular and complex regulation, which involves temporal-, spatial-, and cell-specific cues as well as a combination of mechanisms such as the transcriptional modulation of the *Dio2* gene, post-transcriptional regulation of *Dio2* mRNA stability, and post-translational ubiquitination. Indeed, ubiquitination is a major step in the control of D2 activity, through which T_4 binding and/or T_3 catalysis triggers inactivation of the enzyme; once ubiquitinated, D2 can be either targeted to proteasomal degradation or reactivated by deubiquitination. The mechanism for this D2 regulation has been reviewed elsewhere.[9]

Despite the fact that T_3 is considered the bioactive TH, recent data from pituitary and astrocyte-specific *Dio2* KO mice have disclosed that both T_4 and T_3 are instrumental in the fine-tuning feedback mechanism that regulates the hypothalamic–pituitary–thyroid axis. These studies highlighted the critical role played by tanycytes in mediating this hormone-feedback loop. Tanycytes are unique, specialized ependymal cells that function as the interphase between blood on one side, and the intercellular fluid of the brain parenchyma and the cerebrospinal fluid (CSF) on the other. Located on the lining floor and infralateral wall of the third ventricle adjacent to the thyrotropin-releasing hormone (TRH)-expressing neurons in the paraventricular nuclei (PVN-TRHergic neurons), tanycytes are the main brain cell type expressing D2 and D3. Tanycyte D2 has access to T_4 from plasma or CSF, and it produces T_3 that can then reach the PVN-TRHergic neurons or the pituitary gland via portal blood. Thus, tanycyte D2 is pivotal in the T_4-mediated feedback regulation of TRH and thyroid-stimulating hormone (TSH), as well as in the central hypothyroidism that occurs during non-thyroidal illness. This notion is strengthened by the fact that in thyrotrophs, D2 is co-expressed with TSH and with TRβ2, the dominant isoform that mediates the negative regulation of hypophysial TSH-β and hypothalamic TRH gene expression.[9–13]

Deiodinase Type 3 (D3)

The human D3 gene (*hDio3*) is localized at 14q32.31 and extends for 2.1 kb. Of the three deiodinases, *hDio3* is the only one that does not contain introns and is imprinted by the paternal allele. D3 catalyzes the inactivation of the prohormone T_4 and of the bioactive T_3, producing the inactive iodothyronines rT_3 and 3,3'-diiodothyronine, respectively. In adults, D3 expression and activity are mainly restricted to the brain (neurons), skin, and endothelial cells, as well as the gestating uterus and placenta. However, during embryonic life the enzyme is more abundant and ubiquitously expressed, which is consistent with the crucial role that D3 plays in protecting fetal tissues from premature exposure to surplus THs during critical periods of development.[4,14]

This protective capacity of D3 is preserved in adult life and can be reactivated by various stressful situations and acute or chronic pathologies (fasting, carbohydrate deprivation, surgery, myocardial infarction, sepsis, critical illness, wasting syndrome, etc.). Indeed, recent studies (animal models and human postmortem tissue) have demonstrated the "neo-expression" of D3 in the liver, heart, and skeletal muscle, tissues that normally do not express this deiodinase in adult life. D3 reactivation is accompanied by a reduction in the expression and activity of D1 but without significant changes in D2. Furthermore, besides being tissue-specific (it is absent in endometrial cells and fibroblasts) the reactivation of D3 is dependent on the hypoxia-inducible factor (HIF), a transcription factor that interacts with the hypoxia response element contained in *hDio3*. Recent evidence points to oxygen availability as the major signal to re-locate the enzyme from the plasma to the nuclear membrane or *vice versa*. Indeed, under normoxic conditions D3 is located at the plasma membrane, but in the presence of hypoxia or ischemia, the enzyme is rapidly displaced via a HSP40-mediated shuttle mechanism to the nuclear envelope, where it inactivates T_3. Of note is the fact that hallmarks of all these clinical and/or experimental conditions are a low or undetectable level of circulating T_3 together with normal or low T_4, increased or normal rT_3 and, paradoxically, normal or even a low TSH level.[8,12,15]

CLINICAL CROSSROADS OF DEIODINATION

New Insights About TH Metabolism and Neuropsychiatric Disorders

It is now clear that the critical role of THs in nervous system development and functioning is not limited to fetal and early postnatal periods. Within the past decade, our understanding of the involvement of THs and their derivatives in altered neuropsychological function has experienced remarkable progress. Besides being the backbone of thyronines, Tyr is also the precursor of the biogenic amines or catechol-adrenergic neurotransmitters adrenaline, noradrenaline, and dopamine. Furthermore, Tyr, which derives from the essential amino acid phenylalanine, is also the precursor of the so-called trace amines (TA), a group of endogenous neuroactive sympathomimetic molecules that exert modulatory effects upon pre- and postsynaptic biogenic amine physiology. TA include, among others: β-phenylethylamine (β-PEA), *para*-tyramine, and *para*-octopamine. Structurally, β-PEA is closely related to amphetamine and exerts well-documented amphetaminenergic effects. Indeed, it has been suggested that β-PEA corresponds to an endogenous amphetamine. Likewise, and as part of the reciprocal interplay of both systems, THs are precursors of a novel group of potent endogenous decarboxylated thyronine derivatives named thyronamines (TAM). Although their precise biosynthetic pathway remains elusive, TAM exert, within minutes, short-term negative chrono- and inotropic effects, and depression of metabolism and body temperature, effects opposite to those elicited by THs. Interestingly, both TA and TAM, but not the biogenic amines, are ligands for a new class of recently discovered G protein-coupled receptors called trace-amine associated receptors (TAAR). In addition, it is now recognized that THs exert clear-cut effects on hippocampal and dentate gyrus neurogenesis and differentiation, as well as on the expression of various neurotrophins. Moreover, recent data suggest that THs and TAM, particularly the 3-iodine derivative or T1AM, have neuroprotective, non-genomic effects in the context of ischemic brain damage and/or excitotoxicity. As in the case of biogenic amines, alterations in TA metabolism and/or function, particularly of β-PEA, have been implicated in the pathogenesis of various neuropsychiatric conditions, ranging from schizophrenia and phenylketonuria to affective disorders and migraine. The discovery of TAAR, particularly TAAR1, as well as the engineering of the first synthetic TAAR1 partial agonist, has renewed interest in the proposal that a deficit in β-PEA function and/or turnover is part of the etiology of depression and other psychiatric conditions.[2,16–20]

Although poorly understood, adulthood hypothyroidism is accompanied by various neuropsychiatric complaints and symptoms. Indeed, myxedema historically has been associated with severe mental disorders mimicking melancholic depression and dementia, an entity also known as "myxedematous madness." Current epidemiological studies support a relationship between hypothyroidism and reversible dysfunction in cognition, learning, and mood. Subclinical hypothyroidism, particularly in women, is now considered a predisposing factor for depression, cognitive impairment, and reversible dementia. Similarly, TSH circulating levels have been positively correlated with depression severity in hypothyroid patients. T_3 has been shown to significantly improve (hastening/enhancing) the clinical response to antidepressive pharmacological therapy.[21–23] Consistent with data from experimental animals,

recent neuroimaging, as well as postmortem human brain studies, strengthen the evidence that depression is accompanied by a subtle but significant reduction in gray matter volume. This reduction occurs particularly in the anterior cingulate and orbitofrontal cortex, as well as in the hippocampus, putamen, and caudate nucleus, and most current antidepressant treatments selectively increase neural progenitor cells and angiogenesis in the dentate gyrus. Smaller hippocampal volumes have also been documented in children and adolescents with congenital hypothyroidism and reduced memory function, thereby corroborating the adverse effect of TH deficiency on neurogenesis and hippocampal development. In addition, functional positron emission tomography (PET) scans in hypothyroid patients show a decreased activity in cerebral regions associated with mood, affection, and memory regulation (anterior and posterior cingulate cortex, amygdala, and hippocampus); interestingly, all differences disappeared after TH replacement therapy. Furthermore, several psychoactive drugs ranging from antidepressants (desipramine, fluoxetine) to neuroleptics (haloperidol) clearly increased rat brain D2 as well as tissue concentrations of T_3.[24–26] These data strongly suggest that, besides their well-known effects upon catecholamine- and/or indoleaminergic brain neurotransmission, antidepressants and other neuroactive drugs promote T_4 to T_3 conversion by increasing the activity of D2. Indeed, this possibility is consistent with the finding that distinct single nucleotide polymorphisms (SNP) in *Dio1* and *Dio2* are linked to specific thyroid functional phenotypes (TH circulating levels). Thus, the D1-C785T SNP (rs11206244) is associated with higher rT_3 levels and a lower T_3/rT_3 ratio, suggesting lower enzyme activity in carriers of the T allele. On the other hand, the D1-A1814G SNP (rs12095080) is associated with a higher T_3/rT_3 ratio, suggesting that this variant may result in increased D1 activity.[27] Furthermore, the *Dio2* gene polymorphism Thr92AlaD2 (rs225014), which results in an impaired T_3 production, has been linked to mental retardation and bipolar disorder, as well as other metabolic dysfunctions (see below).

Deiodination and Metabolic Homeostasis

Operating as a single neuroendocrine and neuronal integrative and regulatory system, the hypothalamus controls a vast array of physiological processes including energy balance, feeding, and satiety. Among the different signals (nutrients, hormones, neuropeptides, and neurotransmitters) that participate in controlling metabolic homeostasis, recent experimental evidence indicates that THs, particularly T_3, via the hypothalamus and its sympathoadrenal and parasympathetic output, are major regulators of glucose and lipid metabolism. Thus, although the precise mechanisms are not fully understood, it should be stressed that local hypothalamic T_3 availability is dependent upon tanycyte D2 activity. Indeed the local administration of T_3 into the hypothalamic PVN increases plasma glucose levels independent of plasma T_3, insulin, glucagon, and corticosterone, and the effects of T_3 are prevented by selective hepatic sympathectomy. Interestingly, the systemic or intracerebroventricular administration of T1AM and T0AM also increases plasma glucose as well as glucagon and corticosterone but not plasma insulin. In this context, it has been suggested that the PVN-TRHergic neurons are major metabolic sensor/effectors contributing to the regulation of energy homeostasis. In addition to their negative TH-feedback loop, PVN-TRHergic neurons receive a rich noradrenergic input ($\sim 20\%$ of all synapses) as well as projections, mainly from the arcuate nucleus and the lateral hypothalamus, of several orexigenic (orexin, NPY, MCH, and AgRP) and anorexigenic (POMC, CART, and PACAP) neuropeptides. Recent reviews describing the physiological role played by these neurotransmitters

and neuropeptides in the control of food intake and energy homeostasis can be found elsewhere.[10,11,13,28–30]

Lipid Metabolism and D2

The well-known participation of THs as modulators of energy homeostasis and intermediary metabolism includes, among others, their cross-talk with nuclear receptor ligands like bile acids (BA), cholesterol derivatives, and fatty acids, actions that are mediated by FXR, LXR, and PPARs, respectively. While these mechanisms are beginning to be elucidated, the key participation of the *Dio2* gene has been well established. The installation of adaptive thermogenesis, for which the major effector is brown adipose tissue (BAT), is strictly dependent on the orchestrated interplay of the adrenergic stimulation of D2 to increase local conversion of T_4 to T_3, leading to the expression of BAT uncoupling protein 1 (UCP1). Hepatic gluconeogenesis is also stimulated by catecholamines and thyroid hormones. The UCP1 promoter and the promoter for a rate-limiting enzyme in gluconeogenesis, phosphoenolpyruvate carboxykinase (PEPCK), both have a cAMP and a TRE response element, and both are required to stimulate gene expression. Indeed, the BAT thermogenic response to cold is blunted in mice with BAT-*Dio2* targeted disruption, as is the expression of UCP1 in mice with a mutation in the TRβ isoform. Furthermore, BAT development is impaired in these *Dio2* knockouts. The mechanisms for TH interactions with the adrenergic system involve coordinated regulation of ligand availability, direct gene regulation by TR, as well as co-regulation of some genes by TR and adrenergic-stimulated factors. Likewise, D2 has been found to be a key player in the metabolic effects of BA. For example, mice fed with a high-fat diet containing BA gain less weight than their control counterparts, because the elevation in serum BA activates BAT, inducing key genes involved in the thermogenic process, i.e.,

Dio2 and PGC1α. BA activate the D2 pathway by interacting with the BA G protein-coupled receptor, TGR5, and activating the cAMP–PKA pathway, leading to a dose-dependent increase in *Dio2* expression and D2 activity. Most importantly, *Dio2*$^{-/-}$ mice are resistant to the protective effects of BA against diet-induced obesity, indicating that some of the BA-mediated effects on metabolism depend on the TGR5–D2 pathway.[9,29,31]

Deiodinases and Type 2 Diabetes (T2DM)

T2DM is a polygenic disease usually preceded by the following hallmarks: insulin resistance and/or impaired insulin secretion; increased liver gluconeogenesis and hyperglycemia during fasting, and endoplasmic reticulum (ER) stress. These characteristics can be present even years before the disease is diagnosed. THs and deiodinases, particularly D2 and D3, are instrumental for both the development and functional maturation of pancreatic β-cells. Thus, THs are now considered physiological regulators of functional maturation of β-cells via the induction of specific, TH-regulated transcription factors (i.e., MafA);[32] Furthermore, postnatal rat β-cells express different TR isoforms and deiodinases in an age-dependent pattern as glucose responsiveness develops. Moreover, mice with targeted disruption of *Dio3* gene (D3KO) are glucose intolerant due to impaired glucose-stimulated insulin secretion, without changes in peripheral sensitivity to insulin. THs are likely to have an analogous role in humans, given the similar D3 expression pattern.[33]

Regarding D2, various population-based studies disclosed that Thr92Ala SNP (rs7140952), a polymorphism found in 15% of the population, associates with lower glucose disposal rate and insulin resistance, but they failed to demonstrate an association between this SNP and an increased risk for T2DM. However, a recent case-control study and a meta-analysis of the literature strongly suggest that this D2 SNP is indeed associated with an increase of both insulin resistance and risk for T2DM (~10%). Of note is the fact that this percentage is similar to that found for other genes related to T2DM. These data are consistent with studies showing that D2KO mice are insulin resistant on a chow diet and exhibit a significant increase in body weight and hepatic steatosis on a high-fat diet.[34,35] Interestingly, this same D2 SNP has been linked to central obesity and hypertension, components of the metabolic syndrome that usually precedes T2DM.[36] Other *Dio2* SNPs have been correlated with metabolic impairment, particularly in Pima Indians, an ethnic group with a high frequency of T2DM; however, these findings are not conclusive, and the significance of ethnic components in metabolic disorders has still to be elucidated.[37]

Also associated with obesity, insulin resistance, and T2DM is ER stress, a cellular condition caused by disruption in ER homeostasis. One of the characteristics of ER stress is the accumulation of misfolded proteins in the ER lumen. The elimination of these proteins involves targeting them to the ER-associated degradation pathway, which is mediated by the ubiquitin–proteasome system. It has been demonstrated in D2-expressing cells that undergo ER stress that D2 activity is rapidly lost and that this translational arrest is mediated not by ubiquitination as would be expected, but by the PERK–elF2a pathway. This loss in D2 activity is followed by a substantial and rapid reduction in T_3 production and, therefore, a state of relative cellular hypothyroidism. ER stress can be attenuated or resolved by chemical chaperones which, acting as molecular stabilizing agents, can also reverse the loss of D2 activity. For example, 4-phenyl-butyric acid or tauroursodeoxycholic acid successfully attenuate ER stress in a mouse model of obesity and insulin resistance, restoring glucose homeostasis and sensitivity to leptin and insulin, as well as stimulating *Dio2* expression and D2 activity. These metabolic effects seem to depend on the D2 pathway since the effects are lost in *Dio2*$^{-/-}$ mice.[9]

Tumorigenesis, Cancer, and Deiodination

Given their well-known actions on normal cell proliferation and differentiation, a number of studies have tried to answer the question of whether THs and/or altered deiodinase expression may play a role in tumorigenesis and progression of different types of neoplasms. The results obtained have been sometimes contradictory, and few studies have simultaneously assessed both deiodinase expression and tissue levels of THs. Consequently, it is still not clear if alterations in TH deiodination are a cause or a consequence of tumor onset and progression. A large prospective study showed an increased risk for lung, prostate, and breast cancer associated with subclinical and overt hyperthyroidism, and T_3 is thought to induce cancer cell proliferation, probably by regulating the expression and activity of several tumor suppressors and oncogenes.[38] On the other hand, hypothyroidism has been associated with increased tumor invasiveness and metastatic growth, an increased risk for hepatocarcinoma in women, and a reduced risk of primary breast carcinoma. However, inconsistent results have been found after the administration of T_4, T_3, $3,5$-T_2 and rT_3 to different prostate, breast, and ovarian cell lines, strongly suggesting that there is an heterogeneous response among cancer cells.[39,40] It is important to keep in mind that the hypothyroid patients studied could have subclinical hyperthyroidism due to TH replacement therapy, a situation that has not been ruled out as an explanation of the results obtained.

As summarized in Table 22.1, iodothyronine deiodinase expression can be altered in a variety of tumors. Consistent with their cellular lineage, a common feature of central nervous system (CNS) tumors is a significant increase in the expression and activity of D3 and D2, whereas that of D1 is barely detectable.[40] Although the mechanism remains speculative, a low *Dio1* expression and/or enzyme activity is a common finding in thyroid, kidney, liver, lung, and prostate neoplasias, among others. Furthermore, in thyroid tumors, significant decreases in D1 and D2 mRNA and enzyme activities have been consistently found in all histological subtypes and different clinical stages of samples from papillary carcinoma. Indeed, the decreased expression of D1 has been proposed as a marker of dedifferentiation towards papillary thyroid carcinoma. In contrast, normal or elevated expression and activity of both genes is usually found in tissue samples from thyroid adenoma and follicular cancer, as well as in follicular cancer cell lines. Interestingly, these follicular cell lines retain their normal *Dio1* response to retinoic acid, but they do not respond to T_3 and TSH, thus indicating preservation of their basal regulatory functions, in contrast to less differentiated thyroid neoplasias. At variance with earlier reports, recent studies showed high levels of D1 mRNA and activity in both anaplastic and Hürthle cell carcinoma.[8,41]

Both D1 expression and activity are usually reduced or undetectable in samples from renal clear cell carcinoma, the most frequent renal cancer. Moreover, recent data have revealed the presence of D1 splicing variants in this type of cancer, but not in normal tissue; these variants could be used as unique specific molecular markers for neoplastic kidney cells. Furthermore, while in normal non-lactating mammary tissue, D1 activity is undetectable; in samples from well-differentiated human breast cancers, D1 mRNA and enzyme activity were significantly increased. These findings are consistent with studies on breast cancer cell lines. The well-differentiated MCF-7 cells express D1, and its activity was stimulated by retinoic acid but not by T_3 or by the β-adrenergic agonist isoproterenol. In dedifferentiated MDA-MB-231 breast cancer cells, however, the expression of D1 was undetectable, and the stimulation by retinoic acid was lost. These results led to the suggestion that D1 expression could represent a sensitive differentiation marker of breast cancer cells.[41]

TABLE 22.1   Expression of Deiodinases in Different Types of Tumors

Type of Tumor	D1	D2	D3
THYROID			
Follicular adenoma	↑	↑	
Follicular carcinoma	↑	↑	
Anaplastic carcinoma	↑	↑	
Hürthle cell carcinoma	↑		
Medullary carcinoma		↑	
Thyroid cold nodules	↓		
Papillary thyroid carcinoma	↓		
BREAST			
Differentiated carcinoma	↑	↑	
LUNG			
Lung carcinoma	↓		
Mesothelioma		↑	
LIVER			
Hepatic adenocarcinoma	↓		
KIDNEY			
Clear cell carcinoma	↓		
PROSTATE			
Prostate carcinoma	↓		
CENTRAL NERVOUS SYSTEM			
Oligodendroglioma		↑	
Astrocytoma		↑, ↓	↓
Glioblastoma		↑, ↓	↑
Oligoastrocytoma		↑	
Ganglioglioma		↑	
Neurinoma, cordoma, hamartoma		↑	
Non-functioning pituitary adenoma		↑	
TSH- and ACTH-producing tumors		↓	↑
Pituitary tumors			↑
Gliosarcomas		↑	↑
Glioma		↑	↑
VASCULAR			
Hepatic hemangiomas			↑

(*Continued*)

TABLE 22.1   (Continued)

Type of Tumor	D1	D2	D3
Cutaneous hemangiomas			↑
Hemangiomas	↓		
SKIN AND BONE			
Basal cell carcinoma		↓	↑
Osteosarcoma		↓	

Abbreviations: ACTH, adrenocorticotropic hormone; TSH, thyroid-stimulating hormone.
Sources: Maia et al., 2011; Piekielko-Witkowska and Nauman, 2011; Casula and Bianco, 2012.

In accordance with the findings suggesting an increase in tumor growth and metastasis in hypothyroid patients, D3 mRNA and activity were significantly increased in samples from papillary carcinoma. D3 was also overexpressed in infantile hepatic and cutaneous hemangiomas; this excess D3 activity can generate consumptive hypothyroidism due to excessive TH inactivation. D3 is also overexpressed in basal cell carcinoma (BCC) cells together with an inactivation in D2 expression. Interestingly, when BCC cells are implanted in mice, the implant growth is dramatically reduced by inhibiting D3 expression, suggesting that active THs reduce tumor progression.[41]

Indirect evidence suggests a link between cancer progression and alterations in iodothyronine deiodinase structure and function. Thus, mice with reduced ability to synthesize selenoproteins, including iodothyronine deiodinases, show an increased susceptibility for colorectal, bladder, esophageal, and prostate cancer, among others.[5] In addition, chromosomal deletions in genes coding for iodothyronine deiodinases have been detected in some tumors (meningiomas, ovarian tumors, breast cancer, pheochromocytomas, gliomas, colorectal carcinomas, lung cancer, renal carcinomas).[40] However, so far, no SNP or mutations have been reported in iodothyronine deiodinase genes related to cancer.

Interestingly, mutations of *THRA* and *THRB* have been associated with an increased risk of hepatic, renal, mammary, thyroid, and pituitary tumor development. Furthermore, epigenetic silencing of *THRB* is a common trait in human cancers, and its reactivation inhibited proliferation and migration in thyroid cancer cell lines.[42] These findings led to the proposal that *THRB* is a tumor suppressor gene with potential therapeutic uses.

T_4 and T_3 stimulate angiogenesis by activating the integrin $\alpha\nu\beta3$, a plasma membrane receptor located mainly on endothelial cells, vascular smooth muscle cells, cancer cells, and osteoclasts. This activation of $\alpha\nu\beta3$ is a nongenomic action of TH and is related to a pathway that increases proliferation of various tumor cells as well as tumor-related neovascularization.[17] Interestingly, tetraiodothyroacetic acid (TETRAC), a deaminated analog of T_4, blocks TH cellular (non-genomic) actions by inhibiting the angiogenic effects of vascular endothelial growth factor and basic fibroblast growth factor in human breast cancer cells.[43] Accordingly, TETRAC-covalently linked nanoparticles have been shown to inhibit growth of human renal cell carcinoma xenografts, as well as TH stimulation of lung cancer cells *in vitro* and their growth as xenografts.[44] In this regard, the possibility that TH analogs act as antitumoral agents, inhibiting both cancer proliferation and tumor angiogenesis, is highly encouraging.

CONCLUDING REMARKS

Basic aspects in science move faster than those of clinical practice; thus, whereas basic knowledge of thyroid hormone deiodination is solid, its pathophysiological implications are only beginning to be elucidated. This chapter summarizes recent findings regarding iodothyronine deiodinases as key tissue-specific regulators of intracellular TH availability and signaling, as well as their implications in the pathophysiology of different clinical conditions. Integrated in a complex, but as yet poorly understood, cross-regulatory network of tissue-intrinsic signals, the altered expression and/or activity of D2 and D3 are instrumental in the development and progression of various clinical pathologies, including cognitive and neuropsychiatric disorders, tumorigenesis, and metabolic alterations. In this context, polymorphisms in deiodinase genes are clearly associated with both mood alterations and the clinical response to replacement therapy in hypothyroidism. Likewise, in euthyroid depressed patients, neurogenesis and the clinical response to antidepressants improve with TH adjuvant treatment. Regarding tumorigenesis, the potential roles of TH analogs as well as the modulation of deiodinase expression in tumor cells suggest new therapeutic options. However, the possibility that deiodinases could serve as molecular markers of carcinogenesis and dedifferentiation is complicated by the following aspects, all of which deserve further systematic studies: (i) deiodinases also deiodinate several TH-derivatives known to modulate cellular proliferation and/or migration, rT_3, T_2, TAM and TETRAC, among others; and (ii) the so-far-uncharacterized chemical species of iodine generated as byproducts of deiodination have also been implicated in

antiproliferative or apoptotic effects.[45] The current challenge is to assemble the puzzle with the pieces of information emerging from clinical research and practice. Nevertheless, two facts are conclusive: deiodinase action defines in part the TH transcriptional footprint, and their dysfunction leads to clinical disease.

Acknowledgements

The work of the authors was supported by grants from PAPIIT IN208511 and CONACYT 166357. The authors thank M.S. Patricia Villalobos and Dr. Dorothy Pless for critically reviewing the manuscript.

References

1. Cheng SY, Leonard JL, Davis PJ. Molecular aspects of thyroid hormone actions. *Endocr Rev* 2010;**31**:139–70.
2. Patel J, Landers K, Li H, et al. Thyroid hormones and fetal neurological development. *J Endocrinol* 2011;**209**: 1–8.
3. Heuer H, Visser TJ. The pathophysiological consequences of thyroid hormone transporter deficiencies: Insights from mouse models. *Biochim Biophys Acta* 2013;**1830**:3974–8.
4. Gereben B, Zavacki AM, Ribich S, et al. Cellular and molecular basis of deiodinase-regulated thyroid hormone signaling. *Endocr Rev* 2008;**29**:898–938.
5. Lu J, Holmgren A. Selenoproteins. *J Biol Chem* 2009;**284**:723–7.
6. St Germain St DL, Galton VA, Hernandez A. Minireview: defining the roles of the iodothyronine deiodinases: current concepts and challenges. *Endocrinology* 2009;**150**:1097–107.
7. Williams GR, Bassett JH. Deiodinases: the balance of thyroid hormone. Local control of thyroid hormone action: role of type 2 deiodinase. *J Endocrinol* 2011;**209**: 261–72.
8. Maia AL, Goemann LM, Souza Meyer EL, et al. Deiodinases: the balance of thyroid hormone type 1 iodothyronine deiodinase in human physiology and disease. *J Endocrinol* 2011;**209**:283–97.
9. Arrojo ED, Fonseca TL, Werneck-de-Castro JP, et al. Role of the type 2 iodothyronine deiodinase (D2) in the control of thyroid hormone signaling. *Biochim Biophys Acta* 2013;**1830**:3956–64.
10. Nillni EA. Regulation of the hypothalamic thyrotropin releasing hormone (TRH) neuron by neuronal and

peripheral inputs. *Frontiers Neuroendocrinol* 2010;**31**: 134—56.

11. Rodríguez EM, Blázquez JL, Guerra M. The design of barriers in the hypothalamus allows the median eminence and the arcuate nucleus to enjoy private milieus: The former opens to the portal blood and the latter to the cerebrospinal fluid. *Peptides* 2010;**31**:757—76.

12. Wajner SM, Maia AL. New insights toward the acute non-thyroidal illness syndrome. *Front Endocrinol* 2012;**3**:8.

13. Fonseca TL, Correa-Medina M, Campos MPO, et al. Coordination of hypothalamic and pituitary T3 production regulates TSH expression. *J Clin Invest* 2013;**123**: 1492—500.

14. Charalambous M, Hernandez A. Genomic imprinting of the type 3 thyroid hormone deiodinase gene: Regulation and developmental implications. *Biochim Biophys Acta* 2013;**1830**:3946—55.

15. Jo S, Kallo I, Bardoczi Z, et al. Neuronal hypoxia induces hsp40-mediated nuclear import of type 3 deiodinase as an adaptive mechanism to reduce cellular metabolism. *J Neurosci* 2012;**32**:8491—500.

16. Lin H-Y, Davis FB, Luidens MK, et al. Molecular basis for certain neuroprotective effects of thyroid hormone. *Front Mol Med* 2011;**4**:1—6.

17. Piehl S, Hoefig CS, Scanlan TS, et al. Thyronamines-past, present, and future. *Endocrine Rev* 2011;**32**:64—80.

18. Hackenmueller SA, Marchini M, Saba A, et al. Biosynthesis of 3-iodothyronamine (T1AM) is dependent on the sodium-iodide symporter and thyroperoxidase but does not involve extrathyroidal metabolism of T4. *Endocrinology* 2012;**153**:5659—67.

19. Kapoor R, Desouza LA, Nanavaty IN, et al. Thyroid hormone accelerates the differentiation of adult hippocampal progenitors. *J Neuroendocrinol* 2012;**24**:1259—71.

20. Revel FG, Moreau JL, Gainetdinov RR, et al. Trace amine-associated receptor 1 partial agonism reveals novel paradigm for neuropsychiatric therapeutics. *Biol Psychiatr* 2012;**72**:934—42.

21. Demartini B, Masu A, Scarone S, et al. Prevalence of depression in patients affected by subclinical hypothyroidism. *Panminerva Medica* 2010;**52**:277—82.

22. Bunevicius R, Prange Jr. AJ. Thyroid disease and mental disorders: cause and effect or only comorbidity? *Curr Opin Psychiat* 2010;**23**:363—8.

23. Hage MP, Azar ST. The link between thyroid function and depression. *J Thyroid Res* 2012;**2012**:590648.

24. Cobb JA, Simpson J, Mahajan GJ, et al. Hippocampal volume and total cell numbers in major depressive disorder. *J Psychiat Res* 2013;**47**:299—306.

25. Wheeler SM, Willoughby KA, McAndrews MP, et al. Hippocampal size and memory functioning in children and adolescents with congenital hypothyroidism. *J Clin Endocrinol Metab* 2011;**96**:E1427—34.

26. Pilhatsch M, Marxen M, Winter C, et al. Hypothyroidism and mood disorders: integrating novel insights from brain imaging techniques. *Thyroid Res* 2011;**4**:S3.

27. Cooper-Kazaz R, van der Deure WM, Medici M, et al. Preliminary evidence that a functional polymorphism in type 1 deiodinase is associated with enhanced potentiation of the antidepressant effect of sertraline by triiodothyronine. *J Affect Disord* 2009;**116**:113—6.

28. Hollenberg AN. The role of the thyrotropin-releasing hormone (TRH) neuron as a metabolic sensor. *Thyroid* 2009;**18**:131—9.

29. Fliers E, Boelen A. Type 2 deiodinase and brown fat: the heat is on-or off. *Endocrinology* 2010;**151**:4087—9.

30. Kalsbeek A, Bruinstroop E, Yi CX, et al. Hypothalamic control of energy metabolism via the autonomic nervous system. *Ann N Y Acad Sci* 2010;**1212**:114—29.

31. Yan-Yun Liu Gregory A, Brent GA. Thyroid hormone crosstalk with nuclear receptor signaling in metabolic regulation. *Trends Endocrinol Metab* 2010;**21**:166—73.

32. Aguayo-Mazzucato C, Zavacki AM, Marinelarena A, et al. Thyroid hormone promotes postnatal rat pancreatic β-cell development and glucose-responsive insulin secretion through MAFA. *Diabetes* 2013;**62**:1569—80.

33. Medina MC, Molina J, Gadea Y, et al. The thyroid hormone-inactivating type III deiodinase is expressed in mouse and human beta-cells and its targeted inactivation impairs insulin secretion. *Endocrinology* 2011;**152**:3717—27.

34. Dora JM, Machado WE, Rheinheimer J, et al. Association of the type 2 deiodinase Thr92Ala polymorphism with type 2 diabetes case-control study and meta-analysis. *Europ J Endocrinol* 2010;**163**:427—34.

35. Marsili A, Aguayo-Mazzucato C, Chen T, et al. Mice with a targeted deletion of the type 2 deiodinase are insulin resistant and susceptible to diet induced obesity. *PLoS ONE* 2011;**6**:e20832.

36. Al-Azzam SI, Alkhateeb AM, Al-Azzeh O, et al. The role of type II deiodinase polymorphisms in clinical management of hypothyroid patients treated with levothyroxine. *Exp Clin Endocrinol Diabetes* 2013;**121**:300—5.

37. Nair S, Muller YL, Ortega E, et al. Association analyses of variants in the Dio2 gene with early-onset type 2 diabetes mellitus in pima Indians. *Thyroid* 2012;**22**:80—7.

38. Hellevik AI, AsvoldBO BjøroT, et al. Thyroid function and cancer risk: a prospective population study. *Cancer Epidemiol Biomarkers Prev* 2009;**18**:570—4.

39. Tosovic A, Bondeson AG, Bondeson L, et al. Prospectively measured triiodothyronine levels are positively associated with breast cancer risk in postmenopausal women. *Breast Cancer Res* 2010;**12**:R33.

40. Piekiełko-Witkowska A, Nauman A. Iodothyronine deiodinases and cancer. *J Endocrinol Invest* 2011;**34**: 716—28.

41. Casula S, Bianco AC. Thyroid hormone deiodinases and cancer. *Frontiers in Endocrinology* 2012;**3**:1−8.
42. Kim WG, Cheng SY. Thyroid hormone receptors and cancer. *Biochim Biophys Acta* 2013;**1830**:3928−36.
43. Glinskii AB, Glinsky GV, Lin HY, et al. Modification of survival pathway gene expression in human breast cancer cells by tetraiodothyroacetic acid (tetrac). *Cell Cycle* 2009;**8**:3554−62.
44. Mousa SA, Yalcin M, Bharali DJ, et al. Tetraiodothyroacetic acid and its nanoformulation inhibit thyroid hormone stimulation of non-small cell lung cancer cells in vitro and its growth in xenografts. *Lung Cancer* 2012;**76**:39−45.
45. Aceves C, Anguiano B, Delgado G. The extrathyronine actions of iodine as antioxidant, apoptotic and differentiator factor in various tissues. *Thyroid* 2013;**23**:938−46.

MicroRNAs and Long Non-Coding RNAs in Pancreatic Beta Cell Function

Michael D. Walker

Weizmann Institute of Science, Rehovot, Israel

INTRODUCTION

The mammalian pancreas contains both exocrine and endocrine compartments. The endocrine cells (representing ~1–2% of the organ) are organized within the islets of Langerhans, endocrine "mini-organ" clusters scattered throughout the pancreas. Islets contain five hormone-producing cell types: alpha cells secrete glucagon, beta cells secrete insulin, delta cells secrete somatostatin, epsilon cells secrete ghrelin, and PP cells secrete pancreatic polypeptide.[1] Beta cells constitute the majority of cells in the islet; they are the only cells capable of producing significant amounts of the crucial metabolic hormone insulin, and hence are central for maintenance of lipid and carbohydrate homeostasis.[2]

The two hallmark features of the beta cell are insulin biosynthesis[3] and regulated insulin secretion in strict accordance with physiological needs.[4] In order to perform these functions, beta cells express a characteristic repertoire of proteins, which includes many proteins expressed with high selectivity in beta cells and excludes several "forbidden" proteins whose expression is incompatible with normal beta cell function.[5,6] Insulin biosynthesis in beta cells has been studied extensively. Selectivity of insulin gene expression in beta cells results, in large part, from transcriptional control mechanisms operating through multiple *cis*-elements located in the insulin gene promoter.[7,8] This involves the synergistic actions of several lineage-restricted transcription factors including Pdx-1,[9] NeuroD1 (BETA2),[10] and MAFA.[11] Remarkably, many of these insulin gene transcription factors have subsequently been found to play key roles also at various stages of pancreas organogenesis.[12,13]

Secretion of insulin is regulated primarily by blood glucose levels: the ability of the beta cell to sense glucose depends on metabolic activity, involving specialized proteins including the glucose transporter GLUT2 and the high-K_m enzyme glucokinase, that catalyzes the first step of the glycolytic pathway.[14] As blood glucose levels rise, metabolic flux through glycolysis and tricarboxylic acid (TCA) cycle leads to elevated ratios of adenosine triphosphate/adenosine

Cellular Endocrinology in Health and Disease.
DOI: http://dx.doi.org/10.1016/B978-0-12-408134-5.00023-8

diphosphate (ATP/ADP). This in turn leads to closure of ATP-sensitive K^+ channels, cellular depolarization, opening of voltage-gated Ca^{2+} channels and Ca^{2+} influx to the cell. Increased intracellular Ca^{2+} serves as a critical "trigger" for exocytosis of secretory granules containing processed insulin from granules.[15] Insulin secretion is typically biphasic: the first phase is rapid and transient, typically 2−10 min in duration, whereas the second is much more prolonged, lasting several hours.[16]

In addition to glucose, a range of nutrients and hormones modulate insulin secretion to permit adaptation to changing metabolic circumstances. For example, long chain fatty acids (LCFAs) are well-documented regulators of glucose-stimulated insulin secretion (GSIS). LCFAs have complex effects on insulin secretion; acute exposure of beta cells to LCFAs leads to augmented GSIS, whereas prolonged treatment leads to impaired beta cell function or "lipotoxicity."[17] The actions of LCFAs appear to involve mediators generated by intracellular metabolism of LCFAs as well as activation of the G protein-coupled receptor GPR40,[14,18,19] leading to $G\alpha_{q/i}$-dependent intracellular signaling.[19,20] The incretin hormones glucagon-like peptide 1 (GLP-1) and glucose-dependent insulinotropic polypeptide (GIP) have profound effects on beta cells, including enhanced GSIS and beta cell survival. This is mediated through the activation of G protein-coupled receptors leading to elevated intracellular cyclic adenosine monophosphate (cAMP) levels.[21]

The importance of the beta cell is underscored by the fact that both major forms of diabetes involve beta cell dysfunction: in type 1 diabetes (T1D), the beta cells are destroyed by a T cell-mediated autoimmune process.[22] Type 2 diabetes (T2D), on the other hand, is generally believed to progress in stages: insulin resistance in peripheral tissues, frequently associated with obesity, leads to compensatory insulin production by the beta cells, and eventually to beta

cell failure.[23,24] The basis for beta cell dysfunction remains poorly understood. It has been proposed to be a consequence of chronically elevated glucose (glucotoxicity), lipids (lipotoxicity), or cytokine-mediated damage.[24,25]

Type 2 diabetes has reached epidemic levels worldwide, driven by a parallel obesity epidemic, itself the consequence of "Western" lifestyle, involving lack of exercise and overnutrition. Worldwide, approximately 6.5% of the population is affected (285 million). The incidence is increasing rapidly, particularly in developing countries.[26] Although environmental factors are clearly playing an important role, it has been known for many years that genetic factors are significant in both T1D and T2D.[27,28] Recently, genome-wide association studies (GWAS) have led to the identification of over 65 loci, implicating up to 500 genes as risk factors for development of T2D in humans.[29] Strikingly, many of these genes have been linked with defective beta cell function.[30] Surprisingly, however, the combined effects of these genes account for only some 5−10% of disease risk,[31] perhaps indicating that many relevant genes remain to be identified.

The appreciation of the central role of the beta cell in diabetes pathogenesis, together with the demonstration that islet replacement using the "Edmonton protocol"[32] can restore normoglycemia in diabetic individuals, has led to a major effort to generate new sources of functional beta cells for cell replacement therapy of diabetes. Many approaches are being explored, including expansion of islet cell populations, reprogramming of differentiated cell types, embryonic stem cells and induced pluripotent cells.[33] A major guiding principle in all these approaches is the accumulating information base on the mechanisms governing normal pancreatic organogenesis.[34]

In recent years, our traditional views of the mammalian genome and expression of genetic information have been evolving rapidly. In particular, the discovery that mammalian cells contain thousands of distinct RNA sequences, so called "non-coding RNAs," that do not encode proteins has led to a revolution in our appreciation of the mechanisms controlling gene regulation. Several classes of non-coding RNAs have been distinguished including microRNAs (miRNAs), snRNAs, snoRNAs, piwiRNAs, circular RNAs, and long non-coding RNAs (lncRNAs). Among these RNA molecules, the best studied are the miRNAs. These small (~22 nucleotide) RNAs function as post-transcriptional inhibitors of gene expression, exerting their effects on multiple target mRNAs through interaction with sequences located in the 3′ untranslated region (UTR). Much less is known about the lncRNAs. They have been implicated in the regulation of a wide variety of biological processes. In a number of cases, a mechanism involving chromatin modification has been proposed. Yet the function of the vast majority of lncRNAs remain unknown. This chapter will describe evidence that suggests that both these categories of non-coding RNAs are implicated in beta cell development, function and dysfunction.

miRNAs: BIOSYNTHESIS AND MECHANISM OF ACTION

miRNAs are small (~22 nucleotide) non-coding RNAs originally discovered in 1993 in studies of *Caenorhabditis elegans* by Ruvkun, Ambros and colleagues.[35,36] Their major biological significance has come to be appreciated over the past 10 years, and they are now recognized to play key roles in a remarkable variety of cellular activities, including the development and maintenance of differentiated tissues,[37] response to stress,[38] and cell proliferation, apoptosis, and cancer.[39] Mammalian cells contain hundreds of genes that produce miRNA: as of July 2013, the miRBase repository (http://www.mirbase.

org) listed 1872 loci and 2578 mature miRNA species in humans. These loci show diverse genomic organization: they are often located in intergenic regions, under the control of a dedicated transcriptional control region (promoter). Both monocistronic (giving rise to a single transcript) and clustered, polycistronic arrangements (giving rise to a multiple miRNAs) are commonly observed (Figure 23.1a,b).[40] Such clusters may be functionally relevant, by permitting coordinated expression of miRNAs that function in a single pathway.[41]

The majority of miRNA loci are intragenic, commonly positioned within introns (Figure 23.1c).[42] In this situation, transcription of the miRNA sequence is usually under the control of the host gene promoter; hence, the primary transcript includes both host and miRNA sequences. In about one third of cases, the intronic miRNA sequence is transcribed from a dedicated intronic promoter.[43]

Biogenesis of miRNAs initiates with RNA polymerase II-dependent transcription[44] of the genomic sequence into a capped polyadenylated primary transcript[45] known as the pri-miRNA. This transcript contains a characteristic stem-loop structure encompassing the sequences destined to form the mature miRNA (indicated in red in Figure 23.1). The first stage of processing is mediated within the nucleus by the Microprocessor complex that contains the RNAse III enzyme Drosha and the co-factor DGCR8. The complex crops the primary transcript into an ~60 nt precursor known as pre-miRNA that is exported to the cytoplasm by the Exportin-5/RanGTP complex (Figure 23.2).[46] Subsequent cleavage by the enzyme Dicer is followed by incorporation of the mature miRNA with Argonaute proteins (AGO2) into the RNA-induced silencing complex (RISC).[47]

Base pairing between the mature miRNA and target sites typically located in the 5′ UTR of target mRNA sequences leads to repression

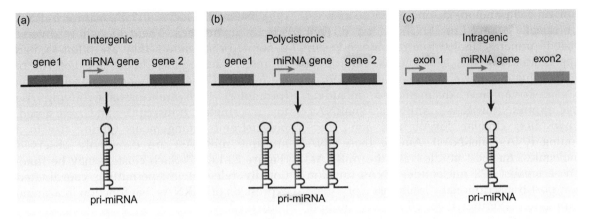

FIGURE 23.1 Diversity of genomic loci generating miRNA. Indicated are two forms of intergenic loci that give rise to a monocistronic (a) and polycistronic (b) pri-miRNA primary transcript under the control of a dedicated promoter region (red arrow). Intragenic loci (c) are generally located within introns and may be transcribed under the control of the host promoter (red arrow on left) or a dedicated promoter (red arrow on right).

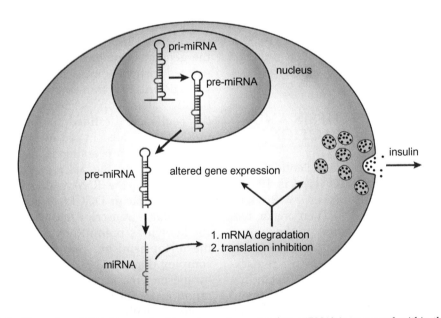

FIGURE 23.2 Stages in miRNA biogenesis. The primary transcript (pri-miRNA) is processed within the nucleus by Drosha to yield pre-miRNA. Following export to the cytoplasm, and processing by Dicer, the mature miRNA (red) is incorporated to the RISC complex, which interacts with target sequences on 3′ UTR sequences of mRNA, guided by base-paring with seed sequences on the miRNA. The result is mRNA degradation and/or inhibition of translation, which may lead to alterations in gene expression and altered secretion of insulin granules from beta cells.

of gene expression. In mammalian cells it was originally believed that this inhibition is mediated primarily through inhibition of translation. More recently, however, it has become clear that reduction of target mRNA levels is an important mechanism.[48,49] The association of an miRNA with a specific mRNA is generally dictated by a "seed" sequence complementary to nucleotides 2—8 in the 5' region of the miRNA.[50]

The effect of a single miRNA on a particular mRNA is often modest.[46] Nevertheless a single miRNA may target many distinct mRNA species, and a single mRNA may be targeted by many different miRNAs. Over 60% of all mammalian mRNAs are predicted targets of miRNAs.[51] Hence, enormous potential exists for generating complex regulatory networks, with a major impact on global gene expression.[52]

ROLE OF miRNAs IN BETA CELL DEVELOPMENT

Pancreas development can be divided into several stages or "transitions."[34,53,54] The first stage (E8.5—12.5 of mouse development) initiates with out-pocketing of the anterior midgut region of the endoderm, giving rise to the dorsal and ventral pancreatic buds. This is followed by initiation of branching morphogenesis to form ductal "trees." In the second stage (E13—E16.5), there is extensive endocrine cell differentiation leading to formation of all major endocrine cell types. In parallel, exocrine acinar cells develop accompanied by formation of acinar structures. The third stage (E16.5 to birth), is accompanied by extensive islet cell proliferation and formation of the islets of Langerhans.

The process is complex and involves extensive permissive and inductive intercellular signaling.[55,56] Genetic analysis has revealed a critical role of intrinsic, cell autonomous determinants of pancreatic organogenesis. This involves hierarchies of transcription factors that orchestrate the complex patterns of gene expression required for development and maintenance of the differentiated state.[12] Among the best characterized and most important of these is Pdx-1, considered to be a pancreas "master regulator," since in its absence pancreas development is arrested at a very early stage both in mouse and humans.[57,58] Pdx-1 expression appears to be controlled by the transcription factors Foxa1 and Foxa2.[59] Neurogenin3 (Ngn3) is expressed transiently and at a later stage in endocrine progenitor cells, where it functions as a crucial regulator of endocrine differentiation.[60] The establishment of mature beta cells is associated with expression of additional transcription factors, including NeuroD1 (BETA2), INSM1, Pax4, Nkx-2.2, Nkx-6.1, Pax6 and MafA/MafB, which are essential for generating the characteristic repertoire of gene expression of the mature beta cell.[12,61]

The first indication of a role for miRNA in pancreas development came from Lynn et al., who used a Pdx1—Cre transgene to generate mice lacking Dicer (and hence lack all mature miRNAs) during early pancreas development. This led to defective development of all pancreatic lineages, but in particular endocrine beta cells, demonstrating an essential role for miRNA.[62] On the other hand, when Dicer was inactivated at later stages of pancreas development, effects were less dramatic. Deletion in endocrine progenitor cells using the Ngn3—Cre transgene, did not affect the specification of hormone-expressing endocrine cells, though neonatal islets exhibited morphological defects and loss of hormone expression.[63] Surprisingly these Dicer-deficient endocrine cells expressed neuronal genes associated with a loss in binding of the neuronal transcriptional repressor REST-1 to target genes. Inactivation of Dicer selectively in beta cells (RIP—Cre) led to little or no disruption of pancreas development.[64] Taken together, these data suggest an important role(s) for miRNAs in pancreatic development, particularly at early stages of organogenesis. Although the identity of the miRNA(s) involved is not yet clear, many miRNAs have been proposed as candidates: these include miR-124a,[65] a regulator of the early developmental factor Foxa2, and mir-15a, mir-15b, mir-16, and mir-195, potential regulators of Ngn3.[66]

miR-7 is expressed at high level in embryonic pancreas,[66] and appears to be localized in endocrine cells.[67] Inhibition of miR-7 during early embryonic development led to downregulation of insulin production, decreased beta-cell numbers, and postnatal glucose intolerance, indicating an important role in generation of beta-cell mass in late embryogenesis.[68] Knockdown of miR-375, also expressed at high levels in developing and mature pancreas,[69] led to aberrant formation of the endocrine pancreas in zebrafish.[70] Consistent with this, ablation in the mouse produced abnormal islets displaying increased alpha/beta cell ratio.[71]

Overall, miRNAs clearly play an essential role during pancreas development, particularly at early stages. Though a number of promising candidates have been identified, dissecting out all the relevant miRNAs remains a challenging task.

ROLE OF miRNAs IN BETA CELL FUNCTION

To directly address the global significance of miRNA in mature beta cell function *in vivo*, Melkman-Zehavi *et al.* used a genetic approach based on the conditional deletion of Dicer using a regulatable transgene controlled by the insulin gene promoter (RIP−CreER).[72] Dicer deficiency resulted in a striking decrease in beta cell content of insulin and insulin mRNA, accompanied by a diabetic phenotype. Although expression of well-characterized

insulin gene transcriptional activators such as Pdx1, NeuroD/Beta2, and MafA was not significantly affected, a striking upregulation of putative repressors of insulin gene expression Bhlhe22 and Sox6 was observed. It was proposed that miR-24, miR-26, miR-182 or miR-148 are involved, since knockdown of these species in cultured beta cell lines or islets led to downregulation of insulin promoter activity and insulin mRNA levels. Thus, this study indicates an important role for miRNAs in maintenance of mature beta cell function through a network involving regulation of insulin gene repressors.

Many reports have appeared describing the effects of individual miRNA on beta cell function. The first of these addressed the role of miR-375. In addition to its role in pancreas development (see above), the high levels of miR-375 in islets and beta cell lines is consistent with a role in mature beta cell function. Indeed, overexpression of miR-375 in beta cell lines led to reduced glucose-induced insulin secretion, whereas, inhibition of miR-375 enhanced insulin secretion.[69] These effects appear to be mediated in part through downregulation of myotrophin,[69] a vesicle transport protein, and PDK1,[73] a key component in the PI3 kinase signaling pathway.

Mice lacking miR-375 (see above) display an aberrant ratio of alpha to beta cells and a diabetic phenotype. Strikingly, the absence of miR-375 in obese mice (ob/ob) led to profound loss of compensatory proliferative capacity, and a much more severe diabetic phenotype,[71] consistent with a role for miR-375 in controlling beta-cell proliferation, and the general notion that miRNAs contribute to the cellular machinery involved in response to stress.[38] Indeed gene profiling of islets derived from these mice has demonstrated altered expression of several negative growth regulators that may play a key role in beta-cell proliferation.[71] Thus miR-375 is essential for normal beta cell glucose homeostasis, intra-islet alpha: beta cell balance, and compensatory beta-cell expansion in response to increases in insulin demand caused by insulin resistance.

The mechanisms directing selective expression of miR-375 in islet cells operate through transcriptional control.[74] The promoter region of the miR-375 gene is located within an evolutionarily conserved region upstream from the transcription start site. Among the proteins that are implicated in regulating this promoter region are several well-established beta cell transcription factors, including HNF1, INSM1, NeuroD1 and Pdx-1.[74,75]

As with miR-375, miR-7 is expressed at high level both in developing pancreas and islet cells.[66] It has recently been shown that miR-7a (the major form of miR-7 in islet cells) targets five members of the mTOR signaling pathway. Thus, inhibition of miR-7a leads to activation of mTOR signaling and enhanced proliferation of mature beta cells. Accordingly, miR-7 is proposed as a "brake" for beta cell proliferation and, thus a potentially attractive therapeutic target in diabetes.[76]

A number of miRNAs exert their effects on beta cells by modulating the activity of key transcription factors. For example, miR-9 has been shown to affect GSIS by targeting the transcription factor Onecut-2, which in turn increases levels of Granuphilin/Slp4, a Rab GTPase effector that inhibits insulin release through action on beta-cell secretory granules.[77] An additional target of mir-9 is Sirt1, an NAD-dependent protein deacetylase, that has been implicated in modifying transcription factor activity and insulin secretion.[78] Thus miR-9 may be required for optimal insulin secretion by beta cells.

Foxa2, previously mentioned as a regulator of the Pdx-1 gene in embryonic development, also plays an important role in maintenance of beta cell function.[59] miR-124a has been shown to directly target Foxa2, leading to inhibition of multiple downstream genes including Pdx-1, Kir6.2 and SUR1 encoding subunits of the ATP-sensitive potassium (KATP) channel, leading in turn to alteration in sensitivity of the insulin secretory machinery to calcium.[65] Consistent with this, miR-124a (and also miR-96), regulates expression of several proteins involved in insulin exocytosis. Over-expression of miR-124a in cultured beta cells led to elevated basal insulin secretion but reduced glucose-dependent secretion.[79] Likewise, miR-19 was shown to target the transcription factor NeuroD1 and, consequently, reduce levels of insulin mRNA.[80]

miRNAs are also implicated in the repression in beta cells of "forbidden genes:" miR-29a and 29b have been shown to directly repress MCT-1 (monocarboxylate/lactate transporter) mRNA,[81] whose enzymatic activity must be kept low to prevent release of insulin under inappropriate circumstances, e.g. during exercise.

MODULATION OF miRNA UNDER PHYSIOLOGICAL AND PATHOLOGICAL CONDITIONS

There has been considerable interest in determining the effects on beta cell miRNAs of exposing beta cells to a variety of physiological and pathological conditions, and attempting to correlate this with the diabetic state. Thus, many studies have examined the effects on miRNA expression of exposure to glucose. For example, exposure of MIN6 cells to varying glucose concentrations led to regulation of the expression of 61 miRNAs.[82] Among the miRNAs upregulated following glucose exposure, was miR-30d. Overexpression of miR-30d, increased insulin gene expression, while inhibition of miR-30d abolished glucose-stimulated insulin gene transcription, indicating a direct role in mediating glucose-dependent regulation of insulin gene expression.[82] A similar study performed using mouse islets revealed upregulation of miR-15a following exposure to high glucose for 1 h, whereas prolonged exposure resulted in reduced expression of miR-15a. These changes correlated with insulin biosynthesis. Uncoupling protein-2 (UCP-2) was shown to be a direct target of miR-15a.[37]

Experiments involving prolonged exposure of human islets to elevated glucose concentrations led to reduced miR-146 but increased miR-133 resulting in downregulation of polypyrimidine tract binding protein (PTB) and decreased insulin biosynthesis.[83] Finally, exposure of cultured rat beta cell lines or islets to glucotoxic conditions led to increased miR-30a-5p expression. NeuroD1 was identified as a direct target and it is proposed that miR-30a-5p-mediated direct suppression of NeuroD1 gene expression is an important step of glucotoxicity-induced beta cell dysfunction.[84]

Since prolonged exposure to fatty acids has been implicated in mediation of beta cell damage in T2D, the effects of lipids on beta cell miRNA have been investigated. Prolonged exposure of MIN6 cells and islets to palmitate leads to increased levels of miR-34a and miR-146. The rise in miR-34a is associated with elavated p53, increased apoptosis, and impaired nutrient-induced secretion. This latter effect may be related to inhibition of vesicle-associated membrane protein 2 (VAMP2), a component of the beta-cell exocytotic machinery. Thus, some detrimental effects of palmitate on beta cells may be caused by alteration in miRNAs including miR-146.[85]

Pregnancy and obesity are associated with insulin resistance and compensatory beta-cell mass expansion. Expression profiling was used to examine the possible involvement of beta cell miRNAs. miR-338-3p, whose expression is reduced in gestation and obesity was identified as a candidate. Inhibition of miR-338-3p in beta cells mimicked gene expression changes in beta cell expansion and resulted in increased proliferation and survival. This implies an important role for miR-338-3p in mediating compensatory beta-cell mass expansion in insulin resistance.[86]

The role of miRNAs in the cytotoxic effects of cytokines on pancreatic beta cells has been examined by microarray profiling of MIN6 beta cell lines, primary islets and prediabetic NOD (non-obese diabetic) mice. IL-1beta and TNF-alpha were found to induce the expression of miR-21, miR-34a, and miR-146a, both in MIN6 cells and human pancreatic islets. These miRNAs also increased in islets of NOD mice during pre-diabetic insulitis. Inhibition of miR-34a and anti-miR-146a protected MIN6 cells from cytokine-triggered cell death.[87] In a follow-up study, these investigators observed increased miR-29 in pre-diabetic islets of NOD mice and in primary mouse and human islets exposed to proinflammatory cytokines. Thus, changes in the level of miR-29 family members contribute to cytokine-mediated beta cell dysfunction occurring during the initial phases of type 1 diabetes.[88] Overexpression of miR-29 led to impaired GSIS, associated with reduced expression of the transcription factor Onecut2 and a rise in granuphilin, an inhibitor of beta-cell exocytosis. Thus, changes in the level of miR-29 may contribute to cytokine-mediated beta-cell dysfunction during early stages of type 1 diabetes.[88]

A similar approach was used to explore the role of miRNA in a type 2 diabetes model. Islets from prediabetic and diabetic db/db mice were compared using miRNA profiling. Differentially expressed miRNAs fell into two classes: one corresponding to changes that occurred before diabetes onset (miR-132, miR-184, and miR-338-3p) and are proposed to have beneficial effects on beta cell function, and a second class observed in diabetic animals (miR-34a, miR-146a, miR-199a-3p, miR-203, miR-210, and miR-383) and are proposed to promote beta cell apoptosis.[89]

A role for miR-21 has been suggested in preventing beta-cell death in type 1 diabetes. Activation of the NF-κB pathway (previously implicated in beta-cell death associated with T1D)[90] increases miR-21 expression which in turn decreases expression of PDCD4 an inducer of apoptotic cell death.[91] A further link between miRNA and diabetes has come from GWAS studies of T2D in humans. Predicted mRNA targets of islet-expressed miRNAs were found to be significantly enriched for signals of T2D association. Six loci with evidence for T2D association (AP3S2, KCNK16, NOTCH2, SCL30A8, VPS26A, and WFS1) contained predicted mRNA target sites for islet-expressed miRNAs that overlapped candidate causal sequence variants. Further analysis will be needed to validate these intriguing observations.[92]

LONG NON-CODING RNAs AND THEIR POSSIBLE ROLE IN BETA CELLS

Major technical advances in high throughput sequencing of RNA (RNASeq) combined with the generation of genome-wide chromatin maps have led to the identification and characterization of a large and previously unappreciated class of RNA, denoted as long non-coding RNA (lncRNA). Defined as non-coding RNA greater than 200 nt in length, the lncRNAs are strikingly heterogeneous. It has been estimated that the genome contains >9000 loci for lncRNAs, and the number will likely surpass the number of protein-coding genes.[93] lncRNAs are transcribed from intergenic regions, or from within protein-coding genes, including sense, antisense, intronic or overlapping with protein-coding regions.

Although the function of the vast majority of lncRNAs is currently unknown, many lncRNAs have been clearly implicated in critical roles within cells. These include transcriptional regulation, imprinting, pluripotency, splicing, and cell-cycle regulation. Intriguingly, lncRNAs show expression patterns that are often highly cell-specific, much more so than for protein-coding genes. In one analysis 29% of lncRNAs were expressed in a single cell line, as compared to 7% for protein-coding genes.[93]

A common functional theme linking many lncRNAs is chromatin modification. As many as 20% of lncRNAs tested by Khalill et al. showed interaction with the Polycomb repressive complex PRC2.[94] Consistent with a role in transcription, si-RNA knockdown of specific PRC2-associated lncRNAs led to alterations in gene expression.[94] One of the best characterized lncRNAs, known as HOTAIR, recruits PRC2 to specific chromosomal loci,[94] and through a separate domain recruits a complex containing LSD1−Co-REST (lysine-specific demethylase−co-repressor for element 1-silencing transcription factor). This leads to concerted histone H3K4 demethylation and H3K27 methylation, both of which participate in transcriptional repression. Thus, HOTAIR is proposed to function as a scaffold for assembly of chromatin modifying enzymes, thereby determining the epigenetic landscape around targeted genes.[95]

A broader attempt to classify the transcriptional roles of lncRNAs has led to the designation of four major classes: signals, decoys, guides, and scaffolds.[96] Thus, activation of a specific lncRNA may serve as a signal of combinatorial presence of distinct transcription factors or activation of a signaling pathway. As decoys, lncRNAs may titrate out transcription factors from chromatin. As guides, lncRNAs may recruit chromatin modifiers to specific loci, or (as a scaffold, see above) bring together distinct protein complexes.[96] Much work is clearly required to determine the relative importance of these categories in mediating the transcriptional effects of lncRNAs.

In addition to effects on transcription, lncRNAs are implicated in post-transcriptional regulatory events. Thus, antisense lncRNAs may base pair with mRNA leading to altered transcript stability.[97] Since this can occur even without perfect base pairing,[98] the potential for regulatory complexity is dramatically expanded. Additional post-transcriptional regulatory mechanisms include assembly of nuclear structures and modulation of splicing efficiency.[99,100]

A recent genome-wide analysis has been performed to identify islet lncRNAs.[101] Over 1100 islet cell lncRNAs were characterized. Their expression pattern is dynamically regulated, consistent with a role in differentiation and maturation of the beta cell. Accordingly, as much as 55% of intergenic lncRNAs is islet specific. One of the islet lncRNAs, HI-LNC25, was shown to regulate expression of GLIS3, an islet transcription factor implicated in diabetes susceptibility.[101] There is an intriguing link between diabetes loci identified by GWAS, and islet lncRNAs. Thus, a number of risk associated polymorphisms localize to ANRIL, a lncRNA expressed from the INK locus that encodes the tumor suppressor genes $p14^{ARF}$, $p15^{INK4b}$, and $p16^{INK4a}$. Since ANRIL can repress $p15^{INK4b}$ by recruitment of PRC2,[102] mutations that affect ANRIL expression or activity might affect beta-cell proliferative capacity and hence predisposition to diabetes.

CONCLUSIONS

During the past decade, the study of non-coding RNAs has led to remarkable progress in our understanding of basic biological processes. miRNAs and lncRNAs, in particular, are now established as key components of complex regulatory networks mediating a wide array of cellular activities. In beta cells, miRNAs are essential for development and maintenance of mature beta cells. Future research in this area will most likely lead to important new and surprising mechanistic insights.

In addition, research on non-coding RNAs in beta cells may help combat diabetes. The ongoing worldwide epidemic of diabetes imposes ever-increasing strains on health care systems. Hence, new approaches to early diagnosis and improved methods of treatment are urgently needed. The unexpected discovery of miRNA in extracellular fluids, including blood, not only implies that non-coding RNAs may participate in novel pathways of cell–cell signaling, but raises the possibility that miRNAs may serve as valuable biomarkers of diseases such as diabetes.[103]

Several studies have indicated that miRNAs affect the efficiency of *in vitro* differentiation of embryonic stem cells.[104] Hence elucidation of the underlying mechanisms may lead to improved protocols for *in vitro* generation of unlimited supplies of functional beta cells for cell replacement therapy of diabetes.

Acknowledgements

Research in the author's lab is supported by the Juvenile Diabetes Research Foundation and the Israel Science Foundation. MDW is the incumbent of the Marvin Meyer and Jenny Cyker Chair of Diabetes Research at the Weizmann Institute.

References

1. Brissova M, Fowler MJ, Nicholson WE, Chu A, Hirshberg B, Harlan DM, et al. Assessment of human pancreatic islet architecture and composition by laser scanning confocal microscopy. *J Histochem Cytochem* 2005;**53**:1087–97.

2. McGarry JD. What if Minkowski had been ageusic? An alternative angle on diabetes. *Science* 1992;**258**:766–70.

3. Steiner DF, Park SY, Stoy J, Philipson LH, Bell GI. A brief perspective on insulin production. *Diabetes Obes Metab* 2009;**11**(Suppl. 4):189–96.

4. Rutter GA. Nutrient-secretion coupling in the pancreatic islet beta-cell: recent advances. *Mol Aspects Med* 2001;**22**:247–84.

5. Pullen TJ, Rutter GA. When less is more: the forbidden fruits of gene repression in the adult beta-cell. *Diabetes Obes Metab* 2013;**15**:503–12.

6. Schuit F, Van Lommel L, Granvik M, Goyvaerts L, de Faudeur G, Schraenen A, et al. beta-cell-specific gene repression: a mechanism to protect against inappropriate or maladjusted insulin secretion? *Diabetes* 2012;**61**:969–75.

7. German M, Ashcroft S, Docherty K, Edlund T, Edlund H, Goodison S, et al. The insulin gene promoter: a simplified nomenclature. *Diabetes* 1995;**44**:1002–4.

8. Karlsson O, Edlund T, Moss JB, Rutter WJ, Walker MD. A mutational analysis of the insulin gene transcription control region: Expression in beta cells is dependent on two related sequences within the enhancer. *Proc Natl Acad Sci USA* 1987;**84**:8819–23.

9. Ohlsson H, Karlsson K, Edlund T. IPF1, a homeodomain containing transactivator of the insulin gene. *EMBO J* 1993;**12**:4251–9.

10. Naya FJ, Stellrecht CMM, Tsai MJ. Tissue-specific regulation of the insulin gene by a novel basic helix-loop-helix transcription factor. *Genes & Dev* 1995;**9**:1009–19.

11. Matsuoka TA, Artner I, Henderson E, Means A, Sander M, Stein R. The MafA transcription factor appears to be responsible for tissue-specific expression of insulin. *Proc Natl Acad Sci USA* 2004;**18**:18.

12. Chakrabarti SK, Mirmira RG. Transcription factors direct the development and function of pancreatic beta cells. *Trends Endocrinol Metab* 2003;**14**:78–84.

13. Edlund H. Transcribing pancreas. *Diabetes* 1998;**47**:1817–23.

14. Prentki M, Matschinsky FM, Madiraju SR. Metabolic signaling in fuel-induced insulin secretion. *Cell Metab* 2013.

15. Henquin JC. The dual control of insulin secretion by glucose involves triggering and amplifying pathways in beta-cells. *Diabetes Res Clin Pract* 2011;**93**(Suppl. 1):S27–31.

16. Curry DL, Bennett LL, Grodsky GM. Dynamics of insulin secretion by the perfused rat pancreas. *Endocrinology* 1968;**83**:572–84.

17. Zraika S, Dunlop M, Proietto J, Andrikopoulos S. Effects of free fatty acids on insulin secretion in obesity. *Obes Rev* 2002;**3**:103–12.

18. Steneberg P, Rubins N, Bartoov-Shifman R, Walker MD, Edlund H. The FFA receptor GPR40 links hyperinsulinemia, hepatic steatosis, and impaired glucose homeostasis in mouse. *Cell Metab* 2005;**1**:245–58.

19. Mancini AD, Poitout V. The fatty acid receptor FFA1/GPR40 a decade later: how much do we know? *Trends Endocrinol Metab* 2013.

20. Shapiro H, Shachar S, Sekler I, Hershfinkel M, Walker MD. Role of GPR40 in fatty acid action on the beta cell line INS-1E. *Biochem Biophys Res Commun* 2005;**335**:97–104.

21. Drucker DJ. Incretin action in the pancreas: Potential promise, possible perils, and pathological pitfalls. *Diabetes* 2013.

22. Atkinson MA, Maclaren NK. The pathogenesis of insulin-dependent diabetes mellitus. *N Engl J Med* 1994;**331**:1428–36.

23. Kahn BB. Type 2 diabetes: when insulin secretion fails to compensate for insulin resistance. *Cell* 1998;**92**:593–6.

24. Rhodes CJ. Type 2 diabetes-a matter of beta-cell life and death?. *Science* 2005;**307**:380–4.

25. Donath MY, Dalmas E, Sauter NS, Boni-Schnetzler M. Inflammation in obesity and diabetes: islet dysfunction and therapeutic opportunity. *Cell Metab* 2013;**17**:860–72.

26. Shaw JE, Sicree RA, Zimmet PZ. Global estimates of the prevalence of diabetes for 2010 and 2030. *Diabetes Res Clin Pract* 2010;**87**:4–14.

27. Poulsen P, Kyvik KO, Vaag A, Beck-Nielsen H. Heritability of type II (non-insulin-dependent) diabetes mellitus and abnormal glucose tolerance--a population-based twin study. *Diabetologia* 1999;**42**:139–45.

28. Todd JA, Bell JI, McDevitt HO. A molecular basis for genetic susceptibility to insulin-dependent diabetes mellitus. *Trends Genet* 1988;**4**:129–34.

29. Scott RA, Lagou V, Welch RP, Wheeler E, Montasser ME, Luan J, et al. Large-scale association analyses identify new loci influencing glycemic traits and provide insight into the underlying biological pathways. *Nat Genet* 2012;**44**:991–1005.

30. Florez JC. Newly identified loci highlight beta cell dysfunction as a key cause of type 2 diabetes: where are the insulin resistance genes? *Diabetologia* 2008;**51**:1100–10.

31. Morris AP, Voight BF, Teslovich TM, Ferreira T, Segre AV, Steinthorsdottir V, et al. Large-scale association analysis provides insights into the genetic architecture and pathophysiology of type 2 diabetes. *Nat Genet* 2012;**44**:981–90.

32. Shapiro AM, Lakey JR, Ryan EA, Korbutt GS, Toth E, Warnock GL, et al. Islet transplantation in seven patients with type 1 diabetes mellitus using a glucocorticoid-free immunosuppressive regimen. *N Engl J Med* 2000;**343**:230–8.

33. Pagliuca FW, Melton DA. How to make a functional beta-cell. *Development* 2013;**140**:2472–83.

34. Bramswig NC, Kaestner KH. Organogenesis and functional genomics of the endocrine pancreas. *Cell Mol Life Sci* 2012;**69**:2109–23.

35. Wightman B, Ha I, Ruvkun G. Posttranscriptional regulation of the heterochronic gene lin-14 by lin-4 mediates temporal pattern formation in C. elegans. *Cell* 1993;**75**:855–62.

36. Lee RC, Feinbaum RL, Ambros V. The C. elegans heterochronic gene lin-4 encodes small RNAs with antisense complementarity to lin-14. *Cell* 1993;**75**:843–54.

37. Sun K, Lai EC. Adult-specific functions of animal microRNAs. *Nat Rev Genet* 2013;**14**:535–48.

38. Leung AK, Sharp PA. MicroRNA functions in stress responses. *Mol Cell* 2010;**40**:205–15.

39. Kloosterman WP, Plasterk RH. The diverse functions of microRNAs in animal development and disease. *Dev Cell* 2006;**11**:441–50.

40. Megraw M, Sethupathy P, Corda B, Hatzigeorgiou AG. miRGen: a database for the study of animal microRNA genomic organization and function. *Nucleic Acids Res* 2007;**35**:D149–55.

41. Kim VN, Han J, Siomi MC. Biogenesis of small RNAs in animals. *Nat Rev Mol Cell Biol* 2009;**10**:126–39.

42. Golan D, Levy C, Friedman B, Shomron N. Biased hosting of intronic microRNA genes. *Bioinformatics* 2010;**26**:992–5.

43. Ozsolak F, Poling LL, Wang Z, Liu H, Liu XS, Roeder RG, et al. Chromatin structure analyses identify miRNA promoters. *Genes Dev* 2008;**22**:3172–83.

44. Lee Y, Kim M, Han J, Yeom KH, Lee S, Baek SH, et al. MicroRNA genes are transcribed by RNA polymerase II. *EMBO J* 2004;**23**:4051–60.

45. Kim VN. MicroRNA biogenesis: coordinated cropping and dicing. *Nat Rev Mol Cell Biol* 2005;**6**:376–85.

46. Ameres SL, Zamore PD. Diversifying microRNA sequence and function. *Nat Rev Mol Cell Biol* 2013.

47. Winter J, Jung S, Keller S, Gregory RI, Diederichs S. Many roads to maturity: microRNA biogenesis pathways and their regulation. *Nat Cell Biol* 2009;**11**:228–34.

48. Fabian MR, Sonenberg N. The mechanics of miRNA-mediated gene silencing: a look under the hood of miRISC. *Nat Struct Mol Biol* 2012;**19**:586–93.

49. Guo H, Ingolia NT, Weissman JS, Bartel DP. Mammalian microRNAs predominantly act to decrease target mRNA levels. *Nature* 2010;**466**:835–40.

50. Bartel DP. MicroRNAs: genomics, biogenesis, mechanism, and function. *Cell* 2004;**116**:281–97.

51. Friedman RC, Farh KK, Burge CB, Bartel DP. Most mammalian mRNAs are conserved targets of microRNAs. *Genome Res* 2009;**19**:92–105.

52. Stefani G, Slack FJ. Small non-coding RNAs in animal development. *Nat Rev Mol Cell Biol* 2008;**9**:219–30.

53. Pictet R, Rutter WJ. Development of the embryonic endocrine pancreas. In: Steiner DF, Frenkel M, editors. *Handbook of physiology.* Washington DC: American Physiological Society; 1972. . p. 25–36. **Sec. 7**

54. Gittes GK. Developmental biology of the pancreas: a comprehensive review. *Dev Biol* 2009;**326**:4–35.

55. Edlund H. Pancreatic organogenesis--developmental mechanisms and implications for therapy. *Nat Rev Genet* 2002;**3**:524–32.

56. Wandzioch E, Zaret KS. Dynamic signaling network for the specification of embryonic pancreas and liver progenitors. *Science* 2009;**324**:1707–10.

57. Jonsson J, Carlsson L, Edlund T, Edlund H. Insulin promoter factor 1 is required for pancreas development in mice. *Nature* 1994;**371**:606–9.

58. Stoffers D, Zinkin N, Stanojevic V, Clarke W, Habener J. Pancreatic agenesis attributable to a single nucleotide deletion in the human IPF1 gene coding sequence. *Nat Genet* 1997;**15**:106–10.

59. Gao N, LeLay J, Vatamaniuk MZ, Rieck S, Friedman JR, Kaestner KH. Dynamic regulation of Pdx1 enhancers by Foxa1 and Foxa2 is essential for pancreas development. *Genes Dev* 2008;**22**:3435–48.

60. Gradwohl G, Dierich A, LeMeur M, Guillemot F. neurogenin3 is required for the development of the four endocrine cell lineages of the pancreas. *Proc Natl Acad Sci U S A* 2000;**97**:1607–11.

61. Hang Y, Stein R. MafA and MafB activity in pancreatic beta cells. *Trends Endocrinol Metab* 2011;**22**:364–73.

62. Lynn FC, Skewes-Cox P, Kosaka Y, McManus MT, Harfe BD, German MS. MicroRNA expression is required for pancreatic islet cell genesis in the mouse. *Diabetes* 2007;**56**:2938–45.

63. Kanji MS, Martin MG, Bhushan A. Dicer1 is required to repress neuronal fate during endocrine cell maturation. *Diabetes* 2013;**62**:1602–11.

64. Kalis M, Bolmeson C, Esguerra JL, Gupta S, Edlund A, Tormo-Badia N, et al. Beta-cell specific deletion of Dicer1 leads to defective insulin secretion and diabetes mellitus. *PLoS One* 2011;**6**:e29166.

65. Baroukh N, Ravier MA, Loder MK, Hill EV, Bounacer A, Scharfmann R, et al. MicroRNA-124a regulates Foxa2 expression and intracellular signaling in pancreatic beta-cell lines. *J Biol Chem* 2007;**282**:19575–88.

66. Joglekar MV, Parekh VS, Mehta S, Bhonde RR, Hardikar AA. MicroRNA profiling of developing and regenerating pancreas reveal post-transcriptional regulation of neurogenin3. *Dev Biol* 2007;**311**:603–12.

67. Correa-Medina M, Bravo-Egana V, Rosero S, Ricordi C, Edlund H, Diez J, et al. MicroRNA miR-7 is preferentially expressed in endocrine cells of the developing and adult human pancreas. *Gene Expr Patterns* 2009;**9**: 193–9.

68. Nieto M, Hevia P, Garcia E, Klein D, Alvarez-Cubela S, Bravo-Egana V, et al. Antisense miR-7 impairs insulin expression in developing pancreas and in cultured pancreatic buds. *Cell Transplant* 2012;**21**:1761–74.

69. Poy MN, Eliasson L, Krutzfeldt J, Kuwajima S, Ma X, Macdonald PE, et al. A pancreatic islet-specific microRNA regulates insulin secretion. *Nature* 2004;**432**: 226–30.

70. Kloosterman WP, Lagendijk AK, Ketting RF, Moulton JD, Plasterk RH. Targeted inhibition of miRNA maturation with morpholinos reveals a role for miR-375 in pancreatic islet development. *PLoS Biol* 2007;**5**:e203.

71. Poy MN, Hausser J, Trajkovski M, Braun M, Collins S, Rorsman P, et al. miR-375 maintains normal pancreatic alpha- and beta-cell mass. *Proc Natl Acad Sci USA* 2009;**106**:5813–8.

72. Melkman-Zehavi T, Oren R, Kredo-Russo S, Shapira T, Mandelbaum AD, Rivkin N, et al. miRNAs control insulin content in pancreatic beta-cells via downregulation of transcriptional repressors. *EMBO J* 2011;**30**:835–45.

73. El Ouaamari A, Baroukh N, Martens GA, Lebrun P, Pipeleers D, van Obberghen E. miR-375 targets 3′-phosphoinositide-dependent protein kinase-1 and regulates glucose-induced biological responses in pancreatic beta-cells. *Diabetes* 2008;**57**:2708–17.

74. Avnit-Sagi T, Kantorovich L, Kredo-Russo S, Hornstein E, Walker MD. The promoter of the pri-miR-375 gene directs expression selectively to the endocrine pancreas. *PLoS One* 2009;**4**:e5033.

75. Keller DM, McWeeney S, Arsenlis A, Drouin J, Wright CV, Wang H, et al. Characterization of pancreatic transcription factor Pdx-1 binding sites using promoter microarray and serial analysis of chromatin occupancy. *J Biol Chem* 2007;**282**:32084–92.

76. Wang Y, Liu J, Liu C, Naji A, Stoffers DA. MicroRNA-7 regulates the mTOR pathway and proliferation in adult pancreatic beta-cells. *Diabetes* 2013;**62**:887–95.

77. Plaisance V, Abderrahmani A, Perret-Menoud V, Jacquemin P, Lemaigre F, Regazzi R. MicroRNA-9 controls the expression of Granuphilin/Slp4 and the secretory response of insulin-producing cells. *J Biol Chem* 2006;**281**:26932–42.

78. Ramachandran D, Roy U, Garg S, Ghosh S, Pathak S, Kolthur-Seetharam U. Sirt1 and mir-9 expression is regulated during glucose-stimulated insulin secretion in pancreatic beta-islets. *FEBS J* 2011;**278**:1167–74.

79. Lovis P, Gattesco S, Regazzi R. Regulation of the expression of components of the exocytotic machinery of insulin-secreting cells by microRNAs. *Biol Chem* 2008;**389**:305–12.

80. Zhang ZW, Zhang LQ, Ding L, Wang F, Sun YJ, An Y, et al. MicroRNA-19b downregulates insulin 1 through targeting transcription factor NeuroD1. *FEBS Lett* 2011;**585**:2592–8.

81. Pullen TJ, da Silva Xavier G, Kelsey G, Rutter GA. miR-29a and miR-29b contribute to pancreatic beta-cell-specific silencing of monocarboxylate transporter 1 (Mct1). *Mol Cell Biol* 2011;**31**:3182–94.

82. Tang X, Muniappan L, Tang G, Ozcan S. Identification of glucose-regulated miRNAs from pancreatic {beta} cells reveals a role for miR-30d in insulin transcription. *RNA* 2009;**15**:287–93.

83. Fred RG, Bang-Berthelsen CH, Mandrup-Poulsen T, Grunnet LG, Welsh N. High glucose suppresses human islet insulin biosynthesis by inducing miR-133a leading to decreased polypyrimidine tract binding protein-expression. *PLoS One* 2010;**5**:e10843.

84. Kim JW, You YH, Jung S, Suh-Kim H, Lee IK, Cho JH, et al. miRNA-30a-5p-mediated silencing of Beta2/NeuroD expression is an important initial event of glucotoxicity-induced beta cell dysfunction in rodent models. *Diabetologia* 2013;**56**:847–55.

85. Lovis P, Roggli E, Laybutt DR, Gattesco S, Yang JY, Widmann C, et al. Alterations in microRNA expression contribute to fatty acid-induced pancreatic beta-cell dysfunction. *Diabetes* 2008;**57**:2728–36.

86. Jacovetti C, Abderrahmani A, Parnaud G, Jonas JC, Peyot ML, Cornu M, et al. MicroRNAs contribute to compensatory beta cell expansion during pregnancy and obesity. *J Clin Invest* 2012;**122**:3541–51.

87. Roggli E, Britan A, Gattesco S, Lin-Marq N, Abderrahmani A, Meda P, et al. Involvement of microRNAs in the cytotoxic effects exerted by proinflammatory cytokines on pancreatic beta-cells. *Diabetes* 2010;**59**:978–86.

88. Roggli E, Gattesco S, Caille D, Briet C, Boitard C, Meda P, et al. Changes in microRNA expression contribute to pancreatic beta-cell dysfunction in prediabetic NOD mice. *Diabetes* 2012;**61**:1742–51.

89. Nesca V, Guay C, Jacovetti C, Menoud V, Peyot ML, Laybutt DR, et al. Identification of particular groups of microRNAs that positively or negatively impact on beta cell function in obese models of type 2 diabetes. *Diabetologia* 2013.

90. Heimberg H, Heremans Y, Jobin C, Leemans R, Cardozo AK, Darville M, et al. Inhibition of cytokine-induced NF-kappaB activation by adenovirus-mediated expression of a NF-kappaB super-repressor prevents beta-cell apoptosis. *Diabetes* 2001;**50**:2219–24.

91. Ruan Q, Wang T, Kameswaran V, Wei Q, Johnson DS, Matschinsky F, et al. The microRNA-21-PDCD4 axis prevents type 1 diabetes by blocking pancreatic beta cell death. *Proc Natl Acad Sci U S A* 2011;**108**:12030–5.

92. van de Bunt M, Gaulton KJ, Parts L, Moran I, Johnson PR, Lindgren CM, et al. The miRNA profile of human pancreatic islets and beta-cells and relationship to type 2 diabetes pathogenesis. *PLoS One* 2013;**8**:e55272.

93. Djebali S, Davis CA, Merkel A, Dobin A, Lassmann T, Mortazavi A, et al. Landscape of transcription in human cells. *Nature* 2012;**489**:101–8.

94. Khalil AM, Guttman M, Huarte M, Garber M, Raj A, Rivea Morales D, et al. Many human large intergenic noncoding RNAs associate with chromatin-modifying complexes and affect gene expression. *Proc Natl Acad Sci USA* 2009;**106**:11667–72.

95. Tsai MC, Manor O, Wan Y, Mosammaparast N, Wang JK, Lan F, et al. Long noncoding RNA as modular

scaffold of histone modification complexes. *Science* 2010;**329**:689–93.

96. Wang KC, Chang HY. Molecular mechanisms of long noncoding RNAs. *Mol Cell* 2011;**43**:904–14.

97. Faghihi MA, Modarresi F, Khalil AM, Wood DE, Sahagan BG, Morgan TE, et al. Expression of a noncoding RNA is elevated in Alzheimer's disease and drives rapid feed-forward regulation of beta-secretase. *Nat Med* 2008;**14**:723–30.

98. Gong C, Maquat LE. lncRNAs transactivate STAU1-mediated mRNA decay by duplexing with 3′ UTRs via Alu elements. *Nature* 2011;**470**:284–8.

99. Mao YS, Sunwoo H, Zhang B, Spector DL. Direct visualization of the co-transcriptional assembly of a nuclear body by noncoding RNAs. *Nat Cell Biol* 2011;**13**:95–101.

100. Tripathi V, Ellis JD, Shen Z, Song DY, Pan Q, Watt AT, et al. The nuclear-retained noncoding RNA MALAT1 regulates alternative splicing by modulating SR splicing factor phosphorylation. *Mol Cell* 2010;**39**:925–38.

101. Moran I, Akerman I, van de Bunt M, Xie R, Benazra M, Nammo T, et al. Human beta cell transcriptome analysis uncovers lncRNAs that are tissue-specific, dynamically regulated, and abnormally expressed in type 2 diabetes. *Cell Metab* 2012;**16**:435–48.

102. Kotake Y, Nakagawa T, Kitagawa K, Suzuki S, Liu N, Kitagawa M, et al. Long non-coding RNA ANRIL is required for the PRC2 recruitment to and silencing of p15(INK4B) tumor suppressor gene. *Oncogene* 2011;**30**:1956–62.

103. Chen X, Liang H, Zhang J, Zen K, Zhang CY. Secreted microRNAs: a new form of intercellular communication. *Trends Cell Biol* 2012;**22**:125–32.

104. Wang Y, Melton C, Li YP, Shenoy A, Zhang XX, Subramanyam D, et al. miR-294/miR-302 Promotes Proliferation, Suppresses G1-S Restriction Point, and Inhibits ESC Differentiation through Separable Mechanisms. *Cell Rep* 2013;**4**:99–109.

Index

Note: Page numbers followed by "*f*" and "*t*" refers to figures and tables, respectively.

Printed and bound by CPI Group (UK) Ltd, Croydon, CR0 4YY

08/05/2025

01864983-0001